AF790801

Photography (Second Edition)

Physics and art in focus

Online at: https://doi.org/10.1088/978-0-7503-3703-8

Photography (Second Edition)

Physics and art in focus

John Beaver
University of Wisconsin Oshkosh, Oshkosh, WI, USA

IOP Publishing, Bristol, UK

© IOP Publishing Ltd 2022

All rights reserved. No part of this publication may be reproduced, stored in a retrieval system or transmitted in any form or by any means, electronic, mechanical, photocopying, recording or otherwise, without the prior permission of the publisher, or as expressly permitted by law or under terms agreed with the appropriate rights organization. Multiple copying is permitted in accordance with the terms of licences issued by the Copyright Licensing Agency, the Copyright Clearance Centre and other reproduction rights organizations.

Permission to make use of IOP Publishing content other than as set out above may be sought at permissions@ioppublishing.org.

John Beaver has asserted his right to be identified as the author of this work in accordance with sections 77 and 78 of the Copyright, Designs and Patents Act 1988.

Multimedia content is available for this book from https://doi.org/10.1088/978-0-7503-3703-8.

ISBN 978-0-7503-3703-8 (ebook)
ISBN 978-0-7503-3701-4 (print)
ISBN 978-0-7503-3704-5 (myPrint)
ISBN 978-0-7503-3702-1 (mobi)

DOI 10.1088/978-0-7503-3703-8

Version: 20221201

IOP ebooks

British Library Cataloguing-in-Publication Data: A catalogue record for this book is available from the British Library.

Published by IOP Publishing, wholly owned by The Institute of Physics, London

IOP Publishing, No.2 The Distillery, Glassfields, Avon Street, Bristol, BS2 0GR, UK

US Office: IOP Publishing, Inc., 190 North Independence Mall West, Suite 601, Philadelphia, PA 19106, USA

For Anne

Contents

Preface

The first edition of this book was published in three volumes in 2018 and 2019 under the title *The Physics and Art of Photography*. Early drafts were written for a course I first taught in the Fall of 2013 at the University of Wisconsin–Fox Valley, in Menasha, Wisconsin (USA). This second edition—now all in one volume—is revised, reorganized, and greatly expanded.

I assume no specific prior knowledge of the reader except for a basic understanding of physical units, dimensions and scientific notation; a brief review can be found in appendix A. The mathematics presented in the text is rudimentary, with only the most basic of algebra (more detailed derivations, or those that require calculus, are relegated to the appendices).

If you have little experience with photography, it is my goal that *Photography: Physics and Art in Focus* will help form a useful foundation from which to learn about photography in whatever way that works best for you. If you are a seasoned pro, but looking to set off in a new direction, then I still hope that you will find much here that is fresh and inspiring. It is my goal that the book will help to open new possibilities.

Photography: Physics and Art in Focus covers some material that is typical of discussions that link physics and photography. But it is also personal; it is very much my own take on the two subjects. I would not say that my personal views regarding science and art are controversial, but they are perhaps somewhat unconventional. There are few details here with which other artists and scientists are likely to strongly disagree. It is, rather, what I have chosen to emphasize—what I have left out all together and the particular connections I point to—that most show my own personal likes and dislikes.

Since my formal training is in physics and astronomy, and I am essentially self-trained in art (with informal mentoring from many others), the science part of this book is perhaps more conventional and straightforward than is my portrayal of art. My choice of physics-related topics should give one a fairly balanced and conventional taste of that subject as it relates to photography. Regarding photography as an art, however, I am surely on shakier ground.

Certainly, I do not pretend to present a comprehensive or balanced overview of art photography; I am unqualified to attempt such a thing. However, I do try to make a case that the particular thin slice that I present here has some merit and is worth spending a little time to consider, even if it turns out not to be your particular cup of tea. Since it is my goal to get you thinking, it does not matter much whether you agree with me or not. Thus, it is fitting that my discussion of art is more personal, since my own art is the wee bit about which I really do know about what I am talking.

One might complain that *Photography: Physics and Art in Focus* is a very long artist's statement, justifying the value and relevance of my own art. That complaint may be partly true, but I do try to approach it in a way that emphasizes broad questions, rather than the particular answers I try to give (tentatively) with my own art. I hope this book does help a little to make you a better photographer, and as such I do spend time on some of the very basic technical aspects of photography that I find important. But in doing so, I try to use these technical issues as points of departure to consider the status of photography as an art, finally exploring some issues relating to this status in the digital age.

This book may also be read as a manifesto of sorts for the aspects of science that have always moved me the most. I am interested in science not for the technological gizmos it has produced or for some notion of inevitable human 'progress.' Rather, science is—for me—part of the study of nature. My interest in Einstein's general relativity, for example, is essentially the same as my interest in bird watching. Because I have spent some time to learn a bit about birds, I can now walk through the woods free of binoculars—looking only at the ground at my feet—and a world is open to me just by the sounds I hear. When I stumble on my way up the stairs, as a physicist, I can take comfort in the idea that my shin in contact with the stair has prevented me from following my normal straight-line path through four-dimensional spacetime.

You will find, throughout the book, illustrations from my own photography as examples. This is convenient; I know my own pictures and the stories behind them, and I do not need permission to use them. Of course I also want you to look at other photography, and so I have included some examples from a few other artists whose work I admire.

A useful companion is *The Photography Book* (Cooke and Kinneberg 2014), which presents hundreds of photographs, spanning the entire history of photography. Each photograph has a short analysis with cross references to other photographs that are related. The photographs, only one per photographer, are arranged in alphabetical order by photographer's name. Thus, the ordering of the pictures is thematically random, which often results in unusual juxtapositions on facing pages. I sometimes refer to pictures in *The Photography Book* as examples, so it is useful to have it handy. However, all of these pictures are famous and can easily be found online as well.

The reader will also find, scattered throughout the three volumes and their appendixes, details and examples from what I call *ephemeral- process (EP) photography*. EP photography is my own invention—sort of—and I spend so much time on it because it is perfect for illustrating many of the concepts in *Photography: Physics and Art in Focus* in a way that I believe goes directly to the heart of the matter. Furthermore, it is accessible. The materials and equipment are inexpensive, it requires no specialized facilities (such as a darkroom), and it is surprisingly versatile. Most importantly, it is a lot of fun. A brief guide to the technique can be found in chapter 48.

My larger concerns are to give the reader some background that is helpful for asking important questions about the nature of art and science. However, the practice of photography is the point of departure for these bigger issues, and *Photography: Physics and Art in Focus* does contain a lot of simply practical information as well. There are five basic goals.

1. To ask basic questions about how photography fits in as an *art* and about the nature of art itself.
2. To ask basic questions about the nature of physics as part of the study of the natural world and about the nature of science itself.
3. To gain some practical knowledge that will allow the reader to more easily learn technical aspects of photography as they are needed.
4. To gain some practical knowledge that will help the reader more easily learn to be a better photographer.
5. To expose the reader to a set of interesting photographic processes and tools that are not usually covered in a beginning photography course.

One of the themes of this book is the meaning of digital technology and what it has to say regarding photography as an art form. This may seem like I am speaking out of turn here, since I have neither formal training in art, nor have I ever been a professional photographer using professional digital equipment. Nevertheless, there is a sense in which I am well-positioned to say something of interest about these issues.

My own photography is almost entirely devoid of the use of a digital camera. I often use equipment and old physical processes that are about as far removed from modern digital photography as one could imagine. However, I use equipment and processes in new ways that depend absolutely on the digital; many of my photographs could not exist without modern digital processing, scanning and printing. This kind of interplay between the old and new is one of the running themes of *Photography: Physics and Art in Focus.*

Despite my collection of old cameras, I am not a knee-jerk hater of digital imaging technology. I am—in fact—one of its early practitioners, having used digital cameras and sophisticated digital-image processing long before most photographers. My formal training is in astronomy, and I was there (in graduate school) for the digital revolution as it transformed astronomy in the 1980s. The charged-coupled device digital detectors used in modern digital cameras were fairly new then and still too expensive (and with insufficient resolution) to be of much practical use for photographers. I am the last person one would want to ask about the latest multithousand-dollar model of digital single-lens reflex camera. However, I do have a decades-long understanding of some of the most basic underlying principles of digital photography.

Acknowledgments

I thank Valeria Sapiain for all-around support and much patience over the several years I spent working on this book. I received much valuable feedback, support and mentoring from many, but I note especially Judith Waller, Doug Fowler, Caroline Geary, Anne Haydock and Teresa Patrick. I also give my heartfelt thanks to the members of Edinburgh LoFi (@edinburghlofi), who kept their *Second Tuesday* meetings virtual after the pandemic lockdown was lifted, in part so I could participate: Gregg McNeil, Alison Millar, Elaine Robson, Graeme Lyall, and especially Brittonie Fletcher, whose crucial support over the past few years has meant more than I know how to say.

The following software applications were used in the writing of this book: GIMP, Gnuplot, Inkscape, Aladin Desktop, SciDAVis, OpticalRayTracer, OptGeo, Twisted Wave, Audacity, TeXstudio, JabRef, ImageJ, VueScan, Darktable and IRAF.

All photographs and illustrations are by the author, except as noted:

Figure 2.2: Copyright 2018 Google (detailed attribution in image)

Figure 3.11, Left: By Geek3—Own work, CC BY 4.0 https://commons.wikimedia.org/wiki/File:Mplwp_dispersion_curves.svghttps://commons.wikimedia.org/wiki/File:Mplwp_dispersion_curves.svg

Figure 3.15 Catslash, Public domain, via Wikimedia Commons

Figure 3.17: By Haade, Wjh31, Quibik—Own work, CC BY 3.0 https://commons.wikimedia.org/wiki/File:Interference_of_two_waves.svghttps://commons.wikimedia.org/wiki/File:Interference_of_two_waves.svg

Figure 3.19: By Nicoguaro—Own work, CC BY 4.0 https://commons.wikimedia.org/w/index.php?curid=49324857https://commons.wikimedia.org/w/index.php?curid=49324857

Figure 3.21: By Arne Nordmann (norro)—Own illustration, CC BY-SA 3.0, https://commons.wikimedia.org/w/index.php?curid=1944668https://commons.wikimedia.org/w/index.php?curid=1944668

Figure 3.22 Public Domain

Figure 3.24: By P. Wormer—Own work, CC BY 3.0 https://commons.wikimedia.org/wiki/File:Electromagnetic_wave.pnghttps://commons.wikimedia.org/wiki/File:Electromagnetic_wave.png

Figure 4.1 Public Domain

Figure 5.3: Public Domain

Figure 5.7: Robert A. Rhohde, CC BY-SA 3.0 https://commons.wikimedia.org/wiki/File:Solar_spectrum_en.svgcommons.wikimedia.org/wiki/File:Solar_spectrum_en.svg

Figure 5.8: CC BY 3.0 https://en.wikipedia.org/wiki/File:Spectrum_of_halophosphate_type_fluorescent_bulb_(f30t12_ww_rs).png https://en.wikipedia.org/wiki/File:Spectrum_of_halophosphate_type_fluorescent_bulb_(f30t12_ww_rs).png

Figure 6.1: By SpoonSpa, Simon Viktória, CC BY 2.0 Generic https://commons.wikimedia.org/wiki/File:Szakkad.jpghttps://commons.wikimedia.org/wiki/File:Szakkad.jpg

Figure 7.2. Eclipse photographs by Fred Espenak, http://www.mreclipse.com, used by permission

Figure 7.6: (Left:) Total solar eclipse prediction diagram by Fred Espenak (http://www.mreclipse.com), used by permission

Figure 7.6: NASA Earth Satellite Geostationary Operational Environmental Satellites image; usage within stated rights permitted by publisher

Figure 8.13: by Anna Atkins, (New York Public Library Digital Collections)

Figure 9.8: Photograph by Teresa Patrick, used by permission

Figure 10.2 WolfWings—Own work, Public Domain

Figures 10.4, 10.5, 10.19: Graphics by the author, using beaver image from: Steve from Washington, DC, USA, CC BY-SA 2.0 Steve from Washington, DC, USA—American Beaver, CC BY-SA 2.0

Figure 10.7 (Wood 1906, public domain)

Figure 10.21 Mliu92Own work CC BY-SA 4.0

Figure 10.22 Morio–Own work CC BY-SA 4.0

Figure 12.4: Public Domain

Figure 13.1: Left: Public Domain. Right: NASA, J. Bell (Cornell U.) and M. Wolff (SSI)

Figure 13.8: Public Domain

Figure 26.11 (graphic by Mix321—Own work, Lewińki, Holak, Biologia dla liceum, wyd. Operon, GFDL) The Chlorofilab.svg image has been obtained by the author(s) from the Wikipedia website where it was made available User: Mix321 under the GNU Free Documentation License, Version 1.2. It is included within this publication on that basis. It is attributed to Lewińki, Holak.

Figure 27.1: graphic by Vanessaezekowitz at en.wikipedia CC BY-SA 3.0

Figure 28.4: graphic by Figure design by Kasuga jawiki; vectorization by Editor at Large; 'The dress' modification by Jahobr, CC BY-SA 3.0.

Figure 29.2: Images made with the Aladin Sky Atlas, Digitized Sky Survey II: (Bonnarel *et al* 2000, Lasker *et al* 1996)

Figure 29.5 Georges Seurat, Art Institute of Chicago, CCO Public Domain Designation

Figure 29.8 By en:User:Cburnett—Own workThis W3C-unspecified vector image was created with Inkscape, CC BY-SA 3.0

Figure 29.9: Janke— CC0, Public Domain

Figure 29.14 By Didier Descouens—Own work, CC BY-SA 4.0

Figure 30.7: Left: graphic by Vanessaezekowitz at en.wikipedia CC BY-SA 3.0

Figure 32.1 By Unknown—illustration from the Scientific American Supplement, January 11, 1879, Public Domain

Figure 32.2 By Fig_retine.png: Cajalderivative work Fig retine bended.png: Anka Friedrich (talk) derivative work: vectorisation by Chris—Fig_retine.pngFig retine bended.png, CC BY-SA 3.0

Figure 32.3 By RicHard-59—Own work, CC BY-SA 3.0

Figure 35.1 Brittonie Fletcher, 2017, use by permission

Figure 35.2 Brittonie Fletcher, 2008, used by permission
Figure 35.3 Gregg McNeill, 2022, used by permission, of a model made by Ian Chrichton
Figure 35.7 Brittonie Fletcher, 2019, used by permission
Figure 36.3 Brittonie Fletcher, 2010, used by permission
Figure 36.4 Right: Brittonie Fletcher, 2022, used by permission
Figure 36.6 Brittonie Fletcher, 2022, used by permission
Figure 37.4: Right: Teresa Patrick, 2005, used by permission
Figure 38.3 Nevit Dilmen, CC BY-SA 3.0
Figure 38.5 By Dougsim—Own work, CC BY-SA 3.0
Figure 39.2 Kim Scarborough, CC BY-SA 2.5
Figure 39.4 Photo by Davepape, released into public domain by its author
Figure 39.6 World Imagingown work, CC BY-SA 4.0
Figure 39.7 CC BY-SA 3.0
Figure 39.8 John Alan Elson, CC BY-SA 4.0 nimslo
Figure 39.8 John Alan Elson, CC BY-SA 4.0 nimslo
Figure 43.5 Photographs by Gregg McNeill, 2022, used by permission
Figure 43.8 Brittonie Fletcher, 2013, used by permission
Figure 43.18 Ralph A. Clevenger, 1999, used by permission of Ralph A. Clevenger
Figure 44.20 Judith Baker Waller, 2019, use by permission
Figure 45.1 Ky Lewis, 2021 (used by permission)
Figure 45.2: Brittonie Fletcher 2020 (used by permission)
Figure 45.3 Brittonie Fletcher 2020 (used by permission)
Figure 45.4 Chrystal Lea Nause, 2011, (used by permission)
Figure 45.5 Chrystal Lea Nause, 2007, (used by permission)
Figure 45.6 Erin Woodbrey, 2021 (used by permission)
Figure 45.7 Erin Woodbrey, 2020 (used by permission)
Figure B.1 Left: Public Domain. Right: By Chris 73, CC BY 3.0

A portion of the work for the first edition was carried out with the support of the University of Wisconsin Colleges sabbatical program.

References

Bonnarel W, Fernique P, Bienaymé O, Egret D, Genova F, Louys M, Ochsenbein F, Wenger M and Bartlett J G 2000 The ALADIN interactive sky atlas: A reference tool for identification of astronomical sources *Astron. Astrophys. Suppl. Ser.* **143** 33–40

Cooke T and Kinneberg C (ed) 2014 *The Photography Book* 2nd edn (London: Phaidon Press Limited)

Lasker B M, Doggett J, McLean B, Sturch C, Djorgovski S, de Carvalho R R and Reid I N 1996 The Palomar–ST ScI Digitized Sky Survey (POSS–II): preliminary data availability *Astronomical Society of the Pacific Conf. Series* 101G H Jacoby and J Barnes (Astron. Soc. Pac. Conf. Ser.), 88

Wood R W 1906 Fish-eye views, and vision under water. *London, Edinburgh Dublin Phil. Mag. J. Sci.* **12** 159–62

Author biography

John Beaver

For over 20 years, John Beaver has used old processes to make new negatives, often in ways that can only be realized as a print with digital scanning and printing. This includes his development of the cyanonegative process, innovative work (in collaboration with Teresa Patrick) with instant film, and most recently his development of an accelerated, unfixed printing-out process he calls (perhaps annoyingly) 'ephemeral-process photography.'

He is Professor of Physics and Astronomy at the Fox Cities Campus of University of Wisconsin Oshkosh, where he teaches physics, astronomy, photography and interdisciplinary courses. He earned his BS in physics and astronomy in 1985 from Youngstown State University and his PhD in astronomy in 1992 from Ohio State University. His published work in astronomy is on the topics of spectrophotometry of comets and gaseous nebulae and multicolor photometry of star clusters.

He has exhibited photographs in many juried competitions in Wisconsin, Minnesota, Ohio, New York, Louisiana, Missouri, Oregon, Colorado, Illinois, and Barcelona, Spain. He has had several solo exhibitions, as well as joint shows with artists Judith Waller, Diana Ludwig, Dawn Patel and Teresa Patrick. Beaver has long been involved in art-science collaborations (many with artist Judith Waller) in the classroom, at academic conferences, and in art galleries and planetaria.

Some of John Beaver's photography can be seen at http://www.JohnEBphotography.com.

Part I

Some preliminary ideas

Chapter 1

What is science? What is art?

One purpose of this book is to use photography as a point of departure to learn a bit about both science and art and how they relate to each other. It may be helpful to take a brief tour of the nature of each in turn before really diving into the material. The difficulty is that there is much more to both art and science than what can be captured in any kind of short description. So I will not really answer the questions 'What is science?' or 'What is art?' in any comprehensive way or with nice one-sentence definitions of these two odd and complex things that humans do.

I will say only a little about the nature of science in general and then almost nothing about the overall nature of art. This is mostly because I am much more qualified to attempt such a 'big picture' description for science. I have been teaching science for a few decades now, mostly at an introductory level to nonscientists. I have had to think long and hard about just what it is that I am teaching and why I am doing it. My relationship with art—on the other hand—is less formal, more personal and more recent. Please note that in what follows I am talking about the natural sciences. The social sciences have much in common but also much that is different, and I leave it to others to talk about them.

Much has been written about the nature of science, and it is a subject of ongoing debate. My brief and very incomplete discussion follows the general approach of Sokal and Bricmont (1998, chapter 4) but is also informed by Chomsky (2000, chapters 4 and 5); it should be taken as only one approach to this complex topic.

1.1 The coherence of our experience

One key aspect of science is that it favors explanations that economically tie together, accurately, a broad range of experience that otherwise would seem to be connected only by coincidence. What may seem on the surface to be a coincidental connection is instead evidence that there is a deeper understanding to be sought, in which all of the facts come together in a natural way.

doi:10.1088/978-0-7503-3703-8ch1

© IOP Publishing Ltd 2022

When a scientific explanation works well at connecting the seemingly unconnected, we are left to consider that either there is something to it or the Universe is conspiring, as a sort of joke, to make it only seem as though the explanation is correct. It is tempting to give a simple, short example right here; the problem is that for any real example, the explanation is always a very long story.

This is what happens when we find a compelling explanation in science: many seemingly-disconnected facts all fit exactly into one simple scheme, called a theory. This theory allows one to correctly predict or explain, again and again, new facts that one has not yet even looked for or considered. So, we are then left with three choices.

1. It is simply a coincidence that our theory gets it so right, so often.
2. There is a conspiracy by some Cosmic Intelligence, either malevolent or at least with an odd sense of humor, to arrange things intentionally so that it only looks like our theory is correct when actually the theory is completely wrong.
3. There is some kind of truth to our theory. It is at least on the right track, at the moment, although details will likely need to be modified as we learn more. Perhaps some day we will see it, not as something that stands alone, but rather as only a part of a larger, more complete theory.

Science chooses the third option whenever it gets to the point that the first option seems too unlikely to take seriously. In science we reject the second option out of hand, with neither argument nor apology. As a human being, one may very well accept option two in a particular case; science is only one of the many things we humans do. But whatever option two is, it is not science.

1.2 Truth in science

We never prove a theory true in science. When a scientist says (quoted in a news story, for example) that a particular theory has been 'proven,' they are being somewhat glib, and they do not mean it in the strictly mathematical sense.

Rather, they mean that the theory is compelling. That is, it is rational to believe the theory has at least a conditional and tentative truth. More importantly, the evidence is great enough that it is, in a sense, irrational to believe the theory is completely wrong.

To say that a scientific theory has been proven is to say that of the options in section 1.1, option one is—in the face of new evidence—too unlikely to take seriously. So we are faced with either option two or option three, and if we are acting as a scientist then we dismiss option two out of hand and choose option three.

Of course, in real life, it is never so clear cut as I have laid out here. It is always the case that evidence is limited and uncertain. This means that for any given theory that gets all of the available evidence correct—that 'gets it right,' so to speak—there will also be an infinite number of other theories that do the same thing. So why then, in a given case, do we pick the particular theory we do?

Well, often, we do not. We disagree, at least for a while. However, we do pretty much agree—as scientists—that the theories that explain more, but still manage to do it with less complexity, are the theories that are more likely to be correct. This principle is sometimes called *Occam's razor*, and it serves as a rule of thumb for choosing between competing theories with equal explanatory power. Even given Occam's razor, it is not always clear which theory is the least complex; there are always different ways to look at it. Therefore, we often disagree about the details in the short term, only approaching overall agreement in the long term as more is learned.

Clearly, this is not really a proof at all; any particular scientific theory, no matter how compelling, is never the only explanation possible for all of the evidence. Since there are other logically possible explanations, the theory has not been proven in the mathematical sense. Rather, the scientist makes a less-strong claim about their theory—simply that it seems unreasonable to disbelieve it.

1.2.1 Proving a theory false

We cannot really prove a scientific theory to be true, but it might seem that it should be easy to prove that an untrue theory is false. In theory (pun intended), a theory is falsified if it predicts something in particular to be true, and then we go check to find that the prediction was incorrect. It would seem that it should be easy to prove a theory incorrect: simply find one fact that it gets wrong.

Surprisingly, even to prove a theory false is, in practice, rarely a straightforward process. Maybe the observed 'fact' is wrong but not the theory. Furthermore, real theories are usually subtle. That is, real theories use simple principles that—when applied to particular circumstances—predict extremely complex results. Since the world is a complicated place, applying any given theory to a particular real-world situation can be surprisingly difficult. In fact, it is often too difficult to do exactly.

Newton's laws of gravity and motion are a good example. These laws are simple enough that, when expressed in the correct mathematical language, they can be written on the side of an envelope (see figure 1.1). However, it can be maddeningly difficult to figure out exactly what they predict in seemingly simple real-world situations. It is so difficult that we are usually forced to make (hopefully) reasonable approximations. This leaves open the possibility, when a theory seems to give the

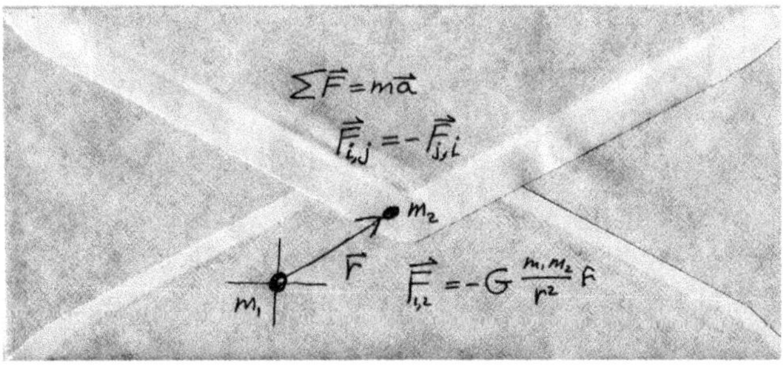

Figure 1.1. Newton's laws of motion and gravity are simple in a mathematical sense. When expressed in the right mathematical language, they can be written on an envelope. However, applying them to real-world situations can be maddeningly complex.

wrong answer in a particular case, that it is not the theory that is incorrect. Rather, the issue may be that we have made invalid approximations when we applied the theory.

In the nineteenth century, it was noticed that the application of Newton's laws to the orbit of the planet Mercury gives answers that are just slightly different from the planet's actual measured orbit. Faced with this discrepancy, should 19th-century scientists have immediately rejected Newton's laws and looked for something better? Certainly not, for there were many imaginable explanations for the discrepancies in Mercury's orbit that did not conflict with Newton's laws (several were suggested at the time). Therefore, since Newton's laws worked so enormously well in so many zillions of cases, it would have been unreasonable to reject it because of this one discordant fact when other explanations could easily be imagined.

I have chosen this example for its irony, because it turns out—in retrospect—that the discrepancy in Mercury's orbit really *is* caused by a failure of Newton's laws of gravity and motion. In 1915, Einstein superseded these laws with a more elaborate and complete theory called *general relativity* (GR). One of the first things Einstein demonstrated was that GR gives the same answer as Newton's laws whenever Newton's laws agree with experiment. But GR also gives the right answer for Mercury's orbit, whereas Newton's gravity gives a slightly incorrect answer.

A very tiny discrepancy in the orbit of one planet seems like a poor reason to discard Newton's laws, a wildly-successful theory for 250 years, and replace it with something much more complex and strange. However, correcting this discrepancy was not Einstein's primary motivation. Furthermore, this successful prediction (postdiction really) was only one of the many reasons (and not the most important one) why most other physicists quickly adopted GR as the new best theory of gravity (Weinberg 1992, chapter 5).

In short, Einstein provided no 'proof' that GR is correct. Instead, he argued that —as a scientific theory—it is compelling. So, what then was his argument? It is a long story![1]

Back to the 'failure' of Newton's Laws regarding the orbit of Mercury; this example is the exception. Every day, facts are discovered that, if they are taken literally, conflict with some tried-and-true theory. Upon closer examination, it is almost always the 'facts' that are wrong, not the tried-and-true theory. Therefore, it is prudent to be conservative and require extraordinary evidence for extraordinary claims.

1.3 Operational definitions

One necessary tool for the physical sciences is the *operational definition*, which is not at all like a dictionary definition. The dictionary appeals to our existing intelligence and knowledge, and most of the words defined have meanings that are somewhat fuzzy and vague by nature.

[1] There are many excellent books on this topic, many of which are accessible to the lay reader; Rucker (1977) is a good example.

In the physical sciences we must define our terms so they are useful not just for describing our existing knowledge, but also for extending it in ways that we cannot predict ahead of time. Additionally, the definition must mean exactly the same thing to everyone.

Fortunately, this is not quite so difficult as it sounds. Instead of tying ourselves up in knots writing a whole book just to carefully define a single word, we instead describe a *procedure*—a set of steps to follow. Basically, the procedure is a recipe. 'Do this, now do that, attach this thing here, read the meter, plug that number into this equation and that thing that comes out in the end—that is what I am talking about.' A given physical quantity is defined by the very method that is used to measure it.

A good example is the quantity mass, which is defined by what is essentially a set of instructions for performing a particular experiment. When this experiment is performed, the concept of mass naturally arises, and is thus named in context. Interestingly, it has long been known that there are actually two very different operational definitions for mass—one in terms of inertia and the other in terms of gravitation. Experimental evidence over 200 years has shown that these two very different definitions give, essentially, the same answer. This was considered to be a strange and mysterious coincidence until Albert Einstein resolved the issue in 1915, in the context of developing his new theory of gravitation: GR.

The use of operational definitions means that terms in physics are defined much more precisely than are the words we use in our everyday speech. This means that, for the most part, if one particular word (velocity, for example) is used correctly in a given physical context, then every other word used in the same context would be incorrect. This is in contrast to the more vague and fuzzy meanings of the words we use in our everyday speech, where the meanings overlap somewhat; one word might be the 'best,' but often several other words are also not wrong. The situation is, unfortunately, made more complicated by the fact that many words in physics are pronounced and spelled just like words we use in our everyday speech, so it is tempting to wrongly apply those same vague meanings in the context of physics. Some common offenders are 'force,' 'energy' and 'momentum.' These words have precise operational definitions in physics. In a given physical context, in a given sentence, only one is correct; the others are wrong.

1.4 Inspiration and perspiration

Einstein's famous phrase that 'genius is 99% perspiration and 1% inspiration' has much truth in it for both science and art. An artist needs many tools to draw from in order to act upon inspiration when it arises. It does no one any good if genius is only in the mind, and it cannot get out.

A good example is *The bird cage*, which can be seen in figure 1.2. It received a little bit of attention; it was selected for a couple of national juried exhibitions—one in New York City. I certainly would not say it is a work of 'genius,' whatever that means exactly. However, I do believe one can say some good things about that particular picture.

Figure 1.2. *The bird cage*, John Beaver 2004.

How did I do it? It is such a long story that I am not sure even where to begin. This is a good thing because if I did know where to start, I might proceed. If I proceeded to recount all that went into that picture, it would be a very boring for the reader indeed. It is not that there is nothing interesting to say about how that picture came to be. But *most* of the story is mundane and what can only be described as a long slog through a lot of work.

This example is particularly illustrative because, in order to make that picture, I had to develop a new photographic technique that I call *cyanonegative photography*, and much of the power of this particular photograph is intimately connected to the technique used to make it.

To take advantage of inspiration when it arises, one needs tools. For a musician, the seemingly endless scale exercises are part of the necessary preparation for creative acts, such as improvisation and composition. A photographer must understand principles of two-dimensional design in order to get more than the lucky shot, and this takes practice. Without having studied many topics very hard, an astronomer is just a person stumbling around in the dark. So, part of the purpose of this book is simply to offer some useful tools. Make of them what you will!

1.5 Criticism and self-esteem

Criticism is an absolutely necessary part of both science and art. To succeed in either, one must have a thick skin. Perhaps I should put it differently; it may sound as though I am saying that one must develop an ability to tolerate abuse, and that is not at all what I mean. Rather, one needs to be able to distance oneself and look at one's own work as if it were seen from the outside.

This ability to look at one's own work from some distance is necessary in order to hear the criticism of others without being paralyzed by it. Do not take it personally!

Notice the exclamation mark. Instead, criticism can be seen as information—sometimes useful, sometimes not—rather than a statement about one's ultimate worth as a human being.

The role of criticism in art is well known, but it is every bit as important in science. A successful scientist internalizes the positive dynamics of criticism early on, and in fact tries very hard to anticipate it. Without looking critically at one's own work, how else can one tell whether or not it is even finished? One must actively look for flaws to find and address them.

As I write this, I am faced with a dilemma[2]. I have been working for three years on a photometric analysis of a single star cluster in the constellation Scutum. In collaboration with a colleague from University of Wisconsin—Oshkosh, I have been working on a photometric analysis of this cluster. Although the cluster is an important and famous one, no one had previously studied this cluster in the particular way we have.

The problem is that our data are not fully cooperating, and it is my part of the analysis that is problematic. Much of the analysis is as one would expect, but some of it makes no sense. Meanwhile, we discover (literally today, as I write this) that someone else will soon present similar data on this very same cluster. What to do?

Well, we make the best of it. I work very hard to get the most coherent, interesting, and honest results I can out of our data—and I do it as fast as possible. Perhaps all of that work will come to something; perhaps it will not[3]. In any event, I have learned a lot by doing it, and I now have many new tools at my disposal. When inspiration arises at some later date, it is more likely that I will be able to make the most of it.

Self-criticism is particularly important for photography because it is so easy to quickly accumulate a lot of photographs. A big part of making a good photograph is the ability to tell the bad from the good—in order to show the good ones to others and throw away (or archive) the bad ones. Self-criticism takes practice because it is easy to become attached to one's own photographs—even when they are awful.

Once, lying in bed awake at 3 A.M., I had a great idea for a series of photographs. I will not go into the details. Suffice to say that it was a clever idea, but pulling it off required a significant amount of work. I finished the first couple of photographs, of which I was very proud, and excitedly began the detailed planning for the rest of them. However, events required me to leave town for a while, and so I had to temporarily set this project aside.

When I returned, I pulled out my notes and the two photographs I had already taken, and I really looked at them. The pictures were pretty bad. The more I stepped back from myself and tried to see the finished product I had planned through the eyes of others, the more I realized that my whole idea was rather lame.

So it goes. Afterward, was it difficult to part with a project with so much time invested? No, not really; I would make other pictures instead. Was the time I spent

[2] I wrote this sentence in November of 2012.

[3] By December 2013 all was well. The research project finally came together and we published a paper (Beaver *et al* 2013).

wasted? Possibly, but I doubt it. Much of the experience of thoughtfully, but wrongheadedly, making those bad pictures will find practical use some day, some way. I am just grateful that I stopped myself before I hurt someone by hanging those dreadful things on the wall.[4]

1.6 Looking at art

How can one look at something and know whether or not it is art? In the novel *Bluebeard* (Vonnegut 1987, p 148), the protagonist gives the following response (attributed to the painter Syd Solomon) when asked how to tell a good painting from a bad one: 'All you have to do, my dear…is look at a million paintings, and then you can never be mistaken.'

To know how to look at art, there is no substitute for looking at a lot of it. Go to museums as often as possible and look at not only photography, but also painting and sculpture. As a simple, practical matter to get much more out of this book, look at every single picture in its companion, *The Photography Book* (Cooke and Kinneberg 2014), read the descriptions and consider the cross-references.

References

Beaver J, Kaltcheva N, Briley M and Piehl D 2013 Strömgren H-β photometry of the rich open cluster NGC 6705 (M 11) *Publ. Astron. Soc. Pac.* **125** 1412

Chomsky N 2000 *New Horizons in the Study of Language and Mind* (Cambridge: Cambridge University Press)

Cooke T and Kinneberg C (ed) 2014 *The Photography Book* 2nd edn (London: Phaidon Press Limited)

Sokal A and Bricmont J 1998 *Fashionable Nonsense: Postmodern Intellectuals' Abuse of Science* (New York: Picador)

von Bitter Rucker R 1977 *Geometry, Relativity and the Fourth Dimension* (New York: Dover)

Vonnegut K 1987 *Bluebeard* (New York: Dell Publishing)

Weinberg S 1992 *Dreams of a Final Theory* (New York: Pantheon Books)

[4] One of the pictures, on further reflection (and input from others) turned out not to be so terrible, and so I did eventually put it in an exhibition. It was better standing on its own and would have been much diminished in the context of the overblown, multiphotograph project I had originally envisioned.

Part II

The nature of light

IOP Publishing

Photography (Second Edition)
Physics and art in focus
John Beaver

Chapter 2

What light is

The properties of light underlie much of photography; it is, after all, the 'photo' in the word itself. Light has several different aspects, and we will look at each of these in turn. In particular, light has a *wave-like* nature, a *particle-like* nature, and a *geometric* nature.

When a photographer speaks generally of light, they often mean a combination of several different things (see figure 2.1). One of these is the intensity of the light on the subject in relation to how the camera will need to be set for a proper exposure. The photographer may also speak of a subtle combination of the intensity, angle and color of the light on the subject. For example, the magic hour of one to two hours before sunset is famous for its 'light.' This reputation comes from a complex

Figure 2.1. *Algae no. 2*, John Beaver 2005. A photographer could mean many different things when talking about 'light' for a photograph such as this.

doi:10.1088/978-0-7503-3703-8ch2

© IOP Publishing Ltd 2022

combination of many aspects of light, only some of which are physical. Just as important are aesthetic concerns that have as much to do with the human mind as with physically definable aspects of light.

We will take up many of these issues throughout *Photography: Physics and Art in Focus*, but in this chapter we concentrate on some strictly physical aspects of light that are independent of photographic concerns. A better understanding of these physical properties will give us a stronger foundation for understanding some of the more subtle ways in which light relates to photography as an art.

There is no perfectly adequate two-sentence definition of light, so I will simply describe—as best I can—its properties. There is a unified, coherent physical description of light, and it is quite a long and complex story to go through in detail. However, there are a few ways to approach the physical nature of light that, although simplified in various ways, still manage to capture the majority of its properties.

The biggest simplification is to describe light as a series of *rays* that travel in straight lines until altered by various devices, such as lenses or mirrors. This is the subject of *geometrical optics*, and we take that up in chapter 6, as it has many practical applications for photography.

In this chapter we go a bit deeper, but—even so—there are two approaches, each of which works best for particular circumstances. One is to model light as a stream of particles, and the other is to view light as a wave phenomenon. It is the latter approach that turns out to be most useful for the purposes of photography, and so we will spend more effort on waves than particles. The particle-like nature, however, is crucial for an understanding of the workings of light *detectors*—a subject we take up in more detail in part VI.

If we can describe light in these three seemingly different ways, then what is light really? Well, it is what it is, and its physical nature is a bit unlike anything in our direct intuitive experience. In order to fully answer that question, we would have to go into a full description of its physical nature, and that means a study of *quantum physics* and much else that is beyond the scope of this book. Suffice to say that the full quantum description of light is a coherent theory that both incorporates and unifies (and thus explains) its wave-like, particle-like and ray-like natures.

We will describe light separately as waves, particles and rays and try to lay out under what circumstances each model is most appropriate. Although this approach is not entirely correct, and strictly speaking is not entirely coherent either, in practice it works most of the time.

2.1 The speed of light

Light carries energy from one place to another, so events in one location can influence events at another location. How fast does this influence travel? Very fast; try to open the refrigerator door before the light comes on.

Many years ago, my friend Doug Fowler bought a new Chevrolet half-ton pickup truck. It was the 'custom deluxe' model, meaning it was so stripped down and featureless that it lacked even a radio. It came off the assembly line in Flint,

Michigan in 1976. Doug drove the car to college at the University of Montana, regularly driving it back and forth between the Rocky Mountains and his home town of McDonald, Ohio. Eventually, Doug moved back to Ohio, and that is where I met him—while I was a student in the physics and astronomy department at Youngstown State University.

I went to graduate school at Ohio State, and Doug went back out West—this time all the way to Bellingham, Washington. Eventually, he ended up back in Ohio again. Although he drove the truck westward many times, including to visit me in Flagstaff, Arizona one summer while I was working at Lowell Observatory.

Finally, around 1991 or so, I ended up with his—by then—much-used truck. When I moved to Wichita, Kansas, in 1993, all of my possessions went West in that truck. It was already on its second engine, and even that one was unhappy about the trip. I had to change the oil twice on the way because the piston rings were so shot. About a month or two after I arrived in Wichita, it went to its final resting place in some salvage yard in Sedgwick County, Kansas.

What did the odometer read after all of that? Very nearly 186 000 miles—the distance light travels in one second.

> The distance light travels in one second is comparable to the total distance traveled by an automobile that has been through a dozen or so years of moderately-heavy driving.

This speed, the speed of light in a vacuum, is such a fundamental quantity that it has its own special symbol, c. It has a value[1] of $2.998 \times 10^8\ \mathrm{ms^{-1}}$, roughly $300\,000\ \mathrm{kms^{-1}}$.

A speed is a distance per time, and with such a large speed as this, one can get a mental grip on either the distance or the time, but not both. If we talk about an intuitive interval of time—the second—then the distance light travels is outside our direct intuitive experience. 300 000 km is nearly seven times around the circumference of Earth.

We can make the light–travel distance manageable: 30 cm, for example, just slightly under one foot. Therefore, we could express the speed of light as a little under one foot in a thousand-millionth of a second (1 ns): a speed of $30\ \mathrm{cmns^{-1}}$. Now it is the time that is far too small for our direct experience. In section 2.1.1 we try to meet both the huge distance and the small time somewhere in the middle.

2.1.1 The speed of light with a shortwave radio

I moved from Wichita in the summer of 1997, again hauling (in a different vehicle, obviously) my belongings halfway across the country—this time eastward, to Appleton, Wisconsin. Because I was starting a new job teaching physics and astronomy at a two-year campus of the University of Wisconsin, I spent much of that summer preparing for my upcoming courses.

---

[1] A brief overview of physical dimensions, units and scientific notation can be found in appendix A.

One of the first things I did upon moving into an apartment was to set the clocks. In my family[2], this had always been accomplished by using a shortwave radio to tune in WWV—a transmitter broadcasting from Fort Collins, Colorado, operated by the National Institute of Standards and Technology (NIST). The NIST uses WWV to broadcast standard time signals on several shortwave frequencies.

On the evening of June 30, 1997 I tuned my radio to 10 MHz, the WWV frequency I most expected to deliver a good signal. Within less than 45 seconds, I heard something I had never heard before on WWV—a female voice. Typically WWV announces each new minute with a male voice. As soon as the female voice announced the minute, the familiar male voice repeated the same announcement.

I knew what was up. The conditions were just right, for the first time in my life, to receive not only WWV from Colorado but also WWVH—its sister station in Hawaii. For just this situation, the two stations use different voices for their every-60 second announcements, and one finishes before the other begins.

I immediately recorded the broadcast, and my quick thinking has paid off. For nearly a quarter century I have used this recording to torture physics and astronomy students. As the first laboratory exercise in many of my beginning courses, the students take a digital sample of my tape recording and use graphical analysis to measure the time of arrival of the two signals (Beaver 2000). Figure 2.2 shows the approximate paths traveled by the two radio waves as they arrived at my receiver in Wisconsin. Shortwave radio waves have the ability to bounce back and forth between the ground and layers of ionized gases (called the *ionosphere*) in Earth's

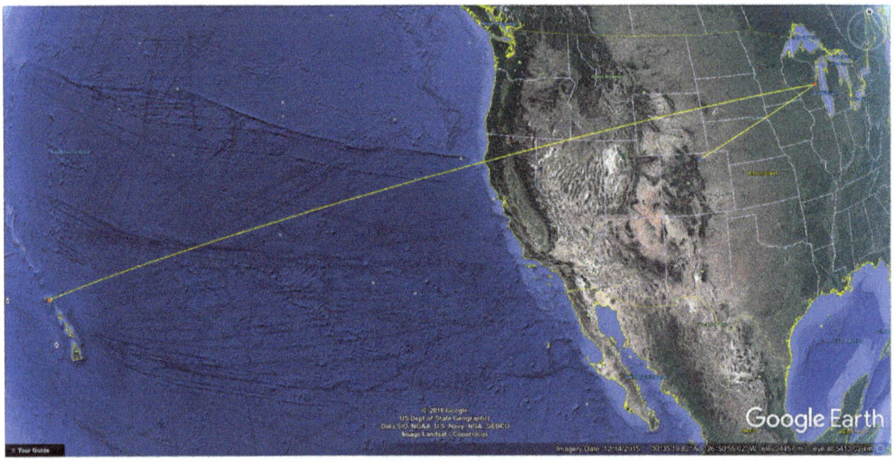

Figure 2.2. The paths of radio waves broadcasting simultaneous time signals from both Hawaii and Colorado and arriving in Wisconsin, as mapped by Google Earth. Since the signal from Hawaii had further to travel, it arrived slightly later. (Image copyright 2018 by Google, U.S. Department of State Geographer, image Landsat/Copernicus, Data SIO, NOAA, U.S. Navy NGA, GEBCO.) Recording available at https://doi.org/10.1088/978-0-7503-3703-8.

[2] My family included many amateur radio operators.

upper atmosphere. They can, in effect, bounce around the curvature of the Earth. This zig-zag route means the radio waves traveled a little bit further than what is shown here, but the difference is less than 10%.

What does this have to do with the speed of light? First, light is an electromagnetic wave, and both visible light and radio waves are but two examples of the many kinds of these waves. To measure the speed of radio waves is to measure the speed of light because radio waves are a kind of light, even though we cannot see them with our eyes.

And so we have the following. Two signals were sent simultaneously from two different locations. Both signals traveled at the same speed—the speed of light. Both arrived at my receiver in Appleton, but one had to travel a greater distance to get there. Thus, the signal from Hawaii arrived later than the signal from Colorado. If one listens carefully to my recording, it is just possible to hear that each of the clicks is doubled, as if a person tapped two fingers on the table but with one finger slightly behind the other.

See figure 2.3 for a graph of two seconds of the data. Just a quick glance at the time axis on the graph (my students do this more precisely) shows that there is a difference of about 0.02 seconds between the arrival of the signals. To calculate the speed, all one needs to do is divide the excess distance the Hawaii signal traveled by the excess time it took to arrive.

One can easily use an online tool such as Google Earth to see that the ground-level distance for the WWVH signal is about 6.9×10^6 m while for the WWV signal

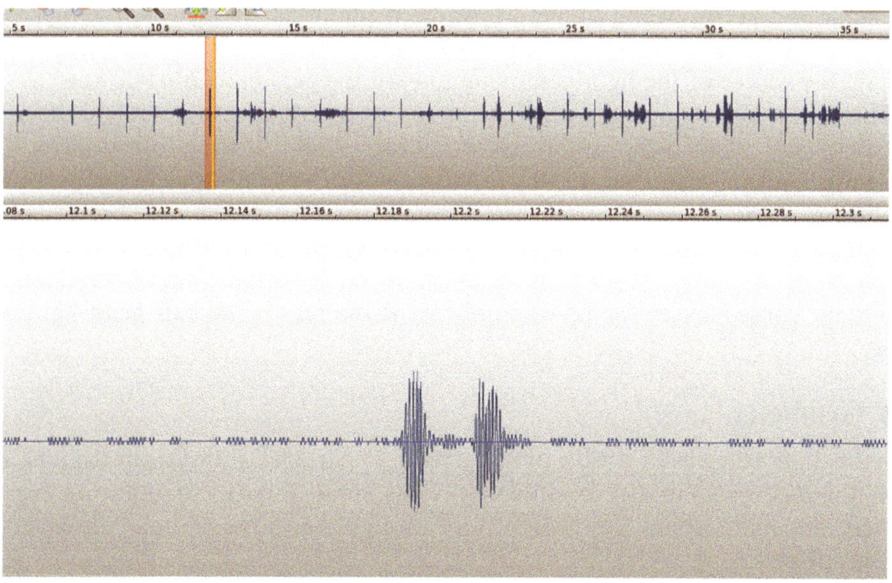

Figure 2.3. A graph of sound level vs. time for signals received from WWV and WWVH. **Top:** The signals arrive as 'clicks' once every second. **Bottom:** a magnified detail of the time of arrival of the first click shows that it is actually two closely-spaced clicks. The first signal arrived from Colorado, and about 0.02 seconds later the signal arrived from Hawaii. (Data graphed with the online twistedwave.com app. Copyright © Thomas Thiriez, Twisted Wave; reproduced with permission.)

it is roughly 1.4×10^6 m. If we divide the difference in those travel distances by the excess 0.02 s for the arrival of the WWVH signal, we have

$$\frac{6.9 \times 10^6\,\mathrm{m} - 1.4 \times 10^6\,\mathrm{m}}{0.02\,\mathrm{s}} = 2.8 \times 10^8\,\mathrm{m\,s^{-1}} \tag{2.1}$$

.

This is a little less than the speed of light, but remember that the radio waves really bounced between the ground and the ionosphere, so the distances used in equation (2.1) are a little too small. Taking that small difference into account gives the correct answer (Beaver 2000).

2.1.2 Relativity and the speed of light

The speed of light is unlike other speeds; all other speeds are relative. An automobile speedometer measures the speed of the car relative to the road and, incidentally, to the tree alongside the road. The speed relative to another car following behind is quite different. This difference is not just philosophical; it has real physical consequences, as would be immediately evident if a car came into contact with one or the other.

The speed of light, on the other hand, is absolute. It is the same for all observers regardless of their relative motion. This fact may seem impossible, but it is well established by experiment. Light always travels (in a vacuum) at the same speed regardless of any relative motion between the source or observer. If I shine a flashlight at a person while running towards them at 200 000 $\mathrm{kms^{-1}}$, the light goes both away from me and toward the person at the same speed of 300 000 $\mathrm{kms^{-1}}$, even though the source of the light is moving toward the person at 200 000 $\mathrm{kms^{-1}}$.

This basic fact about light was first hinted at by Maxwell's complete theory of electricity and magnetism, formalized in the early 1860s. It took four decades until sense was made of that idea; it is the foundation for Einstein's special relativity, first published in 1905.

When light interacts with matter—a piece of glass in a camera lens is a good example—then it may travel at a significantly slower speed. This interaction has important consequences for photography: it is the reason we can bend light with lenses.

2.2 Geometry

Whether light behaves as a wave or as a stream of particles, it travels in straight lines until it interacts with matter in some way. Thus, some of the most important aspects of light are more about the geometry of straight-line rays of light, deflected by various objects.

In many circumstances we can analyze light with geometry alone and apply select details from our knowledge of electromagnetic waves only when it is absolutely necessary. This approach is called geometrical optics, and we will use it often in the chapters that follow. When we talk about a ray of light, we are implicitly using this geometrical approach because it is convenient, even though we know light is really much more complicated than this. When we need to use the more-correct wave or particle models of light, we will. Otherwise, why make things unnecessarily complicated?

2.3 Waves

Wave phenomena allow the transfer of energy from one place to another without actual stuff having to make the trip (see figure 2.4). Move a hand up and down in the water at one end of the bathtub, and eventually the rubber duck at the other end bobs up and down too. Yell, 'Hurry up!' to a roommate in the bathroom and very quickly their eardrums vibrate, even though no air from the yeller's mouth traveled to their ear.

Light has this nature too, although it is not 'stuff' vibrating like most other waves. A light wave alters the very electrical and magnetic properties of space itself, even in a vacuum where there is nothing of substance to move. We call these electric and magnetic properties *fields*, so an *electromagnetic wave* is a changing pattern of the electric and magnetic fields[3]. These changing fields can—in turn—affect matter, and that is essentially what happens when light has some physical influence. Since matter has its own electric and magnetic properties, it is affected by the changing electric and magnetic fields of a passing electromagnetic wave.

One might point out that if no stuff actually makes the trip when a wave moves from one place to another, what do we mean then by the 'speed' of a wave? If the wave is very complex, then this question may have a complex answer. For a simple wave, there is a simple answer—a wave is a repeating pattern in space that moves as time passes. The speed of the wave, then, is the speed at which this pattern moves.

Figure 2.4. *Shoreline and white caps*, John Beaver 2006.

[3] See, for example, Beaver (2020, sections 11.3 and 12.2).

This calls to attention the fact that a wave, by its very nature, is extended in both space and time. It is a pattern spread out in space that moves with time. Even at a given location, the wave changes as time passes. There is no real meaning to assigning a precise location and time to a wave. We can talk about what the wave does at a particular place and time, but we need to consider all of the other places and times in order to describe any particular wave. It is very unlike—say—a stone moving through space, which has a much more precisely definable position at any specific time.

Like all waves, light in its purest form has four basic attributes: speed, amplitude, wavelength and frequency. We will consider each of these in turn.

2.3.1 Amplitude

As a light wave passes a point in space, the electric and magnetic fields change back and forth. The fields themselves actually point perpendicular to the direction of motion of the wave. As the wave passes, they alternately grow in strength, reach a maximum value, weaken to zero, then reverse direction and do the same. The maximum strength of the electric or magnetic field as the wave passes is called the *amplitude* of the wave.

For light, the amplitude is related to the brightness of the light. All else being equal, the larger the amplitude—the greater the maximum strength of the electric or magnetic field as the wave passes—the brighter the light. If one doubles the amplitude, the brightness of the light does not also double—instead, it quadruples. Therefore, the brightness of the light scales with the amplitude squared.

This scale by squaring has another consequence. The electric and magnetic fields alternately switch back and forth in direction, and we can describe this with positive and negative numbers. However, the brightness of the light is proportional to the square of this, and the square of a number is always positive, even if the number being squared is negative. Thus, although the fields reverse directions, it is only the *magnitude* of the strength of the fields that represents the amplitude of the wave. This may seem to be a distinction without a difference, but it has profound and surprising consequences whenever two or more waves interact with each other in a phenomenon called *interference*. We will revisit this idea in section 3.6.

2.3.2 Speed, wavelength and frequency

Let us consider a wave traveling down a string in the way that a wave travels down a garden hose when one end is waved up and down. For now, let us imagine the string to be infinitely long, so we can ignore the interesting complication of waves reflecting off the ends of the string. If one were to take a flash picture of the wave, freezing it in time, a repeating pattern in space would be evident. One can represent this pattern in the form of a graph, as in figure 2.5.

The meaning of the amplitude of the wave is clear from the diagram. However, there is another equally-important measure. The distance over which the wave completes one repetition is called the *wavelength* of the wave. It is a length, so our SI unit (see appendix A) of wavelength is the meter. By historical convention, we use

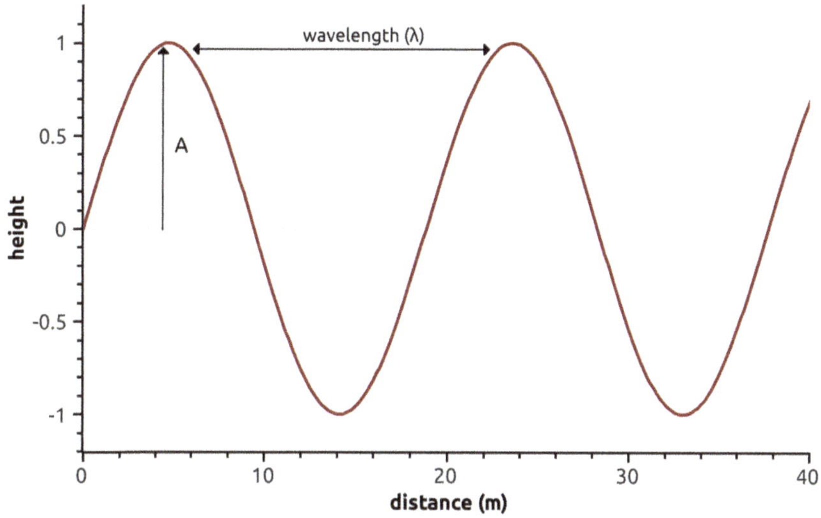

Figure 2.5. Wavelength (λ) and amplitude (A) of a wave. Notice that the horizontal axis of the graph is length. The wavelength is the distance over which the wave repeats itself, at a particular point in time.

Table 2.1. Useful small units for wavelength.

Unit	Symbol	Meters	Notes
Micron or micrometer	μm	1×10^{-6}	1000 μm = 1 mm
Nanometer	nm	1×10^{-9}	1000 nm = 1 mm
Angstrom	Å	1×10^{-10}	10 Å = 1 nm

the Greek letter λ (lambda) as our symbol for wavelength. The human eye is only sensitive to a very narrow range of wavelengths, between about 0.4 to 0.7 millionths of a meter, so it is this range of wavelengths that defines what we call *visible light*.

These tiny lengths make the meter a bit cumbersome, and other smaller units are more commonly used when referring to the wavelength of light. The most common of these are described in table 2.1. Thus, we can represent the range of visible wavelengths as '0.4 mm to 0.7 mm.' If we prefer, we could say the same thing as '400 nm to 700 nm.' As an alternative, if we can find the special symbol in our word processor, we could say with complete equivalence, '4000 Å to 7000 Å.'

We can also represent a wave in terms of changes in time, rather than in space. Figure 2.6 looks, at first glance, the same as figure 2.5, but look at the horizontal scale. Instead of showing the wave at different points in space (at a given instant of time), figure 2.6 shows the wave as time passes (but at some particular point in space). The *period* (T) of the wave is the *time* required for one repetition of the wave to pass a particular point in space, as the wave goes by. In SI units, the period is measured in seconds; however, for most electromagnetic waves of concern to

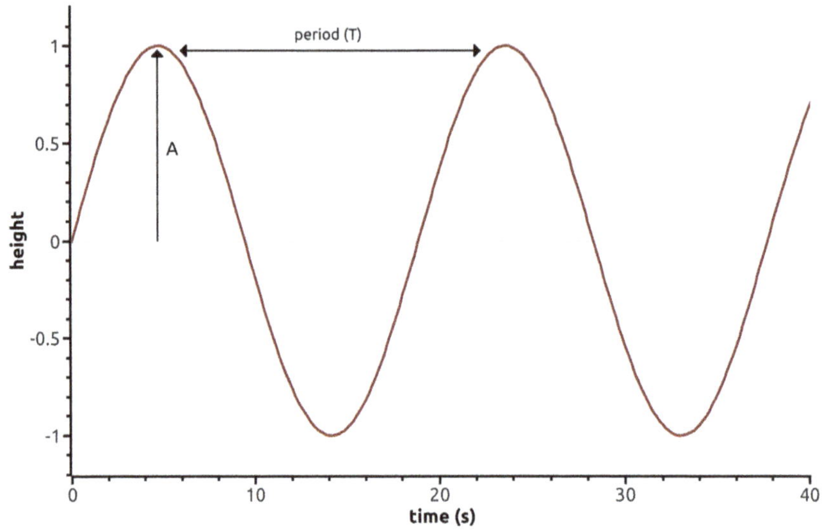

Figure 2.6. Period (T) and amplitude (A) of a wave. Notice that the horizontal axis is time. The period is the *time* over which the wave repeats itself *at a particular point in space*. The *frequency* (f) of the wave is simply one over the period.

photography, it is a tiny fraction of a second. More commonly, we refer to the *frequency* (f), the reciprocal, of the period.

$$f = \frac{1}{T} \tag{2.2}$$

The frequency then would be measured in inverse seconds, labeled s^{-1} or Hertz (Hz). It is the number of wavelengths that go by per second.

While the period for electromagnetic waves is typically a very small number, the frequency is typically a very large number. There are two reasons for this: first, the wavelengths of visible light are very tiny; secondly, the speed of light is very large.

Clearly, the faster is the speed of the wave, the more wavelengths would go by in one second. Also, the shorter the wavelength, the more wavelengths would go by per second, for a given speed. Thus, we have a relationship between the frequency (f), the wavelength (λ) and the speed of light (c).

$$c = f\lambda \tag{2.3}$$

Since the speed of light, c, is a constant, we can see that there is a relation between wavelength and frequency. Thus any given wavelength corresponds to a particular frequency, and vice versa. We can rearrange equation (2.3) as follows.

$$f = \frac{c}{\lambda} \tag{2.4}$$

$$\lambda = \frac{c}{f} \tag{2.5}$$

These are *reciprocal relations*; if frequency is larger, then wavelength is smaller and vice versa. It also means that, for light, we can choose either frequency or wavelength for our description. If given one, the other can be easily calculated.

Given the simple mathematical relation between the wavelength and frequency of light, it may seem practical that we should pick one to stick with and dispense with the other altogether. However, it is not so simple; frequency and wavelength each relate directly to quite different physical phenomena, so both are important.

When light interacts with an individual atom, for example, the atom knows nothing but what is happening locally—where it is. Thus, it is the *frequency* that is relevant for such interactions—how many times per second the wave motion repeats *at any particular point in space*.

There are other interactions where it is the wavelength that is more to the point, in particular the phenomena of interference and diffraction (sections 3.6 and 3.7). Place a wave in a confined space and interesting things happen that have everything to do with how the wavelength compares to the size of the 'box' within the wave is placed.

These observations are true for any wave, not just light. Consider a violin: stretch a string between two fixed points and try to excite waves of all wavelengths to travel along it. The reflections off the end boundaries will result in most wavelengths canceling each other out, leaving only a list of certain specific wavelengths that are 'allowed' by these conditions[4].

This list of allowed wavelengths corresponds to a reciprocal list of allowed frequencies, and the vibrating string thus vibrates air molecules at only those same frequencies[5]. This *sound wave* then propagates through the air, vibrating air molecules next to my eardrum, which causes it—in turn—to vibrate at those same frequencies. It is this set of *frequencies* that we hear as the complex pitch and tonal quality associated with the note from the violin. My eardrum neither knows nor cares about the wavelengths that correspond to these frequencies.

2.3.3 The electromagnetic spectrum

Different wavelengths (or frequencies) of electromagnetic waves interact with matter in different ways. Since both the absorption and emission of light are examples of such interactions, one would need different strategies to produce light of vastly different wavelengths. Likewise, different methods are required to detect light of very different wavelengths.

It was not known that light is an electromagnetic wave until the late 1800s. Furthermore, it was not until the twentieth century that most forms of

[4] In this particular case, an exact integer number of half wavelengths must be equal to the length of the string, or the waves will cancel each other out after several reflections.

[5] In practice, a vibrating string moves little air, so it must be attached to a wooden box—the front surface of which is moved by the string. It is the vibrating top of the box that is able to cause a significant amount of air to vibrate.

Table 2.2. The electromagnetic spectrum.

Name	Typical λ (m)	Typical size	f (Hz)
Gamma ray	$<1 \times 10^{-11}$	Atomic nucleus	$>3 \times 10^{10}$
x-ray	$1 \times 10^{-11} - 3 \times 10^{8}$	Atom	$1 \times 10^{16} - 3 \times 10^{19}$
Ultraviolet	$1 \times 10^{-8} - 4 \times 10^{-7}$	Virus	$7.5 \times 10^{14} - 210 \times 10^{16}$
Visible light	$4 \times 10^{-7} - 7 \times 10^{-7}$	Bacteria	$4.3 \times 10^{14} - 7.5 \times 10^{14}$
Infrared	$7 \times 10^{-7} - 1 \times 10^{-3}$	Protozoa	$3 \times 10^{11} - 4.3 \times 10^{-14}$
Microwaves	$1 \times 10^{-4} - 0.1$	Person	$3 \times 10^{9} - 3 \times 10^{12}$
Radio	>0.1	Building	$<3 \times 10^{9}$

electromagnetic waves were finally identified. Some types had previously been detected, but it was not recognized until much later that they were just different wavelengths of electromagnetic waves. Different ranges of wavelength of electromagnetic waves have different names, in part for historical reasons. Table 2.2 shows the ranges of possible wavelengths, along with their customary names. Taken together, this is called the *electromagnetic spectrum*. Keep in mind that the ranges of wavelengths or frequencies are only approximate; the boundaries are fuzzy and overlap each other. The names really come from the different ways in which we produce or detect them, and the methods to do so have changed over the years as technology has changed.

Let us very briefly consider each of these basic parts in turn. I will start at the long-wavelength bottom of the list; this may seem strange, but remember that long wavelength is the same as low frequency.

- **Radio waves** are made by moving an electrical current back and forth in a wire, and radio waves induce currents to oscillate back and forth in wires through which they pass.
- **Microwaves** can be thought of as very high-frequency radio waves. They can sometimes be made in the same fashion, but other processes (besides electronic circuits) are also used to produce them. Because of their shorter wavelength, they can be focused with special mirrors and more easily guided along paths.
- **Infrared light** is usually produced in ways similar to visible light, but the wavelengths are too long for the human eye to detect it.
- **Visible Light** is the name for the narrow range of wavelengths (about 400–700 nm) to which the human eye is sensitive. Different wavelengths of visible light produce different color sensations: violet for short wavelengths and red for long wavelengths with blue, green, yellow and orange in between.
- **Ultraviolet (UV) light** also is often produced in ways that are similar to visible light, but the wavelengths are too short to be detected by our eyes.
- **X-rays** have wavelengths similar to the size of individual atoms, so they usually interact with matter in ways that involve individual atoms on a one-on-one basis.

- **Gamma rays**, with their enormous penetrating power, interact directly with the tiny nuclei of atoms. Thus, they are associated most strongly with nuclear reactions.

We will discuss different examples in more detail as we go along. However, for most of *Photography: Physics and Art in Focus* we will consider the specific properties of visible light, and wavelengths of infrared and UV that are close to visible light because these wavelengths are of most practical interest for photography.

2.4 Particles

A wave is spread out in space, and it transfers more energy as more time passes; this fact will be very important when we consider the physics of exposure in photography. However, there are situations in which light does not act like this at all. Instead, it acts like a stream of particles (called photons), each of which individually delivers its energy all at once and at a particular place. It was the development of quantum physics, in the early twentieth century, that finally unified these two seemingly-opposed, wave-like and particle-like natures of light.

A particle affects other particles by way of *collisions*, which cause sudden changes at particular places. A wave, on the other hand, gradually causes things to happen over spread-out regions of space. The two ideas—waves and particles—seem on the surface to be utterly contradictory. It is beyond the scope of this book to go into detail as to how quantum mechanics unifies these two seemingly incompatible concepts, but it does.

Suffice to say that sometimes light behaves more like a wave and sometimes more like a particle. The particle-like picture for light is most important regarding how light is actually detected by a light-sensitive material—film, for example, or the charged-coupled device digital detector in a digital camera. By 'detection' we mean that light causes some physical change in the detector that can be recorded in some way, thus eventually producing an image. If one looks carefully enough, this physical detection occurs at particular instants of time and at particular locations on the detector. It is in this way, then, that light behaves like a stream of particles.

This is not to say that when light behaves like a stream of particles that the discussion of light as a wave will be invalid. For even when light behaves like particles, its wave-like nature is still important. In fact, there is always a fundamental connection between the two ways of looking at light.

We describe light as a wave in terms of its amplitude and wavelength (or frequency), whereas a particle would be represented by the energy and momentum it could transfer in a collision. One of the central results of quantum physics is that these wave-like and particle-like natures of light are explicitly connected to each other. If we denote the energy and momentum of a single photon by, respectively E and p, they are always related to the wavelength (λ) and frequency (f) of the light by the following equations.

$$E = hf \tag{2.6}$$

$$p = \frac{h}{\lambda} \tag{2.7}$$

In these equations the letter h represents a very tiny number, one of the fundamental constants of nature, known as the *Planck constant*. The overall consequence is that light of short wavelength or high frequency (when it is acting like a wave) is made of individual photons (when it is acting like a stream of particles), each of which has a high energy and large momentum.

Thus, one can take a source of light and allow it to interact with matter in a wave-like way. From that interaction, one can measure the wavelength. Then one can take the same source of light and allow it to interact in a particle-like way, and in the process measure the momentum of those particles (photons). If one does both, the measured momentum of the photons is related to the wavelength of the light by equation (2.5).

What about the brightness of the light? We have already seen that it is related to the square of the amplitude of the electromagnetic wave. But what is the 'brightness' of a particle of light—a photon? To answer this question for an individual photon is to delve into some of the subtle strangeness of quantum physics (we will consider these interesting issues somewhat more in chapter 4). However, there is a simple answer for a stream of many photons. For the light of a particular wavelength, a brighter light means that there are more photons per second.

References

Beaver J 2000 The speed of light with a shortwave radio *Phys. Teach.* **38** 172–174

Beaver J 2020 *The Big Picture: The Universe in Five S.T.E.P.S. (Synthesis Lectures on Engineering, Science, and Technology)* 2 (San Rafael, CA: Morgan & Claypool Publishers), 1–267

Chapter 3

What light does

3.1 Reflection, absorption and transmission

When matter produces light, it is called *emission*; chapter 5 describes several emission processes. Whatever the origin of a bit of light, many different things may happen when it encounters matter. We call this an *interaction*, and in what follows we describe several different types of light–matter interactions.

When matter travels through space, it carries both energy and momentum; the same is true for light. Unlike matter, light can pass through other light unaffected; light waves from multiple sources can coexist in the same place at the same time and still retain their separate identities. This basic fact leads to the rather surprising phenomena of *interference* and *diffraction*—subjects we take up in sections 3.6 and 3.7.

What happens when ordinary light, traveling through air, arrives at some new material—a piece of glass, for example? To help make sense of this, it is useful to talk about the *plane of interface* between the two materials: the material from which the light came and the different material at which the light has arrived. The plane of interface is the *boundary surface* between those two materials. A good example is the boundary surface one side of a glass window makes with the air next to it. The other side of the glass window would then form a second plane of interface with the air next to it. The plane of interface may be literally flat as in this example, but it could also be curved. Think, for example, of the boundary between a crystal ball and the air around it.

When light arrives at a plane of interface with a new material, three basic things can happen. Some of the incoming (incident) light can bounce off the material (reflection), some can pass through into the new material (transmission) and some can disappear altogether (absorption). Often all three of these happen at once to varying degrees.

The energy carried by light must be conserved. Thus, the energy of the incident light must be equal to the sum of the energies of the reflected, transmitted and absorbed light. That is to say, the energy carried by the incident light cannot just disappear; it must be divided up somehow between these three different things that happen to it.

doi:10.1088/978-0-7503-3703-8ch3

© IOP Publishing Ltd 2022

In the case of reflection and transmission, the light itself does not disappear. The energy from the incident light carries over to the energies of the reflected and transmitted light. Absorption is different; when light is absorbed its energy is transferred to other forms of energy (most commonly thermal energy). The energy as carried by light decreases, while other forms of energy (thermal energy for example) increase by the same amount.

A complication is that all three of these processes—reflection, transmission and absorption—depend on the wavelength of the light. Ordinary window glass is a good example. At a visible wavelength of 500 nm, window glass transmits a lot of light and reflects little. At wavelengths 20 times that, in what we call the thermal infrared part of the electromagnetic spectrum, window glass does the opposite; very little light passes through and much more of it reflects. Keep this in mind as we go along. What is good for the goose wavelength may not be good for the gander wavelength.

3.2 Specular reflection

When light bounces off a smooth and shiny surface, it obeys a simple rule known as the *law of reflection*. Figure 3.1 illustrates the geometry of this type of reflection. Here we have a reflective material on the right and incident and reflected rays on the left, labeled with the Greek letter θ (theta: its use is traditional for angles) with subscripts 'i' and 'r' to represent, respectively, 'incident' and 'reflected.' Specular reflections most commonly occur in two situations.

1. Light encounters a smooth, opaque but shiny (usually metallic) surface. This is the ordinary example of a mirror.
2. Light encounters a smooth interface between two different transparent materials.

If instead of a single ray, a whole bundle of rays, all parallel to each other, arrive at that same shiny surface, they all reflect by the same angle, since they all arrive at the same angle. This organized type of reflection is called a *specular reflection*, and it is the sort that is produced by a mirror. In section 3.4 we consider equally-important diffuse reflections that are not so organized.

We define the direction of these two rays according to an angle made not to the reflective surface, but rather to a line *normal* to that surface. Here the word 'normal' means 'perpendicular to a plane in all directions parallel to that plane.' Place a pencil

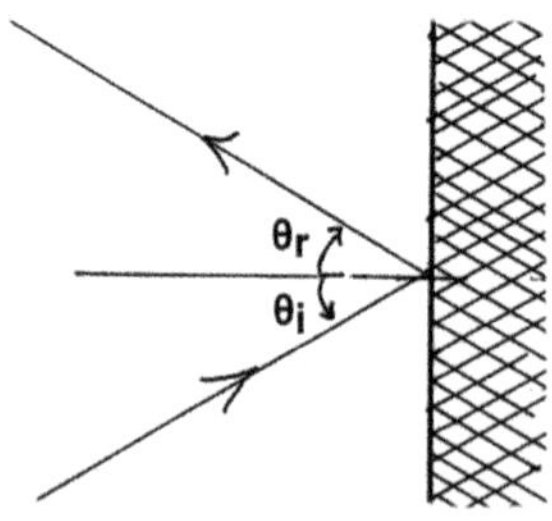

Figure 3.1. The law of reflection. The reflected ray of light makes the same angle with the surface as the incoming (incident) ray. Instead of measuring the angle the ray makes with the surface itself, it is more convenient to measure the angle made with a line normal (perpendicular in any direction) to the surface.

normal to a horizontal desktop, and a 90° angle can then be drawn between the pencil and the desktop in all horizontal directions (see figure 3.2). In this particular case, the pencil points vertically. This is an unambiguous way to describe the relation between a line and a plane, so we will use it whenever we describe the relation of a light ray to some plane (a new material the ray suddenly encounters, for example).

Given figure 3.1, the law of reflection can be expressed very simply: the angle of reflection is equal to the angle of incidence (see equation (3.1)).

$$\theta_r = \theta_i \tag{3.1}$$

But what if the reflective surface is curved? The same relation still holds, so long as we take our angles to mean they are made to the normal of the flat plane locally tangent to the curved surface. A plane tangent to a curved surface means it touches the curved surface at no other nearby points (see figure 3.3). We still have, at any given location, a local law of reflection. It is just that the curve of the surface makes the normal lines (and the reflected rays as well) point in different directions for different parts of the curve. A curved mirror can thus be used to converge or diverge parallel rays of light.

Figure 3.2. A pencil positioned normal to the plane of a horizontal wood floor.

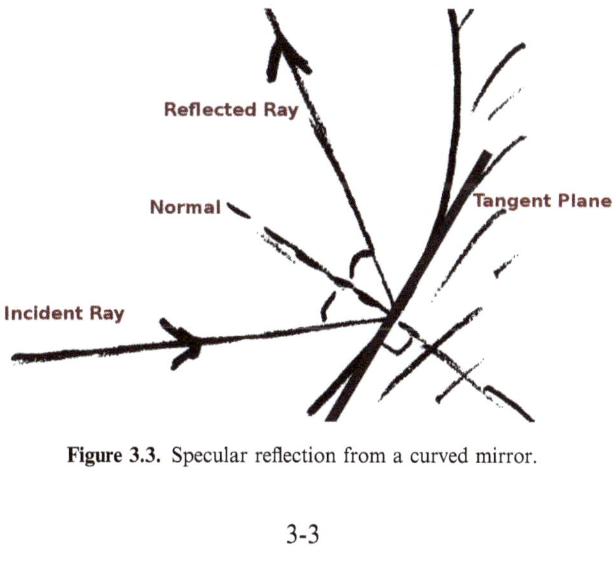

Figure 3.3. Specular reflection from a curved mirror.

3.3 Refraction

While the speed of light in vacuum is a universal constant, the speed is altered when it travels through a transparent material such as water or glass. This is because light is a changing pattern of electricity and magnetism (an electromagnetic wave), and physical materials (glass, for example) alter the electric and magnetic properties of space. In the cases of most interest to photography, materials such as transparent glass or plastic weaken electric fields, and this causes light waves to slow down.

It is helpful to define some terms. Let us say that we have two transparent materials, which we label 1 and 2. They could be water and glass, glass and air, air and acrylic, acrylic and water or any other combination of two transparent materials. As light travels from material 1 to material 2, the plane of interface is the two-dimensional (2D) boundary—the surface—where the two materials meet.

See figure 3.4: according to tradition I have labeled the angles with the Greek letter theta (θ), and I have put subscripts on them to denote about which side we are talking. The key finding is that when this happens, the light makes different angles in the two materials. This bending of light by a sudden change of material is called *refraction*.

The reason the two angles are different is because the speed of light suddenly changes when passing from material 1 to material 2. In the case I illustrate here, the angle with the normal is smaller in material 2 than in material 1. It turns out that this means that light must travel more slowly in material 2. The mathematical relation between the two angles is fairly simple, and it is called *Snell's law*, or *the law of refraction*.

$$n_1 \sin \theta_1 = n_2 \sin \theta_2 \tag{3.2}$$

Here n_1 and n_2 are pure numbers (without units or dimensions), and their values are properties of the two materials themselves. This *index of refraction* of a material is a number greater than or equal to one, and it is usually determined experimentally for a given transparent material.

The specific example shown in figure 3.2 illustrates one of the most important properties of the law of refraction: the angle with the normal is smallest on the side where the index of refraction is largest. As is the case for reflection, we apply Snell's law to a curved surface by measuring the angles with the normal to the plane tangent to the curved surface at each point. See figure 3.5 for an example of light refracted to the bottom of a lake by the wavy surface of the shallow water.

What does the law of refraction have to do with the speed at which light travels between the two materials? A purely geometrical analysis of light provides no

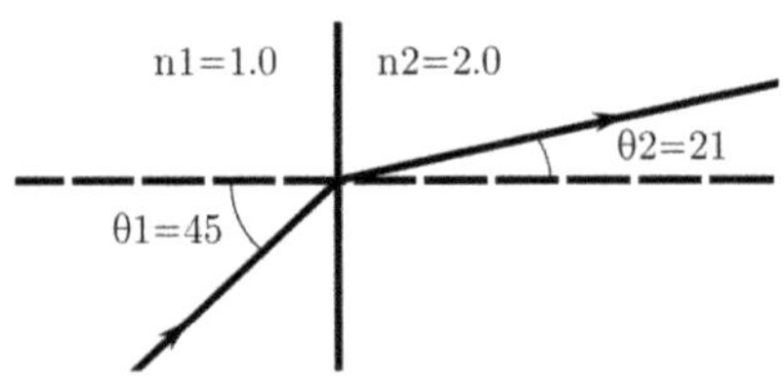

Figure 3.4. Refraction and the plane of interface between two different materials. The angle with the normal is smallest in the material with the largest index of refraction.

Figure 3.5. The wavy surface of the water refracts sunlight into a complex pattern on the lake bottom.

answer except that 'it is what it is,' and this is how we observe light to behave when we perform experiments.

For a physical explanation of Snell's law, we must consider the wave-like nature of light. To do this we will use one of the most important tools of wave optics: *Huygens' construction*, invented by Christiaan Huygens in 1678.

But first, let us define the concept of a *wave front*. If light moves coherently in a particular direction in a given region of space, we can imagine a line that connects all of the places where a particular wave is at a peak. In physical terms, for a light wave, this could be the points in space where, at a given instant of time, the electric field is at its maximum. In three dimensions this region would be a plane (not necessarily flat), but on a 2D drawing this region will appear as only a line (not necessarily straight). An important point is that the direction of travel of a wave is always perpendicular to the wave front. Since our 'rays' point in the direction of travel of the light wave, it follows that the wave fronts are perpendicular to the rays.

Huygens' construction says that if the location of a wave front is known, the next wave front can be found by imagining an infinite number of spherical wavelets emanating simultaneously from different parts of the first wave front—all propagating for the exact same amount of time. The next wave front is then marked by a tangent that connects the edges of all of those spherical wavelets. For a 2D drawing, this means we can imagine circular waves propagating outward from every point on the first wave front. We stop them at a time of our choosing, connect them with a tangent line, and we have found the next wave front.

So let us consider an example such as figure 3.4, but this time drawn with a wave front indicated perpendicular to the incident ray (see figure 3.6). I have chosen to put the first wave front starting right at the boundary between material 1 and material 2. I pick a point on the incident wave front, and I allow just enough time to pass for an imaginary spherical wave coming from that point to make it to material 2. Meanwhile, I allow another imaginary spherical wave to begin at the exact point the ray intersects with material 2.

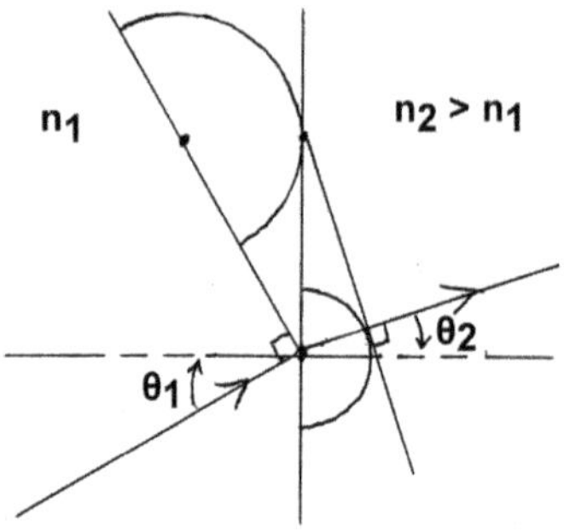

Figure 3.6. Wave fronts in refraction. Because light travels more slowly in the material on the right (higher index of refraction), a wave front in that material travels a smaller distance in the same amount of time. Connecting the wave fronts, according to Huygens' construction, shows that the light must bend in toward the normal as it enters the higher-index material.

Remember that light travels more slowly in material 2, so this imaginary circular wave does not travel as far. So, where is the next wave front? Huygens' construction says that one simply needs to connect those circular waves with tangents. This second wave front is completely within material 2 while the first wave front was completely within material 1. If we recall that the rays are always perpendicular to the wave fronts, we have now found in what direction must be the second ray.

Clearly, the ray in material 2 makes a smaller angle with the axis than does the ray in material 1. In fact, if we analyze figure 3.4 with a little trigonometry, it is not hard to show that there is a fairly simple mathematical relation between θ_1 and θ_2.

$$\frac{c}{v_1}\sin\theta_1 = \frac{c}{v_2}\sin\theta_2 \tag{3.3}$$

where c is the speed of light in a vacuum, v_1 is the speed of light in material 1 and v_2 is the speed of light in material 2. It is easy to see that equation (see equation (3.3)) is Snell's law—equation (3.2)—so long as we make the following identification for each of the two materials.

$$n = \frac{c}{v} \tag{3.4}$$

where n is the index of refraction in a particular material and v is the speed of light in that material.

In purely geometrical optics, the index of refraction is simply a number that is measured by an experiment for a given material—a piece of glass, for example. Wave optics and Huygens' construction reveal its physical meaning. The index of refraction tells us by what factor the material reduces the speed of light as compared to c—its speed in vacuum. For example, if the index of refraction of a particular type of glass is $n = 1.523$, it means simply that light travels 1.523 times more slowly in that particular type of glass than it does in a vacuum.

The second column of table 3.2 gives approximate indexes of refraction for some different materials. Since in all but rather odd circumstances a light wave travels more slowly in a material than in a vacuum, indexes of refraction are greater than 1.0. An important point to remember is that the index of refraction of air is 1.000 3, nearly 1.0 exactly. A perfect vacuum does have an index of refraction of exactly 1.0 (this should be obvious from its definition), and most gases have approximately that value as well.

Notice that Snell's law makes no reference to which side the light was coming from and toward which side it was traveling. Snell's law simply says that the angle is smallest where n is largest. We could reverse the arrows on the rays in figure 3.4 and it would still be a valid diagram.

The same could not be said, however, for figure 3.7, where we have taken into account the fact that when light gets to a sudden change in material, a certain percentage will reflect, instead of passing into the material and refracting. If we reverse the arrows on the top image the result, instead, is something like what is shown in the bottom image—where the ray of light moves from plastic to air instead of from air to plastic. Notice the refracted rays do the same thing in both diagrams: the angle is larger in air and smaller in plastic. But not so for the reflected ray: it always reflects from whichever side the light is incident.

The images in figure 3.7 also illustrate a common misconception about light rays. A light ray is an abstract idea: the ray itself is not *visible* from the side; you can only see the ray if it points directly at your eye. A laser makes a very good approximation of a ray of light. And so why does the 'ray' of laser light show up in these pictures even though the lasers point not toward the camera but rather crosswise to our vantage point?

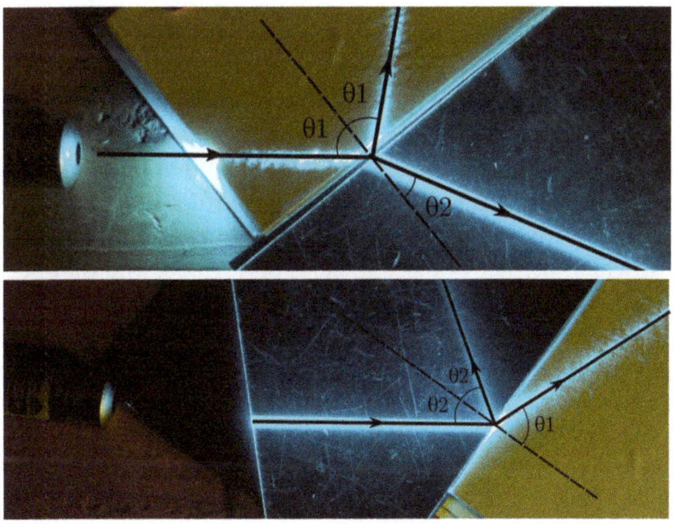

Figure 3.7. **Top:** As light arrives at a sudden change in index of refraction, the light will refract upon entering the new material. But a portion of the light will also reflect off the plane of interface. Here light traveling through air (material #1) refracts as it passes into plastic (material #2). Notice that $\theta 1$ (air) is larger than $\theta 2$ (plastic) because the index of refraction of air is smaller than that of plastic. **Bottom:** The same arrangement as in the top image but with the light instead coming from inside the plastic and refracting as it emerges into the air. Notice that, exactly as in the top image, $\theta 1$ (air) is larger than $\theta 2$ (plastic) because the index of refraction of air is smaller than that of plastic. This time, however, the reflected ray is inside of the plastic instead of in the air.

Table 3.1. Angles of refraction for rays of light entering glass (with $n = 1.5$) from air. The incoming (incident) ray in the air makes an angle θ_i with the normal. It is then refracted and makes a different angle (θ_r) with the normal while inside the glass.

θ_i (°)	θ_r (°)
0	0
10	6.65
20	13.18
40	25.37
60	35.26
80	41.03
85	41.62
88	41.78
89	41.80
89.5	41.81
89.9	41.81

It is because of a trick. I used a particular combination of laser and plastic to take advantage of *fluorescence*—a topic we consider in section 3.8. In this example, a small fraction of the laser light is absorbed by the plastic, and that energy is used by the plastic to emit its own light. It does not emit that light in the direction the laser light was traveling but rather in random directions. The path of the laser light is marked by the fluorescent glow of the plastic sending light in all directions—including toward the camera. For the part where there is no plastic, I let the laser beam partially graze along the surface of a sheet of white paper. You see the little bit of the intercepted beam, scattered toward you by diffuse reflections from the paper, thus it traces out its path. Notice that there is a portion of the laser light's path (on the left side) for which neither of these tricks is performed, and that part of the beam is invisible from the vantage point of the camera.

The amount of refraction depends not only on the angle of incidence, but also on the ratio of the indexes of refraction of the two sides. As an example, table 3.1 gives values for the angles of refraction—given various angles of incidence—for the special case of a ray of light passing from air ($n = 1.0$) into some type of glass ($n = 1.5$).

3.3.1 Total internal reflection

Notice in table 3.1 that for a ray of light passing from air to glass, the angle of refraction seems to approach a maximum value (41.81° in this case), as the angle of incidence approaches 90°. Clearly, the angle of incidence cannot be greater than 90°, simply as a matter of logic. Figure 3.8 shows this limiting case, which can only be approached.

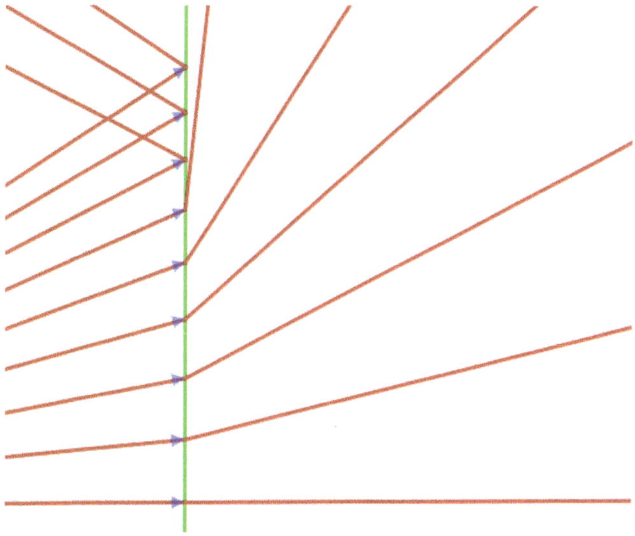

Figure 3.8. If light comes from a material with a larger index of refraction, there will be a particular angle of incidence at which the refracted ray would be 90°. For angles of incidence equal or greater than this there is no refracted ray, and all of the light reflects internally. This is called total internal reflection, and it is often used in place of mirrors made with shiny metal coatings.

Remember that for refraction, one can reverse the arrows and the diagram will still be valid. Table 3.1 shows the angles of incidence and refraction for rays of light in air entering into a piece of glass with index of refraction $n = 1.5$. We could instead imagine the reverse—rays of light coming from the glass and emerging into the air—and all we would need to do is reverse the columns of table 3.1.

However, a moment's thought reveals something odd. The reversed table would only show angles of incidence of 41.81° or smaller, and those would have refracted angles of almost 90°. So, what would happen if one simply made the angle of incidence greater than 41.8°? The numbers in the table seem to imply that a refracted angle of greater than 90° would result. But what would that even mean? An angle with the normal of greater than 90° means the light has not even left the glass.

What does Snell's law say happens in this case? You can find out by using a calculator to plug $n_1 = 1.0$, $n_2 = 1.5$ and $\theta_2 = 45°$ into Snell's law and asking it to solve for θ_1. The calculator will give you an error message. Snell's law is simply invalid for this case, and something else happens that is not predicted by Snell's Law alone.

Let us now consider the reflected ray in the bottom image of figure 3.7. If we were now to gradually increase the angle of incidence, four things would happen.

1. The angle of refraction would increase, according to Snell's law. Since it is in the side with the smaller index of refraction, the angle of refraction would increase faster than we increase the angle of incidence.
2. The reflected ray would increase according to the law of reflection, which is say that the reflected ray would increase so that it is always equal to the angle of incidence.
3. The refracted ray would get dimmer.

4. The reflected ray would get brighter.

As one increases the angle of incidence, the refracted ray gets dimmer and dimmer, disappearing completely as the angle of refraction approaches 90°. Simultaneously, the reflected ray gets brighter and becomes as bright as the incident ray, as the refracted ray disappears. If one then increases the angle of incidence even further there is only a reflected ray, which bounces inside the glass and never gets out into the air. This is called *total internal reflection*, and it is very useful; it is, for example, a way to make a mirror from nothing but clear glass. See figure 3.9 for a demonstration.

From table 3.1 it is clear that, for the case of a ray of light trying to get out of a piece of glass with air around it, the light will reflect inside the glass so long as the angle of incidence is greater than about 42°. This is called the critical angle, θ_{crit}, and we can calculate it as follows[1].

$$\theta_{crit} = \sin^{-1}\left(\frac{n_1}{n_2}\right) \tag{3.5}$$

Table 3.2 shows indexes of refraction and critical angles for several different materials, assuming air ($n = 1.0$) is on the other side of the interface.

Total internal reflection has many practical uses. The right-angle prism of figure 3.9 is, in some cases, a better way to reflect light than the shiny metal surface of a good mirror. Fiber optics employ thin, transparent fibers of glass or plastic. Once a ray of light enters, it inevitably meets the outside edge of the fiber at a very large angle to the normal and undergoes total internal reflection many times, as it bounces from one edge to the other—even if the fiber is tied into knots. Even beads

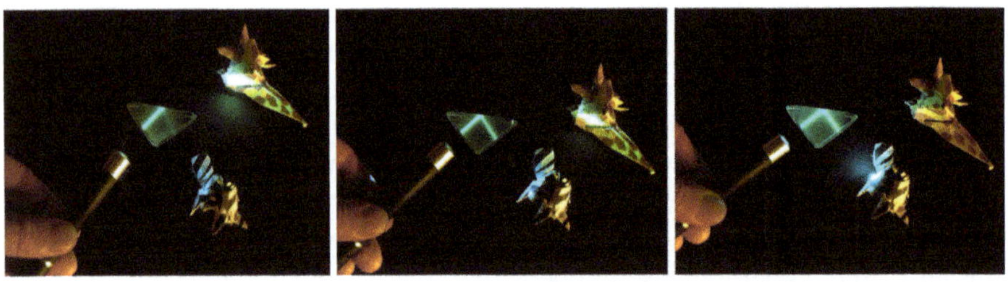

Figure 3.9. A 45° glass prism used to reflect light by 90°. **Left:** The light meets the backside of the prism at less than the critical angle. It mostly passes through, deflected by refraction. Notice, however, that a small portion of the light reflects off the surface internally—like a mirror. **Center:** The light meets the back face of the prism at very nearly the critical angle and refracts almost parallel to the back face of the prism. Notice that the internally reflected portion is now much brighter. **Right:** The ray meets the back surface at greater than the critical angle, and all of the light reflects internally.

[1] In equation (3.5), $\sin^{-1}$ represents the *arcsine* or *inverse sine* of the number. This is the angle one would have to take the sine of in order to get that number, so the $\sin(0°) = 1$ and $\sin^{-1}(1) = 0°$.

Table 3.2. Critical angles for light coming from various materials into air.

Material	Approximate Index of Refraction	θ_{crit} (°)
Water	1.33	48.8
Acrylic	1.5	42
Salt	1.54	40.5
Crown glass	1.5–1.6	38–41
Flint glass	1.6–1.8	35–40
Sapphire	1.77	34.4
Lanthanum glass	1.8–2.0	30–33
Diamond	2.42	24.4

Figure 3.10. Left: Tiny hairs, on the surface of these leaves of American lotus, trap air underneath the water drops. When light rays passing through the water drop encounter this layer of air at greater than the critical angle of 48.8°, total internal reflection occurs. And so the transparent drops of water look like little mirrors reflecting the sky. **Right:** A bundle of optical fibers uses total internal reflection to guide light around tight curves.

of water can act like little mirrors if there is a layer of transparent air trapped underneath (see figure 3.10).

3.3.2 Dispersion

The index of refraction for a typical transparent material is not actually a simple constant; it depends somewhat on the wavelength of the light. Most often, n is slightly larger for shorter wavelengths. This means that the angle of refraction for a given incident ray of light depends on the wavelength of the light—with shorter wavelengths usually refracted by larger angles. This effect is called *dispersion*, and a graph of index of refraction versus wavelength, for a given material, is called the *dispersion curve* of the material.

The left side of figure 3.11 shows examples for some different types of glass. Notice that the graph also shows overall differences in index of refraction. For example at any given wavelength SF10 glass has a greater index of refraction than BK7 glass. However, the blue curve for SF10 is its dispersion curve—it shows a

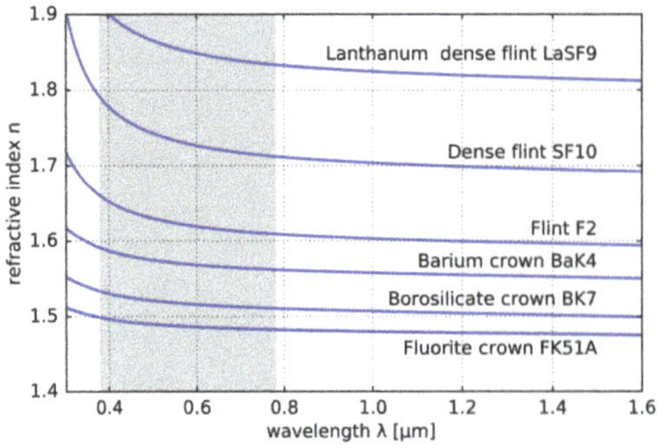

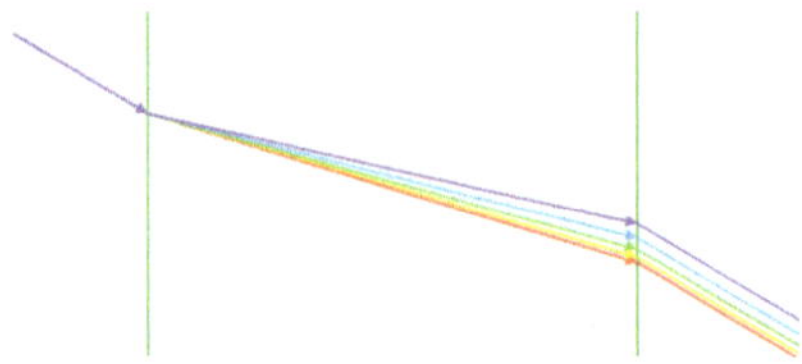

Figure 3.11. **Left:** Dispersion curves for several different types of glass. The vertical axis represents index of refraction, while the horizontal axis is wavelength. The shaded area marks the visible portion of the spectrum. Copyright © Geek3 – Own work, Attribution-ShareAlike 4.0 International (CC BY-SA 4.0), https://commons.wikimedia.org/wiki/File:Mplwp_dispersion_curves.svg, Mplwp dispersion curves.svg **Right:** An example of dispersion by a piece of glass.

different index of refraction for each wavelength. Also, notice that some of the dispersion curves seem to zoom up to huge values at some short wavelengths in the ultraviolet (UV) part of the spectrum. The dispersion curve for F2 flint glass, for example, increases very rapidly at a wavelength of about 0.25 μm (250 nm). What the dispersion curve does not show is that the glass also becomes less and less transparent at these wavelengths. If the glass does not transmit light, then the index of refraction is a moot point.

Most sources of light consist of many wavelengths simultaneously. Thus, if a ray of a typical source of light enters a piece of glass, it is not all refracted at the same angle. Shorter wavelengths are refracted at a greater angle, so the piece of glass disperses light of many wavelengths (but one direction) into many different directions (but each consisting of only one wavelength). See the right side of figure 3.11 for an illustration.

Dispersion can be either useful or a pain in the neck, depending on the situation. If one wants to measure the spectrum of a source of light, then a wedge-shaped piece of glass (a prism) can do the trick. Different wavelengths will refract at different angles, and so one can measure them separately because the prism will deflect them by different angles. If the dispersion curve for the type of glass is known, one can mathematically convert this information into wavelength, and produce a graph of brightness versus wavelength—a spectrum.

If the goal is instead to design a lens to produce sharp images, then dispersion is the enemy. It means that if one designs the shape of the lens so as to focus red light, then blue light will not be in focus because the two wavelengths are refracted at different angles. This basic problem is called *chromatic aberration*, and it is discussed in more detail in chapter 11.

If we consider the phenomena of dispersion, refraction and total internal reflection, we can begin to see why different transparent materials 'look' different. A diamond has facets; light refracts as it makes different angles with the different facets. There is also dispersion, so different colors refract differently, and one can see colors even with white light. Finally, many rays get trapped inside and undergo total internal reflection multiple times before finally meeting a facet at less than the critical angle and escaping. Thus a diamond can look colorful and glittery.

Why does a piece of glass ground to exactly the same shape as a diamond not look as impressive? It is because of the considerably-smaller index of refraction of glass compared to diamond, which means less dispersion, smaller angles of refraction and less internal reflection.

3.4 Diffuse reflections

It is often the case that when light reflects off an object, a single ray will reflect in many directions because of microscopic irregularities in the surface. This is called a *diffuse reflection*, in contrast to the specular reflection already discussed. If one could look closely enough, a diffuse reflection is really a myriad of specular reflections. Instead of one ray, there is really a bundle of many parallel rays very close together. If the surface is rough, then each individual ray in this thin bundle hits a seemingly randomly-oriented surface, and thus goes off in its own peculiar direction.

A mirror is a surface that makes only specular reflections and no diffuse reflections. To be a good mirror, its surface must be smooth down to the level of the wavelength of the light itself. And so, for example, if one grinds glass with successively finer and finer abrasives, it will appear perfectly smooth only once the abrasives (and thus the scratches made in the glass) are smaller than the wavelength of light used to reflect off it. At that point it will produce a good specular reflection and thus appear mirror-like and shiny. With larger scratches, from larger abrasives, the surface will appear dull from the diffuse reflection.

Diffuse reflection is the most common process by which we 'see' solid, opaque objects. A cat (see the left side of figure 3.12) does not create light of its own, at least not in the visible part of the spectrum. We see light coming from the cat because light

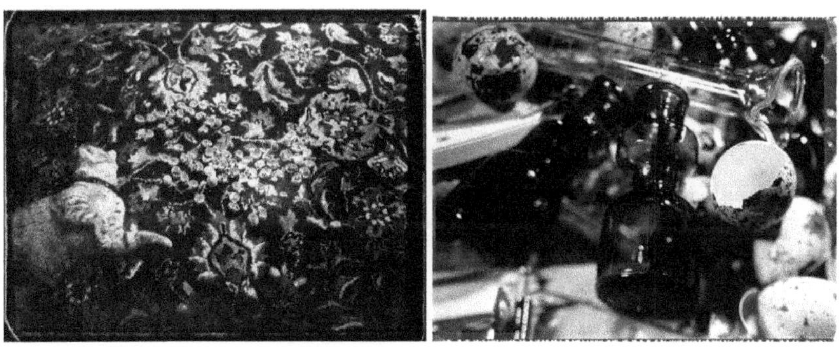

Figure 3.12. Left: *Boris and quail eggs*, John Beaver 2010. A cat does not emit light of its own in the visible part of the spectrum. We see it because of the diffuse reflection of light from other sources. **Right:** *Bottle and quail eggs*, John Beaver 2012. The light from the eggs is due to diffuse reflection while that from the surfaces of the metal and glass is due to specular reflections. In addition, some light from behind has passed through the glass and been altered by refraction.

from some other source has reflected off it, making diffuse reflections with every part of the cat. Thus rays of light go off in every direction from each part of the cat, just as if it were emitting its own light. Our eyes intercept whatever rays, coming from each part of the cat, happen to be going in our direction. The rest go elsewhere (toward a frightened mouse, for example).

We can see the difference between specular and diffuse reflections by looking at the right side of figure 3.12. The rays of light that came from the different objects in the picture, and arrived at the camera, were all due ultimately to reflection from the room lighting. Light coming from the eggs was due to diffuse reflection. In this case, some parts appear light while other parts appear dark because of differing amounts of diffuse reflection; whatever was not reflected was absorbed. Since those diffuse reflections go off in all directions, it is guaranteed that some of the rays from each part of the eggs make it to the camera.

The light coming from the glass surfaces, however, was due to specular reflections. In addition some light reflected diffusely off the eggs in the background has passed through the glass, its path altered by refraction on its way to the camera. In these cases, what appears light and what appears dark is more complex; it depends on the location of the camera, the location of the light source, and the angle and curvature of the surfaces themselves as they both reflect and refract the light.

3.5 Scattering

A light ray is *scattered* if, upon interacting with matter, it splits into many rays in random directions. A diffuse reflection from a solid, opaque object can be thought of as a form of scattering, but the term is usually used to describe other physical processes whereby a distribution of fine particles within a transparent or semi-transparent material deflects light rays randomly as they pass through. As a bundle of rays pass through the scattering material, more and more are successively scattered off in other directions, leaving fewer and fewer rays to travel in the original directions.

In some cases, the light rays mostly scatter at angles less than 90°. This is called *forward scattering*; the left side of figure 3.13 illustrates the concept. But light rays can also scatter at angles greater than 90° to their original directions, resulting in *backscattering* as illustrated on the right side of figure 3.13. Usually, both kinds of scattering are present.

There are many kinds of scattering of light, but we can separate them into two main categories: wavelength-dependent and wavelength-independent scattering. Wavelength-dependent scattering occurs when different wavelengths are scattered

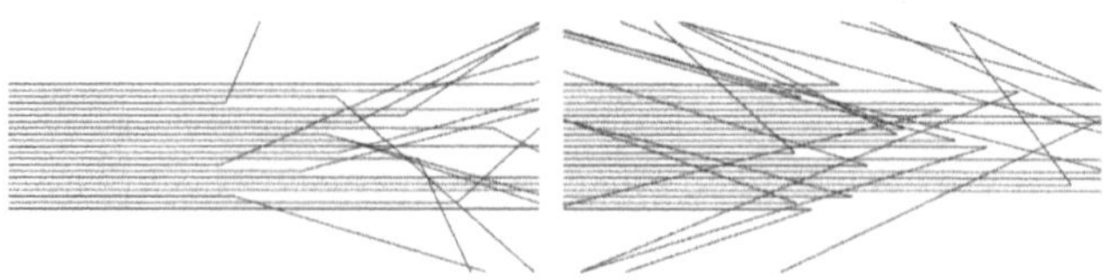

Figure 3.13. **Left:** Forward scattering. **Right:** Backscattering.

by different amounts. Wavelength-independent scattering, on the other hand, affects all wavelengths equally. The physical process of scattering is very much affected by the size of the scattering particles, as compared to the wavelength of light. Roughly speaking, when the scattering particles are much larger than the wavelength of the light, then scattering is independent of wavelength.

The physics of scattering takes many forms, depending on the circumstance. In some cases it must be modeled with quantum physics, and in these cases we think of individual photons of light interacting with individual particles. But in many common circumstances scattering can be modeled by considering electromagnetic waves interacting with the particles, solving the Maxwell equations directly for some simplified (but hopefully realistic) example. In the simplest of cases, scattering can be modeled with the ordinary optics of diffraction. We will briefly consider four cases of the scattering of visible light that are of particular interest to photography. For a more complete overview, see for example chapter 11 of Nassau (1983).

1. **Optical scattering:** the scattering particles are much larger than the wavelength of the light. Ordinary ray optics and wave diffraction theory can be used to model the scattering, and in most cases the scattering is wavelength-independent.
2. **Rayleigh scattering:** the scattering particles are much smaller than the wavelength of the light. A full analysis requires a solution to the Maxwell equations, but for Rayleigh scattering the results turn out to be mathematically simple. Rayleigh scattering is strongly dependent on wavelength, with shorter wavelengths scattered far more than longer wavelengths.
3. **Mie scattering:** the scattering particles are neither much larger nor much smaller than the wavelength of the light. Mie scattering theory is the result of a full solution to the Maxwell equations, usually with simple assumptions about the nature of the particles (for example, that they are conducting spheres). Both Rayleigh scattering and optical scattering can be thought of as special cases of Mie theory. Mie scattering depends on wavelength in a complex way.
4. **Tyndall effect scattering:** The term 'Tyndall effect' is not always used consistently, but it most often refers to a form of wavelength-dependent Mie scattering similar to Rayleigh scattering but arising from much larger particles. The larger particles make the scattering easily observable from even a small amount of material. The term arises from some of the first detailed experimental investigations of scattering by John Tyndall (1820–93). Tyndall described the scattering by the suspension of tiny particles in a fluid (Nassau 1983, p 232).

I emphasize that this is only one way to categorize scattering at visible wavelengths, and many other subtle scattering phenomena can occur that must be analyzed in a more sophisticated manner than what I have laid out here.

3.5.1 Wavelength-dependent scattering

The most common experience of wavelength-dependent scattering is called Rayleigh scattering, and it occurs when light interacts with particles, such as individual atoms or molecules, that are much smaller than the wavelength of the light. Rayleigh scattering of sunlight by air molecules is responsible for the blue of the sky on a clear day. As individual photons encounter individual air molecules (N_2 or O_2), a certain percentage of the light will deflect off in a different direction. But short wavelengths scatter by larger angles than long wavelengths.

Since it is the shorter-wavelength blue light that scatters the most, when you look up at a cloudless daytime sky you sees blue light. This blue sky light comes from the Sun, ultimately, but it is light that would have missed you, had it not been deflected in your direction by Rayleigh scattering.

On the other hand, the setting Sun appears noticeably reddish. In this case you are seeing the light that has *not* been scattered out of your line of sight. Since it is the blue light that mostly scatters off in other directions, you sees sunlight that has had blue light subtracted from it. Thus the Sun appears more reddish.

A kind of wavelength-dependent scattering very much like Rayleigh scattering can also occur with much larger particles—still smaller than, but much closer in size to, the wavelength of the light itself. For visible light, this means particles that are too small to be seen with an ordinary microscope, although still enormous compared to the individual atoms and molecules that cause Rayleigh scattering.

This type of scattering from larger particles is often called the Tyndall effect. Probably the most commonly-seen example is the blue smoke from the tailpipe of an automobile engine that is burning oil and needs an overhaul. The smoke looks blue for the same reason the sky looks blue. You see it by light from other sources scattering off of it, and short wavelengths scatter more than long wavelengths. Not all smoke does this, because in order for the Tyndall effect to be prominent, the smoke particles must be the right size.

These much-larger particles scatter the light—all else being equal—far more than the air molecules of the blue sky. Thus the wavelength-dependent scattering can be observed on a small scale; the Rayleigh scattering by air molecules requires, well, an Earth-atmosphere's worth of air.

The Tyndall effect can also be seen in certain liquids that have submicroscopic globules of one material suspended in another. A tiny bit of milk mixed with water sometimes works and looks bluish when illuminated from the side. Glass can be manufactured with impurities that emphasize the Tyndall effect, and these opalescent glasses are often used for their ornamental properties.

Figure 3.14 shows two photographs of an opalescent glass egg. When white light (of a mix of wavelengths) shines on it, you see only the light scattered toward you. And so it appears blue. When you look *through* it at a source of light behind it, on the other hand, you see the reddish light that has *not* been scattered.

We will consider in later chapters an important photographic example that requires the more-complex Mie theory for an explanation. The silver-gelatin emulsion used in black-and-white film photography and printing consists of tiny crystals suspended in a thin transparent layer of gelatin. Under certain circumstances, silver-gelatin black-and-white printing paper can reflect colors due to wavelength-dependent scattering. But the wavelengths that are most scattered back towards the viewer depend critically on the size of the particles.

Figure 3.14. Tyndall-effect scattering in an egg-shaped piece of opalescent glass. The photo on the **left** was taken with the egg against a black background and illuminated from above. We see the mostly-blue light that is scattered back towards us. The photo on the **right** was taken with the egg illuminated from behind. In this case we see the light that was not scattered by the egg on its way to us. Since mostly blue light is scattered, what is left over appears reddish in color.

Figure 3.15 shows an example of the amount of scattering versus wave frequency for a given particle size, as calculated by Mie theory for the simple case of conducting spheres. The horizontal axis of the graph is in terms of the ratio of the particle circumference to the wavelength of light. Thus unity on the horizontal scale represents scattering by particles that are of similar size to the wavelength of the light. The straight portion on the left is for tiny particles, and represents the case of Rayleigh scattering. The steep upward straight line means that shorter wavelengths scatter more than longer wavelengths.

But when the particle size is near the wavelength of light, the curve turns over, and for a range of wavelengths the opposite is true: longer wavelengths scatter more than shorter wavelengths. Rayleigh scattering and small-particle Tyndall effect give cool-tone blues. But with just the right particle size, warm colors can also be produced by scattering.

3.5.2 Wavelength-independent scattering

Wavelength-independent scattering occurs when light interacts with particles of matter that are significantly larger than the wavelength of light. This can be seen from the Mie theory graph in figure 3.15; at large particle sizes the curve becomes more and more horizontal, indicating no preference for either long or short wavelengths. For even larger particle sizes it is the physics of diffraction theory that is most appropriate, but the overall effect can be similar; an incoming parallel bundle of rays is 'scattered' into many rays traveling in many directions, especially

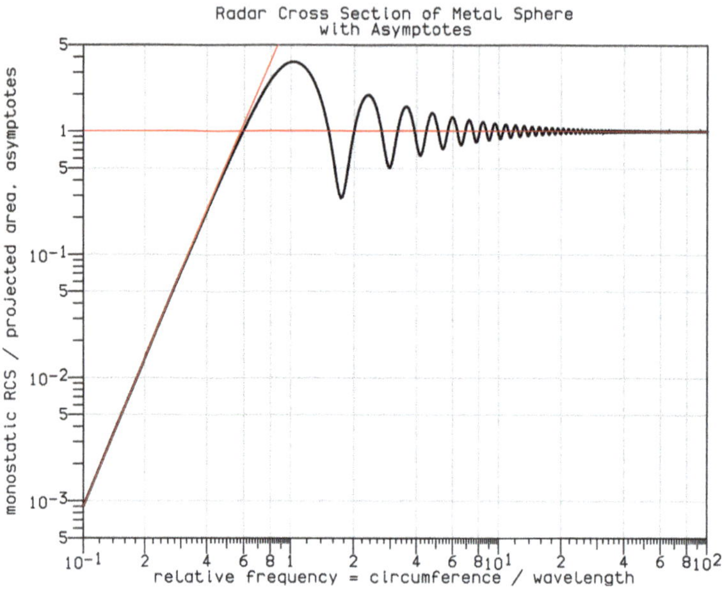

Figure 3.15. A plot of the scattering of light as calculated by Mie theory. The vertical axis is the amount of scattering, while the horizontal axis is the particle size for a given wavelength, or—equivalently—the frequency of the light (shorter wavelength on the right) for a given particle size. Copyright © Catslash, Public Domain, https://commons.wikimedia.org/wiki/File:Radar_cross_section_of_metal_sphere_from_Mie_theory.svg, Radar cross section of metal sphere from Mie theory.svg.

when there are particles of many different sizes. The seemingly-random nature of the scattering comes from the too-complex-to-describe microscopic details of the scattering material.

Light shining on microscopic particles of dust provides a good example. If you take a high-power laser pointer and shines it up into the night sky, it seems as though you can see the beam of light. But the light from the laser is all going in the same direction, and this is not toward you (or at least it had better not be). What you are seeing is light scattered toward you from zillions of microscopic dust particles located all along the beam. Since this scattering occurs most strongly in the backward direction, the beam is more visible if one looks along it from behind.

Fog provides another example of scattering from relatively large particles. Microscopic droplets of water suspended in the air alter the directions of the rays of light by both refraction and total internal reflection. If the fog fills a given space, then a portion of the light that reaches you may have been scattered many times before it gets to you. But fog scatters all visible wavelengths equally, for the most part.

A puffy white cloud in the blue sky appears as it does due to scattering. It is, after all, nothing but countless microscopic, transparent droplets of water or crystals of ice. These tiny particles deflect light rays, each according to the basic laws of refraction and reflection. But the details are far too complex to follow individually, and so the light is scattered in random directions.

Figure 3.16. *Rio from Sugar Loaf Mountain*, John Beaver 2012. Atmospheric perspective, which can give a sense of physical depth to a picture, is here introduced by haze and fog.

In dense fog, light is scattered many times before it gets to you, and it appears as though the air itself is emitting light. Light rays coming from more distant objects have more opportunities to scatter on their way to you, and so the light coming directly from objects is more diminished with greater distance. But while light that would have reached you (had there been no fog) is scattered out of your line of sight, other light—from random directions that would have missed the person had there been no fog—is scattered in your direction. Thus the diminished light from distant objects is filled in by a random glow, and this reduces contrast.

Fog has little effect on one's view of nearby objects, but it has a profound effect on distant objects. For photography, this means the presence of fog can add depth to a picture. Haze causes a mixture of both wavelength-dependent and wavelength-independent scattering, and it too can add depth to a landscape. It makes distant objects appear of lower contrast and slightly more bluish or purplish in color. Painters often use the term *perspective* to mean, in a generic sense, any type of depth cue. And so when a painter adds features that mimic the visual effects of haze scattering in order to show depth in a painting, it is called *atmospheric perspective* (see figure 3.16). But for a photographer, who must deal with whatever haze there happens to be when taking a picture, the lowering of contrast by haze scattering may or may not be desirable.

3.6 Interference

It is possible to put a thin coating of transparent material onto the surface of a glass lens that actually *increases* the amount of light that passes through the lens. It is surprising that one could put something on glass to make it more transparent; would not anything you put on it simply absorb more light and make matters worse? How could the coating cause more light to pass through?

When looking through a typical glass lens, most loss of light is due not to absorption in the glass, but rather because a significant percentage of the light reflects off the glass surface instead of passing through. An *anti-reflection coating* decreases the reflection, while increasing the transmission. It is easy to recognize glass treated in this way from the faint greenish or bluish tint of light that glances off its surface. Anti-reflection coatings work by thin-film interference, a specific example of the phenomenon of interference in general.

If two waves are present at the same place at the same time, they add together. A given wave, at a given point in space and time, causes some kind of disturbance, and this disturbance can be in either one direction or its opposite. We can (arbitrarily) call one direction positive and the other negative. For light it is the direction of the electric and magnetic fields. For sound in air, it is the direction of slight displacements of air molecules.

If one of the two waves is trying to make stuff zig, while the other is trying to make it zag, it is possible for both effects to cancel out, leaving nothing, an effect called *destructive interference. Constructive interference*, on the other hand, means that both waves are trying to cause the same disturbance in the same direction at the same place and time—and so a wave of greater amplitude than either results. There are also all of the possible circumstances in between, where the two waves either partially add or partially cancel each other (see figures 3.17 and 3.18).

There are many ways, in practice, to make this happen. For light, one can split a wave with some kind of obstacle, and then send both waves off on different paths, eventually bringing them back together. Because the waves have traveled different distances, when they recombine they will most likely no longer line up with each other, and so they will interfere.

Thin-film interference works in a different way. Light falling on a very thin layer of transparent material will reflect partly off the top and partly off the bottom; see figure 3.19. These two reflected waves will then interfere, either constructively or destructively depending on the details of the index of refraction of the film, its thickness, and what material is on either side.

And so an anti-reflection coating on a lens is cleverly chosen so that visible wavelengths of light interfere destructively for reflection, but constructively for transmission. Technically, for a single thin layer, this only happens completely for a rather narrow range of visible wavelengths. But if one uses many layers of coatings, all carefully chosen in thickness and index of refraction, then a broad range of

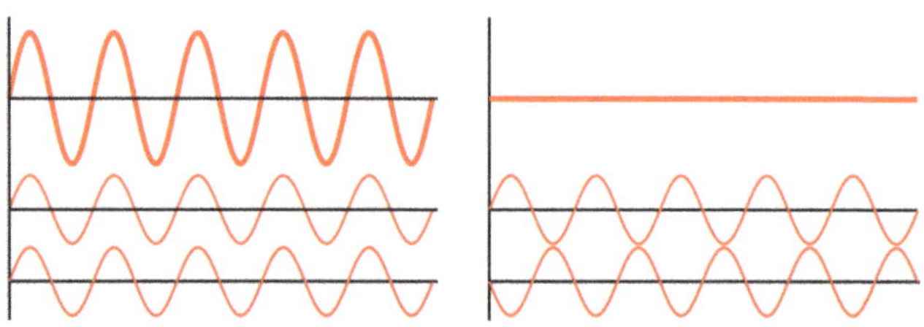

Figure 3.17. When two waves arrive in the same place at the same time, they may undergo constructive interference (**left**) or destructive interference (**right**), or something in between the two. Copyright © Haade, Wjh31, Quibik – Own work, Attribution-ShareAlike 3.0 Unported (CC BY-SA 3.0), https://commons.wikimedia.org/wiki/File:Interference_of_two_waves.svg, Interference of two waves.svg.

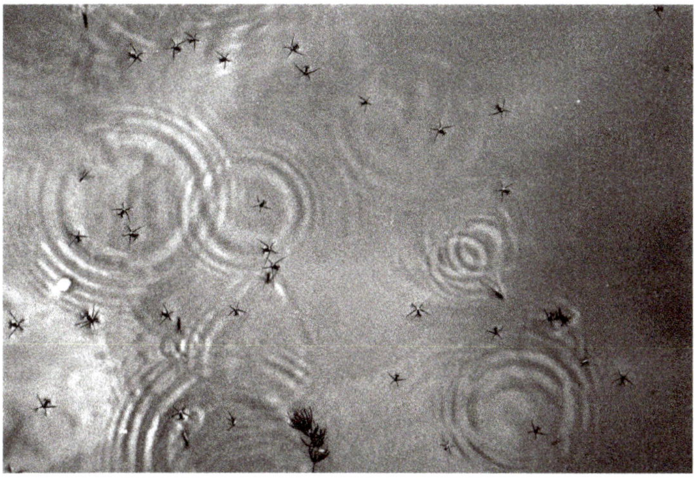

Figure 3.18. The waves made by these water striders show interference. At any given location, the height of the water is given by the sum of whatever waves happen to be at that place and time. In some places, waves can cancel out, while in others they add together.

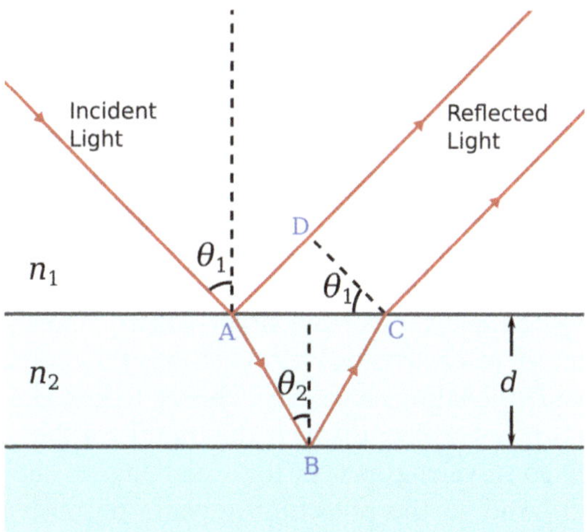

Figure 3.19. A wave can reflect off both the front and back surface of a thin transparent film. The two reflected waves undergo interference due to the fact that they have now traveled by different paths, and so are no longer in phase. Copyright © Nicoguaro – Own work, Attribution 4.0 International (CC BY 4.0), https://commons.wikimedia.org/w/index.php?curid=49324857, Thin film interference.svg.

reflecting wavelengths can be canceled, while transmitted wavelengths are reinforced.

The use of anti-reflection coatings is an important component of modern optics. Thus one can find three types of lenses: uncoated, single-coated and multicoated. A surface of an uncoated lens reflects, typically, about 10% of the light that falls on it, transmitting 90%. This might not seem so bad, but most optical systems (camera

Figure 3.20. The lens on the **left** (from the 1950s) is single coated, while the even-older lens on the **right** is uncoated. Note the slightly-purplish and much-fainter reflections from the coated lens.

lenses included) require many lenses, each reflecting 10% of the light at every boundary, both front and back. And that *is* bad, and not only because less light gets through, resulting in a dimmer image. Worse still, half of those internally-reflected rays *do* make it through eventually, but they are no longer where they should be, and so they add a diffuse glow to the image which reduces its contrast.

Single-coated lenses help with this greatly, but the best lenses are multicoated. Multicoated lenses only reflect about 1% of the light (thus transmitting 99%). Since a real optical system contains many lenses, the best is 'fully multicoated,' which means all lens elements are multicoated at every surface. Figure 3.20 shows examples of the reflections from the surfaces of both coated and uncoated lenses. Notice that each lens shows multiple reflections; what looks like a single lens is actually several lens elements one behind the other, and each surface causes a reflection. But note also that the reflections off of the coated lenses are much dimmer than the reflections off the surfaces of the uncoated lenses.

A purplish color of the reflections from single-coated lenses arises because, while some wavelengths interfere destructively, others interfere constructively. The type and thickness of these lens coatings have been chosen to give destructive interference for reflected light that has wavelengths near the middle of the visible spectrum. But this means that reflected wavelengths near the violet edge of the visible spectrum are not canceled so well. And so the reflection appears purplish in color from these single-coated lenses. Multicoated lenses show reflections that are even dimmer still, and in most versions show a slight greenish tinge. See figure 3.20 for a comparison of uncoated and single coated lenses.

3.7 Diffraction

Recall Huygens' construction, from section 3.3. It says that we can find the location of the next wave front by imagining a bunch of spherical wavelets propagating from the most recent wave front. We can use this to determine what happens when light encounters some kind of obstacle. For example, we can use Huygens' construction to determine what happens to light when it encounters a thin slit (see figure 3.21). The next wave front can be found by imagining spherical wavelets emitted simultaneously from every location along the slit.

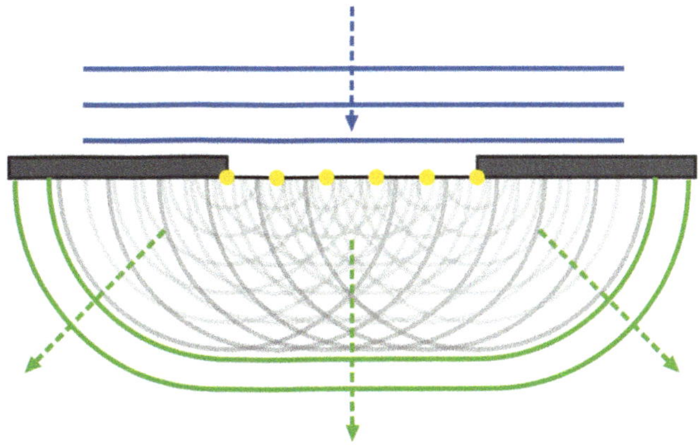

Figure 3.21. Huygens' wavelets propagating from a thin slit and interfering with each other. Copyright ©Arne Nordmann (norro) – Own work (Original text: Own illustration), Attribution-ShareAlike 3.0 Unported (CC BY-SA 3.0), https://commons.wikimedia.org/w/index.php?curid=1944668, Refraction on an aperture - Huygens-Fresnel principle.svg.

But we know that when one adds up light waves, they interfere. So to figure out what would happen, one would need to add up those waves at every point, taking into account the fact that, by the time they all get to a particular point in space, each had traveled a different distance, and so would be in a different part of its wave cycle. The specific part of a wave cycle is called the phase of a wave, and it is the difference in phase between two waves that results in either constructive or destructive interference.

Thus an infinite number of wavelets, sent out from all parts of the slit, arrive together at some distant point in space. But they arrive with different phases, and so they interfere with each other. This situation is called diffraction: the addition of an infinite number of waves, all of different phases (and sometimes also of different intensities). The mathematics of this requires calculus, which is intimately concerned with the business of adding up an infinite number of slightly different things. Here I will simply give the answers for some important cases. Even without working through the mathematics in detail, we can still recognize some basic principles of diffraction theory.

1. When light encounters an obstacle that restricts it to a particular region of space, it will spread out into the region that one would expect there to be a shadow. Sending light through a slit is a good example. It will spread out beyond the confines of the slit.
2. The angle of spreading is greater for a smaller restriction (a narrower slit, for example). And so a more narrow slit causes the light to spread out more.
3. All else being equal, the spreading out of the light is greater for longer wavelengths and less for shorter wavelengths.
4. The spreading of the light is proportional to the ratio of the wavelength to the size of the obstacle. So, for a slit, it is the value of the wavelength divided by the width of the slit that decides how much the light spreads out.

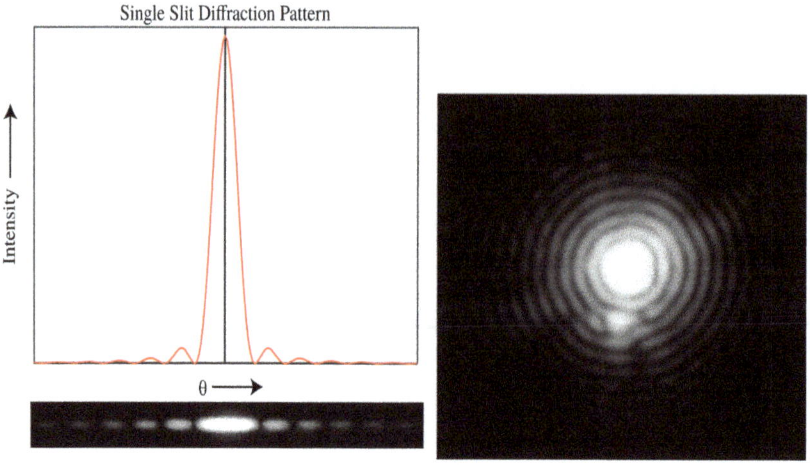

Figure 3.22. Left: The diffraction pattern from coherent light shining through a single slit. The pattern as projected onto a screen is at the bottom. At the top is a graph of the pattern's brightness versus position. Copyright © Nmurdoch – Own work, Public Domain, https://commons.wikimedia.org/wiki/File:Single-slit-diffraction-graph.svg, Single-slit-diffraction-graph.svg **Right:** Diffraction pattern for a circular hole.

5. This process of diffraction will produce an alternating pattern of constructive interference (bright light) and destructive interference (dim light). The size of this pattern when projected onto a screen is proportional to the ratio of the wavelength to the size of the obstacle. Such a pattern is called a *diffraction pattern*, and the bright regions of constructive interference are called *fringes*.

Consider the single slit as an example. The left side of figure 3.22 shows the diffraction pattern for a slit, both as an image and as a graph of intensity versus position. If the slit has a width, a, then θ is the angle by which the diffraction pattern spreads, as measured from its center. Most of the light ends up in one central region of constructive interference. The half-width of this bright central fringe is given by the following equation.

$$\sin\theta = \lambda/a \tag{3.6}$$

Here λ is the wavelength of the light. Notice that it is the ratio of λ to a that counts—not either one by itself. If a is much greater than the wavelength of the light, then the angle between the fringes is very small. If, for example, $a = 1$ cm, then for green light of $\lambda = 500$ nm we have $\sin\theta = 5.0 \times 10^{-5}$, an imperceptibly small angle; the diffraction pattern would be invisible.

If one wants a noticeable diffraction pattern, the slit must be much smaller. For example, if one shines 500 nm light through a slit the width of a human hair, about 50 mm across, then:

$$\sin\theta = \frac{500 \times 10^{-9}\ \text{m}}{50 \times 10^{-6}\ \text{m}} \tag{3.7}$$

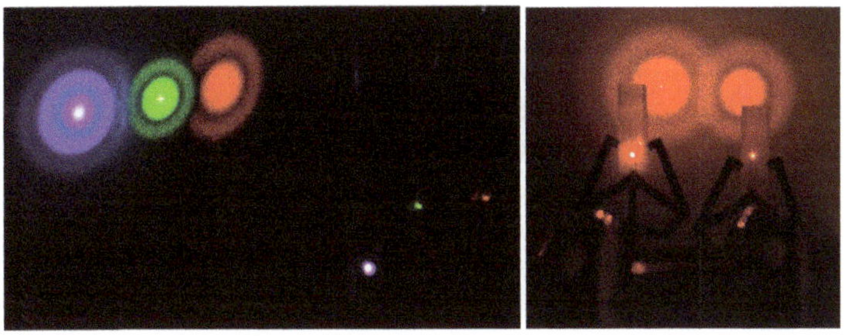

Figure 3.23. Left: Lasers of three different wavelengths shine through smears of human red blood cells. The diffraction pattern is large because the cells are so tiny. The central portion of the bullseye pattern is smallest for the violet (short wavelength) light and largest for the red (long wavelength) light. **Right:** Two lasers of the same wavelength shine through different smears of red blood cells. The diffraction pattern on the left appears slightly larger because the smear was made with red blood cells from a dog, and they are slightly smaller than the human red blood cells used on the right.

$$= 0.01 \tag{3.8}$$

$$\Longrightarrow \theta = \sin^{-1}(0.01) \tag{3.9}$$

$$= 0.57° \tag{3.10}$$

That is still a small angle, but if the screen is far enough from the slit, it would make a noticeable pattern.

What if it is not a slit, but rather some other shape, a circular hole for example? The same basic principles apply, but the detailed formula is different. In the case of a circular hole, the diffraction pattern is something like a bullseye; see the right side of figure 3.22. The half width of the central spot is given by a formula almost the same as (3.6); it differs by only a factor of 1.22.

A bunch of tiny circular dots, all the same size, will also give essentially the same pattern as a circular hole, although with less contrast. Figure 3.23 shows the diffraction patterns made by shining lasers through smears of red blood cells. Clearly, the angle of diffraction is relatively large, and so the blood cells must be small indeed.

Another general feature of diffraction, evident from figure 3.23, is that the size of the diffraction pattern depends not only on the size of the obstacle, but also on the wavelength of the light. Equation (3.6) implies that a longer wavelength of light, all else being equal, will produce a larger diffraction pattern. Equation (3.6) is for the special case of a single slit, but the same rule holds true for any diffraction pattern.

3.8 Fluorescence

Some materials can absorb the energy from light of a particular wavelength, and then use that energy to reemit light of a completely different wavelength, in a process called fluorescence. Most commonly, light of violet and near-UV wavelengths are absorbed, and then light of wavelengths closer to the middle of the visible spectrum is emitted.

Sunlight contains a fair amount of violet and near-UV light, and so fluorescent dyes can make something appear brighter than white. Unbleached paper looks duller than new white copy paper because the latter has a fluorescent dye that absorbs some of the energy of the nonvisible UV, and uses this energy to reemit visible light. This mostly-bluish added light makes the color whiter overall, but it also makes it look brighter. The nonfluorescent unbleached paper can only reflect as much visible light as falls on it. The fluorescent paper does this too, but then adds its own visible light, the extra energy coming from absorbed UV.

3.9 Polarization

There is an additional aspect to electromagnetic waves, apart from the four properties—amplitude, wavelength, frequency and speed—discussed in section 2.3. In an electromagnetic wave the electric and magnetic fields point perpendicular to each other, and the wave travels in a direction perpendicular to both. But this leaves open an infinite number of possibilities. For example, if an electromagnetic wave is traveling directly upward, then the electric field must point perpendicular to that—i.e., horizontally. But does the electric field point horizontally North, South, East, West or one of the infinite possibilities in between? Any one particular choice of orientation of the electric and magnetic fields is called a *polarization* of the wave.

For most sources of light, countless individual atoms and molecules are each producing light in their own fashion, with little coordination between them. Thus many waves are emitted each with its own essentially random polarization. The overall effect is that all polarizations are present at the same time. This is called *unpolarized* light, and it is the light one gets from, say, an ordinary lightbulb. If, on the other hand, a source of light contains mostly only one polarization, we say the light is *polarized*, as in figure 3.24. In practice, this is a more-or-less thing, rather than either-or, and so we find ourselves often describing light as *weakly polarized* or *strongly polarized*. This just means that, while all possible polarizations are present, one of them is more strongly represented than the rest.

There are many ways in which light interacts with matter, for which polarization is a crucial factor. A good example is the Rayleigh scattering discussed in section 3.5. Not only does the amount of scattered light depend on the wavelength, it also depends on the polarization. And so, for a given scattered angle, some polarizations

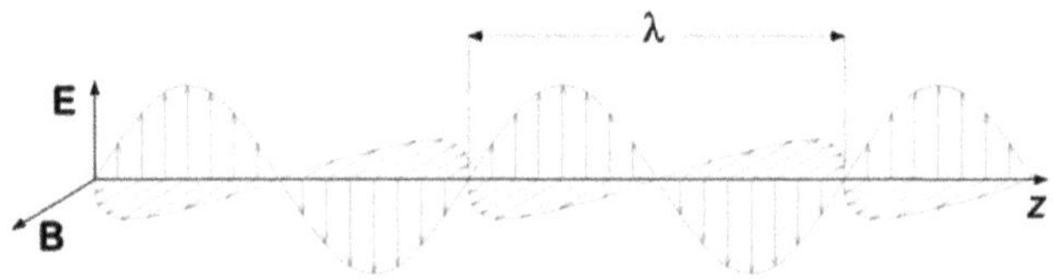

Figure 3.24. A polarized electromagnetic wave traveling from left to right. The electric and magnetic fields (marked **E** and **B**) in a polarized wave maintain a particular orientation. In an unpolarized wave the fields have random orientations perpendicular to the direction of travel (and to each other). Copyright © P.wormer – Own work, Attribution-ShareAlike 3.0 Unported (CC BY-SA 3.0), https://commons.wikimedia.org/wiki/File:Electromagnetic_wave.png, Electromagnetic wave.png.

scatter more than others. For Rayleigh scattering in the Earth's atmosphere, it turns out that light scattered by an angle of 90° is strongly polarized. Why? Because only one particular polarization scatters well at this angle. On a clear, cloudless day of low humidity, the light of the sky comes mostly from Rayleigh scattering. Thus the part of the sky making a circle around the Sun that is everywhere 90° from the Sun is strongly polarized.

Reflections provide another example. A certain percentage of a light wave will reflect when it encounters a sudden change in transparent material, such as when it encounters a piece of glass or the surface of water. This reflected wave may be strongly polarized, depending on the angle of the reflection. At a particular angle, called the *Brewster angle*, the reflected light is completely polarized.

One cannot directly see the effects of polarization because, to the human eye, one polarization looks the same as any other. But there are clever and important ways to distinguish between them with special filters, and we take up this topic again in chapter 30.

Reference

Nassau K 1983 *The Physics and Chemistry of Color: The Fifteen Causes of Color* (New York: Wiley)

IOP Publishing

Photography (Second Edition)
Physics and art in focus
John Beaver

Chapter 4

The weird world of the photon

Let us pause for a moment and think a little about what happens when a 'photon' is absorbed by an 'atom' or 'molecule.' I put these words in quotes because there is an important sense in which photons, atoms and molecules do not exist.

It is likely that our human brains come preequipped to think about the world in particular ways, including the idea that things have solidity. That is—perhaps—we have a common sense, and we naturally see the world as being made of solid objects and immutable particles. When we hear words such as 'atom,' 'molecule' or 'photon' we want to think of a solid thing that either is or is not at a particular location in space at any given point in time.

Just because we naturally think in these terms, it does not follow that the world is really as we imagine. Consider the fact that we have no direct experience—either as individuals or by means of our millions of years of evolution—of a subatomic world at the tiniest scales of existence. It would seem to be an odd coincidence if our minds happened to think in terms of concepts that were directly relevant to that minuscule world.

Quantum physics, developed by many individuals mostly over the years 1899–1929, provides an accurate description of this tiny world. When I say 'accurate,' I mean that quantum physics works; one can use it to correctly predict, with great precision, the outcome of a wide range of experiments. Despite its predictive power, however, it makes little intuitive sense. That is to say, many of the concepts employed by quantum physics disagree with our (perhaps built-in) assumptions about how the world works.

The idea of a photon as a particle is one such concept. We use the word 'particle,' but by that we actually mean something different from what our brains try to tell us a particle should be. A photon does not really exist, in the ordinary meaning of the word, at any particular time or place. Furthermore, a photon is certainly not 'solid.'

In order to learn quantum physics, one must give up the idea that it will make an intuitive sense. Instead, one must use intellect to learn a formal theory that can be

 © IOP Publishing Ltd 2022

applied to many different circumstances. After much effort using the theory for many situations, a physicist gradually develops a second sense that allows the physicist to get a feel for what the theory will say in general. We have a built-in intuition that (in the case of quantum physics) mostly fails us, but we can (with much effort) develop a different intuition to guide us in thinking about the subatomic world[1].

4.1 Young's double-slit experiment and the wave model of light

Light is a wave in the sense that, given the chance, light does the sorts of things that waves do. The phenomenon of interference is a good example. The first utterly-convincing demonstration of the interference of light waves was performed in the early 1800s by Thomas Young.

Young allowed light to emerge from two tiny pinholes that were spaced a small distance apart. The light that emerged from these narrow restrictions was then allowed to fall upon a screen, and an alternating pattern of light and dark was observed. Young successfully explained this effect by appealing to a wave model of light. Since from any particular location on the screen the two holes are at slightly different distances, the two waves may interfere constructively or destructively, depending on the details of the geometry (see figure 4.1).

It may seem that this would be a trivial experiment to perform, but there were several technical obstacles that Young had to overcome because a coherent source of light, such as a laser, was unavailable in Young's time. Young's experiment was key to the acceptance of a theory of light as a wave phenomenon in opposition to the earlier view, expressed most famously by Isaac Newton, that light is a stream of particles. Further developments, especially by Augustin-Jean Fresnel, provided detailed evidence for the wave-like nature of light.

The final nail in the coffin of Newton's particle theory of light was hammered in 1817. Poisson, a skeptic of the wave theory of light, showed mathematically that

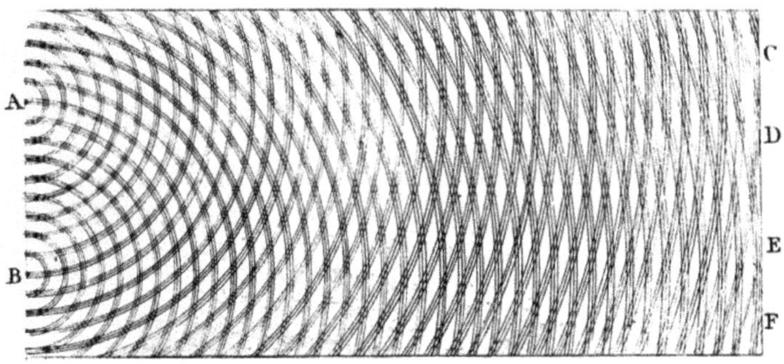

Figure 4.1. Young's diagram that explains the interference of light waves coming from two closely spaced holes. (Copyright © Thomas Young – Own work, Public Domain.)

[1] See Chomsky (2000, pp 108–12) for an interesting historical discussion of a different example (on the topic of gravitation) of this process.

Fresnel's wave theory—if correct—makes a ridiculous prediction: if coherent light is blocked by a round obstruction, there would be a bright spot in the very center of the shadow. When the suggested experiment was immediately performed, it proved to everyone's surprise that 'Poisson's spot' is real.

For Poisson in 1817, the joke was on him. Perhaps in the end, however, he had the last laugh. The success of the wave theory of light demonstrates absolutely that light does the sorts of things that waves do. In that sense, it is a wave. However, it does not mean light is only a wave. Many experiments, beginning around 1900, showed that there is more to it. In certain other circumstances, light does indeed behave like a stream of tiny particles that transfer their action individually at specific times and places. Reconciling these two seemingly incompatible descriptions of light occupied many physicists for the first few decades of the twentieth century, and the result is our modern theory of quantum mechanics.

4.2 The photoelectric effect and the particle model of light

One important historical example of light acting as a stream of particles is the *photoelectric effect.* If light is projected onto certain metallic surfaces, electrons are ejected from the metal. If this is accomplished inside a glass tube evacuated of air, these now-freed electrons can be accelerated by an electric field to make a measurable electric current.

The surprising thing is that careful experiments showed the photoelectric effect to be, in many ways, an either-or effect and not a more-or-less effect. For example, if the wavelength of the light is longer than a particular threshold value (which is different for different metallic surfaces), then no electrons are freed by the light no matter how bright it is. From the point of view of light as a wave, this fact makes no sense. It is true that these longer-wavelength waves, all else being equal, are of lower energy than shorter-wavelength waves. We should, however, be able to compensate for this simply by making the light brighter (so that the waves have a larger amplitude). So, why is there a well-defined particular wavelength, longer than which no effect occurs at all?

Another odd feature of the photoelectric effect is that the energies of the individual electrons that are freed from the metal are not related to the brightness of the light. Instead, they relate only to the wavelength of the light. The brightness, instead, decides how many electrons are freed. From the point of view of waves, this makes no sense because the energy of a wave is related to both the wavelength and amplitude (brightness).

Finally, waves carry energy per time; they carry power. This means the energy transmitted by a wave depends on the length of time over which the wave is absorbed. A low-power wave absorbed over a long period of time transfers just as much energy as a high-power wave absorbed for a proportionally shorter amount of time. Thus, one would expect there to be a time delay for the photoelectric effect. That is, the atoms must absorb the light for a long enough time so that enough energy has been absorbed to free the electrons from the atoms. This also means we would expect a longer time delay for a dim light than for a bright light.

Experiments, however, clearly show that it does not work that way. The photoelectric effect either happens or it does not. When the effect does happen, it is virtually instantaneous.

This mystery was solved by Albert Einstein in 1905. He showed that the experimental facts of the photoelectric effect do make sense if we think of the light shining on the metal not as a wave but as a stream of particles. These particles (called photons) interact one-on-one with individual atoms. From this point of view, either an individual photon has enough energy to knock an electron off the atom, or it does not. If the light shining on the metal is made of photons that—individually—have too little energy, then nothing will happen no matter how many photons there are. There is no time delay because it is like a collision between two particles; all of the energy of the photon is delivered at once.

Einstein went further than these discoveries in an important way; he connected this 'particle model' to the existing 'wave model.' That is, light can act either as a wave or a stream of particles depending on the circumstance, but the two are not independent of each other. There is a correspondence between the two models:

1. A bright light, which corresponds to waves of large amplitude, corresponds to a stream of many photons; while a relatively dimmer light (lower-amplitude wave) corresponds to a stream of relatively fewer photons.
2. Light of relatively long wavelength (low frequency) corresponds to a stream of photons, each of which has relatively low energy. Light of relatively short wavelength (high frequency) corresponds to a stream of photons, each of which has relatively high energy. Einstein proposed that photons have an energy given by the following simple equation.

   $$E = hf \tag{4.1}$$

   Here E is the energy of each individual photon, f is the frequency of the light if it were instead acting as a wave, and h is a very tiny number called *Planck's constant.*

The constant (h) is called Planck's constant because it was first proposed by Max Planck in 1900, in his theoretical accounting for blackbody radiation. Einstein made the additional leap, five years later, that this tiny constant represents the energy per frequency of photons—individual particles of light[2].

The measured value of Planck's constant is 6.626×10^{-34} Js. It has this particular tiny value simply because...it does. It is one of the constants of nature, along with the speed of light and the constant of gravitation. It is unknown why these and other fundamental physical constants have the particular values that they do.

4.3 Young's experiment reconsidered

Let us look at Young's experiment in terms of the particle model of light. How would we do that in practice? That is to say, how could we set up the experiment

[2] The word 'photon' was first coined years later, in 1926, by Frithiof Wolfers and Gilbert N. Lewis.

such that the light is acting like particles instead of waves? All we need to do is to allow the light shining on the two slits to be very dim and to use extremely sensitive light detectors to record the light as it arrives at the screen.

For example, we could place, at the screen, a row of sensitive *photomultipliers.* These are glass vacuum tubes that use the photoelectric effect. A single electron freed from the metal is accelerated (by an electric field) and collides with a second target, and more electrons are liberated by the collision. This process can be repeated several times within the tube. A single photon can result in a shower of many electrons, which causes an easily-measurable pulse of electric current. Thus, a photomultiplier is capable of detecting single photons as they arrive. If we place a row of photomultipliers at the screen location in Young's experiment, we can monitor the arrival of each individual photon at the screen.

Next, imagine that we illuminate Young's double slit with light of such low intensity that on average only one photon at a time arrives at the slit and at the row of detectors. What happens? A moment's thought reveals that this situation makes little sense from the strict point of view of either waves or particles. A single particle can only pass through one slit or the other, so how does it even 'know' that the slit is double? Furthermore, a wave must be spread out in space and time; it is a pattern, and not a thing. So how can a wave arrive at the detector at a single moment in time and at a single place?

In practice, each single photon that passes through the double slit arrives at one of the row of detectors and triggers a signal. For any single one of these photons one cannot predict, even in principle, which of the detectors will be triggered. For a single photon the interaction with the row of detectors is a random event.

Here is the curious thing. Although the arrival of each individual photon is random—if one keeps track of where each photon arrives— after many thousands of photons have been detected, the original interference pattern emerges. More photons arrive at the places where Young's experiment showed the waves to interfere constructively, and very few arrive at the places where the wave experiment produced destructive interference. It is the interference pattern for the double slit that arises, and not the very-different diffraction pattern one sees when light passes through a single slit.

This is very strange, indeed. How does the first photon 'know' where it is supposed to go such that in the end, after many thousands of photons, it is part of the overall 'picture' of bright and dark interference bands? If the photons are arriving only one at a time, the same question could be asked of each photon. How do they, individually, know what to do? Since a wave is spread out in space and time, there is no paradox there. The wave passes through both slits at once, and it stretches all the way from the slit to the screen.

Quantum physics can solve this paradox in different ways. One important example, the *wave mechanics* developed by Erwin Schrödinger in the 1920s, imagines a *wave function* that—although it cannot be observed directly—tells us how to calculate the probability that a particle will be detected. From this point of view, there is a wave that can interfere even when only one photon is detected at a time; it is the wave function. Constructive interference now means, 'there is a high probability of detecting a photon here.' The photons themselves act randomly, according to the laws of probability. Naturally, more of them arrive at the locations where detection is more likely to occur.

One interpretation is that the wave function is—in a sense—'what is really real,' even though it cannot be directly observed even in principle (e.g., see Penrose 2004, p 508). The photon, on the other hand, is just a sudden transfer of energy to the detector. In a sense, it is not a 'thing.' Here, we have an interpretation that what is real cannot be observed, but it does tell us how to predict the likelihood of a particular observation.

Wave mechanics is not the only approach to quantum physics. When other experiments are considered, it all gets weirder still, and the precise connection between our ordinary experience and the strange rules of quantum physics are still a subject of much debate (e.g., see Penrose 2004, chapter 29). I leave it to the interested reader to explore this in more detail in one or more of the many excellent books on quantum physics that are written for the lay reader (e.g., see Feynman 1985).

References

Chomsky N 2000 *New Horizons in the Study of Language and Mind* (Cambridge: Cambridge University Press)

Feynman R P 1985 *QED: The Strange Theory of Light and Matter* (Princeton, NJ: Princeton University Press)

Penrose R 2004 *The Road to Reality: A Complete Guide to the Laws of the Universe* (New York: Vintage Books)

IOP Publishing

Photography (Second Edition)
Physics and art in focus
John Beaver

Chapter 5

Spectra and sources of light

5.1 Light and its spectrum

There are many ways by which matter creates light. Some of these processes are best understood from within the particle model of light, while others are more easily explained in terms of waves. Nonetheless, any particular method for producing light is essentially an interaction between light and matter, and the wavelength (or frequency) of the light is of crucial importance. Any method of producing light will inevitably not create all wavelengths equally. If by 'light' we mean the entire electromagnetic spectrum, then this fact is true in spades.

Let us, for example, use an electronic oscillator circuit to make a current go back and forth in a wire at a frequency of one million Hertz. Electromagnetic waves of that same frequency will then radiate away from the wire. A frequency of one million Hertz is in the radio part of the electromagnetic spectrum, so we have the basis for a radio transmitter.

For visible light with frequencies several powers of ten higher, this strategy simply will not work. If, however, we heat up a tungsten wire to a high temperature, the individual tungsten atoms will vibrate at high frequencies, and visible light (and infrared too) will be emitted.

The point is that any particular method of producing light will make some wavelengths well and others poorly or not at all. To completely describe a particular source of light, we should list separately how much light of each wavelength has been produced. Such a description is called a *spectrum* (plural, *spectra*) of the light source.

For visible and near-visible wavelengths of light, we can use a *diffraction grating* to produce a spectrum. A diffraction grating is a pattern of millions of fine, evenly-spaced parallel lines usually on a transparent substrate. When light of a particular wavelength passes through a diffraction grating, it is deflected by a precise angle. The wavelength (λ) and the angle of deflection (θ) are related by a simple relationship.

$$\lambda = d \sin \theta \tag{5.1}$$

doi:10.1088/978-0-7503-3703-8ch5

© IOP Publishing Ltd 2022

Here d is the spacing between the parallel lines in the grating. For the angle of deflection to be significant, d must be tiny indeed; a typical diffraction grating has hundreds of lines per millimeter.

We can allow light from the source to first pass through a narrow slit and then through lenses to make a narrow, parallel beam. If we allow this to then pass through the diffraction grating, different wavelengths are deflected by different angles. Thus, light is separated into its component wavelengths, which can then be measured separately. Such a device is called a *spectrograph* or *spectrometer*. See figure 5.1 for an example. In this particular case, the light detector is scanned across the spectrum with the wavelengths measured one at a time. One can also simply use a camera to photograph the entire spectrum at once.

The top of figure 5.2 shows a direct photograph of light from a compact fluorescent lightbulb, as recorded by a simple spectrometer, with an

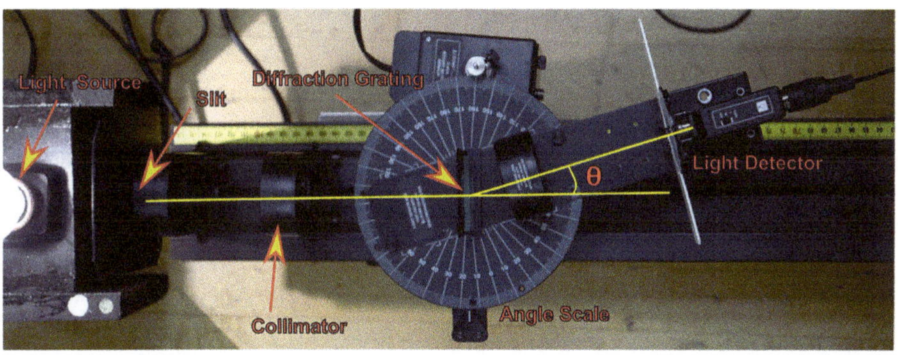

Figure 5.1. One type of spectrometer. A collimated narrow beam from the source on the left passes through a diffraction grating. A light detector on the right measures the brightness of the deflected light as the angle of deflection (θ) is slowly changed by hand.

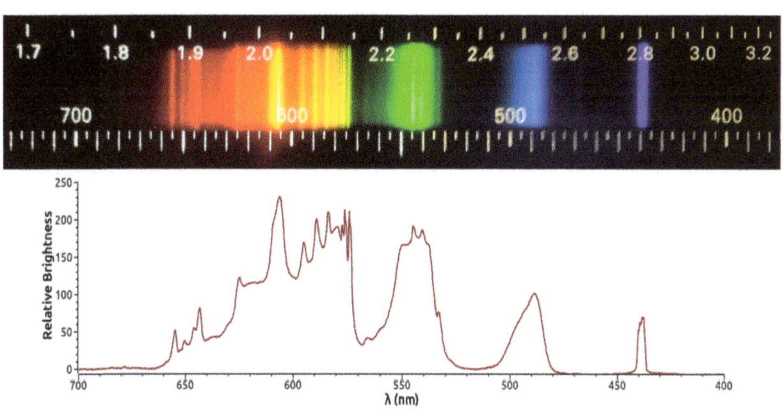

Figure 5.2. Two ways to represent the spectrum of a source of light. **Top:** Wavelength and photon energy scales are superimposed over a direct photograph of the spectrum, produced by passing light from the source through a diffraction grating. **Bottom:** A graph of the same spectrum, plotting brightness versus wavelength.

approximate scale superimposed upon the image. Notice that there are two scales for the spectrum at the top of figure 5.2; the top scale is proportional to the frequency of the light, while the bottom scale is wavelength (in nanometers).

Frequency and wavelength have a simple reciprocal relationship to each other (section 2.3.2), and this means we can represent a spectrum with either. When one goes up the other goes down, thus there is no 'correct' way to order a spectrum. Figure 5.2 represents the spectrum with increasing frequency and decreasing wavelength, but the opposite choice would have been just as valid.

The specific numbers used in the upper scale, although proportional to the frequency of the light, are chosen for an important reason: they measure the energy per individual photon, represented with a tiny unit of energy called the *electron volt*. Many processes whereby matter either creates light or interacts with it are one-on-one encounters between individual photons and individual atoms. Thus, this measure of energy per photon is of crucial importance. In particular, many photographic processes require rather high-energy photons and only work at short wavelengths.

Despite the physical significance of photon energy, most visible spectra are represented in terms of wavelength instead. The reason for this is simple: most spectrometers use diffraction gratings, and a diffraction grating disperses light in a manner that is proportional to wavelength rather than frequency or photon energy. Note that the numbers in the wavelength scale are evenly spaced, while those in the photon energy scale are not evenly spaced.

The apparent colors of different wavelengths of the spectrum, as shown in figure 5.2, are approximately as they would appear if one were to peer directly through the spectrometer with one's eye. The combination of the digital camera used to make the photograph and the color printing of the pages of this book does a fair job at approximating the visual color perception of everyday scenes. However, these are complex combinations of wavelengths—not single wavelengths viewed individually.

The graph at the bottom of figure 5.2 shows a more useful way to represent the same data. A graph allows for a quantitative measurement of the brightness at each wavelength and can show detail that may not be apparent in the direct visual representation of the spectrum.

In the best of circumstances, the graph of a spectrum is not only wavelength calibrated but also *flux calibrated*. This means that when the graph indicates that one wavelength is three and a half times brighter than another, it really is. In practice it is always difficult to flux calibrate a spectrum because no light detector is equally sensitive to all wavelengths, and the optics of the spectrometer itself will inevitably transmit different wavelengths by different amounts. The spectra shown here, for example, are uncalibrated.

We will use this idea of the spectrum of a light source many more times throughout this book; it is one of the most important tools for understanding light. In the sections that follow, the spectra of light produced by several different physical processes is described.

5.2 Thermal radiation

The term *thermal radiation* refers to electromagnetic waves emitted due to the normal motions of atoms and molecules in some material. Whether it be a liquid, solid or gas, its atoms and molecules are constantly in motion. Although the individual particles are moving with a wide range of energies (called kinetic energy), there is a meaningful average in many circumstances. This average of the kinetic energies of the individual particles in a material is directly related to what we call *temperature*. Thus, the atoms and molecules making up a hot potato are moving, on average, with greater energy than are those of a frozen potato.

These motions of the individual atoms and molecules generate electromagnetic waves, and the waves produced have a wide range of frequencies just as the atoms and molecules have a wide range of energies. Thus, a spectrum of frequencies (or wavelengths) is produced. If all of the individual atoms and molecules are in perfect balance with each other—and with the light they produce—then a particular kind of spectrum is produced called a *blackbody spectrum*. That perfect balance of matter and light, called *thermodynamic equilibrium*, ensures that the details of the black-body spectrum depend only on temperature.

Figure 5.3 shows three different blackbody spectra, with temperatures of 3000 K, 4000 K and 5000 K, where the 'K' stands for kelvin—the SI unit of absolute

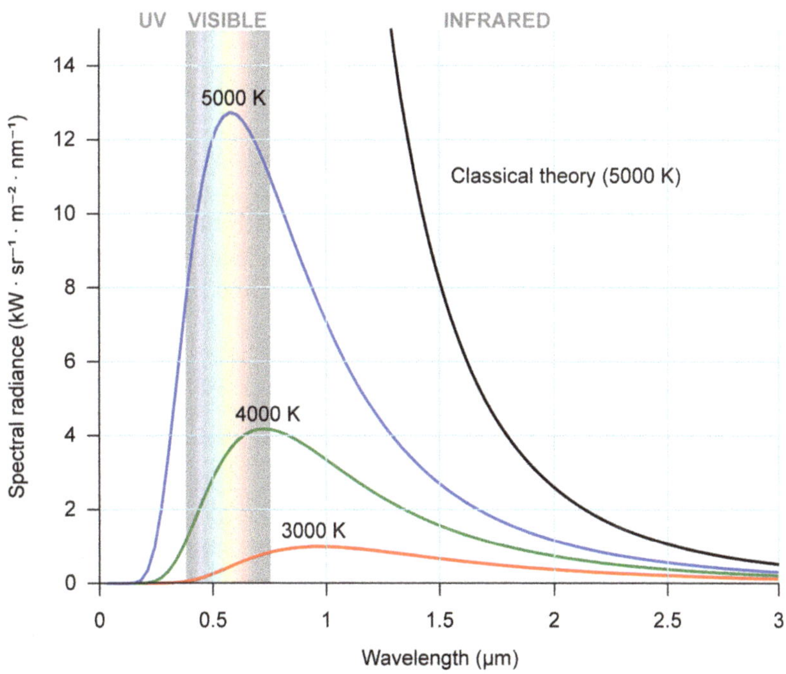

Figure 5.3. The blackbody spectra for several different temperatures. The range of wavelengths corresponding to the visible spectrum is highlighted in rainbow colors. The curve marked 'classical theory' shows how poorly theory fit the data before the development of quantum physics. (Copyright © Darth Kule – Own work, Public Domain.)

temperature. The kelvin scale starts at absolute zero, and room temperature is roughly 300 K. The surface temperature of the Sun is 5770 K, so the temperatures represented in figure 5.3 are very high indeed. Nearly all of the chemical elements are vaporized at 5000 K.

The graph is silent about what type of material made these spectra for a very simple reason—it does not matter. A 5000 K blackbody is a 5000 K blackbody, whether it is produced from calcium, cobalt, carbon or chocolate-chip cookies[1].

The overall shape of a blackbody spectrum is that of a broad hump with a peak at some particular wavelength. There is a particular mathematical formula that describes the exact shape of a blackbody spectrum, but there are two basic features that are obvious from the graph alone. First, more energy is emitted (the hump is bigger) if the temperature is higher. This is a huge effect, as the brightness is related to the total area under the graph. Notice how much larger the area is under the 5000 K graph as compared to the 3000 K graph. It turns out that the brightness of a blackbody scales with the fourth power of the temperature. Doubling of the temperature increases the brightness by $2^4 = 16$ times.

It is also clear from figure 5.3 that a higher temperature produces a blackbody peak at a shorter wavelength. This is described by a simple inverse proportionality. As an example, figure 5.3 shows that a 3000 K blackbody 'peaks' at about one micrometer (1000 nm). A room-temperature 300 K blackbody with one tenth the temperature peaks at ten times that wavelength, or ten micrometers, in a region of the spectrum known as the *thermal infrared* (TIR). Table 5.1 shows the peak wavelength for different temperatures.

A blackbody the temperature of the Sun, a little less than 6000 K, peaks right in the middle of the visible spectrum. The part of the Sun's atmosphere that makes the visible surface of the Sun is not in perfect equilibrium (if it were, the light could not escape). Therefore, the visible spectrum of the Sun is not quite the same as a blackbody, but it is a rough approximation of one with the same overall shape.

Figure 5.3 also shows the spectrum of a 5000 K blackbody as predicted by the electromagnetic theory of the late nineteenth century, and it is a catastrophic failure

Table 5.1. Peak wavelengths (λ_{max}) for different temperature (T) blackbody spectra.

T (K)	λ_{max} (μm)	Part of Electromagnetic Spectrum	Example
2.7	1100	Microwave	Cosmic microwave background
300	10	TIR	Room-temperature objects
3000	1	Near infrared	Lightbulb filament
6000	0.5	Visible light	Visible surface of Sun
10 000	0.3	Near ultraviolet	Visible surface of Vega

[1] A 5000 K chocolate-chip cookie would be a hot cookie indeed. It would be of necessity in the gaseous state and would have to be held together under enormous pressure in order to produce a blackbody spectrum.

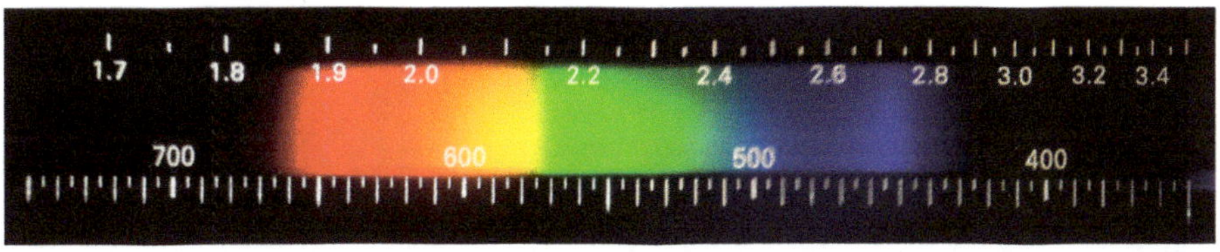

Figure 5.4. The spectrum of an incandescent lightbulb emitting approximately like a blackbody of about 2500 K.

of agreement between theory and experiment. At the dawn of the twentieth century, Max Planck demonstrated that the observed shape of the blackbody spectrum could be explained only if one assumed that energy came in discrete clumps or *quanta*. This *Planck quantum hypothesis* was the first step in the development of quantum physics, and the idea was extended by Einstein and others to form the modern concept of the photon—a particle of light. Although we observe the spectrum emitted by a blackbody as a wave phenomenon, the fact that light can act as individual particles is essential to its creation (see figure 5.4).

5.3 Atomic spectra

Thermal motion of atoms and molecules is not the only way to produce light, and not all sources of light emit with a spectrum approximately like that of a blackbody. When a low-density gas is excited by a high voltage, it glows with an *emission-line spectrum*. Only certain, specific wavelengths are emitted, which is characteristic not of the temperature but of the type of gas. Most wavelengths are devoid of light altogether, while at certain, quite-specific wavelengths, a lot of light is emitted. Each atom, ion[2] and molecule has its own characteristic list of wavelengths at which it emits light in this way.

A nonthermal source of light, such as a gas tube, produces light at the atomic level. Each atom creates its own light, emitting individual photons; the particle-like nature of light must be considered to understand these sources. They are one-on-one interactions between light (in the form of individual photons) and matter (in the form of individual atoms). As such, the spectrum of the light produced depends critically on the type of atoms emitting the light. This dependence is the opposite of the case of thermal radiation. For purely thermal radiation, only the temperature matters; the type of atom is irrelevant. Many real sources of light, when considered in more detail, involve a combination of both thermal and nonthermal emission.

Figure 5.5 shows the emission-line spectra of three neutral gases: hydrogen, helium and neon. The spectrum of hydrogen is clearly the simplest of the three. Not coincidentally, neutral hydrogen is also the simplest atom—a single proton for a nucleus, 'orbited' by a single electron.

I put the word 'orbit' in quotes because it is not so simple as that. Yes, there is an electrical attraction between the positively-charged nucleus and the negatively-charged electron. And yes, this electron, acting under the influence of that force and

[2] An ion is an atom that is either missing electrons or has extra electrons, and thus it has an overall electric charge.

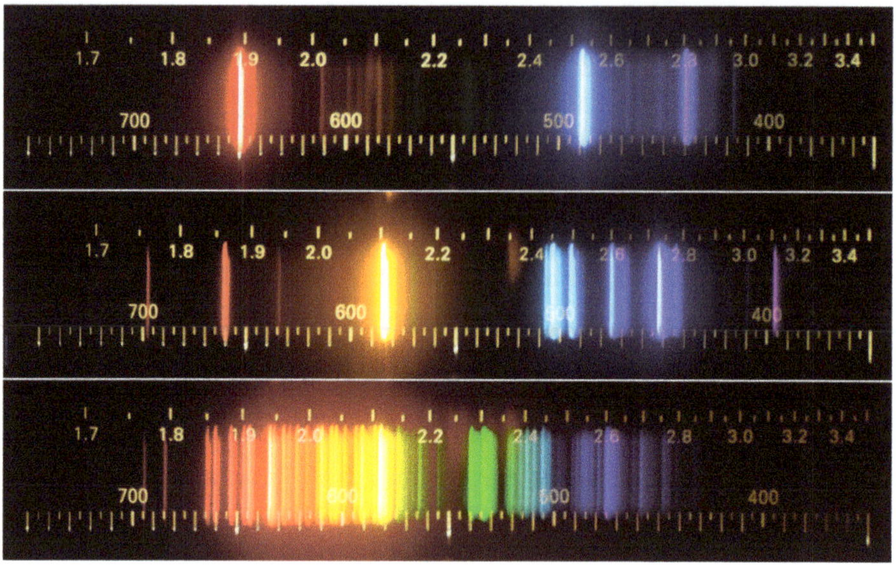

Figure 5.5. Atomic spectra of hydrogen (**top**), helium (**center**) and neon (**bottom**).

obeying Newton's laws of motion, would literally orbit the nucleus in much the same way Earth orbits the Sun[3].

The electron, instead, acts in a manner very much at odds with the Newtonian laws of motion that so successfully explain the Earth's orbit and most everyday experiences. Instead of a point-like particle orbiting the nucleus, the electron acts more like a spread-out wave that is trapped in a fixed region of space. And this changes everything.

Waves passing down the G string of a violin, for example, must have only certain wavelengths, or else repeated back-and-forth reflections off the nut and the bridge will quickly cancel each other out. These allowed wavelengths result in a set of resonate frequencies that produce the particular pitch and sound[4] of the string, even though the rubbing of the resin and horsehair bow tries to vibrate the string at essentially all frequencies at once.

In similar fashion, the electron in a hydrogen atom acts as a wave trapped by the electrical attraction of the atom's nucleus, and it must resonate with itself in that tiny space. This leads to a specific list of possible energies that the electron may have. The different emission lines in the hydrogen spectrum arise then from transitions of the electron from one allowed energy level to another.

[3] Both the gravitational force responsible for Earth's orbit and the basic electrical force act as an inverse square with distance, and this leads theoretically to similar kinds of orbits for both.

[4] In practice, the string vibrates at a combination of these allowed frequencies, and we usually associate the lowest of these with the perceived pitch. A particular combination of the other, higher allowed frequencies (called *harmonics*) alters the quality of the sound.

For example, the red line in the hydrogen spectrum, with a precise wavelength of 656.3 nm, arises when the electron makes a sudden transition from the third allowed energy level downward to the second. The electron's energy decreases but is not lost; a photon of light is created with precisely the right energy to make up the difference. In this way an isolated hydrogen atom creates light, but it can only create wavelengths that correspond to differences in its quite-specific list of possible energy levels.

Neutral helium has two electrons instead of one, and each feels not only the electrical attraction from the nucleus but also repulsion from the other electron. This situation leads to a quite different 'space' for the electrons, so the list of allowed resonate energy levels is different and more complex than those for hydrogen. Thus, the spectrum of neutral helium is completely different from the spectrum of neutral hydrogen and so too for other atoms[5].

It was the quantum revolution in the early twentieth century that first allowed for a successful physical explanation of the emission spectrum of gases in spectral tubes. Part of that story is that light has a particle-like nature in addition to its wave-like nature. The other side of the story is that matter—the physical stuff we like to think of as solid—also has a wave-like character.

5.4 Sunlight

Sunlight is the source of light for most daytime outdoor photography, so it is important to understand its spectrum. It is also one of the most common sources of ultraviolet (UV) light used for alternative and historical photographic processes that require it. The spectrum of the Sun—the visible-wavelengths portion is shown in figure 5.6—can be thought of as arising from three different effects.

1. The interior of the Sun is an opaque, ionized gas. Its spectrum is a near-perfect blackbody, and hottest in the center. The upper-most layer of the Sun's opaque interior has a blackbody spectrum of roughly 6000 K.
2. For the light from the Sun's interior to escape into space, it must first pass through layers of gas that are cooler and less dense—the Sun's atmosphere. This gas absorbs light according to the types of atoms, ions and molecules in the Sun's atmosphere. The process produces dark lines in the Sun's spectrum.
3. For sunlight to reach Earth's surface, it must pass through Earth's atmosphere. The molecules that make up Earth's blanket of air further absorb sunlight, making dark bands in the spectrum.

Figure 5.6 shows the visible part of the Sun's spectrum. A continuous spectrum, much like a blackbody, is evident but with narrow dark lines at specific wavelengths that correspond to absorption of light by atoms and molecules in the atmospheres of both the Sun and Earth.

[5] If helium is ionized to have only one electron, its spectrum is similar to that of hydrogen but with all of the frequencies scaled by a different factor because of the greater electrical attraction from the two protons in the helium nucleus.

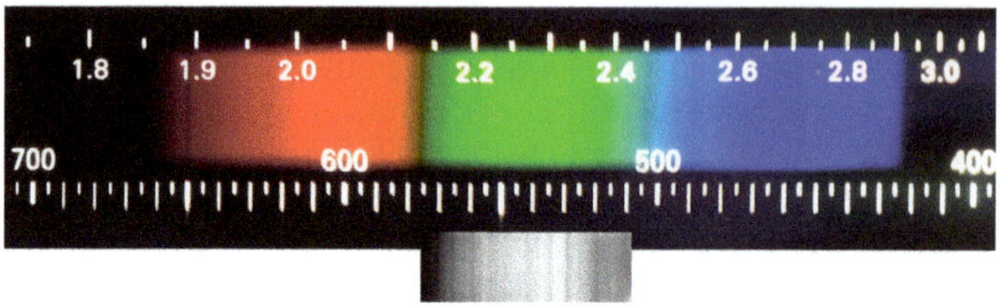

Figure 5.6. Part of the visible spectrum of sunlight as recorded at Earth's surface with a simple spectrometer. Faint dark vertical lines are visible in the spectrum, marking wavelengths partially blocked by atoms, ions and molecules in the Sun's and Earth's atmosphere. The inset at the bottom is a selection of the spectrum with the contrast enhanced to better show the faint absorption lines.

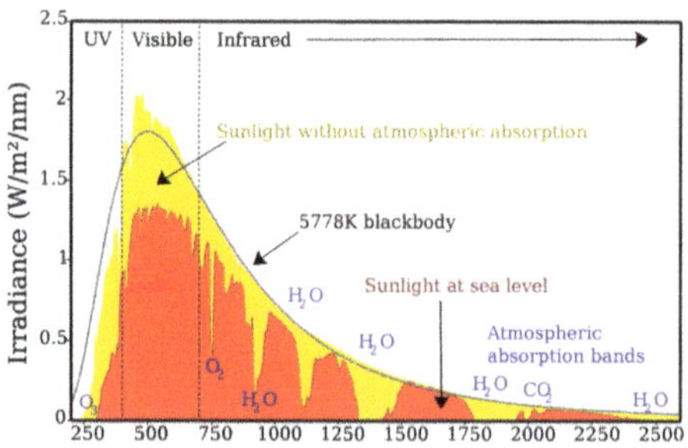

Figure 5.7. The UV, visible and infrared spectrum of the Sun. Yellow represents the solar spectrum as it enters the Earth's atmosphere. Many wavelengths are absorbed by our atmosphere to produce the sea-level spectrum shown in red. A blackbody, at the same temperature as the Sun's visible surface, is also shown for comparison. (Copyright © Robert A. Rohde – Own work, Attribution-ShareAlike 3.0 Unported (CC BY-SA 3.0), https://commons.wikimedia.org/wiki/File:Solar_Spectrum.png, Solar Spectrum.png.)

Figure 5.7 illustrates the different factors that contribute to the Sun's spectrum. The solid smooth curve is a blackbody of 5780 K, which approximates the Sun's overall spectrum. The yellow curve is the actual spectrum of the Sun with the effects of many absorption lines from the Sun's atmosphere. There is so much absorption in the near-UV, that the spectrum is significantly fainter there than a pure blackbody would be. The red curve shows the actual spectrum of sunlight at the Earth's surface with significant dimming at many wavelengths because of absorption by molecules in the Earth's atmosphere. In particular, water vapor and carbon dioxide absorb over broad bands of wavelength in the infrared, and ozone (O_3) absorbs a lot of UV light.

5.5 Fluorescent light

A lamp to produce a mercury emission spectrum is inexpensive to produce. But some of mercury's brightest emission lines are in the UV part of the spectrum, which would do us little good if we were to use such light as a source by which to see. The property of *fluorescence* (see section 3.8), however, allows us to reprocess the UV light to visible wavelengths. Before it gets out of the lamp, it strikes a powder of small crystals, called phosphors, that line the inside of the glass lamp. These crystals

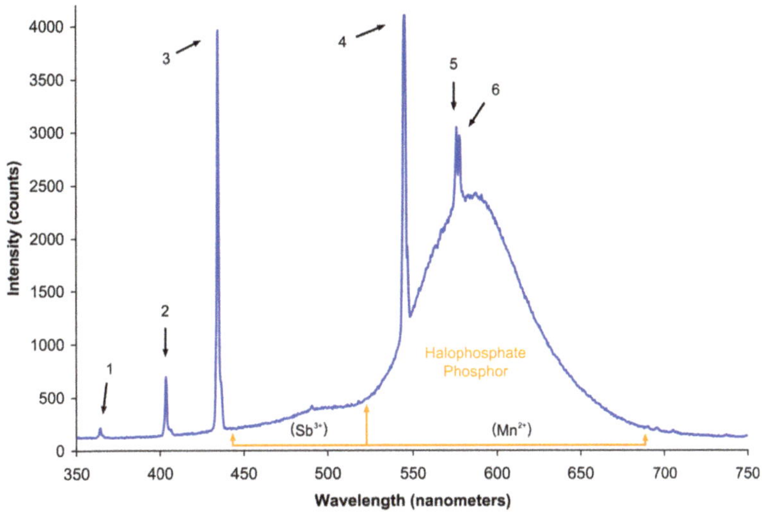

Figure 5.8. The spectrum of a fluorescent tube light. The numbered peaks are emission lines from the spectrum of mercury. The smooth parts of the spectrum are due to the glowing phosphors excited by the UV emission lines in the mercury spectrum. (Copyright © Д.Ильин, CC0 1.0 Universal (CC0 1.0).)

have the ability to absorb UV light and use its energy to emit its own visible light. Since the phosphors emit a range of many wavelengths after being excited by UV light, the overall effect is that the invisible UV light is converted to visible light. If phosphors with the right properties are chosen, the overall light emitted may appear whitish. However, it is not produced by high temperatures, and the spectrum is very different from that of sunlight.

Figure 5.8 shows the spectrum of a typical fluorescent light. One can see the mercury emission spectrum poking through, but superimposed on this is the broad spectrum of the glowing phosphors. Although the overall perception by the eye of fluorescent light is roughly similar to that of white light, it is a very different spectrum.

5.6 LED light sources

A *light-emitting diode* (LED) is a decidedly nonthermal source of electromagnetic waves. In certain carefully-chosen cases when a voltage difference is applied to a junction between two dissimilar semiconductors, electrical current flows and light is emitted. Although silicon is usually the base semiconductor, small amounts of other elements are interspersed throughout the silicon crystals. These 'doped' impurities are carefully chosen for the properties of their energy bands. Practical LEDs have been engineered to emit at many wavelengths, from the infrared, through the visible spectrum and into the UV. Something like the reverse of this process can also be engineered; it is the foundation for a silicon solar cell (Nassau 1983, chapter 8).

Since most LEDs naturally emit over a relatively narrow range of wavelength, those that emit in the visible part of the spectrum tend to produce highly saturated spectral colors. LEDs of different wavelengths can combine to make a whiter-looking source of

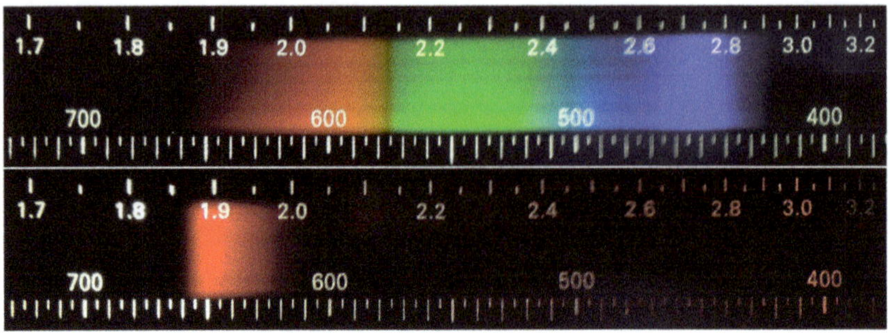

Figure 5.9. The spectrum of a white LED lamp (**top**) and a red LED lamp (**bottom**). The broad spectrum of the white LED arises from a complex combination of relatively narrow wavelength emission from solid-state band gaps and broad wavelength fluorescence.

light, and fluorescent materials can be used to reprocess short wavelength emission to a more broad range of wavelengths. Figure 5.9 shows, at top, the spectrum of a 'white' LED and, at bottom, the much-narrower spectrum of a deep-red LED.

5.7 Lasers

A laser emits *coherent light*. Most sources of light are, by contrast, incoherent. The glowing filament in an incandescent lightbulb is an example; each tiny portion of the metal filament emits light with no concern for–and no knowledge of—the rest. In a laser, however, all of the physical light-emitting portions work in concert with each other through the process of *stimulated emission*. The word 'laser' is in fact an acronym for 'light amplification by stimulated emission of radiation.'

When an electron in an atom or molecule is excited to a higher energy level, there is a tendency for it to spontaneously de-excite and emit a photon of light in the process. This usually happens so quickly that there is too little time for another process to interfere. However, some materials have energy levels that are *metastable states*—once excited to that energy level, they linger somewhat and allow for some other event to trigger a de-excitation and emission of a photon.

Stimulated emission is one such process. If a photon passes by an atom excited to exactly the energy of the passing photon, there is high probability the electron will de-excite and an identical photon emitted. Unlike the ordinary de-excitation–emission process, this photon is emitted in the same direction as the stimulating photon. It is connected to the stimulating photon in *phase*—the overall shift of the wave. Since it is by necessity the same energy as the stimulating photon, it also shares the same wavelength. See, for example, Nassau (1983, p 58ff) for a more detailed description.

A laser is a device that engineers such stimulated emission of light. The material (sometimes a mixture of gases) must be carefully chosen for its metastable energy transition—this decides the wavelength of the laser light emitted. To amplify the effect, a laser usually is constructed of a cylinder with a mirror at each end. One mirror is only partially reflective, so the light can finally escape. Some outside source of energy causes the excitation, and the stimulated emission produces a light source that is not only coherent but also *monochoromatic*—all of the same wavelength. This is because the particular metastable transition, caused by the very material it is made of, guarantees the emission of a particular photon energy and thus a particular wavelength of light. The two opposing mirrors, between which the stimulated emission is trapped, have another desirable effect: the source of light is unidirectional. The unidirectional nature of the light is the 'beam' in laser beam.

The first laser was made from a ruby crystal, but they can also be made from glass tubes filled with carefully chosen low-density gases. In recent years, most lasers are made from inexpensive semiconductor materials. In this case—in the so-called diode laser—the metastable state is provided by an energy band gap that arises from the contact of two dissimilar semiconductors. The technology is related to the LED described in section 5.6. For laser light to be emitted, the two materials, the impurities within them and the physical dimensions of the junction between them must all be carefully engineered. Diode lasers of many wavelengths are now available, from the infrared through the UV.

5.8 Many ways to see a cat

We have seen how emission processes produce a spectrum of light—different amounts of light emitted at different wavelengths. We have described different ways in which matter can both create light and also alter light that interacts with it: absorption, reflection (diffuse and specular), refraction, scattering and emission (thermal and nonthermal). These processes depend critically on the wavelength of the light. Their dependence means that, for one particular region of the electromagnetic spectrum (visible light, for example), one or more of these processes may be important while others are negligible. However, at some other region of the electromagnetic spectrum (infrared, for example) the situation could be entirely different.

Let us reconsider, for example, Boris the cat in figure 3.12, from section 3.4. In the visible part of the spectrum, we see Boris by diffuse reflection of light that has arrived at him but was created by other means. In this particular case, Boris was illuminated by an incandescent lightbulb—a white-hot tungsten filament that emitted light by means of (mostly) thermal radiation.

The light emitted by the lightbulb and arriving at Boris had a spectrum similar to the red curve in figure 5.3. The spectrum was mostly visible light and the infrared part of the spectrum right next to visible light—what is sometimes called the *near infrared* (700–5000 nm). In this case, the light detector that recorded the image ignored whatever near-infrared light reflected from Boris and only recorded the little bit that was in the visible part of the spectrum. The picture, therefore, shows visible light emitted by a lightbulb and reflected diffusely off of Boris.

Every part of Boris also absorbed a portion of the visible light that arrived there. His stripes absorbed a little more, so they reflected a little less and thus appear slightly darker in the picture. Figure 3.12 shows only what happened to visible light; it says nothing about other wavelengths. What about that near-infrared light that also illuminated Boris but was not detected, and so is not represented, in the picture? What would have happened if we had instead taken the picture with a camera that is sensitive not to visible light, but instead only to near-infrared light? Would the picture have been different?

See figure 5.10 where I have photographed a different cat[6], Tobias, with two different cameras: one sensitive to visible light (left) and the other sensitive to near-infrared light (right). Because the striped part of the fur absorbs more (and so reflects less) visible light, it does not follow that it must also do so for near-infrared light.

For both visible light and near-infrared light, we see what is reflected by Tobias; the source of the light is elsewhere. For TIR light, however, Tobias is the source. In figure 5.11 I photographed him with an ordinary visible-light digital camera on the left. The right-hand image was made with a *thermal imaging camera*, sensitive mostly between the wavelength range 8–14 μm.

TIR is called 'thermal infrared' because it is the range of wavelengths emitted by objects at roughly room temperature. The amount emitted depends strongly on temperature and—since Tobias is warmer than his surroundings—he emits more TIR than anything in the picture.

The window glass is transparent to visible light, but it is opaque to TIR light; it does, however, reflect a significant fraction. Thus none of the outdoors for which Tobias pines is visible in the TIR view, and it looks as though he is sitting in front of a mirror admiring his own handsome reflection.

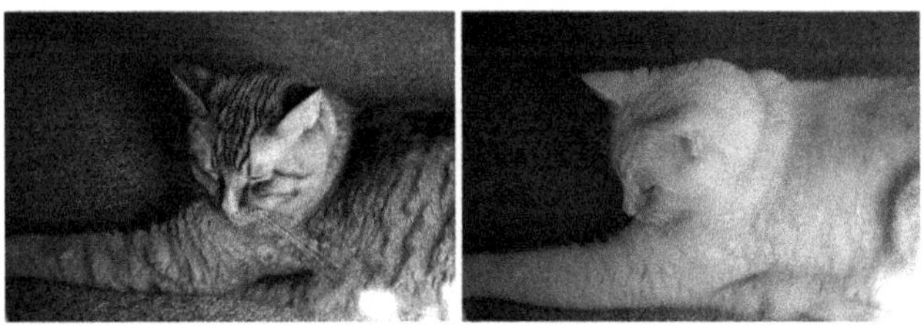

Figure 5.10. Tobias, the cat, illuminated by sunlight and photographed with visible light (**left**) and near-infrared light (**right**).

Figure 5.11. Tobias illuminated by daylight and photographed in visible light (**left**) and TIR light (**right**).

[6] Rest in peace, Boris, under the peonies!

The TIR image looks almost as though it was a negative, but it is not; the brightest parts of the picture are those that emitted the most TIR light. The gaps and crevices in Tobias's fur reveal areas that are shadowed from reflected visible daylight, so they appear dark in the visible-light picture. They also reveal the skin of Tobias, which is warmer than the outer parts of his fur, making those areas some of the brightest parts of the TIR image.

Reference

Nassau K 1983 *The Physics and Chemistry of Color: The Fifteen Causes of Color* (New York: Wiley)

Part III

Geometry and two-dimensional design

IOP Publishing

Photography (Second Edition)
Physics and art in focus
John Beaver

Chapter 6

Geometry and the picture plane

The history of geometrical optics is tied intimately to the history of the visual art that predates photography. Look at paintings from the middle ages, and they are unlike photographs in many ways. The sizes of forms in the paintings were not necessarily connected in a representational manner, as they are in a typical photograph, to the sizes and distances of objects in the real world. Instead, the size and positioning of forms in a medieval painting often had more to do with what was more or less important (see e.g., Duby 1992, chapter 1). It was not until the Renaissance that painters really worked out the details of *linear perspective*, allowing one to duplicate a scene on canvas akin to what would later appear in photographs (Fichner-Rathnus 1992, pp 333–4).

Photography clearly has much to do with the connection between the three-dimensional (3D) world and two-dimensional (2D) art. At its most basic, a camera takes rays of light from the 3D world and redirects them to a 2D surface. The 'fourth dimension' of time is also important. It is now common to see photography as an act of freezing a moment of time, and photographers often take advantage of this. See, for example, the disturbing use of this power in *Execution of Vietcong Prisoner* by Eddie Adams (Cooke and Kinneberg 2014, p 10). A case can be made that the idea that an instantaneous moment of life can be captured with paint on canvas came into its own with the Impressionists, who were very much influenced by then-new photography.

6.1 From 3D to 2D

Part of the essence of photography is that elements of the inherently 3D world are directed onto a 2D surface. At first glance, one might think that, as with a camera, this is exactly what the human eye does. It is true that rays of light from the world are focused by the eye's lens and cornea onto specific locations on the 2D surface of the retina. However, 'seeing' is far more complex than that.

doi:10.1088/978-0-7503-3703-8ch6 © IOP Publishing Ltd 2022

Part of the reason that 2D art works at all is that our brains seem to be specifically structured such that we can easily attach worldly, 3D meaning to patterns on a flat surface. Presumably, this is because the brain must make sense of the essentially 2D information coming from the eye. How the eye–brain accomplishes this is extraordinarily complex and only partially understood, but much of it seems to have little direct counterpart in the operation of a camera. To cite one rather-obvious example, the image from a camera (or eye) lens literally represents only directions in space—not distances. An image is only a 2D representation, and yet we construct a 3D Universe out of that image.

This fact is particularly (and painfully) obvious to an astronomer. Take a picture of the night sky, and one sees a pattern of bright dots—stars. There are an infinite number of possible 3D arrangements of those stars consistent with that same picture. To determine the third dimension of distance requires much additional information that is not available in the photograph alone, and this task is a central preoccupation of astronomers.

The human brain, in its second-by-second workings, does not really operate in this mathematical way. The brain adds its own interpretation to the mix of data coming from the eye; it 'makes stuff up,' in a sense. It is possible to take advantage of this fact and trick the brain; the phenomena of *optical illusions* provide evidence and clues to the actual mechanisms at work.

6.2 The brain's construction of 3D reality

Experiments have shown that the image on the retina of the human eye is very different from what we actually perceive as sight. In its construction, the human eye has much in common with a camera; a lens focuses an image onto a surface (the retina) that is sensitive to light. The similarity, however, ends there. Much is sometimes made of the fact that the image on the retina is—like in a camera—upside down, but that is not such a big deal. Most of 'seeing' is in the brain and not in the eye; flipping the image right-side up is the least of it.

Some of the research carried out in the 1950s and 1960s by the Russian psychologist Alfred Yarbus provides a good example (Yarbus 1967). Yarbus recorded the eye movements of observers as they performed certain tasks. He found that, for even something so simple as looking at a motionless face, the human eye darts around all over the place (e.g., see figure 6.1). It seems that we tend to look at the eyes and mouth a lot, as can be seen from the eye movement trajectories he recorded. If one pointed a video camera at the same face, and moved it according to such a trajectory, the resulting video would likely be unintelligible.

Somehow the brain makes sense of the seemingly-confusing information coming from the eye. The brain constructs an image from the information, rather than simply recording it, and that image is of a 3D world with objects in it. What we 'see' not only ignores much of the irrelevant information coming from the retina, but it also includes elements not even present in that stimulus. This means that we can attach meaning to even the simplest marks on a flat surface. For painters and photographers this is a lucky break! A photograph is just a flat piece of paper with marks on it, but we can see the world when we look at a good one.

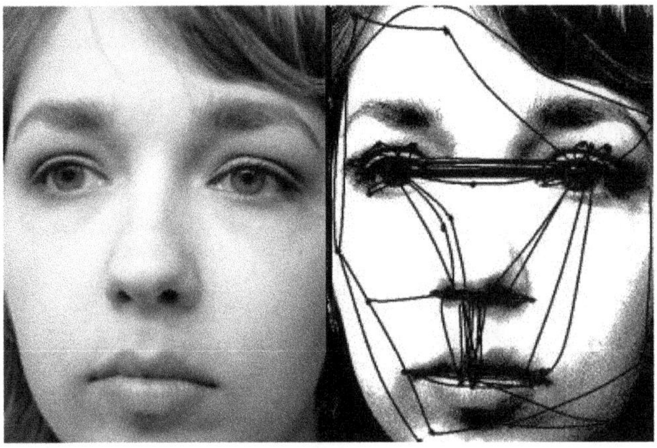

Figure 6.1. An illustration tracking the eye movements of a subject looking at a photograph of a face—work pioneered by the Russian psychologist, Alfred Yarbus. Even when looking at something so simple, the human eye is constantly darting. We, however, still perceive a motionless image. (Copyright © SpooSpa – Own work and Derivative: Simon Viktória, Attribution-ShareAlike 2.0 Generic (CC BY-SA 2.0), https://commons.wikimedia.org/wiki/File:Szakkad.jpg, Szakkad.jpg.)

6.3 Linear perspective and the camera obscura

Which came first, the camera or the light-sensitive material that goes in the camera to make the photograph? It is a surprise to most people that the correct answer is the camera, which preceded the invention of photosensitive materials by a couple of hundred years. However, when one realizes that the word 'camera' is simply the Latin word for 'an enclosure,' then maybe it is not so strange. Long before cameras were used to record images, they were used to view them. The *camera obscura* is simply a 'dark box.'

It has been known for centuries that one can use geometry to produce images. If a small hole is placed on one side of a dark box, each ray of light coming from objects in the world will be restricted by the hole to only one spot on the opposite side of the box. Thus, an image of the outside world is automatically reconstructed on the inside of the box. This basic idea underlies both the ancient camera obscura, and the modern pinhole camera we discuss in detail in chapter 9.

A glass lens allows one to (among other things) brighten an otherwise dim image, and this too has been known for centuries—at least since the 1500s. When light-sensitive materials were invented in the 1800s, the already-existing camera obscura developed into the photographic camera (Marien 2002, pp 3–7).

One simple way to view the image in a camera obscura is to make the dark box very large and put oneself inside. The photographer Abelardo Morell has been doing this with great success. He uses paint to black out the windows of a room with a view and then scratches a small hole in the paint. An upside-down image of the outside world appears on the opposite wall. This is, in effect, a giant pinhole camera with the photographer inside. Very little light enters through the small hole, so the image is very dim—barely visible to the naked eye. He photographs this image with a

Figure 6.2. My physics classroom photographed from the inside. A small hole in the blinds produced the image of what is outside the laboratory. The picture has been turned upside down, so the room appears inverted, but the camera-obscura image is upright.

separate camera, set up inside the room, using a long-time exposure to make up for the dim light.

A not-very-impressive example of this same technique is shown in figure 6.2. The darkened room of my physics classroom was photographed with a long-time exposure. The image on the wall is from a small hole in the opposite blinds. The photograph has been inverted so the camera-obscura image, of the grounds outside the lab, is upright.

The camera obscura and related optical and sighting devices were instrumental in the discovery of mathematical rules for laying out a realistic perspective in paintings —what art historians call *linear* or *mathematical perspective*. Some have speculated (the thesis is still controversial) that some of the Flemish painters of the seventeenth century (Johannes Vermeer, in particular) used a camera obscura to properly locate the details in their paintings (Steadman 2001). In fifteenth century Italy, Filippo Brunelleschi performed experiments with a sighting device much like a camera obscura in order to work out the first formal methods for incorporating precise linear perspective into paintings and drawings.

6.4 The picture plane

Photographs are often *representational*; forms in the photograph are meant to directly represent things in the real world. However, as is the case for all 2D art, there is more to it than that. An important element is the *picture plane*—the flat plane representing the 2D surface of the picture itself.

There are really two ways to look at this idea, one more abstract and the other more physical. From one point of view, the picture plane consists of shapes, lines, colors, and contrasts—whatever the physical means by which they are embodied. We will further consider these formal aspects of 2D art in chapter 13. However, there is also the physical nature of the picture surface itself, and it too plays a role. While the concept of the picture plane as a physical surface is often of crucial importance for painting, it is sometimes downplayed or ignored in photography—much of which strives to be purely representational.

This can be seen in the way many photographs are displayed. A mat is used to overlap the image area of the print, so the opening in the mat appears as a window with the picture behind it. The intent seems to be to hide the fact that one is looking at a piece of paper; only the representation of the image matters.

Many photographers, however, who align themselves somewhat more with the history of painting and the early history of photography, go out of their way to draw attention to the physical surface of the picture plane. One method is to use alternative printing processes that introduce their own elements in the form of random detail and textures (Rexer 2002). We will explore some of these printing processes in more detail in part VII of *Photography: Physics and Art in Focus*. Often the goal is similar to that of traditional printmaking: the image is the same, but each print is slightly different—and thus unique—due to the handmade process. The more-interesting and less-uniform surface of the paper itself is often evident. It is not just an image; it is an image on a flat piece of interesting paper, and the paper is the picture plane.

The negative process of film photography allows for another interpretation of the picture plane. One does not normally view a photographic negative directly; rather one looks at a positive print made from it. The negative also represents a flat plane, and it is what was actually inside the camera at the moment the photograph was taken.

In negative-based film photography, we actually have a possibility for two picture planes: the surface of the print itself and the surface of the negative that was used to make the print. Details in each can draw attention to these two flat surfaces, forming an additional element in the formal construction of the picture. This is an especially important point in the modern age, when many pictures never exist as a physical print. The picture may be an image on a large computer screen or on a tiny smart phone, depending on who is looking at the picture. Where, then, is the picture plane? Is the concept still meaningful when no physical image surface exists?

Figure 6.3 shows an image made from an old type of Polaroid large-format negative film. This type of film made its own instant-print and a high-quality negative that could then be printed in the darkroom. No darkroom was needed to produce the negative itself; one just needed to clear off the developer goo, in daylight in the field, with a simple sodium sulfite solution (often carried around by the photographer in a bucket).

In this case, I allowed the goo to remain on the negative, and in fact rubbed it in the dirt to make it even more yucky. I then allowed the negative to dry without washing it. The result was a bad negative that would have been difficult, if not impossible, to print in a traditional darkroom. It was possible, however, to scan it with a high-quality scanner and then use digital techniques to reverse it to a printable positive image.

Figure 6.3. *Proof of life on Earth*, John Beaver 2008.

The result is that the surface of the negative is introduced as a picture plane, independent of the manner in which the image is printed or displayed. In this particular case the flat plane of the negative plays off the seemingly-curved plane of the table, introduced by the distortion and vignetting of the particular lens used.

In the chapters that follow we explore this 2D nature of photography and how it relates to our 3D world. We will see that even a virtual image that never exists as a physical object still has abstract relations between the parts of its 2D 'surface.' In part VI of *Photography: Physics and Art in Focus* we revisit some of these ideas in consideration of the physical processes by which light from the world interacts with, and changes, a flat surface in order to form an image. We also explore some methods by which nature can have its own say in the making of photographic art—ways that are only partially under the control of the artist.

References

Cooke T and Kinneberg C (ed) 2014 *The Photography Book* 2nd edn (London: Phaidon Press Limited)

Duby G 1992 *Medieval Art: Europe of the Cathedrals 1140–1280* (Geneva, Switzerland: Skira)

Fichner-Rathnus L 1992 *Understanding Art* 3rd edn (Englewood Cliffs, NJ: Prentice-Hall)

Marien M W 2002 *Photography: A Cultural History.* (Englewood Cliffs, NJ: Prentice-Hall)

Rexer L 2002 *Photography's Antiquarian Avant-Garde: The New Wave in Old Processes* (New York: Harry N. Abrams)

Steadman P 2001 *Vermeer's Camera: Uncovering the Truth Behind the Masterpieces* (Oxford: Oxford University Press)

Yarbus A L 1967 *Eye Movements and Vision.* (New York: Plenum)

IOP Publishing

Photography (Second Edition)

Physics and art in focus

John Beaver

Chapter 7

Light and shadows 1: eclipses

Some of the first photographs were made with shadows; the light that would have fallen upon a light-sensitive surface is selectively blocked by intervening objects. The practice of capturing shadows long predates photography: hand-traced silhouette portraits made from candle light is a good example. We will explore the intentional making of shadow prints—or photograms—in chapter 8. Here we consider the geometry of some of nature's most famous uses of shadows—lunar and solar eclipses. This leads us to consider some of the subtleties of shadows in an outdoor scene or a subject illuminated with artificial light.

7.1 Angular size of the Sun and Moon

The fact that the Moon orbits about Earth, and both together orbit a far-more distant Sun, was not firmly established until the late 1500s. It was not even clear that the Moon was a sphere, like Earth[1], until Galileo's telescopic observations of 1610. Although Tycho Brahe estimated the distance to the Moon in the late 1500s, the distance to the Sun would not be successfully estimated until the late seventeenth century and not measured accurately until 1877 (North 1995, p 253).

Although precise prediction of eclipses had to wait at least for Newton's law of gravity, a basic understanding of the cycles of lunar and solar eclipses stretches back over 2000 years. Part of the reason is that, regarding much of what happens during an eclipse, the absolute scale of sizes and distances was unimportant; it was mostly how these compare to each other that was of importance. A key concept is that of *angular size*; it is how big something appears rather than how big it is. Clearly, this is related to the actual size, but also to the distance. Figure 7.1 illustrates the concept.

The radian measure of an angle provides the most appropriate tool for measuring angular size. If we sweep out an angle in a circle, the angle in radians is simply the

[1] The ancient Greeks knew Earth was a sphere, and Eratosthenes accurately measured its diameter in 240 B.C.E.

doi:10.1088/978-0-7503-3703-8ch7

© IOP Publishing Ltd 2022

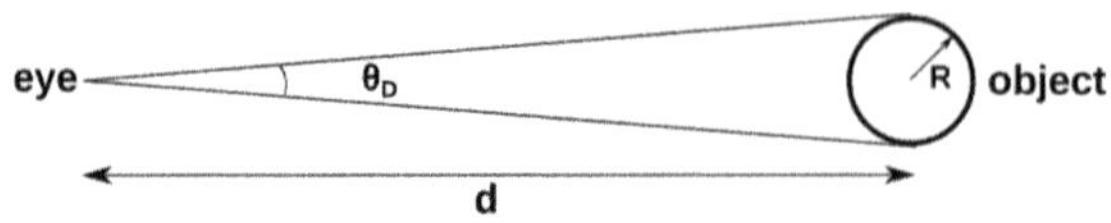

Figure 7.1. The geometrical relationship between angular size (θ_D), radius (R) and distance (d) for a spherical object.

length of the arc swept out divided by the circle's radius. We can think of the diameter, $D = 2R$, of the object in figure 7.1 as the tiny arc of a large circle with a radius equal to the distance *(d)* to the object. The angular size is, thus, simply the ratio of the object's size to its distance.

$$\theta = \frac{d}{D} \tag{7.1}$$

Even if we have no idea either how big or how far away the Sun is, we can measure its angular size directly—simply by how big it appears in the sky[2]. The angular diameters of the Sun and the Moon are both approximately 0.5° (0.008 7 rad). We can reproduce this same angular diameter by picking any particular object and placing it at the distance needed such that the ratio of its distance to its diameter is the same as for the Sun and Moon. This is roughly the equivalent to a US dime or UK five-pence coin held at arm's length. The angular diameter of the source of light is one of the key factors that determines the 'fuzziness' of a shadow.

7.2 The kinds of eclipses

One of the central facts about eclipses is this: the angular diameters of the Sun and the Moon are, by coincidence, very nearly the same as each other[3]. This leads to a particularly unique geometry for solar eclipses; the situation for lunar eclipses is different. To help with our discussion, let us outline the basic types of eclipses and define some helpful terminology. The following lists all of the possible variations of eclipses. We will look at each of these in more detail, regarding the geometry of shadows and the arrangement of Earth, Moon and Sun.

- **Solar Eclipse:** The shadow of the Moon is cast upon Earth. When one sees a solar eclipse one is located within the shadow of the Moon. For this to occur the Moon must be on the same side of Earth as the Sun. This only occurs at the new Moon phase.
  - **Total solar eclipse:** As seen from a particular location on Earth, all of the Sun's bright surface is covered by the intervening Moon. This can happen for only a small region of Earth at once. As the Moon orbits and Earth rotates, this region moves along to make a *path of totality*. One will see a total eclipse only if one is located along that path, and even

[2] Do not look directly at the Sun!

[3] This is indeed a coincidence; it has not always been so, and it will not be in the future.

then only for several minutes at most. When one sees a total solar eclipse, one is in the *umbra* of the Moon's shadow—the shadow's darkest part.
  - **Partial solar eclipse:** As seen from a particular location on Earth, only part of the Sun's bright surface is covered by the intervening Moon. When one sees a partial solar eclipse, one is within the *penumbra* of the Moon's shadow—the shadow's fuzzy, not-completely-dark part.
  - **Annular solar eclipse:** The alignment of Earth, Moon and Sun is like that of a total solar eclipse. The angular size of the Moon is slightly smaller than that of the Sun. Instead of completely covering the Sun, a bright ring—or *annnulus*—of sunlight is seen surrounding the silhouette of the intervening Moon. The angular sizes of the Sun and Moon are only approximately the same, since they both vary somewhat in their distance from Earth. When one sees an annular solar eclipse, one is within the *antumbra* of the Moon's shadow.
- **Lunar Eclipse:** The shadow of Earth is cast upon the Moon. When one sees a lunar eclipse, one is looking at the Moon while Earth's shadow is cast upon it. For this to occur, the Moon must be located on the opposite side of Earth as the Sun. This only occurs at the full Moon phase.
  - **Total lunar eclipse:** All of the Moon is within the dark part of Earth's shadow, its umbra. To see a lunar eclipse, one must be on the side of Earth that is facing the Moon when it happens. Since the Moon must be opposite the Sun in the sky for a total eclipse to occur, if the eclipsed Moon is visible, then it is night; if the Moon is not visible, then it is day. In contrast to the case of a total solar eclipse, the entire night side of Earth gets to see it. A total lunar eclipse may last for more than an hour rather than only a few minutes.
  - **Partial lunar eclipse:** Only part of the Moon is in the umbra of Earth's shadow. Just as with a total solar eclipse, a total lunar eclipse will have partial phases both before and after totality, as the orbiting Moon moves into and then out of Earth's umbra. In some cases, the Moon only grazes Earth's umbra, and only a partial eclipse occurs.
  - **Penumbral lunar eclipse:** The Moon passes through only the penumbra of Earth's shadow, missing the smaller and much darker umbra. The shadowing during a penumbral eclipse is often barely noticeable, betraying itself only as a seeming uneven illumination of the Moon.

Figure 7.2 shows the visual appearance of all six types of eclipses, photographed by Fred Espenak. In the sections that follow, we will explore the geometry involved and some important complications.

At first glance, the image of the total solar eclipse may appear similar to the image of the annular eclipse. In the case of the total eclipse, however, it is the relatively dim outer atmosphere of the Sun—its *corona*—that is revealed. For the annular eclipse, it is the blindingly-bright *photosphere* of the Sun that the Moon has failed to

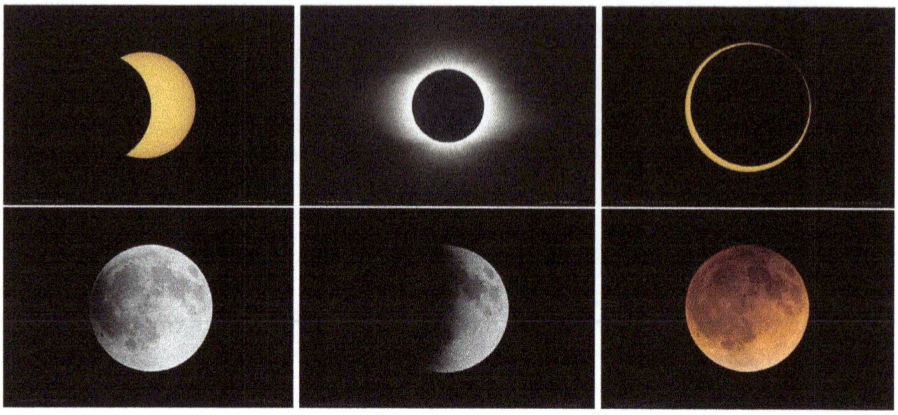

Figure 7.2. From **left** to **right, top**: Partial solar eclipse, total solar eclipse and annular solar eclipse. From **left** to **right, bottom**: penumbral lunar eclipse, partial lunar eclipse and total lunar eclipse. (Eclipse photographs by Fred Espenak, http://www.mreclipse.com, used by permission.)

completely block. A special filter that rejects all but a tiny fraction of light must be used to safely photograph a partial or annular eclipse. The corona of the total eclipse, on the other hand, is dim enough to photograph with an unfiltered camera lens.

7.3 The geometry of simple shadows

A shadow is more than a simple matter of light here and dark there. It is possible to create a shadow that is a perfect sharp-edged darkness cut out of the light. To do so, we would need to engineer a very specific environment.

> There must be only a single source of light, and the light source must be a perfect point. Furthermore, there can be no reflections off the surfaces of walls or other objects.

Clearly, this happens rarely in the real world, where objects are typically illuminated by multiple, often-large light sources and secondary reflections. The left side of figure 7.3 illustrates the geometry of this simplest of shadows in which one is either in the shadow (region two) or one is not (region one).

A more-complex and realistic case is to allow the source of light to be large, rather than just a point, as diagrammed on the right side of figure 7.3. Even this case is not often realized in our daily lives, but it is a useful point of departure for shadows in general. There are really two variations for the large source of light: the light source may be either bigger or smaller than the object casting the shadow. Figure 7.4 illustrates the latter case.

While there are only two distinct regions for the simple case of the left side of figure 7.3, we now see that there are three different regions for an object larger than (or the same size as) the light source, and four different regions for the case of a

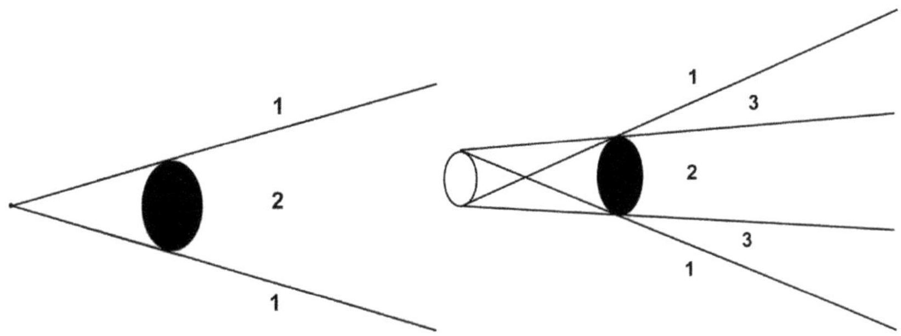

Figure 7.3. Parts of a shadow for the cases of (**left**) point light source, and (**right**) an extended light source that is smaller than the obstructing object. The different regions are marked as follows: (1) full illumination (no shadow), (2) umbra and (3) penumbra. For both illustrations, the source of light is to the left.

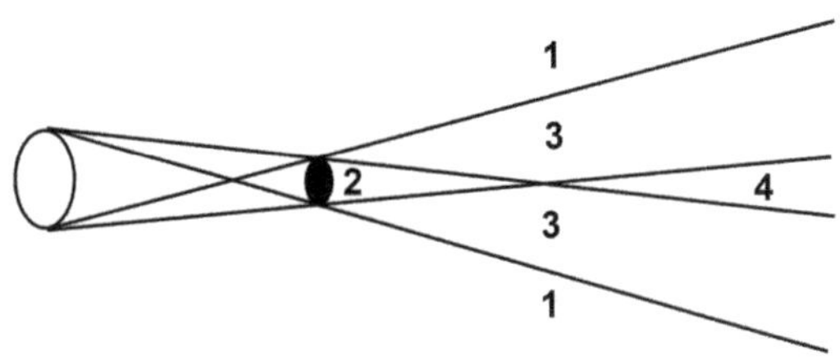

Figure 7.4. Parts of a shadow for the case of an extended light source that is larger than the obstructing object. The different regions are (1) full illumination (no shadow), (2) umbra, (3) penumbra and (4) antumbra.

shadow-casting obstruction that is smaller than the light source. These four regions of the shadow are named as follows: (1) no shadow, (2) umbra, (3) penumbra and (4) antumbra. These names should be familiar from section 7.2. The geometry of solar and lunar eclipses is similar to that of figure 7.4.

For a lunar eclipse, the obscuring object is Earth. The Moon is located in region one before the eclipse, region two during a penumbral eclipse, entirely in region two for a total eclipse, and only partly in region two if the eclipse is partial. The Moon cannot be in region four because the Moon is closer to Earth than is Earth's antumbra.

For a solar eclipse, the obscuring object is the Moon, and a given part of the Earth's surface is located in region three for a partial eclipse, region two for a total eclipse, or region four for an annular eclipse. By coincidence, the average distance of Earth from the Moon is very nearly the same as the length of the Moon's cone-shaped umbra. The distances vary since the Earth–Sun and Earth–Moon distances change as both Earth and Moon go about their orbits. Thus, sometimes a part of Earth's surface may be within the umbra (total eclipse); for a different eclipse, it may be slightly too far away and lie instead within the antumbra (annular eclipse).

The Sun is much larger than either the Earth or the Moon, but it is also very far away. Figure 7.5 shows a computer-generated diagram like that of figure 7.4 but drawn with all sizes and distances to scale for the case of a total solar eclipse.

Figure 7.5. The Sun, Earth and Moon with the construction lines for the Moon's umbra, drawn to scale. The Sun is on the left. The Earth and Moon are too tiny to see at this scale: they would appear only about 0.25 mm apart on the right.

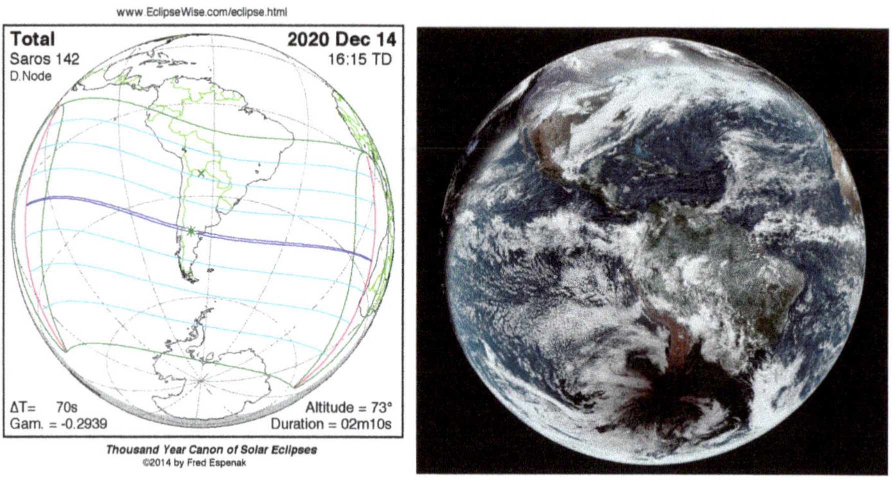

Figure 7.6. Left: Total solar eclipse prediction diagram by Fred Espenak (http://www.mreclipse.com, used by permission) for the 14 December 2020 total solar eclipse. The narrow blue path shows the progress of the Moon's umbra as it passed across Earth's surface due to Earth's rotation and the Moon's orbital motion. **Right:** NASA Earth Observatory GOES satellite picture (https://eoimages.gsfc.nasa.gov/images/imagerecords/147000/147659/eclipse_goes_2020349_lrg.jpg) of the Moon's umbra at mid-eclipse. The umbra and surrounding darkest parts of the penumbra appear irregular in shape because of varied cloud cover.

The angle made by the Moon's umbra is only about 0.5°. To fit the Earth–Sun distance on the page, the Sun must be less than 2 mm across; the Earth and Moon at this scale would require a microscope to see, and would be spaced only about 0.25 mm apart.

Since Earth's surface is situated almost exactly at the tip of the Moon's umbra, figure 7.5 can also be taken to represent just the umbra of the Moon with the Moon on the left side of the diagram rather than the Sun. Because the rays from the Sun are so close to parallel, the penumbra would then expand to about the twice the diameter of the Moon by the time it reaches Earth. Since Earth is about four times the diameter of the Moon, the Moon's penumbra during a solar eclipse is roughly half the diameter of Earth. Figure 7.6 shows the limits of the Moon's umbra and penumbra for the 14 December 2020 total solar eclipse (calculated by Fred Espenak) and a geostationary operational environmental satellite (GOES) image of the Moon's shadow during mid-eclipse. The umbra sweeps across the Earth, forming a path as time passes, due to Earth's rotation and the Moon's orbital motion about Earth. This motion, combined with the tiny size of the Moon's umbra, means that a given location experiences a total solar eclipse for only seven minutes or less. A video of the shadow's path for the 2020 eclipse can be seen at the NASA Earth Observatory website.

7.4 Make your own eclipse

When the morning sun shines through a window onto the opposite wall, make a shadow with your hand. How does this relate to an eclipse? Clearly your hand is considerably smaller than the Sun, so it would seem that the example of figure 7.5 would apply.

If we think about this a bit more, it is clear that the nature of such a shadow must not depend upon the real size of the Sun. If it did, then we could figure out the Sun's diameter simply by making shadow puppets. The reason is that the Sun is very far away compared to the distance between your hand and the wall. This means light rays from any given part of the Sun's surface all arrive at your hand very nearly parallel to each other.

Light rays from *different* parts of the Sun's surface, however, do arrive in slightly different directions; otherwise the Sun would appear as only a point in the sky. That small range of directions to different points on the Sun's surface is exactly what we mean by angular size as discussed in section 7.1. A shadow made by sunlight, blocked by an object only inches from a screen, has a geometry as illustrated in figure 7.7. Here I have drawn parallel rays coming from opposite sides of the light source. The angle between these rays from opposite sides of the light source is the angular diameter of the source.

We see that this everyday shadow geometry also has penumbra, umbra and antumbra regions. In this case the details are determined by the physical size of the obstructing object and its distance from the screen, and the angular size of the light source. For the Sun, that measure is a fixed 0.5°—much smaller than the example shown in the figure 7.7.

Any rays coming from some other part of the light source (the center, for example) also arrive parallel to each other. These rays partly fill in the penumbra and antumbra, as seen in figure 7.8. Here I have traced many parallel rays coming from three regions of the light source: the top (green rays), bottom (blue rays) and center (red rays).

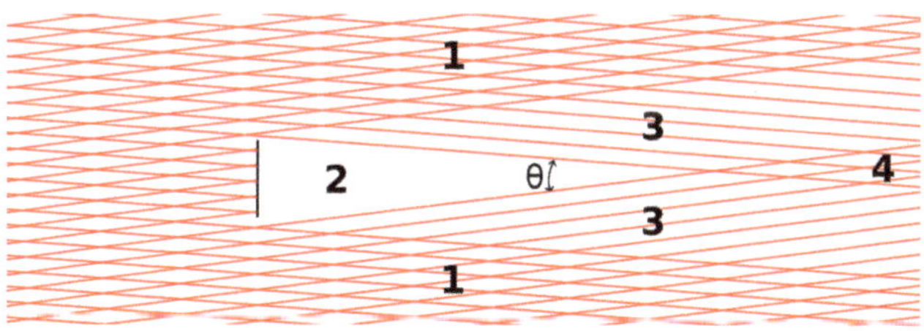

Figure 7.7. When a light source of significant angular size is very far away, rays of light arrive parallel from each part of the light source. Parallel rays arriving from both the top and bottom of the light source still trace out regions of umbra, penumbra and antumbra. A shadow cast by a small object in sunlight is a familiar example, though the angular size is much smaller than depicted in this diagram.

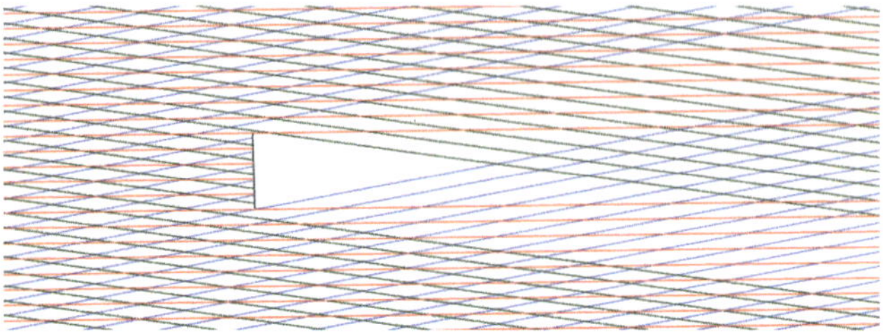

Figure 7.8. The geometry shown in figure 7.7 for a distant object of significant angular size but with many overlapping rays traced from the top (green rays), bottom (blue rays) and center (red rays) of the light source. When rays from all parts of the light source are included, the penumbra and antumbra become more and more filled with light at increasing distances from the light-obstructing object.

Even with only three sets of rays from the light source, it is clear that the penumbra and antumbra parts of the shadow are darkest where they are closest to the umbra. If sets of rays from all parts of the light source are included, the shadow fades out from complete darkness to full illumination as one goes from umbra to the outer cone of the penumbra (see figure 7.7). The same is true as one moves through the antumbra, farther and farther from the tip of the umbra.

Let us allow figures 7.7 or 7.8 to represent a shadow cast by a small object on a sunny day, here at Earth's surface, onto a sheet of white paper for example. For this familiar and important circumstance, the angle θ is too small to effectively illustrate with a diagram (see figure 7.5). However, it is easy to see what would happen.

If the shadow-casting object is close to the paper, then the shadow will be mostly umbra with a narrow penumbral boundary where it fades to full sunlight. The greater the distance between the paper and the object, as compared to the object's diameter, the smaller the umbral region as compared to the penumbral region. If the object is too far from the paper, there will be no umbra at all—only antumbra surrounded by penumbra. As distance (relative to the size of the object) increases further, these antumbra and penumbra regions become more and more filled with sunlight, and the shadow becomes less and less distinct.

The key measure is the angular diameter (θ) of the Sun. We can easily calculate the average[4] value of θ in radians by dividing the diameter of the Sun (1.393e6 km) by the Sun's average distance from Earth (1.496e8 km, the *astronomical unit*). Thus, $\theta = 0.00931$ rad $= 0.533°$. This tiny angle represents a skinny triangle that is about 107 times longer than its base. Table 7.1 lists different shadow circumstances for an object 21 mm in diameter—the approximate size of many coins (e.g., the US, Canadian and Euro five cent, and the UK penny).

From figure 7.7 it is easy to see that the diameters of the umbra (D_U), penumbra (D_P) and antumbra (D_A) are given by equations (7.2)–(7.4), where D is the diameter of the obstructing object, d is the distance to the screen, and θ is the angular diameter of the Sun, expressed in radians. We have also used the small angle approximation that $\sin\theta \approx \theta$ when $\theta \ll 1$ rad, a very good approximation in this case.

[4] Earth's orbit is elliptical, so the distance to the Sun varies throughout the year.

Table 7.1. Diameters of the umbra (D_U), penumbra (D_P) and antumbra (D_A) for a 21 mm coin casting a shadow a distance (d) in sunlight.

d (cm)	D_U (mm)	D_P (mm)	D_A (mm)
5.0	20.5	21.5	—
20	19.1	22.9	—
50	16.3	25.7	—
75.2	14	28	—
150	7.0	35	—
225.6	0.0	42	0.0
300	—	49	6.9
400	—	58	16.2

Figure 7.9. Comparison to the same scale of obstructing object (**left**) and its shadow (**right**) in sunlight, for a distance of $d = D/3\theta$, where θ is the angular diameter of the Sun and D is the diameter of the obstructing object (e.g., a 20.3 mm British penny). At this distance, the umbra is 2/3 the diameter of the coin, and the penumbra is twice the diameter of the umbra.

$$D_U = D - \theta d \tag{7.2}$$

$$D_P = D + \theta d \tag{7.3}$$

$$D_A = \theta\left(d - \frac{D}{\theta}\right) \tag{7.4}$$

From equations (7.2)–(7.4) we can see that the umbra ends and the antumbra begins at a distance of $d = D/\theta$ (2.256 m for our $D = 21$ mm coin in sunlight). At this distance, the penumbra is twice the diameter of the shadow-casting object. We can also see that at a distance of $d = D/3\theta$, or 75.2 cm, for our sunlight-illuminated coin, the umbra is 2/3 the diameter of the coin and 1/2 the diameter of the penumbra. Figure 7.9 illustrates this last case. I glued a penny to a sheet of clear plastic,

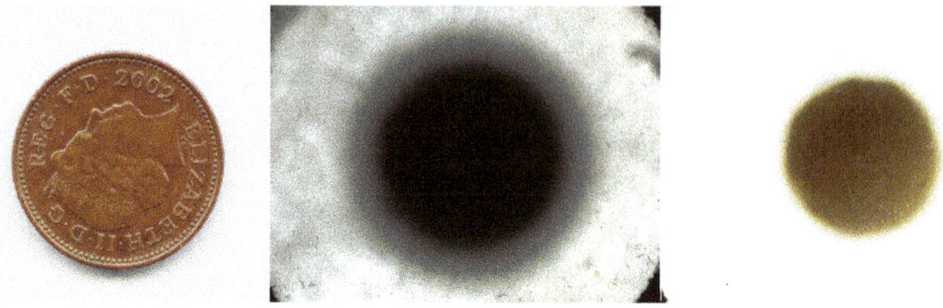

Figure 7.10. The same shadow photograph as in figure 7.9, but with the levels adjusted to better show the outer regions of the penumbra (**center**) and the umbra (**right**).

suspended it over the end of a cardboard tube, and pointed it at the Sun. I used ephemeral-process photography (chapter 48) to directly record the shadow with a 3 s exposure[5]. The coin is shown to the same scale.

The penumbra is larger than the coin, but only its inner edge—closest to the umbra—is readily visible in figure 7.9. It is difficult to display the full, yet subtle, range of values in one single image. For figure 7.10, I have made two different versions of the same image, adjusting the brightness levels to better show the limits of the penumbra (center) and umbra (right).

We now have everything we need to make our own solar eclipse, suitable for viewing by a tiny bacterium on the surface of a globe that is about 57 mm in diameter. Simply suspend a 21 mm coin a distance of 2.256 m from its surface[6]. You can choose whether the bacterium sees a partial, total or annular eclipse simply by moving the coin a little closer (for a total eclipse), a little farther (for an annular eclipse) or a bit to one side (for a partial eclipse).

There is, however, an obvious difference between our scaled-down–coin solar eclipse and a real solar eclipse. For even in the umbra of the coin's shadow on a clear and sunny day, the bacterium will still be illuminated by the rest of the bright, blue sky. A real total solar eclipse is different; the sunlight is blocked by the Moon before it reaches Earth's atmosphere. We consider these complications further in section 7.5.

7.5 Atmospheres of the Sun and Earth

7.5.1 The Sun's atmosphere and solar eclipses

The total solar eclipse illustrated in figure 7.2 shows a glowing haze around the silhouette of the Moon. This is the Sun's corona, its superheated but low-density outer atmosphere. The corona is only visible from Earth during a total solar eclipse, when the far-brighter photosphere of the Sun is completely blocked by the Moon.

The Sun's photosphere seems to have a sharp edge to it, and this would give the visual impression[7] that the Sun is a hot and perfectly spherical solid. This, however,

[5] The corners of the dark circle in the right-hand shadow image are the edges of the cardboard tube.

[6] The diameter of Earth is about 3.67× the diameter of the Moon.

[7] Do not look directly at the Sun!

is an illusion. The Sun is all gas, and the gas in the mostly-invisible layers just above the blinding photosphere are only somewhat cooler and less dense than the gas that makes up the intensely bright part of the Sun's visible 'surface.'

The cause of this illusion is that most of the gas of the Sun is opaque, and it quickly transitions to transparent over a surprisingly narrow range (compared to the diameter of the Sun). Thus, the light of the Sun is trapped within until it reaches this narrow range of layers that make up the photosphere. At that point most of the light escapes, passing through the mostly transparent layers above, on its way to Earth and beyond. Nearly all of the light of the Sun comes from a thin spherical shell with a diameter of 1.392 7e6 km—the diameter of the Sun's photosphere.

Above the photosphere, the gas in the corona[8] is mostly transparent and emits very little light; the photosphere is blindingly bright in comparison. The corona is far hotter than the photosphere, so it emits mostly ultraviolet light and even x-rays. Because the coronal gas is so thin, it emits very little light in total compared to the photosphere. Although even a brief unprotected view of only a part of the Sun's photosphere can easily damage the retina of the eye, the Sun's corona can be viewed briefly with the unaided eye.

7.5.2 Earth's atmosphere and eclipses

Surrounding the solid, opaque Earth is a gaseous layer of varying density—the atmosphere. The atmosphere both refracts and scatters light. The Rayleigh scattering (section 3.5) of sunlight by air molecules means that we see light from the Sun even when looking in other directions. Rays of sunlight that would have missed us are instead scattered towards us. Since it is the shorter-wavelength light that undergoes the largest amount of Rayleigh scattering, the sky is blue.

Because of *atmospheric refraction,* sunlight from a low angle, grazing close to Earth's surface, is bent by Earth's atmosphere. This angle of refraction is much less than the typical scattering angles from Rayleigh scattering. The angle is, however, enough that we have a slightly-curved line of sight to the Sun as it rises and sets, and this can add up to several minutes to the hours of daylight.

7.5.2.1 The atmosphere and solar eclipses

The scattered blue daylight is far brighter than the Sun's corona or chromosphere, so only the Sun's brilliant photosphere is evident from Earth's surface. In order for us to see these fainter parts of the Sun's atmosphere, light from the photosphere must be blocked in space—by the Moon, for example, during a total solar eclipse—so it never arrives at the scattering nitrogen and oxygen molecules of the part of Earth's atmosphere over our heads.

During even the deepest of total solar eclipses, the Moon's umbra is less than 300 km across. How does this compare to the 'size' of the part of Earth's atmosphere that is over my head right now? Because of both the curvature of Earth and the relatively small thickness of the atmosphere, the answer is clearly only a tiny fraction

[8] There is also a thin, irregular layer between the photosphere and corona—called the *chromosphere*.

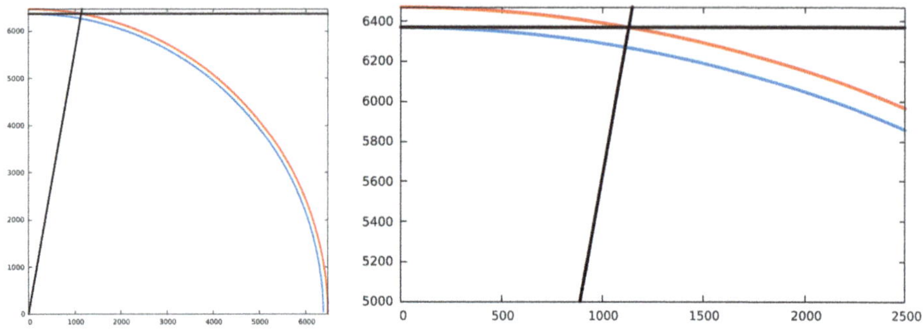

Figure 7.11. Left: A diagram representing 1/4 of a 6370 km radius Earth (blue curve) surrounded by a 100 km thick atmosphere (red curve). The horizontal black line at the top is a tangent line from Earth's surface that shows where it intersects the top of Earth's atmosphere. **Right:** An enlargement of that part of the diagram.

of the diameter of Earth. But how does this compare to the size of the Moon's umbra during a total eclipse?

A simple (and simplistic) model is shown in figure 7.11. The blue curve represents 1/4 of Earth's circumference, while the red curve represents the top of its atmosphere. The horizontal black line near top is a tangent from Earth's surface, and where it intersects the top of the atmosphere is the radius of the circle of atmosphere directly visible above us. The right-hand side of the diagram is an enlargement of this part. The length of the horizontal black line until it intersects the red curve is, then, half the diameter (D) of our locally-visible atmosphere.

We note that this forms a right triangle with the vertical radius line, of length equal to Earth's radius ($R_\oplus$) and the diagonal radius of length ($R_\oplus + T$), where T is the thickness of the atmosphere. From the Pythagorean theorem we have the following equation.

$$D = 2\sqrt{(R_\oplus + T)^2 - R_\oplus^2} \tag{7.5}$$

Of course it is not so simple. Atmospheric refraction means we actually see a bit around the curvature of Earth. The atmosphere does not really have a sharp boundary; it becomes less and less dense with altitude. For the diagram I have picked an atmospheric thickness of 100 km, but most of the atmosphere's mass is contained within much less than that. One would expect most of the scattering that leads to the brightness of the daytime sky to occur at relatively low altitudes.

Oumbe and Wald 2009 calculate the effect of elevation on both direct and diffuse scattered sunlight. From their figure 3, at sea level scattered skylight is about 15% of the total, whereas by an altitude of 10 km the diffuse component has reduced to less than 2.5%. As a crude estimate, then, we can assume that $T = 10$ km. Given that $R_\oplus = 6370$ km, we have from equation (7.5) that $D \approx 700$ km, which is clearly larger than the Moon's umbra during a total solar eclipse. Therefore, it is unsurprising that it is not completely dark during a total eclipse, and the sky is noticeably brighter at large angles from the Sun.

7.5.2.2 The atmosphere and lunar eclipses

Over small distances our atmosphere affects the directions of light rays very little. Regarding light rays that pass closely by the limb of Earth during a lunar eclipse, Earth's atmosphere may have a profound effect on whether these rays do or do not reach the Moon.

Consider the images of a not-quite–total lunar eclipse in figure 7.12, taken from Appleton, Wisconsin on the night of 18–19 November 2021. The left-hand image was taken in the early evening, when the Moon still lay outside Earth's penumbra. The right-hand images were taken in the early-morning hours, near the time of the Moon's greatest immersion into Earth's umbra.

The Moon's umbra is never completely dark because atmospheric refraction bends some light around Earth's surface and into its own umbra. It is mostly the lower atmosphere that does this; that is where the air is most dense and has the greatest index of refraction.

Although it is primarily refraction that redirects sunlight slightly around the opaque Earth and into Earth's nominal umbra, it is scattering that most determines which light is refracted. Light is scattered by Earth's atmospheres over large angles, most of which miss Earth's relatively small umbra. For a clear atmosphere, it is mostly Rayleigh scattering by nitrogen and oxygen molecules. Most of the blue light is scattered away from the narrow paths that graze Earth's surface and are refracted into Earth's umbra. Therefore, when the eclipsed Moon is deep inside Earth's umbra, it appears reddish. Figure 7.13 shows a slice of the right-hand image from figure 7.12 with the dark areas brightened so as to match the brighter areas. Clearly, there is a gradation of color, with the reddest part coming from deepest within the umbra.

If much of the Earth's atmosphere is not clear, but is instead filled with dust or smoke, then Rayleigh scattering will likely be unimportant compared to the wavelength-independent scattering of the much-larger particulates. This can happen, for example, after a major volcanic eruption or significant dust storms. A total lunar eclipse under those conditions typically has an umbra that is both darker and more gray.

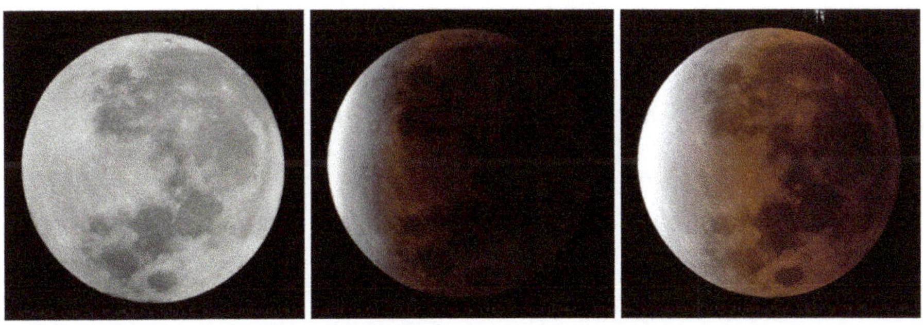

Figure 7.12. The Moon from Appleton, Wisconsin on the early morning of 19 November 2021. **Left:** Fully-illuminated Moon before the beginning of the eclipse. **Center:** Maximum eclipse, with the umbra covering most of the Moon's disk. **Right:** The same image, but adjusted to better show the dark red parts of Earth's umbra.

Figure 7.13. A slice of the Moon's umbra from the right-hand image in figure 7.12 with the dark areas brightened. The color becomes more red deeper into the umbra.

7.6 Shadows on a sunny day

A shadow of an object in direct sunlight, on a clear day, naturally produces a penumbra and either an umbra or antumbra, depending upon the geometry. If it were only the direct light of the Sun, then our discussion in section 7.4 would be the end of the story. However, there are usually two other significant sources of natural light that can greatly alter the look of shadows:

1. The blue sky adds significant blue light to the shadows; this is especially true for *open shade*. The skylight both brightens and colorizes the shadows. In contrast to the direct sunlight, which has had much blue light removed by Rayleigh scattering, the shadows appear bluer still.
2. There is likely direct sunlight reflected into the shadows by objects in the landscape. This brightens shadows with light that has the color of the reflecting surfaces. A familiar example is the warmer and softer shadows and lowered contrast in photographs taken at the beach because of added light from the (typically) warm-colored and highly-reflective sand.

References

North J 1995 *The Norton History of Astronomy and Cosmology* 1st edn (New York: Norton)

Oumbe A and Wald L 2009 A parameterisation of vertical profile of solar irradiance for correcting solar fluxes for changes in terrain elevation *Earth Observation and Water Cycle Science Conf. (Frascati, Italy)* S05

IOP Publishing

Photography (Second Edition)
Physics and art in focus
John Beaver

Chapter 8

Light and shadows 2: photograms

A *photogram* is a camera-less photograph. An object is placed in direct contact with some light-sensitive material and then exposed to light, directly recording the object's shadow. Many of the earliest photographs were photograms for the simple reason that the earliest light-sensitive materials were of such low sensitivity that their use in a camera was impractical. Some of the first photographs using light-sensitive silver salts on paper, by William Henry Fox Talbot, were photograms of botanicals; he called them 'photogenic drawings.' The first book to be illustrated with photographs was *Photographs of British Algae: Cyanotype Impressions* by the botanist and photographer Anna Atkins. She used John Herschel's newly-invented *cyanotype* process to illustrate her 1843 book with direct photograms of the specimens.

Early photographic detectors were far less sensitive to light than those of the modern era, and so they lent themselves naturally to the making of photograms. Many contemporary photographers have found old photogram techniques to be useful tools for making new art, so there is now a strong revival of these old processes (Rexer 2002).

Most photochemical-photographic processes directly produce a negative; the image is darkest where it was most exposed by light. We shall later explore some of the artistic and practical motivations behind leaving a negative as a negative or—instead—reversing the values to a positive. There is a tradition of leaving photograms in their original state as negatives. For the sake of clarity, however, the examples in this chapter have, except where noted, been inverted to positive images; the dark parts of these examples represent places of shadow.

8.1 Shadows and the source of light

A photogram is, essentially, a capture of a shadow cast directly onto a light-sensitive material. To experiment with photograms is to explore the nature of shadows. For the right-hand image in figure 8.1, I used ephemeral-process photography (chapter 48) to directly record the shadow cast by the object shown in the left-hand image.

 © IOP Publishing Ltd 2022

Figure 8.1. **Left:** A cordial glass illuminated by a single source of light located to the upper right. **Right:** A photogram made with this same illumination. The shape of the photogram is distorted by the angle of projection.

When the shadow is seen in context, it is clear that it is illuminated by a single source of light positioned to the upper right of the cordial glass. However, the photogram itself seems distorted; the circular base looks exactly as it does on the object, but the rest looks very different.

The seeming distortion arises from the fact that the photogram is a projection of light rays from the source through the object. So, the shadow of the top of the glass is to the side of the shadow of the bottom of the glass. It is also evident that the shadow of the top of the glass seems a bit disproportionately large when compared to the proportions of the different parts of the object that cast the shadow. Since the bottom of the glass was in direct contact with the light-sensitive paper, it appears the same shape and size and in the same position as the object itself; this is not so for the top of the glass, which is some inches above the surface of the paper.

Careful inspection of the photogram in figure 8.1 also reveals that the blurriest portion is from the top of the glass, the part most distant from the paper, while the sharpest portion is from the base that was in direct contact. We have already seen in chapter 7 that objects more distant from the shadow surface cast blurrier shadows.

Figure 8.2 shows four photograms, all made with the object in the same arrangement as in figure 8.1. However, I have changed the light source for each photogram. For the first and third photograms, the source of light was positioned relatively close to the object. For the second and fourth photograms, the light was positioned far away. For both the first and second examples I used a very tiny source of light (shining through a small hole); while, for the third and fourth examples, I used a light that had a large surface area.

Let us first think about the relative distortions of size and put the issue of sharpness or blurriness of the shadows aside for the moment. Comparing the first image to the second, we see that when the source of light is close by (as in the first image), objects more distant from the detector (and thus closer to the light source)

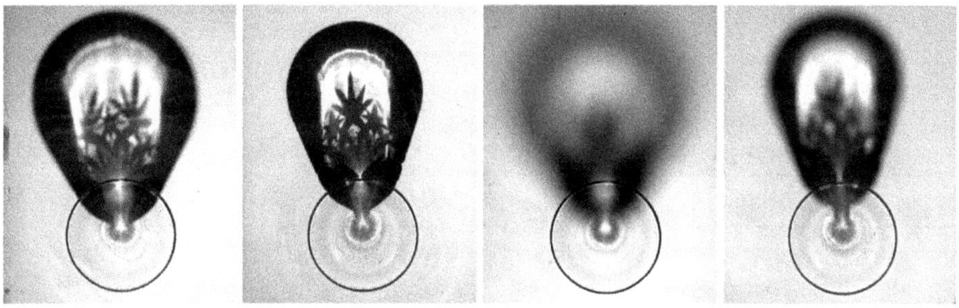

Figure 8.2. Four photograms of the same crystal cordial glass. From **left** to **right**: small light, close up; small light, far away; large light, close up; and large light, far away.

appear larger than life. The same is clearly true when comparing the third and fourth images. As compared to the third image, the light source was moved farther away for the fourth image, so the shadow of the top of the glass appears closer to its real proportion when compared to the bottom.

For the first and third pictures, which show considerable distortion in the size of the top of the glass, the light source was only a few times further away than the distance between the top of the glass and the light-sensitive paper. Thus, the light rays were spreading out significantly when they reached the nearby top of the glass. Following these same rays to the light-sensitive paper, they continue to spread, and so make an enlarged shadow of the top of the glass.

If the source of light is very far away, on the other hand, the light rays are nearly parallel, and so this effect is much less noticeable. Our most common source of light for outdoor photography, the Sun, is very far away indeed. For shadows cast by the Sun the distortions in size, illustrated by the first and third image in figure 8.2, do not occur.

The first and third images were both made with the light positioned at roughly the same distance. However, one is very blurry while the other is very sharp. The same is true for the pair of the second and fourth images. Here the difference results from the physical size of the light source. For both pairs, the smaller light source produced the sharper shadow. Furthermore, the distance of the light source also affects the sharpness of the shadow. The first and second shadows were both made by the same small light source; however, the second, made with the light source farther away, is clearly sharper. The same relation holds true for the third and fourth images. The same source of light moved farther away produces a sharper shadow.

A more distant light source produces a sharper shadow, while a physically larger light source produces a blurrier shadow. So, what if the light source is both larger and more distant? Will the shadow be blurrier or sharper? The answer is that it depends: we must look at the details for the particular case in question.

Consider, again, the four examples in figure 8.2. If we rank them from sharpest to blurriest, it is clear they would be in the order two, one, four and three. Table 8.1 gives the relevant data for the four examples—the size of the light source and its

Table 8.1. In order of increasing blurriness from top to bottom, column one lists the four examples from figure 8.2. Columns three and four list the distance and size of the light source that illuminated the cordial glass to make these four photograms. From either of those columns alone, it is unclear why the rows of the table should be ordered in this way. However, by dividing distance by size (or size by distance), one can see that neither the size nor the distance alone is the crucial issue but rather how they compare to each other.

Example	Description	Distance (cm)	Size (cm)	Distance/Size	Size/Distance
2	Sharpest	132	0.64	206	0.004 8
1	Less sharp	41	0.64	64	0.016
4	Blurrier	132	7.6	17	0.058
3	Blurriest	41	7.6	5.4	0.185

distance from the light-sensitive paper—in order of increasing blurriness. By looking at each of the two variables individually, it is far from obvious why they would be in this particular order. The fifth column of the table provides the answer. Here I have divided the distance to the light source by its size. It is clear that the smaller this ratio, the blurrier the shadow.

If we invert this ratio, instead dividing the size of the light source by its distance, we get numbers (fractions in this example) that are smaller with increased sharpness and larger with increased blurriness (sixth column). This measure, the size of a light source divided by its distance, is closely related to the important concept of *angular size* or *angular diameter*. The angular size of an object is how large it appears as seen from a particular vantage point. With this concept of angular size, we can now state the following simple rule for shadows.

> As seen from the point of view of the object casting the shadow, a source of light with a relatively small angular size will produce relatively sharp shadows, while a source of light with a relatively large angular size will produce comparatively blurry shadows.

Photographers often refer to sharp shadows as *hard* (especially if they are also of high contrast), while referring to blurry shadows as *soft* (especially if they are also of low contrast). Hard shadows may give a sense of drama to a photograph, while soft shadows tend to have a more relaxed feel. Portrait photographers often work in the controlled environment of a studio in part to exert detailed control over the hardness or softness of shadows cast by one part of a subject's face onto other parts. To greatly soften shadows, a compact source of light may be enclosed in a large *soft box* that spreads the source of light over a large area.

8.2 Photograms with sunlight

The Sun is an important source of light for many types of photograms. It is intense and compact, and it has the right spectrum of light for many types of alternative photographic processes. We explore these issues more fully in later chapters; here we consider the geometry of printing with sunlight.

The Sun, as seen from Earth, has an angular diameter of about 0.52°, which corresponds to a size-to-distance ratio of about 110. When using sunlight to make a photogram, this means that the edges of shadows will be blurry by that same ratio as compared to the distance of the shadow-casting object from the paper.

Another way of looking at this is to consider the smallest possible detail that the photogram shadow can contain, given the size of the object and its distance from the paper. For a piece of paper at a comfortable viewing distance of 250 mm, the human eye can discern detail down to a width of about 0.1 mm. The standard for photographic detail on a final print, however, is less stringent than this—about 0.25 mm (Ware 2019, p 60). If the photogram is to be used as a negative, however, to be enlarged by some factor to make a final print, then the photogram should have detail finer by that same factor.

Let us imagine that we want such a tiny detail to cast a visible shadow on our photogram. By way of example, let us consider some strands of human hair—my own, measured to be 0.064 mm in width. This would allow for an enlargement of roughly a factor of four, while still maintaining fine detail in a print held at a comfortable viewing distance.

We may note that if the hair width is at its antumbral distance, then the shadow has no umbra. At just beyond this distance, however, there is still a mostly-dark shadow with only a little light peaking around the edge of the hair. Given the 110 ratio of the Sun's size to distance, the antumbral distance of the strand of hair would have that same ratio with its 0.064 mm width—a distance of about seven millimeters[1]. Figure 8.3 shows three photograms made with direct sunlight of a few strands of this hair stretched across a small, opaque frame. The left-most image was made with the hair in direct contact with the paper, while for the middle and

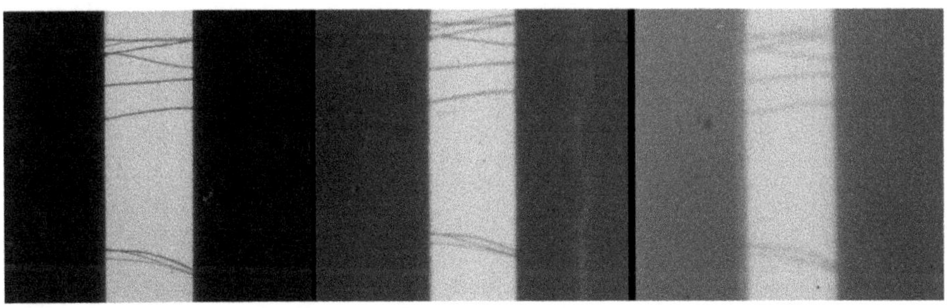

Figure 8.3. Photograms of human hairs of approximate width of 0.064 mm, stretched across a small metal frame. **Left:** In direct contact with the light-sensitive paper. **Center:** At 28 mm above the paper. **Right:** Fifty-six millimeters above the paper.

[1] If the object is separated from the light detector not by air but by glass, refraction at the glass surface extends the antumbral point to a distance greater by a factor of n—the index of refraction.

right-hand images, the hair was suspended 28 mm and 56 mm, respectively, above the paper.

It is evident from figure 8.3 that even at a distance eight times greater than the antumbral distance, a shadow is still visible—albeit barely. Even at less than 3 cm from the paper, the shadows of the hair are noticeably blurry and of significantly lower contrast than the exposure made with direct contact.

Notice too, when comparing the three images in figure 8.3, that even the opaque frame holding the strands of hair no longer casts a dark shadow when held centimeters above the paper. This is due not to the geometry of the shadow cast by the Sun but rather to open skylight leaking underneath, partially exposing the paper and filling in its shadow.

It is significant that the shadows in figure 8.3 are of long, thin objects. For the example of a 0.064 mm wide hair, it will completely cover the Sun if centered at distance of less than about seven millimeters. Even at much greater distances, though the apparent width is much smaller than the angular size of the Sun, the long dimension still stretches completely across the Sun.

This is not so for the shadow of a circular object, which covers the Sun in both dimensions less and less as one increases the distance to the paper. Figure 8.4 shows brightness of the center of the shadow versus distance for both a thin line such as a strand of hair (green curve) and a circle (violet curve). In the scale on the left 'one' represents full illumination, while 'zero' represents full shadow. The horizontal distance scale is in terms of the antumbral distance. Therefore, a distance of 'one' means the object is placed at a distance 110 times greater than its width (for the hair) or diameter (for the circle) and completely covers the Sun.

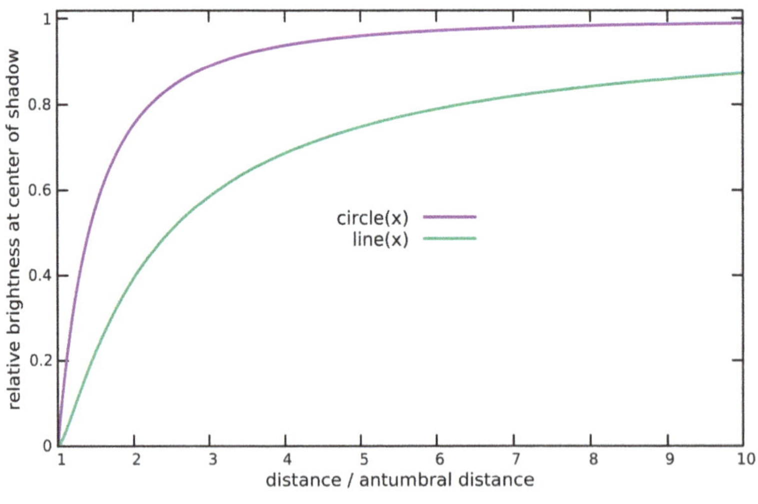

Figure 8.4. Relative brightness of a shadow vs. distance from the screen in units of the antumbral distance. The violet curve is for a circular shadow-casting object, while the green curve is for a line. A line of a given width casts a significantly darker shadow than a circle of the same diameter.

Figure 8.4 shows that the shadow of the hair is much darker at its center when compared to that of a circle with a diameter the same as the hair's width. Even at eight times the antumbral distance, the center of the hair's shadow is darkened by nearly 16%, while the circle's shadow is darkened by only 1.8% at its center.

We can easily see the reason for the difference. For both the hair and the circle at the antumbral distance, the Sun is completely covered by the object, and the shadow is completely dark. Let us now imagine moving the light-sensitive paper ten times farther away. The circle will then cover a region that has 1/10 the diameter of the Sun; that is only $1/10^2 = 1/100$ the Sun's area. The antumbral region of the shadow is decreased in brightness by a fraction of only 0.01 or 1%; the violet curve in figure 8.4 shows a shadow brightness of $1.00 - 0.010 = 0.99$ at ten times the antumbral distance.

The geometry for the hair, however, is different. As one moves the paper ten times farther away, the apparent width of the line covering the Sun decreases by a factor of ten. The full length, however, still covers the Sun. Thus, the shadowing area decreases by a factor of only ten, not 10^2, leading to a 10% decrease in brightness compared to the only 1% decrease for the circle.

It is, of course, more complex than this; the Sun is a circle, so the shadowed area produced by the hair is approximately a rectangle only if it covers a tiny portion of the Sun. The green curve in figure 8.4 takes the circular shape of the Sun into account, and it shows that moving the paper to ten times the antumbral distance decreases the covered area of the Sun to 12.7% (a shadow brightness of $1.000 - 0.127 = 0.873$) rather than 10%. However, the basic argument still holds; the shadowed area from the hair decreases approximately linearly with distance, while the shadowed area from the circle decreases with the distance squared and thus decreases much more rapidly.

Intricate objects placed several times farther than their antumbral distance from the paper may still show detail at a fine enough scale to be interesting, especially if they possess highly-elongated features. Consider the two photograms in figure 8.5 of

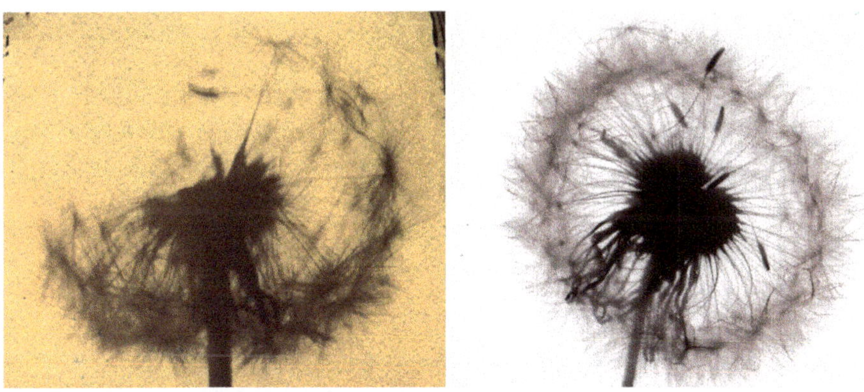

Figure 8.5. **Left:** a cyanonegative photogram of a dandelion seed head that was held loosely against the paper for the exposure with direct sunlight. **Right:** an accelerated lumen photogram of a similar seed head shows finer detail because it was flattened by a piece of glass to about 1/3 of its diameter.

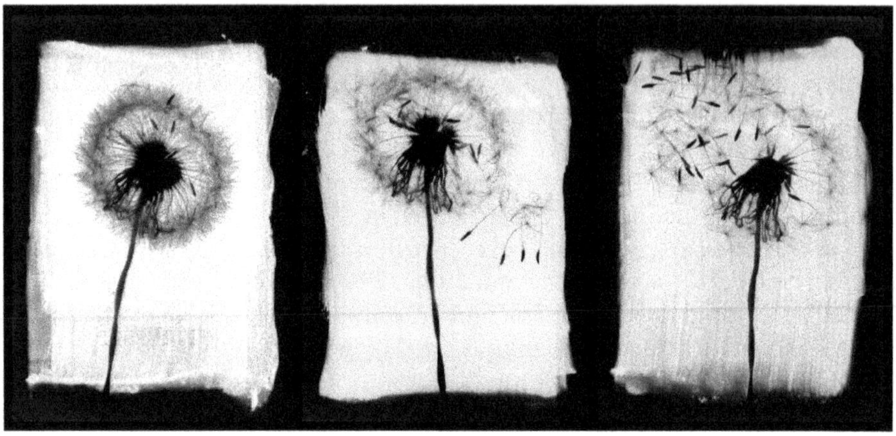

Figure 8.6. *Three reasons to leave*, John Beaver 2019. Digital pigment print, 89 cm × 43 cm.

the roughly 40 mm head of a dandelion. For the left example, the dandelion head was held lightly touching the paper for the exposure. For the right-hand example, the dandelion was pressed gently against the paper, flattened somewhat by a sheet of glass. Clearly the more-flattened example shows more detail, but the other still shows enough fine detail to be interesting.

The tiny radiating structures (pappi) are finer than human hair, only about 0.025 mm across at the middle of their length (Meng *et al* 2014). This gives an antumbral distance of 2.8 mm, nearly 15 times less than the roughly 40 mm diameter of the seed head. Therefore, it is no surprise that many pappi in the left-hand image of figure 8.5 are indistinct. The more-flattened right-hand example, on the other hand, contains enough detail for significant enlargement (see figure 8.6).

Objects at distances from the light-sensitive paper many times their antumbral distance make blurry shadows, but we can sometimes use this to our advantage. Carefully examine the cyanonegative photogram 'Paloverde' in figure 8.7. Some of the branches appear very sharp, while others are blurry. This gives the photogram a look of depth, as if it were a photograph taken with a camera using a shallow depth of focus. However, the photograph was simply a piece of light-sensitive paper held next to a tree branch while it was exposed by the Sun. The branches touching the paper made sharp-edged shadows, while the branches in front of and further from the paper made fuzzier shadows.

As described in section 7.6, sunlight on a clear day includes not only the direct light from the Sun but also the blue light scattered by the sky. This diffuse light shining on the photogram, from all directions, can partially fill in the shadows of an object that is not in direct contact with the paper. This effect may result in decreased contrast, but is likely to have only a modest effect on a crisp, clear day; the entire blue sky taken together is still significantly dimmer than the direct light of the Sun. On a hazy day, however, when the sky is brighter and the direct sunlight dimmer, the loss of contrast can be significant.

Figure 8.7. *Paloverde*, John Beaver 2005. In this photogram, the shadows of the branches that were in direct contact with the light-sensitive paper appear sharp, while those that were some distance away produced blurry-edged shadows.

8.2.1 Photogram sharpness and exposure time

Photograms are often made with historical and alternative photographic processes that are of very-low sensitivity. Even in bright sunlight, exposures of many minutes or even several days[2] may be necessary in some cases.

Imagine that we set up a photogram in sunlight of a thicker object with the Sun at just the right angle to create exactly the desired shadow. We then expose the photogram for—say—15 minutes. This would all be well and good but for an important fact: the Earth rotates, so the angles of the shadows change as time passes. This is a good effect for sundials but not for our photogram if sharp detail is wanted. The shadows will slowly creep across the face of the photogram, blurring the image[3].

The Earth rotates 2π rad in 24 hours, or 4.36×10^{-3} rad min^{-1}. For an object that forms a sharp-edged line oriented crosswise to the Sun's apparent motion, the shadow edge will thus drift at this same angular rate. If the edge feature of the object is located a distance (d) from the light-sensitive paper, the shadow edge will then drift sideways over a length (x) in time (t [in minutes)]), given by an equation.

$$x = 4.36 \times 10^{-3} d \times t \tag{8.1}$$

As an example, let us imagine a square object with sides of length $l = 1$ cm, placed a distance of $d = 10$ cm above the photographic paper. Most of the shadow then is

[2] See, for example, the anthotype process, in section 36.4.

[3] For a useful but somewhat different exposition of many of these ideas, see Ware (2019) in section 4.6.

Figure 8.8. A ten minute exposure photogram of a 1 cm square suspended 10 cm above the light-sensitive paper. The left and right edges are greatly smeared because of the Earth's rotation (oriented left to right in the image) during the exposure.

in umbra; the extension of the penumbra at this distance would be only about 10 cm/110 = 0.9 mm.

If the square is oriented such that one dimension is parallel to the Sun's apparent motion while the other is crosswise to it, then the photogram of the square will stretch out over time along only one dimension. The square will appear, instead, as a rectangle. The elongated dimension is blurred at its edges because those parts of the paper were in shadow for only part of the exposure time and are—therefore—not fully dark.

From equation (8.1), a ten minute exposure would stretch the shadow of the 1 cm square by an additional 0.44 cm. Only the central 0.56 cm (1 cm − 0.44 cm) would have been in the umbra for the entire exposure. Figure 8.8 is a photogram made under these very conditions.

We can see that, for the dimension of the object that was parallel to Earth's rotation, the shadow is sharp. The other dimension is blurred by an amount consistent with equation (8.1).

8.2.2 Sharpness and scattering

We have so far assumed that rays from the Sun (or other light source) are either blocked on their way to the light-sensitive paper or they are not. If a light ray is only partially absorbed, so that it arrives at the paper diminished in brightness, our analysis would still apply.

But what if the light-blocking medium—a leaf, for example—also scatters light rays passing through it? If a part of the leaf deflects light rays by large angles, then nearby shadow regions may be filled in even if the leaf is very close to the light-sensitive paper.

Consider figure 8.9, a tiny detail of a photogram of a grape leaf. The thin leaf was held in place against the light-sensitive paper during the exposure. Fine detail is

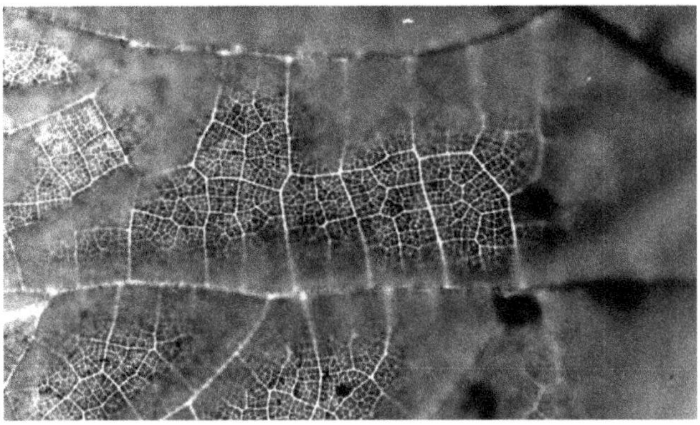

Figure 8.9. A small portion of a photogram of a grape leaf. Where the bottom of the leaf was in direct contact with the light-sensitive paper, fine detail is evident. Parts of the leaf that were only millimeters above the paper surface produced a image blurred by scattering.

evident, but only in certain places; the small-scale veins are blurred to invisibility in other regions.

These differences in detail arose from tiny variations in the distance between the leaf and the paper—a couple of millimeters at most. This distance is too small to account for the blurriness by our analysis in section 8.3. The loss of detail is not from the geometry of light rays arising from the source but because of scattering of light by the leaf. No matter how the light rays enter the leaf, each is scattered over a wide range of angles as it exits.

We can use as a reasonable benchmark that the majority of light exits a particular point of the leaf in a scattered cone that has a diameter equal to its height (corresponding to an angle of 53°). This leads to the following rough rule-of-thumb:

> If the light-blocking object strongly scatters light, then in order to capture detail on the scale of—for example—0.5 mm, the object must be placed at most no further than 0.5 mm from the light-sensitive paper. To preserve the maximum sharpness, the object must be placed in direct contact with the paper.

The example of a leaf is an important one; leaves are thin, flat and have always been among the most common subjects for photograms. Leaves aside, most translucent objects scatter light. When we make a photogram that is more than just black and white—that also shows gradations of value because the subject is translucent—then scattering is likely to be the most important fact of life. In section 8.4, we consider a practical method for preserving sharp detail in a thin subject even when it scatters light.

8.2.3 Cloudy days

Except for portions of the subject that are in direct contact with the light-sensitive paper, photograms made on a cloudy day are blurry. The light source, after all, has

Figure 8.10. A photogram made with small objects but exposed to diffuse skylight on an overcast day. Only the parts of the objects that directly touched the light-sensitive paper are in focus.

as large an angular diameter as possible. We can, however, take this as an opportunity; what if one wants blurry shadows? Figure 8.10 shows an example of a photogram taken on an overcast day.

Note that there are sharp parts of the image in figure 8.10; they are the parts where the objects were in direct contact with the paper. Although they were in the lowermost part of the photogram setup, the sharp details seem to float on the top. The shallow depth of focus is exaggerated, and it is as if the objects are emerging from a dense fog.

8.3 Contact prints

An important example is to allow light to pass through a very thin object or a transparent medium with dark marks on it. To achieve the most detail we press the object or transparency in direct contact with the light-sensitive material before making the exposure, which reduces the distance to effectively zero.

The most common example of 'a thin transparent medium with dark marks on it' is a print-sized photographic negative, made in a large-format camera. Although one could think of this as a special type of photogram, it is usually called a *contact print* given that the intent is to produce a positive print on paper from a transparent, negative image.

Often, a special contact-printing frame is used that presses the film against the light-sensitive paper with a piece of glass. To make a contact print, one needs a transparency that is, of course, the same size as the print. Figure 8.11 shows an example of a contact print made with sunlight.

A contact-printing frame can also be used to press thin objects—leaves for example—against the light-sensitive paper to achieve photograms with sharp detail.

Figure 8.11. A contact print made by direct exposure through a large-format negative.

The first photographically illustrated book—*British Algae: Cyanotype Impressions* by Anna Atkins—was a collection of photograms of botanical specimens pressed under glass and contact printed with the recently-invented cyanotype process. Like most photochemical processes, it directly produces a negative—a blue-toned negative in this case. Thus, Anna Atkins's seaweed contact prints render white shadows on a blue background.

A contact-printed photogram of a very thin object or transparency can show sharp detail even if the source of light is completely diffuse. One can make a sharp-contact print even on an overcast day when the light of the Sun is diffused over the entire sky, although a much longer exposure time is required to make up for the dimmer light.

Many historical photographic processes require ultraviolet (UV) light. Although it is possible to use artificial UV lighting to expose a photogram, such lighting can be expensive and bulky when used to evenly expose a large print.

8.3.1 Cliche verré

An important technique, dating back to the earliest days of photography, is sometimes referred to as *cliche verré*. One makes a drawing or illustration on a thin sheet of clear glass or plastic, using an opaque medium such as black paint. A contact print is then made with the hand-drawn transparency. Figure 8.12 shows a

Figure 8.12. A cliche verré print. A sheet of acetate was painted black, and a drawing scratched into it. This was then contact printed onto the light-sensitive material.

cliche verré contact print; the acetate transparency was painted with a comical lack of skill by the author. Although a negative process was used to make the print, figure 8.12 has not been inverted to a positive image. It looks like an ordinary drawing—dark marks made on a light background—because I first coated the acetate completely with black paint and then scratched in the drawing with a stylus.

If glass is used instead of a very thin sheet of acetate, the thickness of the glass sheet can raise the drawing above the photogram enough that details are softened. It should be printed with the drawing on the bottom, touching the paper; the print, then, is a mirror image of what was originally drawn.

All of the text and figure captions in Anna Atkins's cyanotype volumes were made with the cliche verré technique. Even the title and preface pages are blue with white text (see figure 8.13).

8.4 Shadows and diffraction

So far I have used a simple, geometrical-ray model to analyze shadows of obstacles placed between the source of light and a surface of interest. Due to its wave-like nature, when light encounters obstacles, it does not really propagate along straight-line rays. The phenomenon of *diffraction* (see section 3.7) means that light will, instead, bend around obstacles in complex ways. Thus, we might expect our ray-optics analysis of shadows to be only an approximation. To determine what really happens we must consider the interference of all waves that pass by the obstacles and arrive at our light-detecting surface.

So long as the obstacles and the spaces between them are much larger than the wavelength of the light, the ray model gives a good approximation to reality. If we

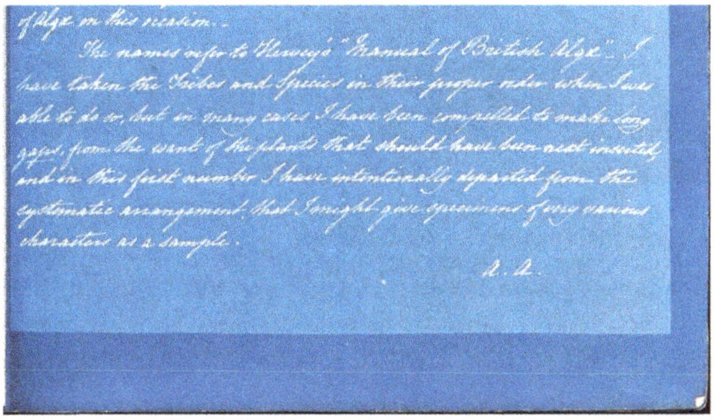
of Algae in this vicinity..
The names refer to Harvey's "Manual of British Algae". I have taken the Genera and Species in their proper order when I was able to do so, but in many cases I have been compelled to make long gaps, from the want of the plants that should have been next inserted, and in this first number I have intentionally departed from the systematic arrangement that I might give specimens of very various character as a sample.
A. A.

Figure 8.13. A detail from the introduction to the first volume of *British Algae: Cyanotype Impressions*, by Anna Atkins. (From The New York Public Library – Anna Atkins, The New York Public Library Digital Collections, https://digitalcollections.nypl.org/items/510d47d9-4b3b-a3d9-e040-e00a18064a99.)

do care about the precise nature of shadows, down to very fine detail, then we should think a bit about how diffraction might alter the picture we have sketched out so far. It is important to distinguish between three different cases.

1. **Simple shadows:** We ignore the wave-like nature of light altogether and simply consider whether or not rays of light from the source are blocked by some obstacle, allowing for the possibility that rays may have been refracted by—for example—the atmosphere of the Earth. This is the path we have taken so far in chapters 7 and 8.1.
2. **Fraunhofer diffraction:** The 'screen' is very far away from the obstacles as compared to their size, and the obstacles or spaces between them are typically less than one thousand times bigger than the wavelength of the light.
3. **Fresnel diffraction:** We add up all of the waves passing by the obstacles, allowing them to interfere with each other but taking into account their different amplitudes as well as their phases. To accomplish this mathematically, Fresnel diffraction assumes that the obstructions are much larger than the wavelength of the assumed point source of light but much smaller than the distances to either the screen or the light.
4. **Full diffraction theory:** All of the light rays are added together, taking phase and amplitude into account, but with no approximations. This can only be done in general by direct numerical calculation with a computer, separately for each situation.

For most cases of Fraunhofer diffraction, we ignore the geometry of shadows. If a light-detecting screen is very distant from a tiny pinhole, for example, the Fraunhofer diffraction pattern is significantly larger than the shadow region determined by ray optics; that is, in fact, the point! Light traveling through a tiny hole does not produce a tiny spot of light on a distant screen but instead produces a much-larger diffraction pattern.

Fraunhofer diffraction assumes that all rays arriving at the screen and interfering with each other have traveled close enough to the same distance that we can assume they all have the same amplitude. The interference arises only from differences in their *phases*—how the positions of the peaks of the interfering waves line up with each other. Furthermore, Fraunhofer diffraction assumes that the source of light is so far away that the wave fronts are parallel planes rather than expanding circles.

Fraunhofer diffraction is invalid, for example, for calculating the diffraction only 1 cm from a 0.5 cm hole. The waves arriving at a given part of the screen are not only different in phase, but their amplitudes are significantly different because waves from the edges of the hole have traveled a much greater distance than waves coming from the center.

The more-complex Fresnel diffraction, on the other hand, takes differences in amplitude into account, and it allows for spherical wave fronts so that the source of light can be much closer. Seemingly by magic, the proper addition of all the waves reproduces the shadow geometry of ray optics but with significant differences in detail if one looks close.

The most famous example of Fresnel diffraction goes back to Augustin-Jean Fresnel himself. It was pointed out at the time by his contemporary Siméon Denis Poisson, a critic of the wave theory of light, that if Fresnel's procedure for adding light waves was correct, then—due to symmetry—the waves passing around a circular obstacle would all interfere constructively at the center, producing a bright spot at the middle of the shadow. Experiment quickly showed[4] this seemingly nonsensical prediction was in fact correct, much to Poisson's embarrassment and the vindication of Fresnel and the wave theory of light. Figure 8.14 shows a photogram of a penny, suspended by a sheet of glass about 2 m from the light-sensitive paper and illuminated by the light of a laser about 6 m from the penny. Both images in figure 8.14 are the same, and they are from a direct scan (inverted from a negative to

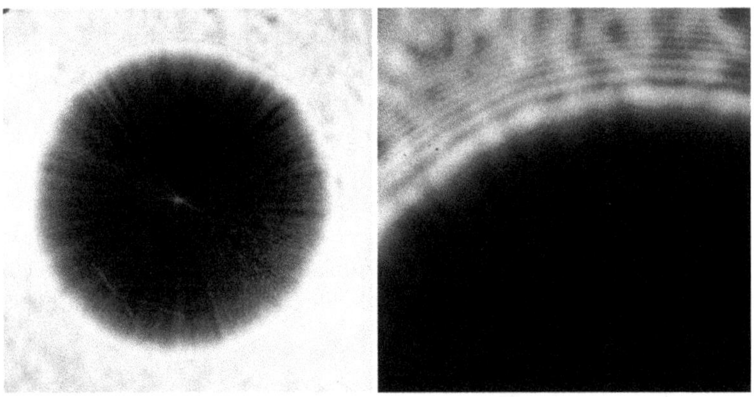

Figure 8.14. A photogram of a penny suspended about halfway between a sheet of light-sensitive paper and a laser about four meters away. **Left:** Poisson's spot is visible at the center of the shadow. **Right:** A small detail of the same image is shown but with the contrast adjusted to better reveal the series of Fresnel diffraction fringes that ring the circular edge of the shadow.

[4] The experiment was first performed by Arago, so the phenomenon is often called the Arago–Poisson spot.

a positive) of the photogram. I have simply adjusted the contrast differently to reveal different details.

The mottled background is due to the poor optics of my inexpensive laser pen and the lens I used to diverge its beam, and the center-pointing rays arise from slight irregularities in the edge of the penny. However, one can clearly see Poisson's spot in the center of the left-hand version. The contrast of the right-hand version has been adjusted to reveal the series of increasingly-narrow circular-interference fringes just as predicted by Fresnel diffraction theory.

References

Meng Q, Wang Q, Liu H and Jiang L 2014 A bio-inspired flexible fiber array with an open radial geometry for highly efficient liquid transfer *NPG Asia Mater.* **6** e125

Rexer L 2002 *Photography's Antiquarian Avant-Garde: The New Wave in Old Processes* (New York: Harry N. Abrams)

Ware M 2019 Argyronomicon: Silver Photographs on Paper: Chemical History of Their Invention, Deterioration and Conservation https://www.mikeware.co.uk/downloads/Argyronomicon.pdf

IOP Publishing

Photography (Second Edition)
Physics and art in focus
John Beaver

Chapter 9

Ray optics 1: pinhole photography

A lens uses refraction to redirect many light rays to the same place, and this allows for the creation of both bright and detailed images. We will consider lenses in more detail in chapter 11, but there is much about the formation of an image that has nothing to do with the lens per se. Rather, the formation of an image is about simple geometry. Before we analyze lenses, it is helpful to consider the geometrical aspects of the simplest of all ways to record an image: the pinhole camera.

Let us consider that light undergoes diffuse reflection off of some object in the world —a cat, for example, such as in figure 3.12. The diffuse reflection means that every part of the cat reflects light in all directions at once. From the point of view of someone looking at the cat or taking its picture, it is as if every point on the cat is emitting light in all directions. We know, of course, that the cat is only reflecting visible light, but if it were emitting its own light a ray diagram of that process would be very similar.

We will adopt this approach as we go along in the next several sections. We will describe an object as if it is emitting, from each point on it, rays of light pointing in all directions. Of course, we cannot literally draw all of these rays, as there would be an infinite number of them. Instead we choose to depict only those rays that, strategically, allow us to figure out what we want.

Consider, for example, rays of light coming from an object that are then focused to an image by a lens. When analyzing this situation, we gain nothing by depicting the plethora of light rays coming from the object that miss the lens all together. In many circumstances, by being only a little clever, we will be able to disregard most of the rest of the rays as well, choosing only those that are necessary in order to determine whatever it is that we want to know. In fact, we will often be able to accomplish our task by considering only two out of the infinite number of rays coming from the object.

A *pinhole camera* is essentially a dark box with a hole in it. A small hole in one side of a box is used to restrict light rays so as to form an image on the opposite side of the box. The image is then recorded with some light-sensitive material.

 © IOP Publishing Ltd 2022

One way to accomplish this is to use black-and-white enlarging paper, such as would be used for making prints in a darkroom, as the 'film.' While in the dark, place the paper on the inside of the box and put a piece of black tape over the pinhole. Then set the camera up outside on a tripod and uncover the hole. After enough time has passed (typically a minute or so), the print paper will be exposed. Cover the hole, carry the entire camera into the darkroom, remove the paper and develop it. A positive can be made from the negative print that results from placing it in contact with a second sheet of print paper and shining diffuse light through the paper[1].

Figure 9.1 shows the geometry of a pinhole camera and how an image is formed. Since light can only enter through a small hole, only one ray from a given part of the object makes it into the camera. The geometry does the rest; the ray can only go to one specific place on the light detector. Each part of the object exposes its own place on the detector, and an image is formed. The image is, of course, upside down.

The left side of figure 9.2 shows the same diagram as figure 9.1 but with a larger hole. The big disadvantage of a pinhole camera is that the light enters only through a pinhole. This means the image is very dim and requires either very bright light or very long exposure (or both). One might very much want a larger hole for a brighter image. However, a large hole means that rays from one point on the object go to a range of positions on the image. In other words, a larger pinhole should mean a blurrier image. There is an inevitable trade-off for a pinhole camera; a brighter image means a blurrier image.

We can compensate for a dimmer image by taking a longer exposure. If the subject is motionless, and the camera is on a stable mount, then perhaps one could make the image as sharp as one likes simply by using the tiniest possible hole even if one had to wait all week for a proper exposure to build on the film. This will only work up to a point, however, because of the phenomenon of *diffraction*.

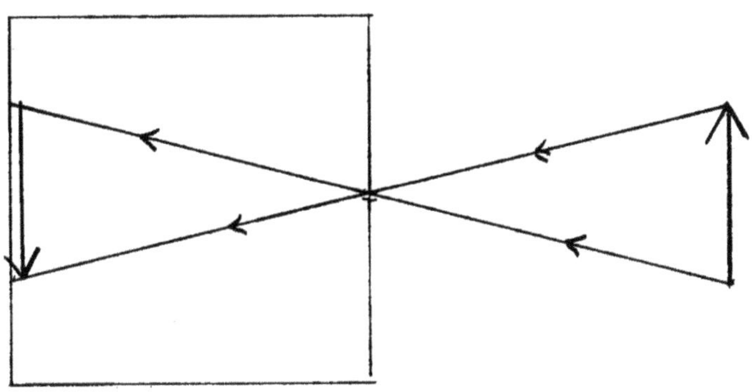

Figure 9.1. The formation of an image in a pinhole camera. Light rays reflect from the tip of the arrow in all directions, but only one can make it through the small hole in the pinhole camera. Thus, rays from each part of the object are guided each to their own separate place to form an upside-down image of the object.

[1] There are many how-to guides available in print and online to fill in practical details for this process and its many variations, if one wants to try it for oneself.

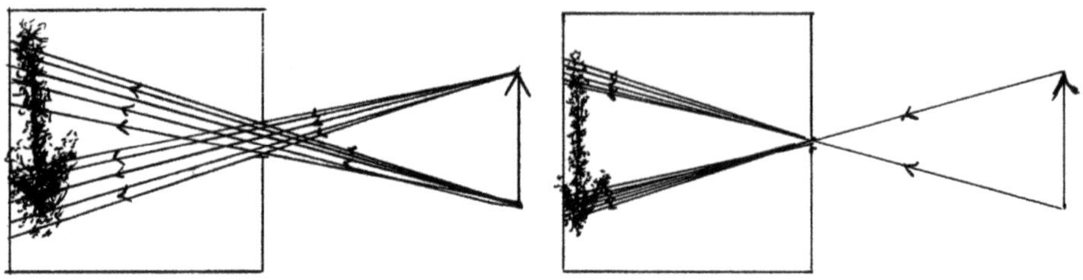

Figure 9.2. Left: A larger hole in a pinhole camera would let in more light and thus produce a brighter image. However, the image would be blurry because many different rays from any given point on the object are allowed to strike many different locations on the detector at once. **Right:** If the hole is too small, the image in a pinhole camera will also be blurry; diffraction at the pinhole will spread the rays out on their way to the detector.

Recall that because of the wave-like nature of light, a ray of light passing through a tiny hole spreads out to form a bullseye pattern. The tinier the hole, the more the light spreads. The light passing through a tiny pinhole will not simply travel in a straight-line ray; the light will spread out, and dramatically so if the pinhole is only a little bigger than the wavelength of the light. See the right side of figure 9.2 for an illustration.

Imagine taking picture after picture with a pinhole camera and each time using a slightly smaller hole. For large holes, diffraction has little effect. At first, because of simple geometry, smaller holes would yield sharper images. However, as one tried smaller and smaller holes, eventually the images would stop getting sharper and sharper. Instead, because of diffraction, smaller holes would make the image blurrier.

Therefore, there must be a particular size hole that gives the sharpest image. Images from pinholes both larger and smaller than this optimum size would be blurrier—large holes would give a blurry image because of geometry, while tiny holes would give a blurry image because of diffraction. This is the best one can do with a pinhole camera; we are up against the very nature of light itself.

With a little bit of calculus and some algebra, one can derive an equation for that optimum pinhole size. The details of the calculation and its motivation are in appendix C, but the result is very simple and given by equation (9.1).[2]

$$D \approx \sqrt{2.44 F \lambda} \tag{9.1}$$

D is the best diameter of the pinhole and λ is the wavelength of the light. If we choose $\lambda = 550$ nm (the middle of the visible part of the spectrum), and convert the units so that D comes out in millimeters, we have the following equation.

$$D(\text{mm}) \approx \frac{\sqrt{F(\text{mm})}}{27}. \tag{9.2}$$

[2] There are slightly different ways to define the 'sharpest' image, so there are other versions of equation (9.1) that are slightly different. When applied to specific numerical examples, however, they give similar results.

I also define the *focal length* (F) of the pinhole camera as the *distance between the pinhole and the light detector*. Therefore, a pinhole camera with a focal length of eight inches (about 200 mm) should have a pinhole about half a millimeter across in order to produce the sharpest possible images.

9.1 Focal length and angle of view

The distance between the pinhole and the light detector is the focal length, but what about the other dimensions of a pinhole camera? What is the difference between a large focal length on the one hand and a small focal length on the other? To answer these questions, we must distinguish between four different concepts: focal length, detector format, image size and angle of view. There is another quantity, *focal ratio*, that relates to focal length in a different way; we consider focal ratio in section 9.4. All of these concepts apply not only to pinhole cameras but to all cameras in general.

Focal length: The focal length, F, is the distance between the pinhole and the light detector.

Image size: The image size refers to the size of the image on the detector, which is directly measured in some unit of length.

Detector format: The detector format refers to the physical size of the light detector in the camera. Is the detector 6 × 9 cm or is it 8 × 10 inches?

Angle of view: The angle of view is the range of directions that can be imaged at once, as seen from the pinhole looking toward the subject.

We consider each of these, in turn, and the relations between them.

9.1.1 Image size

The size of the image (s_i) on the detector of a pinhole camera depends on three factors: the size (s_o) and distance (d) of the subject and the focal length (F) of the camera. The relation between these four quantities is simple and follows directly from the law of similar triangles, as applied to figure 9.1.

$$s_i = s_o \frac{F}{d} \tag{9.3}$$

For a given subject at a given distance, a camera with a longer focal length will produce a larger image of that subject. We shall see in chapter 11 that equation (9.3) is strictly-speaking incorrect for the case of an image formed by a lens, although it is often at least approximately correct even then.

9.1.2 Detector format

Most cameras are built around the particular light detector they employ. Some cameras are designed to use a roll of light-sensitive film, each frame of which is a rectangle that measures 36 × 24 mm. Therefore, the detector format of such a camera is 36 × 24 mm. Such a camera has long been called '35 mm' as a shorthand.

There have been literally dozens of different detector formats in widespread use over the history of photography. For any given camera, however, it is usually the one thing that is fixed and unchangeable by the photographer. A 35 mm camera uses a format of 36 × 24 mm, and only that format, by way of its fundamental design.

The typical pinhole camera is the exception. It can be made by taping a piece of darkroom-enlarging paper to the inside of a box. The light-sensitive paper can be purchased in many sizes, and it can be easily cut smaller. The photographer can put whatever size piece of paper they want in their pinhole camera, so long as it fits in the box. If it does not fix the box, they can simply find a different box. Equation (9.3) for the image size holds whatever the detector format. It may be that a particular desired image does not fit on the detector because the format is too small; even then, the part of the image that does fit will follow equation (9.3) for the part of the subject that is imaged.

In the end, we are often concerned with neither the image size nor the detector format per se. We may care, instead, about how the two compare to each other. Consider two cameras of the same focal length situated right next to each other and both pointing at the same subject. The image size of the subject, as measured in inches, would be the same in both cameras. However, if one camera has a larger detector format than the other, then that image will be a smaller percentage of the full picture. Does the picture only include a friend's head or does it include their whole body as well?

9.1.3 Angle of view

Figure 9.3 illustrates the concept of angle of view (θ) for a pinhole camera. The angle of view depends on two factors: the focal length and the detector format. It is clear from these diagrams that, for a given focal length, a larger detector will produce a larger angle of view. For a given detector size, however, a longer focal length will produce a smaller angle of view.

Since many cameras have a detector format that is rectangular, a camera's angle of view may have different values for the horizontal and vertical parts of the picture. As such, the angle of view is sometimes described by that which is given by the diagonal of the format. If one bisects the angle of view in any of the diagrams in figure 9.3, two right triangles are formed inside the camera. It is easy to show with a little trigonometry that the relations between the detector size (S), the focal length (F) and the angle of view (θ) are given by equations (9.4)–(9.6).

$$F = \frac{S}{2\tan(\theta/2)} \tag{9.4}$$

$$S = 2F\tan(\theta/2) \tag{9.5}$$

$$\theta = 2\tan^{-1}\left(\frac{S}{2F}\right) \tag{9.6}$$

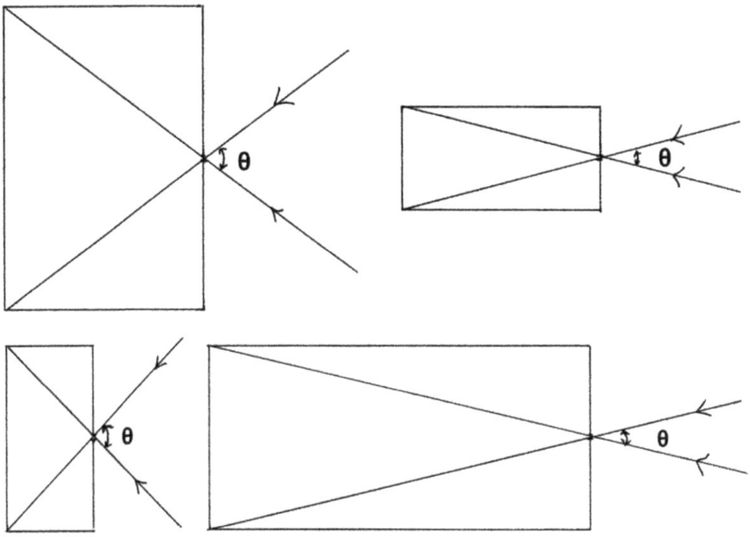

Figure 9.3. **Top:** Two different angles of view, both with the same focal length. The larger detector on the left produces a larger angle of view. **Bottom:** Two different angles of view, both with the same size detector. The shorter focal length on the left gives a larger angle of view.

A large angle of view is often called *wide angle*, while a small angle of view is sometimes called *telephoto*. This terminology is more clear if one thinks of it in terms of most professional cameras; the cameras have a fixed detector size but use interchangeable lenses of different focal lengths.

Imagine that I build two pinhole cameras, both to hold an eight-inch square piece of darkroom paper, but one of them has a longer focal length than the other. The one with the longer focal length would then have the smaller angle of view as shown in the bottom of figure 9.3. For that camera, the longer focal length also means that the image of a particular object—say, a squirrel—would be larger. On the other hand, the camera with the shorter focal length, thus the larger angle of view, would show more than just a single squirrel. It would also show the squirrel's friends gathered around it in a large crowd. With my telephoto pinhole camera, I can see only the one squirrel albeit in much greater detail. Therefore, unless I am able to read the sign it is carrying, with my smaller angle of view I would not know that it is only one of a mighty crowd of squirrels in open rebellion—angry and dangerous.

Pictures taken with an angle of view of about 35°–40° seem approximately 'normal' to most people, so that is often the dividing line between what is called wide angle (or short focus) and what is called telephoto (or long focus). From equation (9.4), one can see that an angle of view of 40° means the camera's focal length is about 1.4 times greater than the size of its detector.

9.2 Distortion and angle of view

In section 10.1 we consider a formal and geometrical meaning of the term *distortion*. Here we use the same term in a more generic sense, regarding some photographic concerns that are as much psychological as geometrical. Our, perhaps, built-in concepts of what a picture should look like can be subverted simply by the choice of subject distance and angle of view. Figure 9.4 illustrates two different ways to produce an image of a given object to be a specific size on the detector. One can either use a long focal length from far away or a short focal length from up close. In this example the sizes of both the image and object are exactly the same, but the angle of view and the distance between the subject and the camera are different. If the object is simply an arrow on a flat wall, then both situations would produce identical images.

A picture taken in the real world, however, usually includes a variety of subjects of different sizes, distances and three-dimensional shapes that have depth. This means very-different results can arise when using different angles of view, even if they produce the exact same image size for the main subject.

Let us consider, as an example, two ways of making an image of the head of Tobias—the cat. Figure 9.5 shows the geometry of Tobias as imaged by a short focal

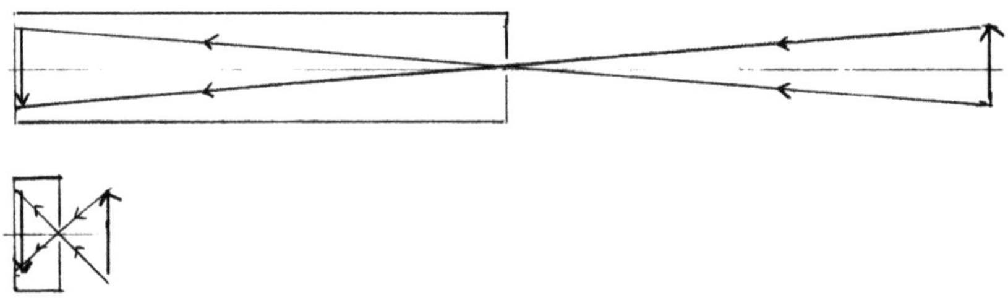

Figure 9.4. Two different ways to achieve the same size image of an object. On the **top**, a long focal length is used from far away. On the **bottom,** a short focal length is used from up close. Although the object portrayed in the diagrams will appear the same in both cases, objects at both nearer and farther distances will be portrayed differently.

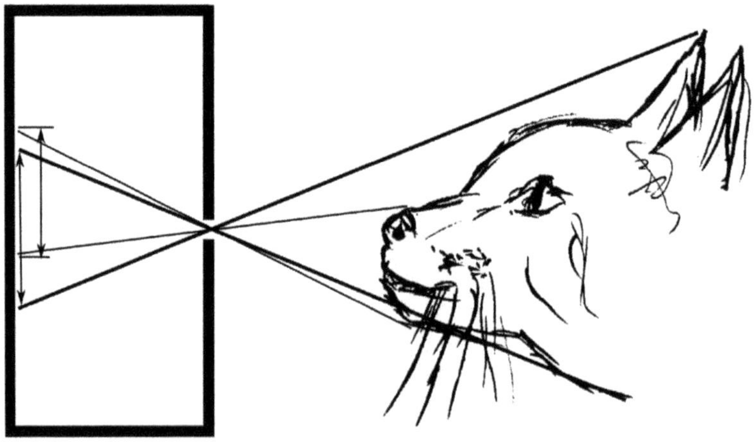

Figure 9.5. If the head of a cat is imaged at a large distance with a long focal length (telephoto), the sizes of the images of the nose and the back of the head are nearly proportional to their real sizes. However, if instead an image of roughly the same size is made by moving close up with a short focal length (wide angle), then the image of the nose is disproportionately large compared to the image of the more-distant back of the cat's head.

length from up close. Although his nose is only about 1/3 the width of his head overall, in this image his nose would appear almost as big as the rest of his head.

An image of Tobias of roughly the same size could be made, instead, by using a long focal length from far away like the top example in figure 9.4. In that case the image of the nose, as compared to the head overall, would be much closer to its real proportions.

From far away, the ears are only a tiny percentage farther away than the nose. From close up, the back of the poor kitty's head is twice as far from the camera as is his nose. This means that the image that results shows a much larger nose as compared to the size of his head. See figure 9.6 for the comparison of the two cases.

Using a long focal length from far away will form an image such that the different parts of the image are in proportion to the actual sizes of the objects. Using a short focal length from close up, on the other hand, has the effect of making the nearer objects look larger in comparison to more-distant objects.

Using a long focal length (and narrow angle of view) tends to make things look flat and with little depth; distances seem compressed together. Using a short focal length (and wide angle of view), on the other hand, exaggerates depth and makes nearby objects look disproportionately large.

As humans, we are used to looking at things from a 'moderately-close' distance, so pictures taken from far away with a long telephoto seem unnaturally flat. Likewise, pictures taken with a very short focal length from very close up seem unnatural as well but in the opposite sense. In practice, photographers have devised guidelines for what is the most useful angle of view for common situations. An example is the rule that for a 'natural-looking' head shot, the angle of view should be roughly 20° (this corresponds to a 100 mm lens with 35 mm film). As always, it is one's own picture. It is up to the photographer to decide what is the most appropriate angle of view for the subject matter and the effect that is desired.

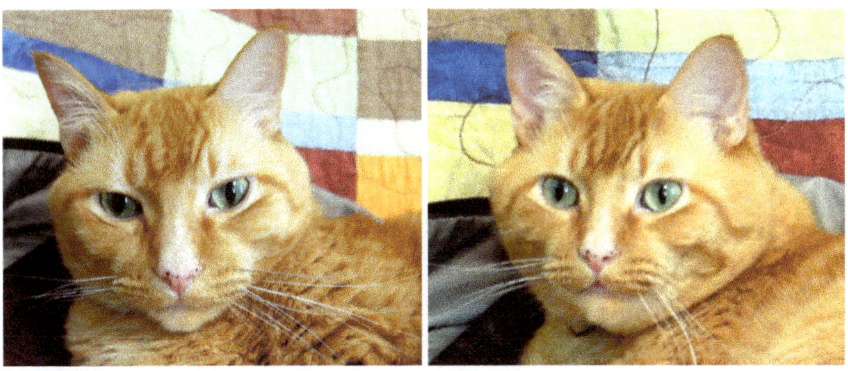

Figure 9.6. The results of taking pictures of Tobias with two very-different angles of view. **Left:** A short focal length lens was used from close up, as in figure 9.5. **Right:** A long focal length lens was used from far away. In the wide-angle, close-up view, the nose was significantly closer to the camera than the ears and so looks disproportionately large.

9.3 Vignetting

For a given value of the focal length (F), how large of a light detector could one use in a pinhole camera? Would a pinhole camera with a focal length of only two inches, for example, produce an image that would cover a detector that is 8 × 10 inches?

Light coming straight in along the axis of the pinhole passes through a circle of a particular diameter. The brightness of the light that enters along that path is proportional, not to the diameter of the pinhole, but rather to its area (πr^2) for a circular hole of radius (r).

Light rays passing through the hole at an angle do not appear as a circle, but an ellipse, as in figure 9.7. This ellipse has the diameter of the pinhole as its long axis, but it is smaller than that along its short axis. Clearly, this off-axis ellipse has a smaller area than the pinhole itself, so less light passes through at an angle.

The image, therefore, will be brightest in the center—exactly opposite the pinhole—and progressively dimmer farther from the center. This effect is called *vignetting*, and the apparent narrowing of the pinhole with angle is only one of several factors involved. We will consider vignetting in detail in chapter 20, but figure 9.8 shows an example in practice.

9.4 Focal ratio

The ratio of the focal length (F) of the camera to the diameter (D) of the pinhole is called the focal ratio (f) of the camera.

$$f = \frac{F}{D} \tag{9.7}$$

The focal ratio of a pinhole camera is an important measure for a simple reason: it is one of the factors that determines the exposure when the image is recorded by the detector. This topic is considered in detail in chapters 18–20, but from the definition alone one can see that a smaller focal ratio means a larger hole and results in a brighter image.

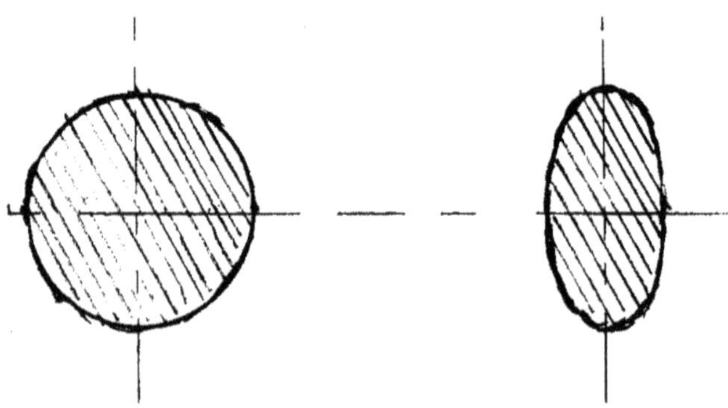

Figure 9.7. A circular hole appears elliptical with smaller area (and thus admitting less light) when seen from an angle. Thus, the center is the brightest part of an image formed by a pinhole camera.

Figure 9.8. *Polar(oid) bear*, Teresa Patrick 2005, used by permission. This wide-angle image from a pinhole camera shows vignetting at the corners.

For a camera with a lens instead of a pinhole, there is a second reason that the focal ratio is important; we consider that topic in chapter 11.

IOP Publishing

Photography (Second Edition)
Physics and art in focus
John Beaver

Chapter 10

Ray optics 2: a fish's eye

10.1 Rectilinear geometry and distortion

The pinhole cameras described in chapter 9 have a very particular geometry: a ray of light from the world passes through the pinhole undeflected and arrives at the flat light detector, which is positioned perpendicular to the axis of the pinhole. Such an optical system is called *rectilinear*; angles formed by light rays inside such a camera are the same as those in the world outside the camera.

We will later consider cases in which the light rays interact with a detector that is angled relative to the image axis or even curved. For these cases, the angles the rays make inside the camera are still unaltered; there is no way for a pinhole to change the direction of a ray. The repositioning of the detector, however, can have a profound effect on the geometry of the image.

Here we consider a different case, bringing us to the next level of complexity—somewhat between that of a pinhole camera and the lenses we will consider in chapter 11. We allow *refraction* from a transparent medium to alter the directions of the light rays as they enter the camera. Lenses also use refraction to alter the directions of light rays, but the intent is mostly different. The main purpose of most lenses is to gather a bundle of light rays and redirect them to the same location on an image that the single ray of a pinhole would have taken. This allows for an image that is both brighter and sharper, as we shall see. Here we consider a different use of refraction—to alter the geometry of the rays in what is still, essentially, a pinhole camera.

Such an optical arrangement produces *distortion*. Some examples are shown in figure 10.1. The left-hand image is rectilinear; rays from the world make the same angles inside the camera as outside. For the middle example, on the other hand, rays make larger angles outside the camera than they do inside. The opposite is true for the example on the right; rays inside the camera make larger angles than those coming from the world. For both cases, rays making only small angles with the axis are nearly undeflected; even with distortion, the center of the image appears approximately rectilinear.

 © IOP Publishing Ltd 2022

The center example in figure 10.1 produces what is known as *barrel distortion.* It is most common when imaging large angles of view—that is, wide-angle lenses. *Pincushion distortion,* on the other hand, arises from the geometry shown on the right in figure 10.1. Pincushion distortion most often occurs when imaging small angles of view—a telephoto lens, for example. The names for these two basic types of distortion arise because of how a regular grid of squares is imaged in each case (see figure 10.2).

Figure 10.3 shows two images made with lens systems. The image on the right was made with a rectilinear lens, while that on the left was made by a lens with extreme barrel distortion.

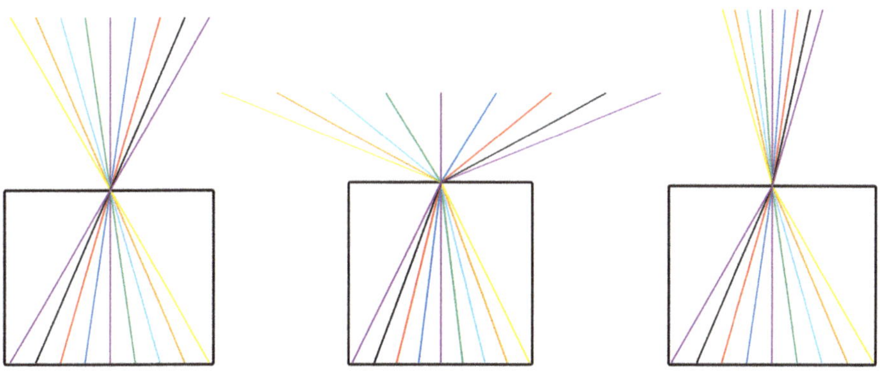

Figure 10.1. Rays from the world as imaged by hypothetical pinhole cameras that introduce distortion. **Left:** A rectilinear system with no distortion. **Center:** A system that produces barrel distortion. **Right:** A system that produces pincushion distortion. To produce either type of distortion, the light rays must be bent somehow by refraction. So, a true pinhole camera is rectilinear.

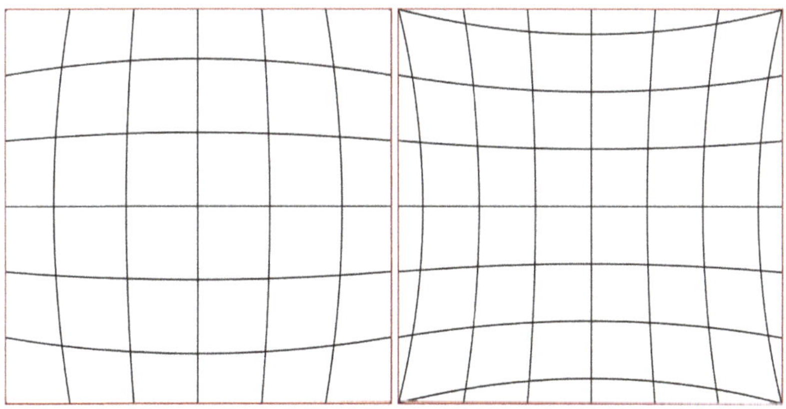

Figure 10.2. Two types of distortion of a flat, square grid. **Left:** Barrel distortion. (Copyright © WolfWings – Own work, Public Domain, https://commons.wikimedia.org/wiki/File:Barrel_distortion.svg, Barrel distortion.svg.) **Right:** Pincushion distortion. (Copyright © WolfWings – Own work, Public Domain, https://commons.wikimedia.org/wiki/File:Pincushion_distortion.svg, Pincushion distortion.svg.)

Figure 10.3. Left: *The world is flat*, John Beaver 2010. This image, taken with a fisheye lens, shows considerable barrel distortion, which curves the horizon line. **Right:** *What happened*, John Beaver 2012. This image, taken with a wide-angle rectilinear lens, preserves straight lines.

10.2 A beaver's-eye view

There is an easy way to produce a rather extreme version of barrel distortion. Note that the center example in figure 10.1 looks very similar to simple refraction of light rays from a lower index of refraction above to a higher index below. The rays all seem to bend inward towards the axis. A natural example has been described for centuries, involving the simple refraction of rays of light passing between air and a body of water. In theory, this produces a greatly distorted view of the world above when seen from below the water's surface. All of the world appears within a much-smaller circular opening, often called *Snell's window* after Snell's law of refraction.

Recall the concept of angle of refraction and total *internal reflection*, described in chapter 3. Consider an underwater aquatic animal—a beaver for example[1]—looking up out of the water at the world beyond (see figure 10.4). Light rays that arrive at the beaver's eye from greater than the critical angle do not arise from the air above the pond; they instead arise from a reflection off the under-surface of the water. The entire 180° of the world above the pond's surface is compressed by refraction to only twice the critical angle as these rays arrive at the beaver's eye.

For the interface of air and water, with index of refraction $n = 1.33$, this critical angle is given.

$$\theta_c = \sin^{-1}\left(\frac{1}{1.33}\right) = 48.8° \tag{10.1}$$

And so in the beaver's-eye view, 180° of world is compressed to only $2 \times 48.8° = 97.5°$. It is as if the entire world is visible through a round window, outside of which one only sees a reflection of the bottom of the pond.

This basic insight has been understood for centuries. It was the physicist Robert W Wood who first tried to photograph the phenomenon (Wood 1906). Before we

[1] Most authors use as an example a fish, which is a member of a vague paraphyletic grouping of organisms. I have improved upon this by invoking the obviously far more superior *Castor canadensis* of the Mammalian order Rodentia.

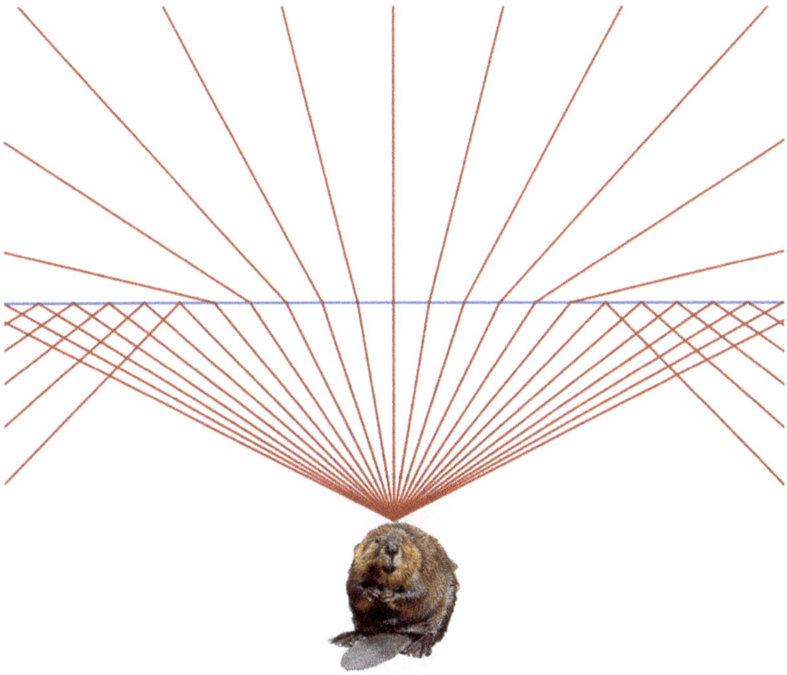

Figure 10.4. A (rather blurry) beaver's-eye view from under water, demonstrating 'Snell's window.' A full 180° of angles above the pond is compressed by refraction to a much-smaller window of only 97.5° across as viewed from below. Outside that window, only the bottom of the pond is seen because of total internal reflection at the water-air interface. (Beaver image from Steve Hersey, CC BY-SA 2.0.)

consider Wood's clever camera, however, let us ask the obvious question: can a human see this effect directly, simply by diving under the water and looking upward?

Wood considered this question, and there are really two answers depending upon how one attempts the feat. One could, for example, literally dive under water, turn so as to face upward and open the eyes. Apart from the fact that many find it uncomfortable to open one's eyes under water, it is well known that humans have very-poor underwater vision.

The reason is clear, if we recall that the angle of refraction made by a ray of light going from one medium to another depends upon the indexes of refraction of both media. The lens and cornea of our eyes (hopefully) bend the light rays by just the right amount to form a clear image, under the normal circumstance of air on the outside. If we instead immerse the human eye in water, then incoming rays refract by angles that are too small, and we see only blurry outlines of objects.

For this reason, those humans who want to see under water often wear a diving mask or goggles; they allow for a layer of air over the eyes, so the lens–cornea combination functions normally. But this trick does not quite work to see the beaver's-eye view, as pointed out by Wood. Once a ray exits the water and goes back to the air inside the diver's mask, the law of refraction bends the ray back to its

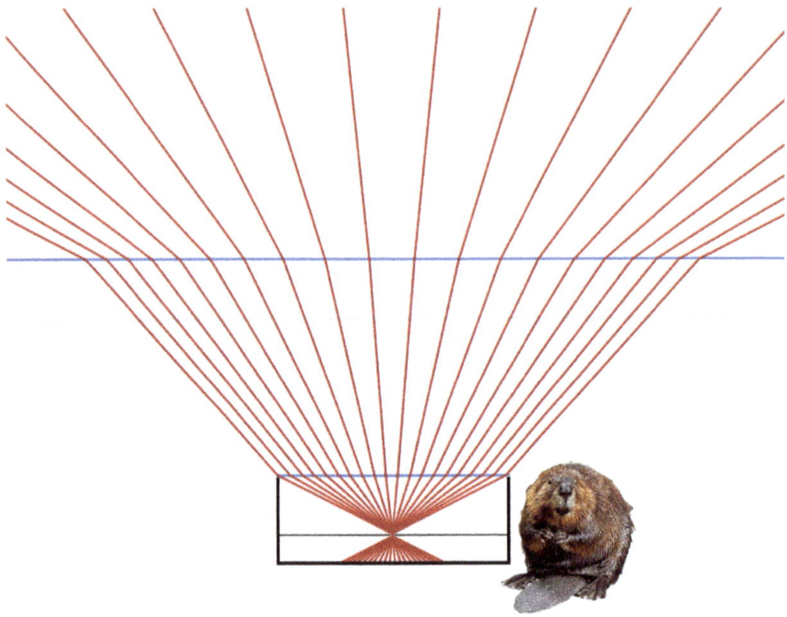

Figure 10.5. A beaver trying in vain to photograph Snell's window with a waterproof, glass-topped underwater pinhole camera. Sadly, the rays refract back to their original angles as they enter the air-filled camera. (Beaver image from Steve Hersey, CC BY-SA 2.0.)

original angle. This same fact also nixes the idea of placing an ordinary underwater camera in a transparent waterproof box, pointing face-up, to take a picture of the effect (see figure 10.5).

There is, however, a sense in which one can still experience Snell's window with swimmer's goggles or an underwater camera: simply abandon the idea of experiencing it all at once. One can point the eye and diver's mask in a direction near that of the critical-angle ray, for example. The ray, upon entering the air inside the mask, now arrives at 0° to the normal axis and no refraction occurs. Thus one can point in different directions, surveying different parts of the beaver's view; it just cannot all be seen at once in a single view (or in a single picture) with such an arrangement.

Speaking of *Castor canadensis*, what is the beaver's-eye view? They are only semiaquatic rodents, so how can they see well in both air and water? It turns out that they do not; their vision is poor on land and even worse under water. Beavers do have a second, transparent eyelid they can close under water, but there still remains the issue of their eye lens focusing differently under water than in air. It seems they split the difference, being somewhat nearsighted on land and very farsighted under water (Rosell and Campbell-Palmer 2022, pp 89–90).

10.3 R W Wood's camera

Robert W Wood (1868–1955) was an iconoclastic US physicist with wide-ranging interests. I first became aware of him while an undergraduate, from his book of charming nonsense poetry, *How to Tell the Birds from the Flowers.* His own illustrations accompanying the poems prove once and for all that carrots and parrots, for example, differ in only superficial and unimportant ways.

While in graduate school I read an account of the famous incidence of his debunking of N-Rays. I was struck by the combination of a rational and skeptical intellect and both the humility and humanity implicit in his approach to telling the world, delicately, that an eminent colleague was publishing paper after paper full of measurements of a phenomenon that simply did not exist.

Perhaps it was this to-the-point way of looking at the world that led Wood to photograph the 'fisheye view' of Snell's window in the most direct way possible. He simply made a pinhole camera with everything under water, including the pinhole and the light-sensitive photographic plate. Figure 10.6 illustrates Wood's arrangement.

Wood made his camera in a bucket, with a sheet of glass resting on the top surface of the water so as to suppress ripples. The pinhole was at the midpoint, and he used a clever remote control shutter mechanism so the pinhole could be open and closed from outside the bucket in order to make an exposure.

In Wood's original version the dimensions are the same above the pinhole to the surface of the water and below the pinhole to the light-sensitive photographic plate.

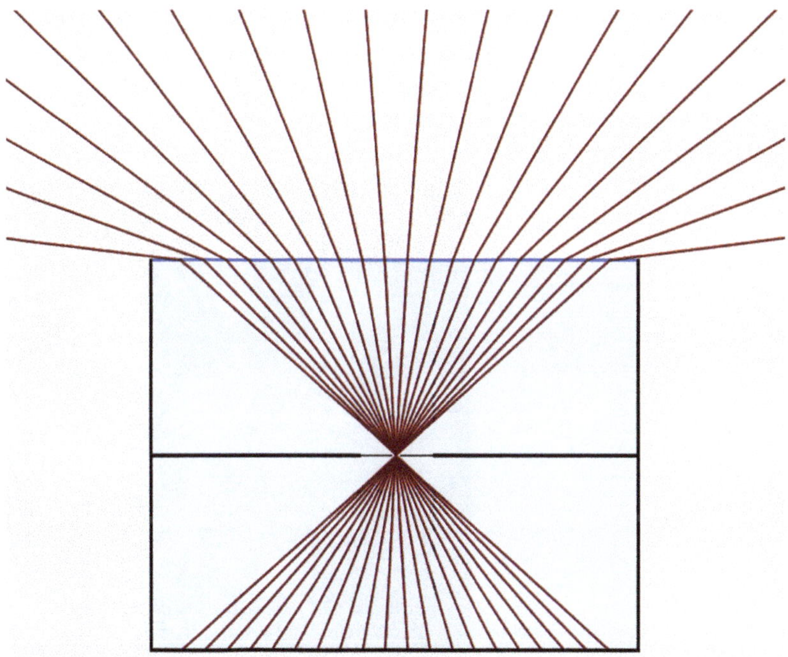

Figure 10.6. Optical design of the 'Fisheye view' camera of R W Wood. Light rays refract as they enter the top of the entirely water-filled pinhole camera. The light-sensitive photographic plate is placed at the bottom, facing the pinhole. The entire 180° of the world is then imaged onto only 97.5° inside the camera.

But there is no reason that it needs to be symmetrical. It is the angles that matter, not the distances per se.

Wood made his camera in 1906, and although no longer in its infancy, photography was still young; it was a technical challenge at the time to take such a photograph. The photographic plate would have been a gelatin-silver emulsion coated onto a sheet of glass. The exposed plate when developed produced a glass negative, which could then be used in a darkroom to make a positive print onto paper, as we shall describe in some detail in later chapters.

The photographic plate, water and the separating layer containing the pinhole and shutter mechanism had to be assembled in the darkroom under a red or deep-orange lamp. The remote shutter mechanism had to be both light and water tight. The loaded and water-filled bucket camera (with the shutter closed) could then be transported outdoors and put in place to take a picture. At that point the top layer of water could be added, a sheet of glass carefully placed on top to prevent ripples and the shutter momentarily opened to expose a picture. Figure 10.7 shows the result—the first known 'fisheye' photograph (Wood's term), published in 1906. It is of several people standing around the bucket camera, and the entire horizon line can be seen circling the picture. Apart from the ahead-of-its-time technical prowess, I love this photograph for many reasons. The sky is overexposed, an almost inevitable consequence of the blue-sensitive photographic emulsions available at the time. This makes the figures' necks nearly disappear, giving them an alien-like appearance worthy of Steven Spielberg. I also love the profile of the shapes that form a backdrop for the figures' knees: the pole conveniently situated in a gap, and the triangular forms creating interesting negative spaces. Little wonder that Wood had some success exhibiting these photographs for their artistic merits.

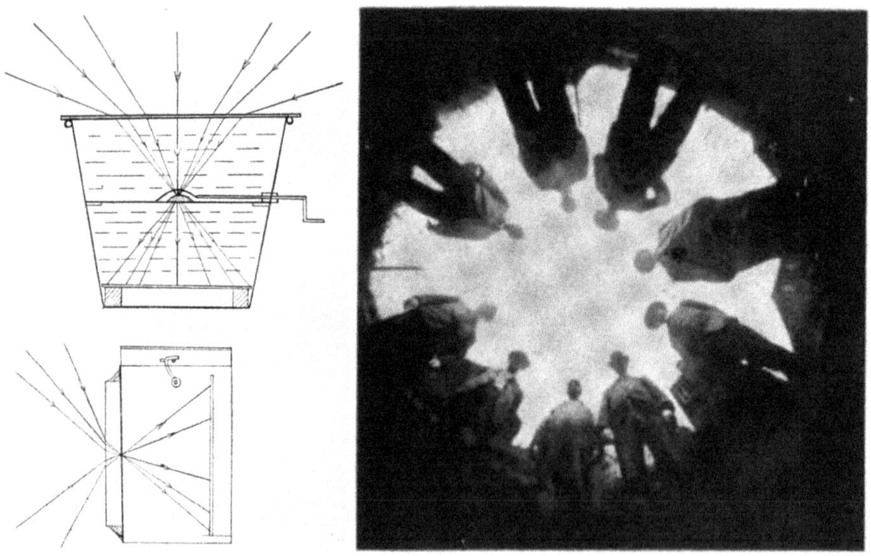

Figure 10.7. Robert W Wood's design for a pinhole-fisheye camera, and one of the first images made with it. (Graphic: Robert W Wood.)

A fisheye camera produces what is technically a form of barrel distortion. It is so extreme that it is usually called, simply, fisheye distortion. The fisheye effect is really more obvious when the camera points in directions other than vertical; we will see examples in later sections. Of course, by its very design, Wood's bucket camera could only take pictures pointing straight upward. He later made cameras that pointed horizontal instead of vertical; imagine an opaque cube-shaped tank with one vertical side made of transparent glass (see figure 10.7).

10.4 Some variations on Wood's camera

Having taught a class on the physics of photography for several years and having already published the first edition of this book, one would think I should know all about the history of R W Wood's water-tank–fisheye cameras. Much to my embarrassment, I knew none of this in early December, 2019, when I was approached for an interview on the subject of the 'mechanics of a fisheye lens' by the online magazine *Vox*.

The writer and producer of the *Vox* video magazine article, Estelle Caswell, was way ahead of me, having already completed much research on the subject when she interviewed me for the video article. The focus of the *Vox* story is on the common use of fisheye photographs for the covers of classic record albums, but Caswell also provides an excellent overview of both the history and the physical principles behind the fisheye lens. The story can be watched here: How the Fisheye Lens Took Over Music.

And so Estelle Caswell of *Vox* is responsible for filling a gaping hole in my physics education. It was also her idea that I try to make my own reconstruction of Wood's camera with the hindsight of more than a century's technical progress in photography. During our conversation, I realized that I might know of a particularly easy way to pull it off.

10.4.1 An 'ephemeral-process' R W Wood camera

In chapter 35 I describe what I call 'ephemeral-process' (EP) photography. This technique, although it uses traditional darkroom photographic paper, requires no darkroom or special equipment to make pictures. The easily-available black-and-white photographic paper is not chemically developed after exposure, as it would normally be. Instead a nondeveloping 'accelerator' is applied immediately before exposure, which causes it to much more rapidly print out—turn dark by the action of light alone rather than through the chemical process of development. The sensitivity is still so low that no darkroom is needed, but the sensitivity is enough that a photograph can be made with a long-time exposure.

After the exposure, the accelerator is simply washed off the paper and dried, at which point its sensitivity lowers enough that it can be scanned and a digital negative image captured. We will use this accessible technique in later chapters for illustrative examples of several photographic and physical concepts.

It occurred to me that I could simply reconstruct something like Wood's bucket camera, of course using a modern ready-made photographic paper emulsion instead of an early twentieth century glass plate. If I fill the tank not with plain water, but rather with my 'EP' accelerator solution, a long enough time exposure will allow the paper to print out with no developer.

The practical advantages are that I do not need a shutter, and the whole thing can be put together in dim room lighting; no darkroom is needed for any step of the process. The disadvantage is that since this is essentially a pinhole camera, there is very little light illuminating the emulsion, so the exposure time needed is long indeed.

My first, very-poor effort can be seen in figure 10.8; it was an all-day-long exposure onto a three inch diameter sheet of photographic paper. Even with such a long exposure, there was barely an image to scan. The tank was an ordinary film-developing tank, filled with a solution of sodium sulfite (which acts as an accelerator for printing out). The pinhole is significantly larger than it should be, given the focal length, and for this reason the image is blurry even for a pinhole camera. The angle of view is somewhat less than 180°, but it does show a fisheye-type distortion of the vertical arbor that the camera was placed under.

My next attempt was both better and worse. To decrease the exposure time, I cheated; I replaced the pinhole with a simple lens. We will consider lenses in detail in chapter 11, and so we will revisit this particular example. But one of the principle reasons for using a lens is that it allows for more light to be gathered, so exposure times can be shorter. Figure 10.9 shows the result: an exposure time of only a few hours resulted in much more visible image detail. The center of the image is now sharpened by the lens, compared to the too-large pinhole of figure 10.8. But now the

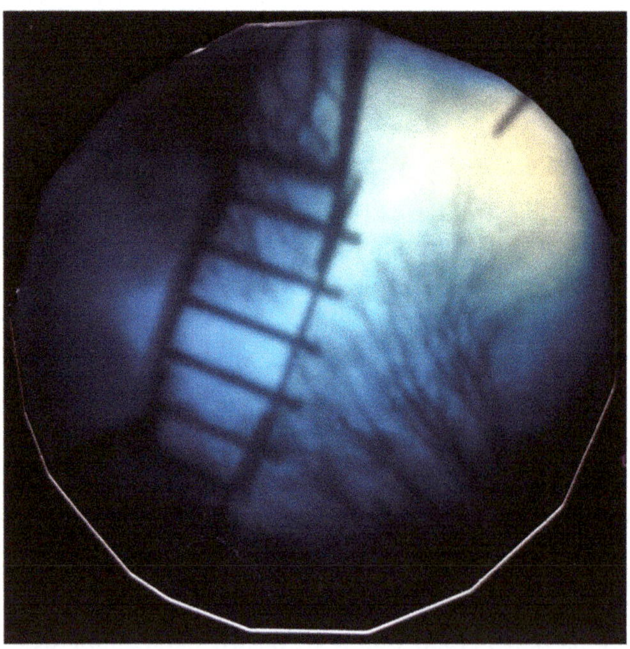

Figure 10.8. My first attempt at an EP Wood pinhole photograph.

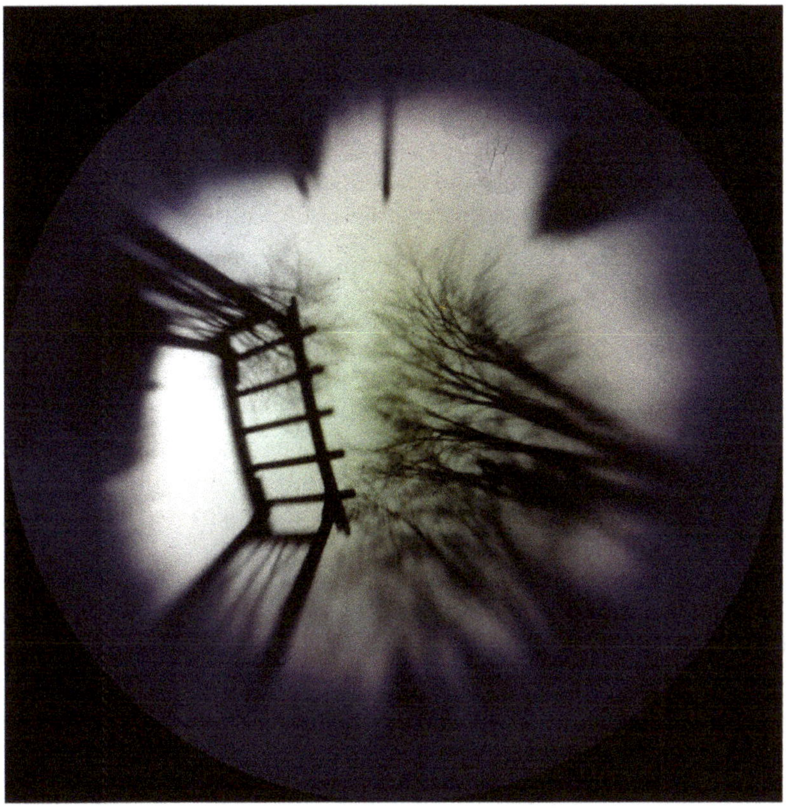

Figure 10.9. My second attempt at an EP Wood pinhole camera, replacing the pinhole with a simple lens. Now the edges are blurry.

edges are blurrier still. Plus, the lens itself is somewhat restricting the field of view, so a full 180° is still not visible, although one can clearly see at least the tops of trees and buildings circling the entire horizon.

10.4.2 A practical R W Wood camera

I next set out to make something more like Wood's original camera; I would use a pinhole rather than a lens, and I would employ far more sensitive developed-out photographic emulsions that must be handled and processed in the darkroom. But I wanted to make a camera that could be loaded and unloaded outdoors with the exposed emulsions stored in an opaque envelope for later darkroom processing. I also wanted the camera and its liquid to be sealed and self-contained, and not dependent on gravity, so it could be pointed in any direction. With a century's hindsight, I had several advantages over Wood.

1. I can make use of readily-available photographic paper that is waterproof, stays flat when wet, can be cut with scissors, and is easily handled with a special orange 'safe light,' since it is not sensitive to long wavelengths. No glass photographic plates for me!

2. I can go to the local hardware store and purchase a wide variety of parts, including plastics easy to drill or cut and that are designed to glue or thread together with waterproof joints. Available in many shapes and sizes, they are called 'plumbing supplies.' It was easy to find an arrangement that could serve as the basis for a waterproof camera.
3. Waterproof silicone sealers and epoxies are easily available and aid greatly in the construction of a camera that not only must be light tight but also water tight.

My camera contains two plastic waterproof chambers, separated by a pinhole-containing thin sheet of brass. A smaller chamber is covered by a clear sheet of acrylic plastic and points toward the subject. The back of the other chamber is removable with large threads (water sealed with Teflon plumbing tape) and holds the light-sensitive paper. A strategic set of internal holes and chambers allows water to flow in, via a hose, and completely fill both chambers while allowing air to escape out a second hose—all while remaining perfectly light-tight (see figure 10.10). More details on its construction are given in chapter 47.

The light-sensitive paper must be loaded and removed in the dark (or under a safe light). But the emulsion-loaded camera can be filled or emptied of water in full daylight. Thus one loads the light-sensitive paper while the camera is dry and sets it up on a tripod for the exposure; a removable cover keeps light from entering until the exposure begins. The camera is then filled with water via the attached hose and funnel; air escapes out the second hose. The cover is briefly removed to make the exposure, and the hoses are then used to empty the tank. The camera can then be taken to the darkroom, and the paper can be removed for processing.

If one wants to take multiple pictures, but there is no ready access to a darkroom, all is not lost. The entire camera, after the paper is exposed and the water has been

Figure 10.10. My fully pointable fisheye pinhole camera.

Figure 10.11. A changing bag. The light-proof cloth bag unzips, and light-sensitive materials can be placed inside. By putting one's arms through the elastic sleeves, the materials can be manipulated safely by feel.

drained, can be placed inside a *changing bag* (see figure 10.11). The exposed paper is then removed by touch alone and placed in an opaque envelope for later darkroom processing. With everything inside the changing bag, a new piece of photo paper can then be loaded into the camera. Once sealed back up and the cover put in place, the loaded camera can be removed from the changing bag and the whole process repeated for a second exposure. A changing bag is often used for pinhole photography in general. Many pinhole cameras are simply some kind of a box with photographic paper held in by magnets, tape or photo corners.

Four of my pictures with this camera can be seen in figure 10.12. The angle of view is very nearly 180°, and the typical fisheye barrel distortion is obvious.

10.4.3 A 'dry' R W Wood camera, version one

I have already pointed out that one cannot simply put a camera with a wide-angle lens inside a waterproof box and place it under water, pointing upward, in order to photograph Snell's Window. The light rays will simply bend back to their original directions when they enter the air-filled camera box.

One could, however, place a sheet of glass at the bottom of Wood's camera, and tightly press the light-sensitive paper to it from the outside. Even though the light rays alter their direction once they exit the glass, they have nowhere to go if they instead immediately encounter the photographic emulsion. Although some smearing of the image may be expected, the exiting rays would expose essentially the same parts of the emulsion as they would have inside Wood's camera (see figure 10.13). Instead of placing the photographic emulsion inside the water at the bottom of the camera, it is place outside the camera on the other side of a thin sheet of glass.

I have tried this, and it works—sort of. It does produce what is technically a fisheye distortion, but the image is significantly blurrier and vignettes severely at the edges. Thus it does not produce the near-full 180° image made by Wood's water-filled camera.

The reason for this more-limited field of view is clear, if we consider the added step of the rays emerging from glass to air. The key problem is internal reflection, as

Figure 10.12. Four photographs from the camera shown in figure 10.10.

the light rays try to exit into the air and to the light-sensitive paper. The same concept of critical angle that produces the fisheye effect in the first place also applies to these exiting rays. Rays that enter from the air at very near 90° exit at that same large angle. Thus they arrive at the bottom of the camera at nearly the critical angle; if they had arrived at more than this angle, they would have undergone total internal reflection and never have escaped at all.

The trouble is that, as we saw in section 3.3.1, internal reflection is not simply an either-or effect. As the critical angle is approached, the exiting refracted ray becomes dimmer, while the internal reflected ray grows brighter, finally reaching 100% at the critical angle.

Even Wood's original camera experiences significant vignetting at the edge. The internal rays near the critical angle arise from the external light rays that are at nearly 90° to the normal of the upper window. That is, the rays arrive nearly parallel, and a significant fraction of the intensity of these rays are reflected too and so never even make it into the camera. This is only made worse by the internal

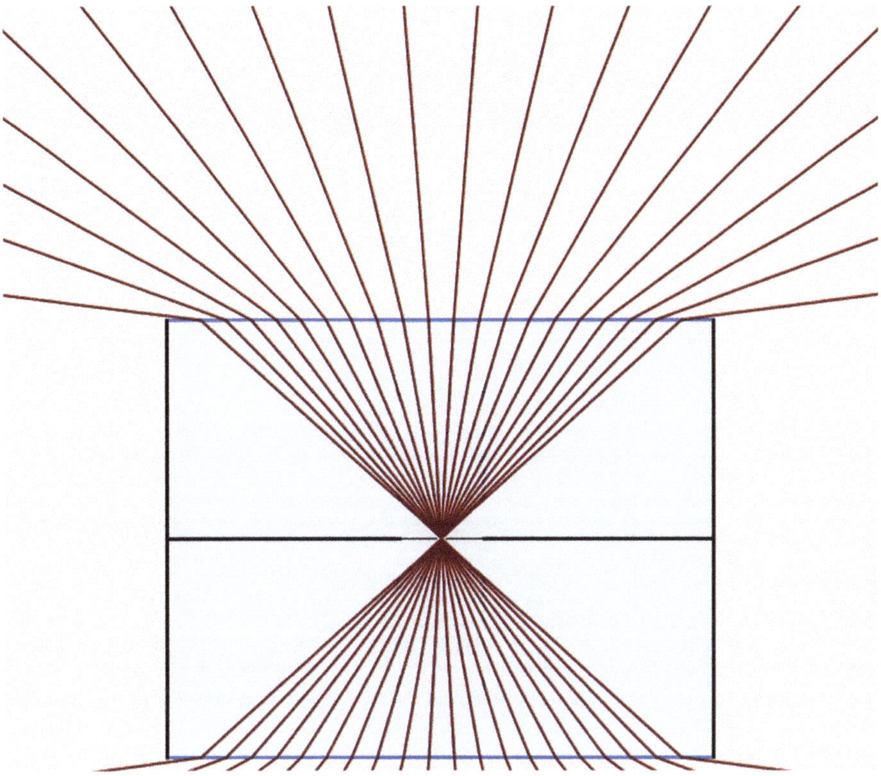

Figure 10.13. The geometry for my first attempt to make a 'dry' pinhole fisheye camera. The rays pass through a second sheet of glass at the bottom and directly expose the light-sensitive paper, which is pressed firmly against the bottom of the glass.

reflections of the would-be exiting rays at the bottom of my 'dry' version of his camera.

But what about those sheets of glass on the top and bottom? The light is now refracting both into, and out of, glass—not water. Does this make a difference? Let us examine this case in more detail because it brings up an obvious complication that, interestingly, turns out not to be so complicated as it might at first seem.

Consider figure 10.14, where I show the complex example of two different materials in layers. The top layer is glass, with $n = 1.6$ covering the tank of water ($n = 1.33$) containing the pinhole at its center. On the bottom is a second layer of glass, to make a sealed container for the water, and then air again is underneath.

All of these layers are flat and parallel, so regarding the direction of the rays it turns out to be just like Wood's camera. Yes, as a light ray from the world passes through air and arrives at the upper surface of the top glass sheet, it refracts at a different angle than it would have if it had encountered only water instead. But the ray refracts again as it passes from the bottom surface of the glass sheet into the water below. This second refraction brings the ray back to the same direction it

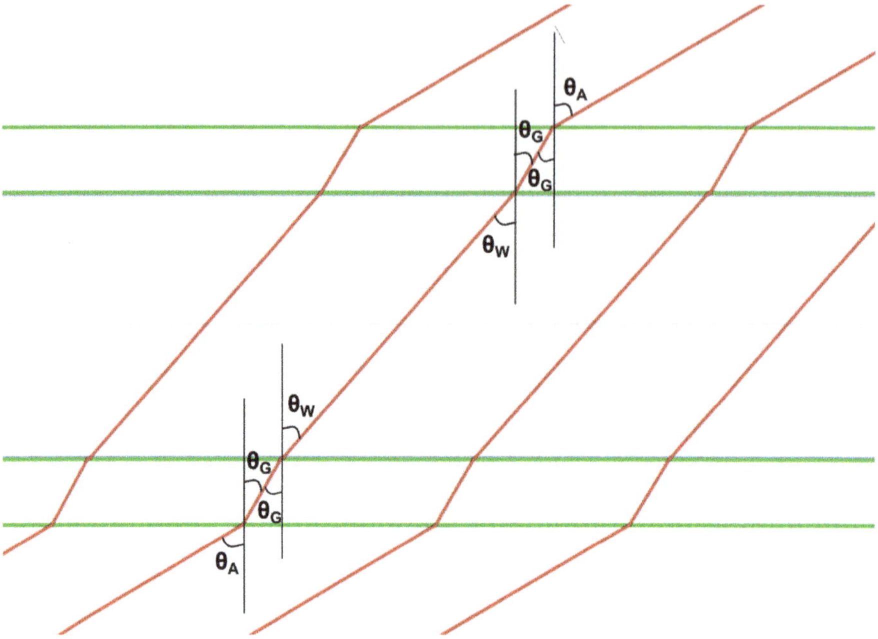

Figure 10.14. If light refracts from air to glass, then to water, then to another layer of glass and then finally back to air—if the layers are flat and parallel—the final rays are in the same direction as the original rays.

would have had if it had refracted directly from air to water with no intervening layer of glass.

We can write Snell's law for the first two refractions, from air to glass, and then from glass to water.

$$n_A \sin \theta_A = n_G \sin \theta_G \tag{10.2}$$

$$n_G \sin \theta_G = n_W \sin \theta_W \tag{10.3}$$

Clearly, we can combine these two equations.

$$n_A \sin \theta_A = n_W \sin \theta_W \tag{10.4}$$

But this is simply Snell's law written for the ray coming from the air directly to the water! So the presence of the glass sheet does not alter the directions of the rays entering the water and passing through the pinhole, and the same must be true for the layer of glass at the bottom of our tank.

But what about the critical angle for total internal reflection? A light ray entering the top sheet of glass at θ_A near 90° produces a ray inside the glass at near the critical angle for the glass–air interface.

$$n_A \sin 90^\circ = n_G \sin \theta_G \tag{10.5}$$

$$\theta_G = \sin^{-1}\left(\frac{n_A}{n_G}\right) \tag{10.6}$$

For glass of $n = 1.6$ this is about $\theta_G = 38.7°$, which is considerably smaller than the 48.8° critical angle for an air–water interface. But this same ray then enters the water at angle θ_W.

$$n_G \sin\theta_G = n_W \sin\theta_W \tag{10.7}$$

$$n_G \sin\left[\sin^{-1}\left(\frac{n_A}{n_G}\right)\right] = n_W \sin\theta_W \tag{10.8}$$

$$n_G\left(\frac{n_A}{n_G}\right) = n_W \sin\theta_W \tag{10.9}$$

$$\theta_W = \sin^{-1}\left(\frac{n_A}{n_W}\right) \tag{10.10}$$

This is simply the equation for the critical angle of an air–water interface. Thus the intervening layer of glass does not alter the fact that this largest-possible ray from the air still enters the water at the critical angle for an air–water interface at about 48.8°. The same argument holds in reverse for the rays at the bottom of figure 10.14. The light rays that enter the camera's top at near $\theta_A = 90°$ arrive at the bottom-most glass–air interface at very near the critical angle for that interface.

The largest-angle exiting rays are the very ones in which we are most interested if we want a fisheye effect. Since these arrive near the critical angle at the boundary from glass to air, they are now subject to increasingly strong internal reflection as they try to leave the camera; vignetting is the result.

There is a way out of this dilemma, although it is rather unappealing. Find a liquid that has the same index of refraction as glass, and simply spread a thin layer between the bottom-most sheet of glass and the photographic emulsion. There are many oils that fit this bill; ordinary corn oil, for example, has nearly the same n as borosilicate glass.

If the rays exiting the bottom of the glass encounter not air, but instead a liquid of the same n, then neither refraction nor internal reflection occurs. The rays simply pass to the emulsion undeflected. This strategy is called an *optical transfer*. Although a useful concept in many circumstances, it is an unappealing choice in this case. The whole point of this exercise was to avoid immersing our emulsion in liquid!

For my very crude attempt, I used acrylic sheets, and filled the camera not with water but with acrylic epoxy casting resin, which hardens to a clear solid of very near the same n as the clear plastic sheets containing it. I then used a thin layer of vegetable oil as an optical transfer from the bottom acrylic sheet to the photographic emulsion; the result can be seen on the left side of figure 10.15.

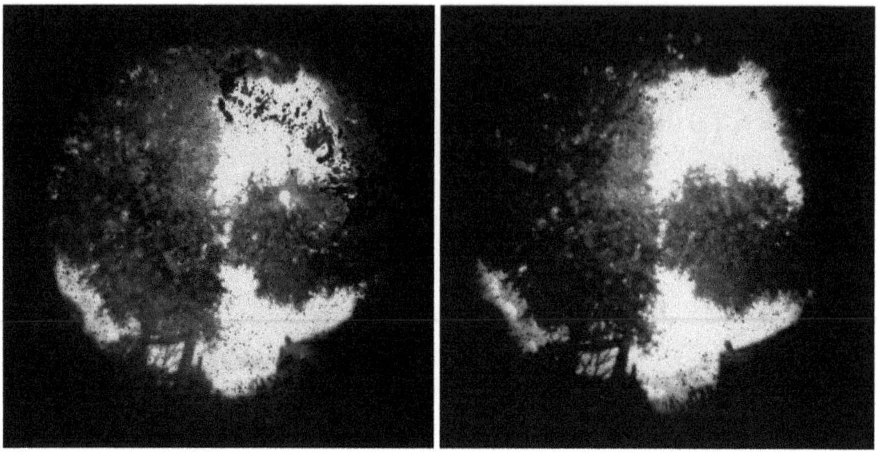

Figure 10.15. Photographs made with the pinhole fisheye camera arrangement of figure 10.13. **Left:** An oil optical transfer was used. **Right:** No oil optical transfer was used.

The image on the right was made without the oil, so there was nothing but a tiny air gap between the acrylic and emulsion. It clearly shows significant vignetting by comparison. The small dark spots in both pictures arise from my unskilled, beginner's effort with casting resin; they are caused by a plethora of tiny bubbles. The odd dark smudge in the left-hand image is from not completely washing off the vegetable oil before putting the emulsion in the developer solution.

10.4.4 A 'dry' R W Wood camera, version two

The 'dry' fisheye pinhole camera described in section 10.4.3 is not very practical because one must still use some kind of oily liquid for an optical transfer to the emulsion. There is a simple variation on Wood's idea, however, that allows one to make a true fisheye 'lens' that can attach to the front of an otherwise ordinary pinhole camera of any size.

Recall from section 10.2 why Wood put the emulsion inside the water in the first place: if one allows the light rays to return to air, refraction brings them back to their original directions, and so the fisheye effect is lost. This is true if we assume the rays traveling through the water to the pinhole encounter a flat interface with the air—a sheet of glass, for example. But what if we arrange a solid transparent material above the pinhole such that the rays traveling through the water to the pinhole always arrive at the material normal to its surface? These rays, with angle of incidence of 0°, cause no refraction at all.

The solution is obvious. Simply place the pinhole at the center of a thin, transparent hemisphere at the bottom of the flat tank of water. The refraction occurs at the surface as in the ordinary Wood camera. The rays then pass undeflected through the hemisphere into dry air and finally through the pinhole to form an image in the completely dry camera. See figure 10.16 for a diagram of the optical arrangement.

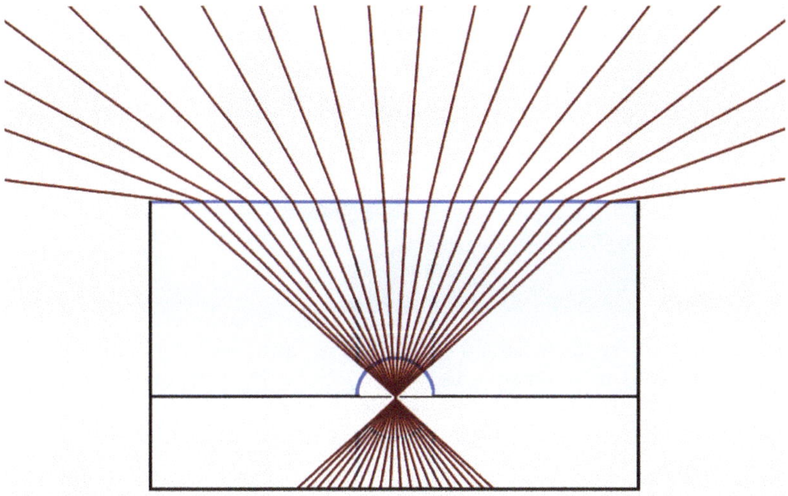

Figure 10.16. My simple design for a 'dry' version of Wood's camera. The rays pass unaltered through a small, transparent plastic hemisphere into air and through the pinhole located at the center of the hemisphere.

Such a fisheye pinhole lens can be placed on any size pinhole camera with some stipulations. First, the camera must, of course, still have an angle of view of at least $2 \times 48.8°$ if it is to capture the entire fisheye image circle. Second, recall that there is an optimal size for a pinhole when given the focal length of the pinhole camera. Thus it is best to construct it such that the pinhole itself is replaceable.

The water-tank part of this lens could be made much, much smaller than the camera itself. But here too, there are some considerations. In particular, both the transparent hemisphere and the water tank itself should be many times larger than the pinhole—not a difficult constraint for most pinhole cameras!

Since a pinhole camera only redirects tiny bundles of light rays, the transparent hemisphere need not be optically perfect. My best example uses an acrylic hemisphere over the pinhole that is 30 mm in diameter. It was sold as a weather cover for a tiny outdoor camera and cost less than ten dollars. I have also tried small acrylic spheres that separate into hemispheres so that they can be filled with objects, sold as a way to make one's own decorations. These 30 mm diameter hemispheres also cost less than ten dollars—but for four dozen of them. They worked nearly as well as the more expensive example. More detailed instructions for making such a camera can be found in chapter 47.

Two of my dry Wood fisheye cameras can be seen in figure 10.17. Both are made from inexpensive small black plastic jars with flat lids—with the pinhole and hemisphere mounted in the lid.

The bottom of the jar was cut off to a proper depth to give a wide enough angle of view for the fisheye image. For the example on the left, a piece of clear acrylic was glued in place of the original jar bottom. It was then filled with water by simply submersing it (having first taped over the pinhole) while open and then sealing the lid under water.

Figure 10.17. A version of Wood's water-tank pinhole fisheye optical system can be made as a standalone 'lens' that may be attached to an ordinary dry pinhole camera. The parts can be seen on the **left:** An inexpensive transparent plastic hemisphere, some clear acrylic sheet, and a black plastic jar, the bottom of which can be cut off to the proper depth. **Right:** Two pinhole cameras using these lenses.

For the example on the right, I simply filled the entire container with acrylic resin —right up to the cut-off edge. I then allowed the resin to cure undisturbed. This example is thus nearly a solid block of clear acrylic. The top surface of the acrylic is not perfectly flat, but this seems to have only a modest effect on the image quality. Some examples of photographs taken with both cameras can be seen in figure 10.18.

The concept shown in figure 10.16 can be used to make a waterproof wide-angle (or even fisheye) camera to take under water and photograph Snell's window. Simply surround the front of the lens with a transparent hemisphere instead of a flat window (see figure 10.19). Dome-shaped underwater housings are available from several manufacturers for those who want to photograph under water with their wide-angle or fisheye lens.

10.5 The birth of the fisheye lens

At its most basic, a camera lens uses the same geometry as a pinhole to make an image. While a pinhole redirects only single rays from the world, a lens is able to capture many at once and still (hopefully!) redirect them to their proper places on the image plane. This produces an image that is both brighter and sharper. We have seen how to use a liquid 'lens' to alter the rectilinear geometry of a pinhole camera to an extreme fisheye barrel distortion. Can we produce a true lens that does the same thing, redirecting significant bundles of light rays to their proper places, in addition to the single rays imaged by our pinhole versions?

In one sense, the way forward to a fisheye lens is clear. We simply shape a piece of glass like the liquid part of the 'dry' pinhole fisheye camera diagrammed in figure 10.17. To image the fish's eye view, we need then place what amounts to a tiny wide-angle camera at the center of the inner hemisphere that can redirect the rays onto a light detector (see figure 10.20).

But once we try to properly collect and redirect bundles of rays rather than the single rays of a pinhole camera, everything becomes far more complex. The outer surface of the lens must be greatly over-sized and at least slightly convex. Furthermore, both the inner and outer curvatures of the glass are often more

Figure 10.18. Photographs taken with the 'dry' fisheye pinhole cameras shown in figure 10.18.

mathematically complex than sections of simple spheres. Additional lenses are then needed to take the place of the pinhole—that is, to relay the fisheye-distorted geometry of these rays to an actual image. Finally, to properly redirect rays of different colors, we must use compound systems of combinations of lenses made of different types of glass working together to correct for each other's mistakes. Figure 10.21 shows diagrams for a couple of early designs for true fisheye lenses. Note the similarity of the top-most lens element to our pinhole version shown in figures 10.16 and 10.20. Missing from the diagrams is the important fact that all of the rays ultimately pass through an aperture, near the center of the arrangement of lenses, that is dwarfed by the large curved outer element.

It was not until the early 1920s, nearly 20 years after Wood's fisheye pinhole camera, before real fisheye lenses were successfully manufactured out of glass. Even then, they were so expensive that they were used mostly for research. Nikon is usually credited with producing, starting in 1962, the first practical fisheye lenses made to mount on an existing 35 mm camera system.

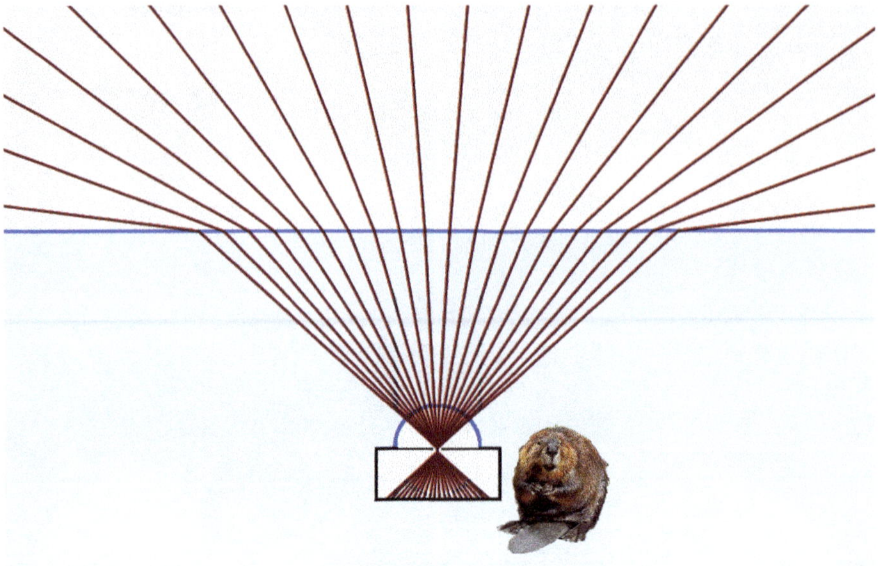

Figure 10.19. A beaver at home in the pond, happily photographing Snell's window after purchasing a transparent hemisphere for the front of their underwater pinhole camera. (Beaver image from Steve Hersey, CC BY-SA 2.0.)

Figure 10.20. The basic concept for a fisheye lens is to shape a piece of glass something like the liquid part diagrammed in figure 10.17, and then place a tiny wide-angle camera lens at the center of the open hemisphere.

The very expensive lens (from 1972), shown in figure 10.22, is able to image angles greater than 180°. That is to say, the lens can image angles that point behind the camera. Only about 200 were made, and an individual example can sell for more than $10 000 on the rare occasion one comes up for auction. Although it gathers a bundle of rays only a few millimeters across, the glass needed to produce a sharp fisheye image over this huge range of angles dwarfs the camera. Very few were produced, mainly for industrial use and scientific research.

With modern techniques for making lenses, a fisheye lens is now within range of even the casual photographer. One company even makes an inexpensive plastic

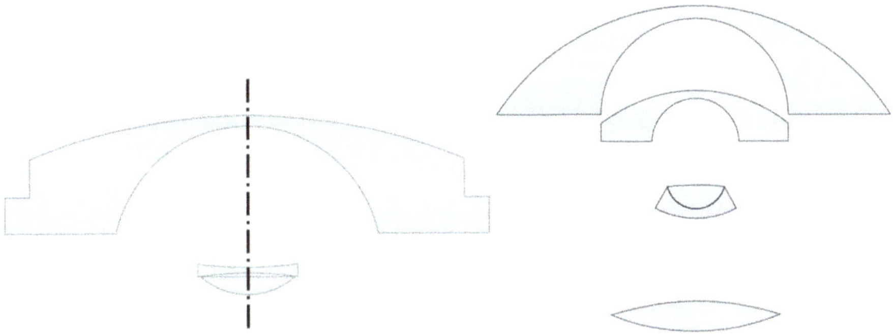

Figure 10.21. Diagrams that show two early designs for fisheye lenses. (Copyright © Mliu92 – Own work. Redrawn from https://worldwide.espacenet.com/publicationDetails/originalDocument?CC=GB&NR=225398A&KC=A&FT=D&ND=&date=19241204&DB=&locale=, Attribution-ShareAlike 4.0 International (CC BY-SA 4.0), https://commons.wikimedia.org/wiki/File:Hill_Sky_Lens_(1924).svg, Hill Sky Lens (1924).svg.)

Figure 10.22. An early Nikon fisheye lens (1972) made to attach to an ordinary 35 mm camera. (Copyright © Morio – Own work, Attribution-ShareAlike 4.0 International (CC BY-SA 4.0), https://commons.wikimedia.org/wiki/File:Fisheye-Nikkor_Auto_6mm_f2.8_lens_2015_Nikon_Museum.jpg, Fisheye-Nikkor Auto 6mm f2.8 lens 2015 Nikon Museum.jpg.)

35 mm film camera with a fisheye lens made from plastic elements (e.g., see figure 10.23). The lens on the left images a full 180° but only from corner to corner in the rectangular image frame. The camera on the right has an image scale that is considerably smaller than the film format, and so most of the circular image is visible.

Figure 10.23. **Top:** A contemporary, modestly-priced fisheye lens for a 35 mm film camera (left), and a simple plastic camera with a fisheye lens made from plastic elements (right). **Bottom:** Examples of photographs made with each camera.

References

Rosell F and Campbell-Palmer R 2022 *Beavers: Ecology, Behaviour, Conservation, and Management* (Oxford: Oxford University Press)

Wood R W 1906 Fish-eye views, and vision under water *London, Edinburgh Dublin Phil. Mag. J. Sci.* **12** 159–62

IOP Publishing

Photography (Second Edition)
Physics and art in focus
John Beaver

Chapter 11

Ray optics 3: lenses

For a pinhole camera there is a necessary trade-off between image sharpness and image brightness, and the image sharpness has a natural limit set by the wave-like nature of light. We could, however, use the property of refraction to redirect light rays from the edges of a large hole so that they end up at the same place on the image as do light rays passing through the center of the hole.

A wedge-shaped piece of glass—a prism—will do the trick. Near the edge of the hole, the rays must be deflected at a greater angle (see the left side of figure 11.1), so a more-angled wedge is required. Nearer the center of the hole, the light rays need not be deflected as much, so a more flat piece of glass would be needed. All of these little prisms can be put together into one smooth shape, called a lens (see the right side of figure 11.1). One may have noticed that a lentil has basically the same shape and that the words 'lens' and 'lentil' have much in common (see figure 11.2).

A lens allows us to get around the inherent geometric restrictions of using a pinhole to make an image. We can, in principle, use as large of a hole as we want, so

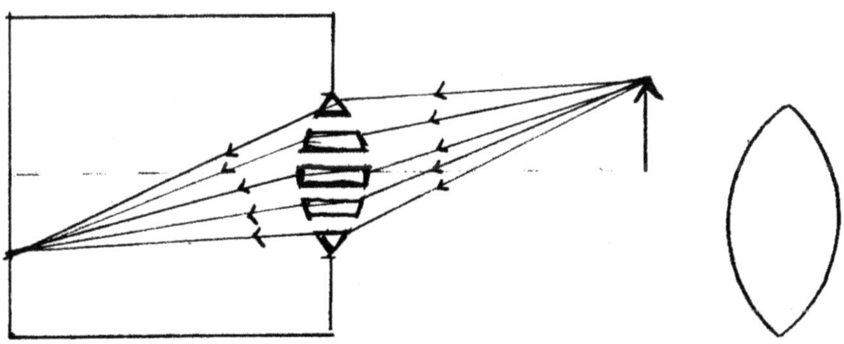

Figure 11.1. Light from a large opening could, in principle, be redirected to a common focus by appropriately-placed prisms, as illustrated on the **left**. On the **right** is a cross section of a double-convex lens. Each part of the lens acts as a prism to refract light by different angles such that they all reach a common focus.

 © IOP Publishing Ltd 2022

Figure 11.2. Lenses and lentils.

long as we can design a lens to deflect all of the light rays to their correct places on the image. In practice, there are—of course—limitations, and we will talk some more about those later. For now, let us put a lens on our camera and see what happens.

11.1 Focus

One consequence of using a lens instead of a pinhole to form an image is that, for a given distance to the object, there will be only one lens-detector distance for which the light rays from one point on the object converge to a single point on the image. This should be clear from figure 11.1. If one were to move the detector closer to or farther from the lens, the light rays would not all meet at one spot.

This is not the case for a pinhole camera. It is true that, because of the effects of diffraction and ray geometry, there is an optimum focal length for a given pinhole size. But this is a subtle effect, and it does not depend at all on the distance to the subject. So, for a pinhole camera, one can simply use the right size pinhole for one's focal length, and everything at all distances will be equally in (crude) focus.

Introduce a lens, however, and for a given object distance there will be only one image distance for a perfect focus. This means that one cannot, in principle, get objects of different distances to all be in focus at once. It also means there must be some mechanism for adjusting the lens-detector distance to achieve perfect focus for a given object. This is called, of course, focusing the camera, and there are many strategies for accomplishing this—all of them greatly complicating the construction of a camera. One of the appeals of pinhole photography is that no focus mechanism is needed, thus making the construction of a pinhole camera simpler.

We can use the most basic simple lens as an example to illustrate some important points about the focus of any lens. Figure 11.3 shows that light rays coming from a point on a nearby object diverge at a steep angle as they arrive at both sides of the camera lens. Light rays coming from a distant object, on the other hand, are more-nearly parallel.

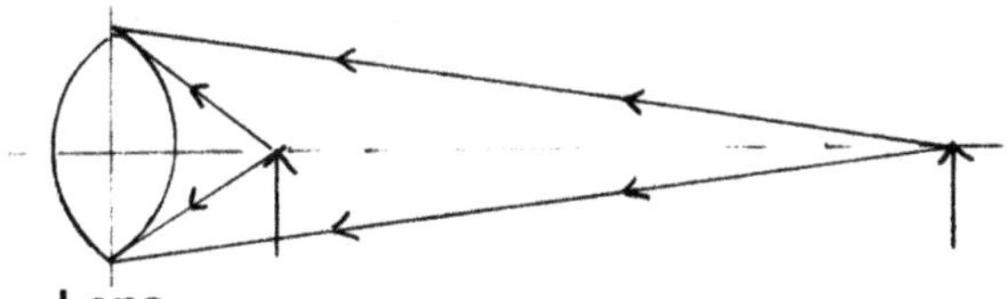

Figure 11.3. As they reach the edges of a given lens, rays of light from nearby objects diverge at a greater angle than those from objects far away.

It should thus be unsurprising that a given lens, which must bend the light rays inward to get them to converge, will bend those coming from the distant object at a steeper angle than those coming from the nearby object. For the nearby object with greatly diverging light rays, some of the refractive power of the lens must be used simply to straighten the rays before it can bend them inward. Thus the image distance for a nearby object is greater than for a more distant object.

What if we slightly change the distance of the two objects in figure 11.3? For the nearby object, a small difference in distance will make a big difference in how much the light rays are diverging when they arrive at the lens and will cause a significant difference in image distance. And so for nearby objects, small differences in distance lead to large differences in focus. For the distant object on the other hand, a slight change in distance will make much less difference in the angle of the light rays as they arrive at the lens. So the focus changes little between two objects that are at different, but both very-distant, locations.

If the object were an infinite distance away, the light rays would be exactly parallel. But even if the subject is only very far away, the light rays arrive practically parallel to each other. For most lenses, put the subject one mile away or two miles away, and the light rays are still so close to parallel that one cannot tell the difference between them. And so every lens has an 'infinity' setting, but do not take the meaning literally.

11.2 Focal length

For a pinhole camera we referred to the focal length (F) as simply the distance between the pinhole and the detector—the image distance. But for a camera with a lens, the focal length is the image distance for an object at infinity. Said another way, the focal length is the distance required for a lens to bring parallel light rays to a focus. Since light rays coming from infinity are parallel, these two statements say the same thing. For a pinhole camera, the image distance and the focal length are the same. For a camera with a lens, this is only true if the object is very distant. For nearby objects, the image distance is greater than the focal length.

We can use the ray diagram in figure 11.4 to investigate the geometry of a lens in focus. The lens is represented by the vertical line in the center. The horizontal dashed line is the axis of the lens, an imaginary line everywhere perpendicular to the plane of the lens. The arrow on the right represents an object to be imaged, and the arrow on the left side is the focused image of that object.

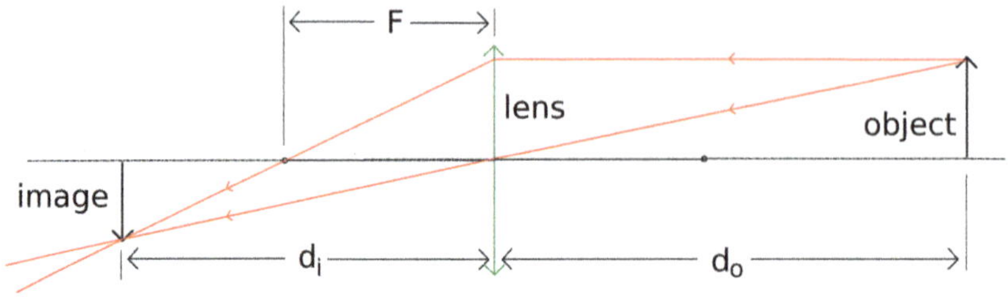

Figure 11.4. A ray diagram showing two principal rays for a converging lens focusing an image at left of an object at right. F is the focal length, d_i is the image distance and d_o is the object distance.

In this diagram we only consider light rays from the tip of the arrow. There are an infinite number of such rays scattering off in all directions, but we only need to consider two of them—called *principal rays*. The *first principal ray* passes through the exact center of the lens. Since the center of the lens is flat, this ray passes undeflected. The *second principal ray* passes parallel to the axis of the lens. Since all light rays arriving at the lens parallel to the axis are—by the definition of F—focused at a distance from the lens equal to the focal length, then this ray must also cross the axis at that point.

If we then project these two principal rays farther, we see that they cross each other. If the lens is shaped correctly, any other ray emitted from the tip of the arrow and intercepted by the lens would also cross at that same point. Thus, the two principal rays show the location of the image of the tip of the arrow.

It is easy to confirm that if we do the same thing for a point halfway down the arrow, the rays converge halfway between the lens axis and the image of the arrow's tip. Thus an image of the arrow is reconstructed as shown in the diagram. And so we have a lens, an object and an image of that object. We call the distance between the object and the lens the *object distance* (d_o), and we call the distance between the lens and the image the *image distance* (d_i). The other distance that is relevant is the focal length (F). This is a property of the lens itself; ultimately it depends on the shape of the lens and the type of glass from which it is made.

The two principal rays and the three distances, d_o, d_i and F, form a set of triangles. It only takes only a bit of geometry and trigonometry to see that there must be some mathematical relationship between these three quantities. I will not work through the details, but I have supplied the answer.

$$\frac{1}{F} = \frac{1}{d_o} + \frac{1}{d_i} \tag{11.1}$$

This is called the *thin lens equation*, and it says that the reciprocals of the image and object distances add to give the reciprocal of the focal length. As we will discuss later, real lenses are almost always more complicated than this. Even for those cases, the thin lens equation is still a good starting point, as it is often approximately true even then.

So let us consider some implications of this equation. To help with this, it is useful to solve equation (11.1) for each of the three quantities separately.

$$F = 1\bigg/\left(\frac{1}{d_o} + \frac{1}{d_i}\right) = \frac{d_o d_i}{d_o + d_i} \tag{11.2}$$

$$d_i = 1\bigg/\left(\frac{1}{F} - \frac{1}{d_o}\right) = \frac{F d_o}{d_o - F} \tag{11.3}$$

$$d_o = 1\bigg/\left(\frac{1}{F} - \frac{1}{d_i}\right) = \frac{F d_i}{d_i - F} \tag{11.4}$$

These three relations give five implications.

1. For a pinhole camera, the focal length is simply the image distance. But for a camera with a lens, the focal length is a property of the lens, whatever the image distance happens to be.
2. Equation (11.2) shows that one can determine the focal length of a lens if one knows even one combination of image distance and object distance. Measure the distance between lens and subject. Then find the best focus (by, for example, moving a screen back and forth near the lens) and measure the image distance. Finally, use equation (11.2) to calculate the focal length.
3. For an object at infinity—or so far away that it might as well be at infinity—the image distance is equal to the focal length. Thus another way to determine the focal length of a lens is to simply measure the image distance for an object that is very, very far away (the Moon, for example).
4. Pick a focal length and then plug different values for the object distance into equation (11.3), and it soon becomes clear that if the object distance is smaller then the image distance is larger. Move the object closer to the lens, and the focus moves farther from the lens.
5. Equation (11.3) shows that if the object distance is less than the focal length of the lens, a negative image distance results. This means that there is no image at all of the kind we have been talking about. The light rays diverge so much that the lens is unable to even bring them parallel let alone bring them together to a focus. Thus there is no *real image*—an image that can be focused onto a screen, piece of film or digital light detector. A simple lens cannot focus an image of a subject that is closer to the lens than its own focal length.

11.3 Spherical lenses: the lensmaker's equation

The simplest type of lens is made from the intersection of two spherical surfaces. A spherical surface is by far the easiest to manufacture; it is the only curved surface that is the same everywhere. Since every other type of surface is far more difficult to grind to precision, nonspherical surfaces are lumped together under the category *aspherical*.

We can describe a lens with two spheres that—for the converging lenses we have so far described—overlap each other. We can also make a different kind of lens using the space between two spheres that are next to each other but do not overlap. This produces a *diverging lens* instead, as described in section 11.5.

The *lensmaker's equation* allows us to calculate the focal length of any lens that has a spherical surface on each side. We will not explore all of its details here; excellent descriptions are widely available (here for example.) Instead we apply it to the simple case of a symmetric, biconvex-converging lens with each side having the same curvature. For our purposes in this book, we can also use the *thin lens approximation*: we assume the lens is much thinner than the radius of either of the two overlapping spheres.

Note that a larger radius of curvature means a flatter lens surface. Given the radius (R) of both sides of the lens[1], and an index of refraction (n) for the glass, the focal length (f) is given.

$$\frac{1}{f} = \left(n - 1\right)\left(\frac{2}{R}\right) \tag{11.5}$$

Since both the focal length and the radius of curvature are in the denominator, a larger radius of curvature results in a longer focal length. The lens converges light less-strongly if its surfaces are more-nearly flat. Equation (11.5) also implies that a larger index of refraction for the lens material leads to a shorter focal length.

11.3.1 Focus of a lens in water

One physical difference between the eye and a camera is that the interior of the eye, through which the light rays pass to form an image onto the retina, is filled with liquid rather than air. As this is reminiscent of R W Wood's fisheye pinhole camera from chapter 10, let us consider the simple case of the focus of a lens when it is placed in water instead of air. How does this change the focal length of the lens?

For our simple version of the lensmaker's equation, we assumed the lens is thin compared to R. But we have made another assumption as well—that the lens, object and image plane are all surrounded by vacuum or air. What if the lens were immersed in a liquid instead? If we have a lens of index of refraction (n_{lens}) immersed in a liquid of index of refraction (n_{liquid}) equation (11.5) changes (Jenkins and White 1950, p 50).

$$\frac{1}{f_{\text{liquid}}} = \left(\frac{n_{\text{lens}}}{n_{\text{liquid}}} - 1\right)\left(\frac{2}{R}\right) \tag{11.6}$$

Note that if we replace n_{liquid} with the index of refraction of air ($n = 1.00$), then we are back to equation (11.5).

Let us now compare these two focal lengths to each other by dividing equation (11.5) by equation (11.6).

[1] The full version of the lensmaker's equation makes a distinction between the radii of each side with a complex sign convention for the two radii, and it applies to both converging and diverging lenses.

$$\frac{f_{\text{liquid}}}{f_{\text{air}}} = \frac{n_{\text{lens}} - 1}{\left(\frac{n_{\text{lens}}}{n_{\text{liquid}}} - 1\right)} \tag{11.7}$$

$$f_{\text{liquid}} = f_{\text{air}}\left[\frac{n_{\text{liquid}}(n_{\text{lens}} - 1)}{n_{\text{lens}} - n_{\text{liquid}}}\right] \tag{11.8}$$

If we consider, for example, a lens made of glass with $n_{\text{lens}} = 1.55$, comparing its focal length in air with its focal length in water ($n_{\text{liquid}} = 1.33$), equation (11.7) gives $f_{\text{liquid}} = 3.33 f_{\text{air}}$. Thus placing the lens in water instead of air greatly increases its focal length. The denominator of equation (11.7) implies that if the two indexes of refraction are the same, the focal length is infinite. This should make sense; no refraction at the lens surface occurs unless there is a difference in index of refraction.

11.4 Real images and focus screens

Figure 11.4 depicts the formation of a real image. Light coming from each part of the object actually converges on particular points in space. Can we view such an image directly with the eye, simply by placing oneself in the path of the light rays converged by the lens? Figure 11.5 shows what happens when one attempts to view such an image directly.

If the eye is located very close the rays (shown in blue), coming from a particular image point, diverge greatly before entering the pupil. The eye is unable to focus such widely-diverging rays to a point on the retina. If, on the other hand, the eye is located much farther from the image, the rays (shown in red) entering the pupil are more-nearly parallel and can be focused to a point on the retina. Thus it is possible

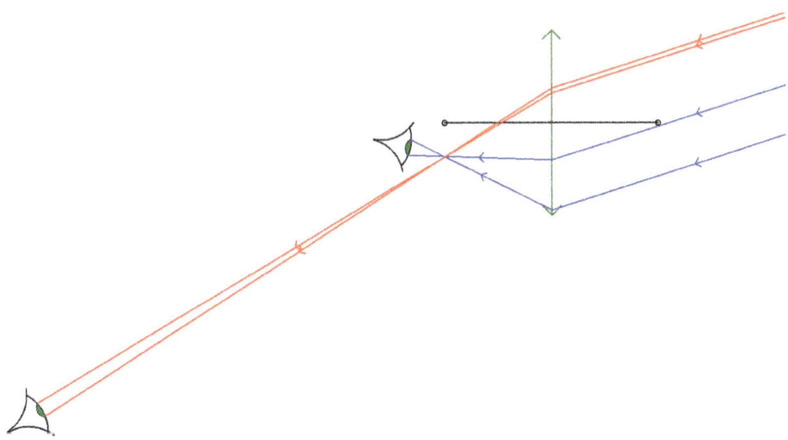

Figure 11.5. A converging lens (green vertical line) produces a real image from a distant object. If the eye is placed close to the focal point, the light rays (shown in blue) spread out at a large angle as they reach the eye's pupil, and the real image cannot be focused by the eye. Placing the eye far enough from the image allows for more-nearly parallel light rays (shown in red) to enter the eye.

to directly view the real image produced by a converging lens, so long as the eye is not located too close to the lens.

But this introduces its own problem. The lens presumably focuses much more than only one image point; the goal, after all, is to make an *image plane*, with different parts of the image corresponding to rays coming from different directions in the world.

The situation is diagrammed in figure 11.6. The lens (small green vertical line on the right) focuses bundles of light rays coming from many different directions to the image plane, represented by the vertical black line at center. A distantly-located eye only intercepts a tiny portion of these, and so most of the image plane is invisible.

We can remedy this situation by placing at the image plane a *ground-glass screen* (GGS). A flat piece of thin glass is finely roughened on one side. As light rays converge on one side of the glass, the light rays then scatter in many directions as they exit the other side of the glass, as shown in figure 11.7.

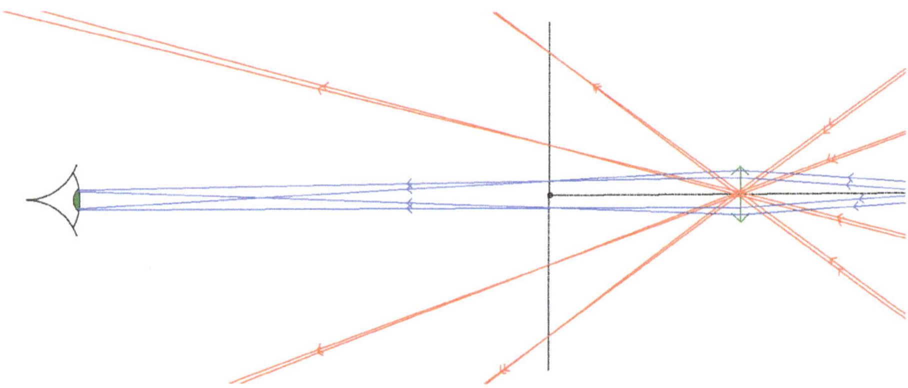

Figure 11.6. A small lens (green vertical line on right) produces a real image with the focal plane located at the vertical black line in the center. A narrow range of light rays (shown in blue), coming from the world and representing only a tiny part of the image, can be viewed by the eye on the left. But light rays focused to most parts of the image miss the eye altogether.

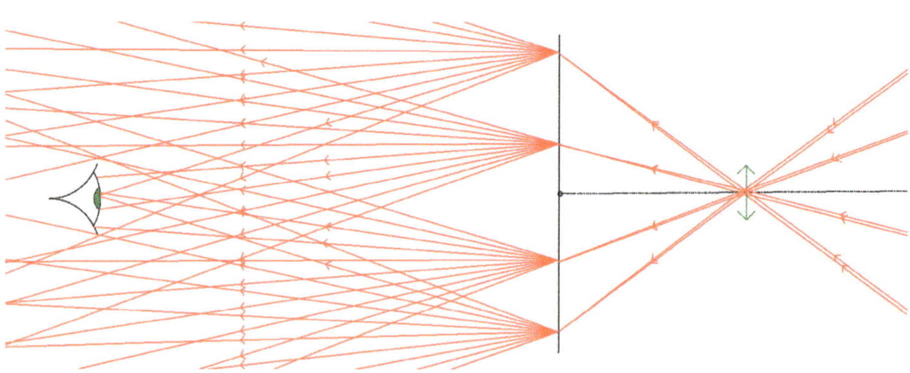

Figure 11.7. The situation depicted in figure 11.6 is repeated but with a GGS located at the image plane. The scattering of light by the GGS makes it possible for the eye to see all parts of the image.

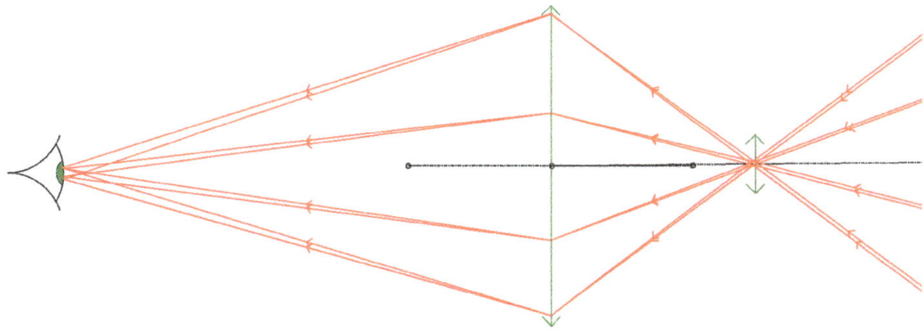

Figure 11.8. If a second, much larger short-focus lens is placed at the focal plane of figure 11.6, light rays from different parts of the image can be redirected to be viewed by the eye.

Now light rays from every point on the image plane make it to the eye, so the entire real image focused by the lens onto the screen may be viewed. We can accomplish something similar by locating a large, short-focus lens at the image plane instead of a GGS (see figure 11.8). Such an arrangement is sometimes called a *brilliant viewfinder*, and we will consider its use for photography in chapter 14.

If the image plane is physically large, then so too must be this second lens. If both a large image plane and a comfortable viewing distance is desired, a lens must be used that strongly converges light rays as compared to its dimensions. Such a lens must be thickly curved and may be impractical in some situations. The arrangements shown in figures 11.7 and 11.8 may be combined with both a GGS and converging lens located at the image plane. There are important situations in photography where this is an advantage.

For example, consider again the GGS shown in figure 11.7. Although light rays from every part of the image make it to the eye, most of the light rays focused by the lens miss the eye altogether because they are scattered off in other directions. The addition of the image-plane lens can redirect most of those scattered light rays towards the eye as in figure 11.8. This means the GGS need not scatter the rays so widely in order for them to reach the eye because the big lens is doing most of the redirecting. Thus more of the total light passing through the GGS makes it to the eye.

But of course, for the brilliant viewfinder arrangement shown in figure 11.8, all of the image rays reach the eye. So why bother with the GGS at all? One of the principle reasons to directly view a real image is to locate the exact physical location of the image plane—this is what we do when we focus a camera.

The arrangement of figure 11.8 works poorly for the task of accurately locating the image plane, as the formation of a sharp image onto the retina also depends upon the optics of the viewer's eye. With the arrangement of figure 11.7, on the other hand, the real image is either located at the GGS or it is not. If it is not, the image will appear blurry to the eye no matter how the GGS is viewed from the other side.

Thus, a GGS combined with a properly chosen large, short-focus converging lens gives us the best of both worlds. The problem is that for large image planes, the

Figure 11.9. Left: A combination GGS and Fresnel lens. The GGS side is shown on the left, attached to a camera. The Fresnel side of another example is on the right. A viewing hood to block out stray light while focusing is at the top. **Right:** An image focused by a lens onto the GGS/Fresnel combination. The Fresnel lens makes all parts of the image uniformly bright.

thickness of the lens needed to sufficiently redirect the light rays greatly complicates the situation.

A solution is to use instead a *Fresnel lens*—essentially a greatly curved convex lens that has been collapsed in concentric rings to a thin flat sheet. A Fresnel lens can be made inexpensively out of molded plastic, but it will not focus light rays precisely. But if the image is already defined by the GGS, the Fresnel lens can be acceptable for the much less critical task of redirecting light rays from different parts of the GGS to the eye. See figure 11.9 for a picture of a GGS with Fresnel lens and an image focused onto it.

11.5 Virtual images and diverging lenses

We have, so far, described what is called a real image; light rays spreading out from some object in the world are converged to a focus by a lens onto some surface. But there is another important type of image as well. Consider the ray diagram depicted in figure 11.4. Figure 11.10 is a modification of the same diagram, but I have moved the object so that it is closer to the converging lens than its own focal length. A very different kind of image results.

The rays coming from the object diverge too greatly to be brought together by the lens; they still spread out, although less so. These diverging rays coming from the lens seem to arise from a different place above and behind their actual origin. From the point of view of the eye, it is as if the rays come from an object that is larger and behind the real object. And so a magnified image results.

This is called a *virtual image* because no light rays actually pass through it. Such an image cannot, for example, be directly focused onto a screen. But so long as it is viewed from a comfortable distance, it can be seen directly with the eye. A reflection in an ordinary mirror also produces such a virtual image.

Our basic relation between image distance, object distance and focal length—equation (11.1)—still works for this case. But we must interpret the virtual image distance as negative instead of positive. Notice that the virtual image that results is

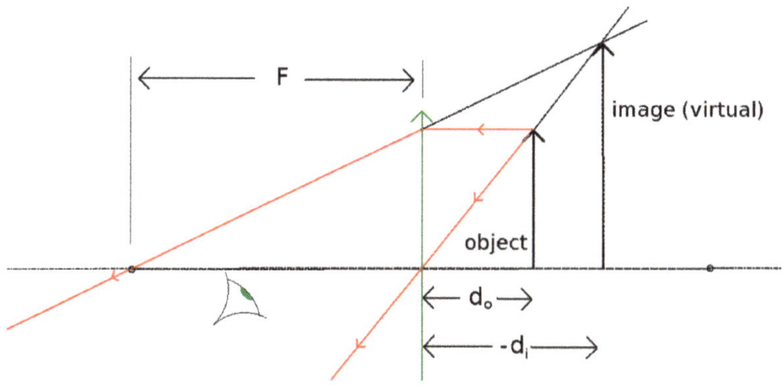

Figure 11.10. When an object is placed closer to a converging lens than its focal length the rays do not converge, and a virtual image results.

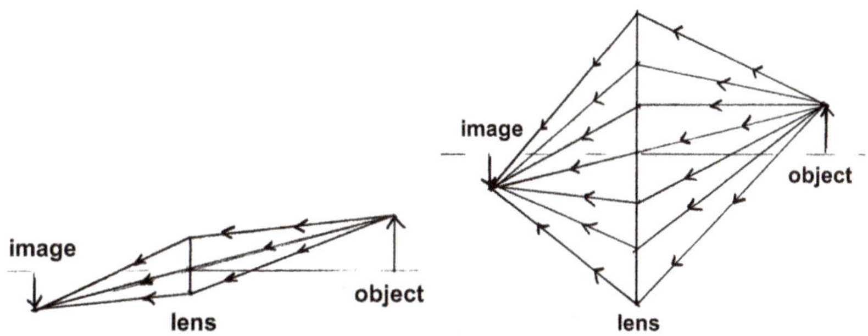

Figure 11.11. Left: Ray diagram for a small-diameter lens. **Right:** Ray diagram for a large-diameter lens. Since the rays converge to focus at a greater angle for the larger-diameter lens, the plane of focus must be located more precisely. This means one must focus the larger lens more carefully than the smaller lens.

on the same side of the lens as the object and that it is right-side up—in contrast to the upside-down real image from figure 11.4.

A virtual image can also be made by a diverging lens; concave surfaces are said to have a negative focal length. This produces a virtual image with a negative focal length no matter what the object distance.

11.6 Depth of focus and focal ratio

In practice, careful focusing is sometimes very critical and other times not all that important. Consider the difference between the two examples in figure 11.11. Both have the same object distance, focal length and image distance. But the lens on the right has a larger diameter (*D*). Clearly, if one moves the detector in the right-hand example only slightly, there is a large error in how well the rays come together. But for the left-hand example with a smaller diameter lens, the rays are less angled. And so the detector would have to be moved much farther to cause an equal error in focusing.

We can ask a different but related question. If the detector is located so as to form a perfectly focused image of a particular subject, by how much could I move that subject closer or farther away and still achieve a not-perfect but still-acceptable focus? This range of subject distances that still produces an acceptable focus is called the *depth of focus* or depth of field.

For the same reason that careful focus adjustment is more critical for the right-hand example in figure 11.11, the larger lens also has a smaller (more shallow) depth of focus. For a given focus setting, there is a smaller range of subject distances that are in acceptable focus for the large lens than for the small lens.

Since depth of focus is about the angle the rays make at the edge of the lens, it cannot be about lens diameter alone. For what if I made a lens with twice the diameter but also twice the focal length? Clearly, everything would scale together, and so the angles would be the same (so long as the object were also twice as far away). Thus depth of focus is primarily about not lens diameter (*D*) but rather how the lens diameter compares to the focal length (*F*). We have already defined a quantity that makes this comparison—the focal ratio: $f = F/D$. A smaller focal ratio means a larger-diameter lens as compared to its focal length. And so we have the following rule:

> A smaller focal ratio means a more shallow depth of focus. A larger focal ratio means a larger depth of focus.

For all lenses, these effects of depth of focus are more pronounced at small image distances simply because the rays coming from the subjects are less parallel for nearer subjects. So one must change the focus a lot when focusing on objects six inches and then 12 inches away. But for practically any lens one might ever use, the difference in focusing at ten miles compared to focusing at 20 miles is tiny indeed.

Most real camera lenses include a set of thin metal leaves that can slide against each other to make a hole of adjustable size in order to vary the *aperture*—the size of the hole where the light enters. And so one can vary the aperture, the effective size of the lens, even though the glass lens itself is fixed. This is called an *iris* or *aperture stop*, and the adjustments are labeled with the focal ratio that results. For this reason focal ratio is often referred to as '*f*-stop.' Sometimes, to emphasize that *f* is a ratio, it is instead written as '*f*/.'

For most lenses, the *f* settings come in discrete increments, called *steps*. Recall that there is another reason to adjust *f* apart from depth of focus: the focal ratio is one of the factors that determines exposure when a picture is taken. And so the focal ratio steps are chosen to be equal steps in exposure and not depth of focus. This topic is considered in detail in part IV, but I note a key result here: equal steps in exposure are related not to focal ratio directly but rather to its square.

This means that doubling the focal ratio changes the exposure by a factor of four rather than a factor of two. Since photographers choose exposure steps such that they result in successive doubling (or halving) of exposure, then our steps in *f* must be square roots of successive factors of two. Successive steps in focal ratio are, for example, $\sqrt{1}$, $\sqrt{2}$, $\sqrt{4}$, $\sqrt{8}$, $\sqrt{16}$, $\sqrt{32}$, $\sqrt{64}$, etc. Approximating these square roots, the *f* settings on a typical camera lens is some subset of 1.4, 2, 2.8, 4, 5.6, 8, 11, 16, 22, 32, 45, 64, 90.

11.6.1 Zone focusing

The *infinity setting* on a camera lens is set for the exact focus at—literally—infinity. But because any given lens has a certain depth of focus, it means that things even 'beyond infinity,' if there could be such a thing, would be in focus as well. The landscape photographer often wants a large depth of focus with not only the most distant objects but also subjects as close as possible to be in acceptable focus. We have already seen that a large focal ratio is the best choice for this situation. But one can do even better by putting the best focus somewhat closer than infinity such that infinity is still in acceptable focus. Now the range of acceptable focus has been moved even closer. Such a setting is called *hyperfocal infinity*, and many manual-focus camera lenses have marks to allow one to set them that way.

With an understanding of depth of focus, one can predict ahead of time what range of distances will be in acceptable focus for a particular combination of focus and f. This leads naturally to the idea that one can prefocus the camera lens for the situation. Whenever only a restricted range of distances are the center of interest, one can preset the lens focus to the middle of that range of distances. If the focal ratio is then picked so that the depth of focus extends for the full range of distances of interest, then all of our pictures will have acceptable focus. This process of carefully choosing a preset focus and focal ratio to give the required depth of focus is called *zone focusing*.

Zone focusing was common before the arrival of cameras with electronic autofocus, and many old cameras included markings to make zone focusing easier. But even now zone focusing is sometimes useful. The autofocus process is not instantaneous, and so there is an inevitable slight delay for the camera to focus. Zone focusing, when used appropriately, is faster than autofocus. Zone focusing may also be a good choice when the scene includes many subjects moving rapidly in different directions; an autofocus mechanism may be unable to quickly and correctly identify a proper focus subject.

11.7 Aberrations

A camera lens has a lot of work to do. It must bring light rays of many wavelengths, coming from many directions, all to a proper focus at the right part of the detector. In practice, no single lens can do all of this, and so a real camera 'lens' incorporates many different *lens elements* to correct each other's mistakes. See figure 11.12 for an example of a photograph taken with only a simple double-convex lens. In this section we consider the three most important *aberrations*—errors in bringing the image to a proper focus—that are inevitable with a simple lens.

11.7.1 Ray tracing

For the design of a real camera lens, the thin lens approximation is far too crude. For the simplest type of lens, with two surfaces that are simply sections of spheres rather than some more-complex shape, the lensmaker's equation allows one to find the focal length (F) given the curvature of each side and the dimensions of the lens.

Figure 11.12. This photograph was taken with a simple single-element magnifying glass attached to a 35 mm camera. It resulted in large amounts of distortion and aberrations.

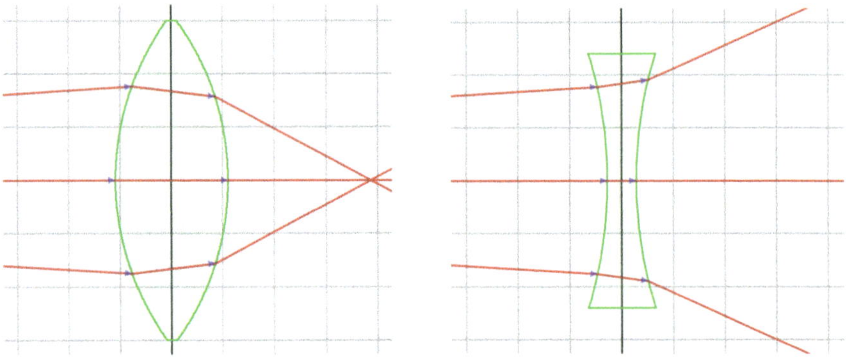

Figure 11.13. The thin lens approximation is too crude for precise optical design. Instead, individual rays are 'traced' from the source to the image with Snell's law applied at each interface.

But even this is too crude as anything but a starting point for the complex situations faced by lens designers

Instead, optical systems are designed with the aid of *ray tracing*. Rays coming from the object are followed geometrically until they reach an interface—a sudden change of material, such as from air to the first surface of a glass lens or from one type of glass to another. At that point of contact with the new surface, geometry is used to find the angle of incidence, and then Snell's law is used to calculate the angle of refraction. The ray is then followed geometrically until it reaches the next interface, and on and on—until the image point is finally reached. See figure 11.13 for an example of rays traced through two very simple but decidedly nonthin lenses. It should be obvious that the tedious calculations of ray tracing are much easier now than they were before the advent of the computer.

11.7.2 Spherical aberration

A spherical lens surface is one that is shaped like a slice off the edge of a sphere. This is the easiest type of surface to make, by a long shot, for the simple reason that all parts have the same curvature. Lenses that have nonspherical surfaces are called aspherical, and they are much more expensive to manufacture. The problem is that spherical lenses do not form perfect images because different parts of the lens have different focal lengths; it is not really the correct shape needed to bring all of the rays to a common focus.

Figure 11.14 shows parallel rays traced through two lenses. They look almost identical to the eye, but they are not. The left-hand lens is spherical, while the right-hand lens has a more-complex hyperboloidal curvature. *Spherical aberration* means that rays of light passing through the edges of the lens focus more closely than rays passing through the center. The (difficult to manufacture) hyperboloidal curvature eliminates spherical aberration in this particular example, but other aberrations arise when the light rays are not parallel or arrive at the lens at an angle to the lens axis.

11.7.3 Coma

Even if a simple lens has the correct aspherical shape to bring on-axis light rays to a proper focus, it will still misfocus light rays coming in from an angle to the lens axis. The result can be seen in figure 11.15. Severe *coma* spreads out light from point

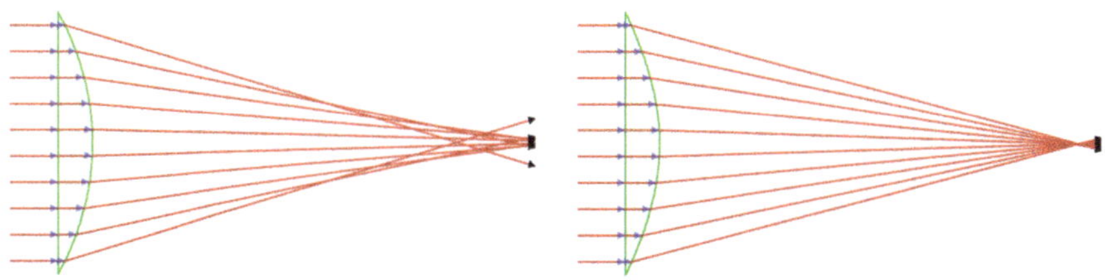

Figure 11.14. Left: Spherical lens, showing severe spherical aberration. **Right:** A lens with hyperboloidal curvature.

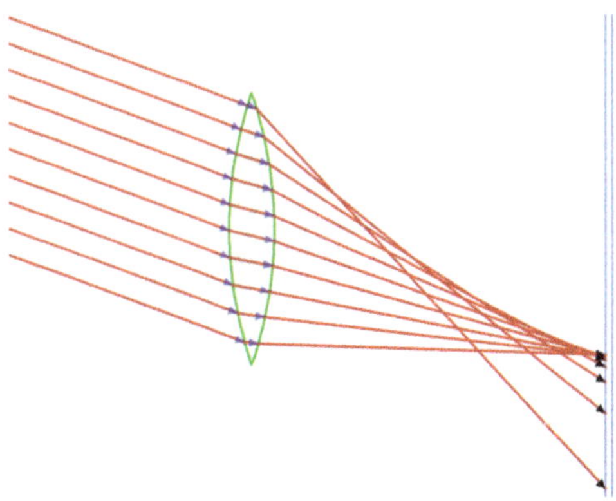

Figure 11.15. Parallel rays entering a simple lens at an angle focus in different places and making a point spread out into a comet-shaped blob.

sources into fan shapes that point away from the center of the image. The effects of coma are greater farther from the image center and with a wide-angle lens.

11.7.4 Chromatic aberration

The focal length of a thin lens is related to the index of refraction of the glass from which it is made. Since a larger index of refraction means the light rays coming from air bend at a larger angle, a larger index of refraction results in (all else being equal) a shorter focal length. But the index of refraction of glass is not a simple number; it depends on wavelength. This is the *dispersion* discussed in section 3.3.2.

The examples of dispersion curves in figure 3.11 show that shorter wavelengths have a larger index of refraction. Thus an ordinary lens has a smaller focal length for shorter wavelengths and a larger focal length for longer wavelengths. If light of many wavelengths must be focused simultaneously, a blurry image is thus inevitable.

See figure 11.16 for an example of rays traced through a simple lens, with dispersion taken into account. Notice that the short-wavelength violet rays focus more closely, while the long-wavelength red rays focus at a larger distance. This error in focus is known as *chromatic aberration.*

Even though it is about wavelength, and thus color, chromatic aberration is important for black-and-white photography as well as color photography; even black-and-white detectors are sensitive to many wavelengths at once.

One way to reduce chromatic aberration is to use a filter to restrict the light entering the camera to only a narrow range of wavelengths. See figure 11.17 for an example. But this is usually a somewhat impractical solution for two reasons. First, a more narrow range of wavelengths means less light enters the camera, and so a proportionally longer exposure is needed. Also, if only a narrow range of wavelengths is admitted, then color photography is impossible.

The best way to reduce chromatic aberration is to abandon the use of a single, simple lens. We combine a strong—but low dispersion—converging lens with a weaker—but higher dispersion—diverging lens. The two lenses are made of different types of glass so they have different amounts of dispersion. The trick is to choose the types of glass and the shapes of the lenses such that the combination still acts as a converging lens; the diverging lens is weaker than the converging lens regarding the

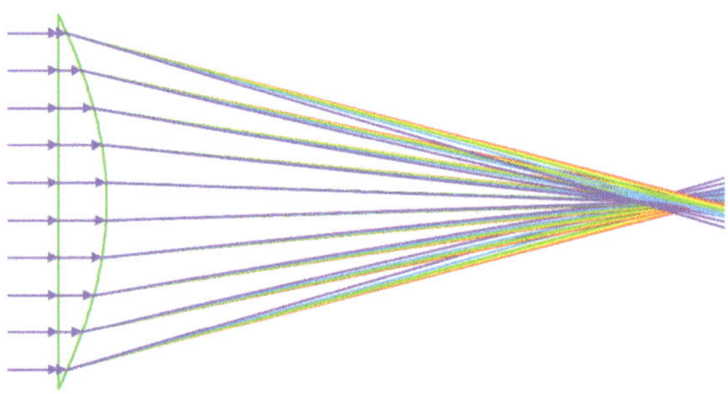

Figure 11.16. Even if spherical aberration is corrected, a simple lens will focus different wavelengths of light differently because of the dispersion of the glass.

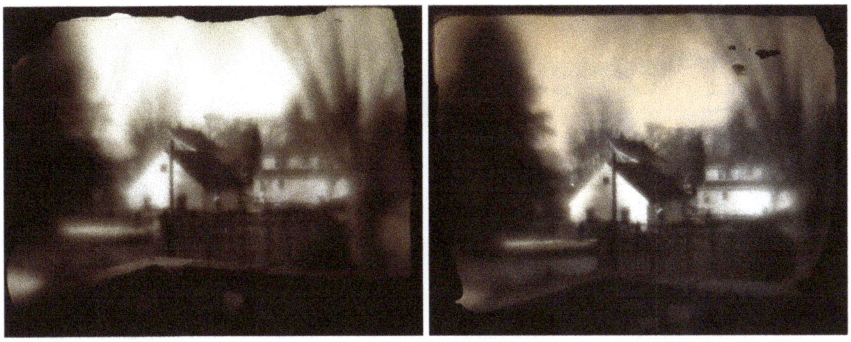

Figure 11.17. Both of these images were taken with the same simple double-convex lens, and they show pronounced chromatic aberration. But the detail is sharper in the example on the right because a filter was used to greatly restrict the range of wavelengths of light. The image is still blurry, especially at the edges, because spherical aberration and coma still remain.

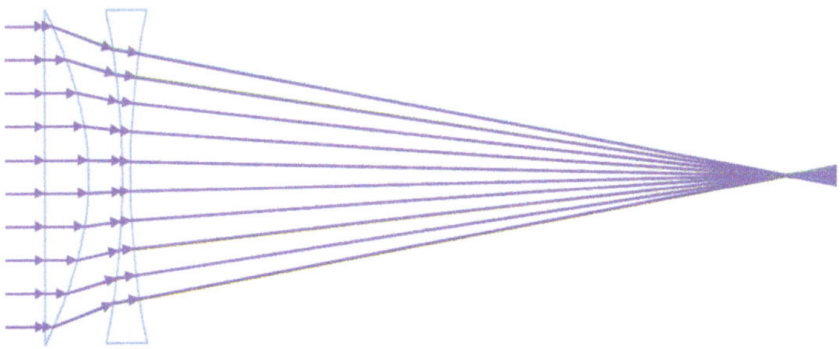

Figure 11.18. Chromatic aberration can be partially corrected if at least two lenses, of different dispersions and indexes of refraction, are strategically combined. In this example, the positive lens on the left is stronger than the negative lens on the right, so the overall effect is that of a positive converging lens. But the negative lens has a higher dispersion and cancels out the dispersion of the positive lens.

overall bending of light rays, but the dispersion of the diverging lens does cancel the dispersion of the converging lens.

See figure 11.18, where I have traced rays of many wavelengths through such a combination—called an *achromatic lens*. Chromatic aberration can be reduced even further by combining more than two lenses. A lens combination that effectively eliminates nearly all detectable traces of chromatic aberration over the range of visible wavelengths is often called an *apochromatic lens*.

11.7.5 Aperture and aberrations

One way to reduce aberrations in general is to use a lens with a very large focal ratio. Regarding chromatic aberration, for example, one can see from figure 11.19 that the rays entering the edges of the lens are refracted by a greater angle, but those rays also show greater dispersion. The rays entering through near the center of the lens, on the other hand, are bent very little and show much less dispersion.

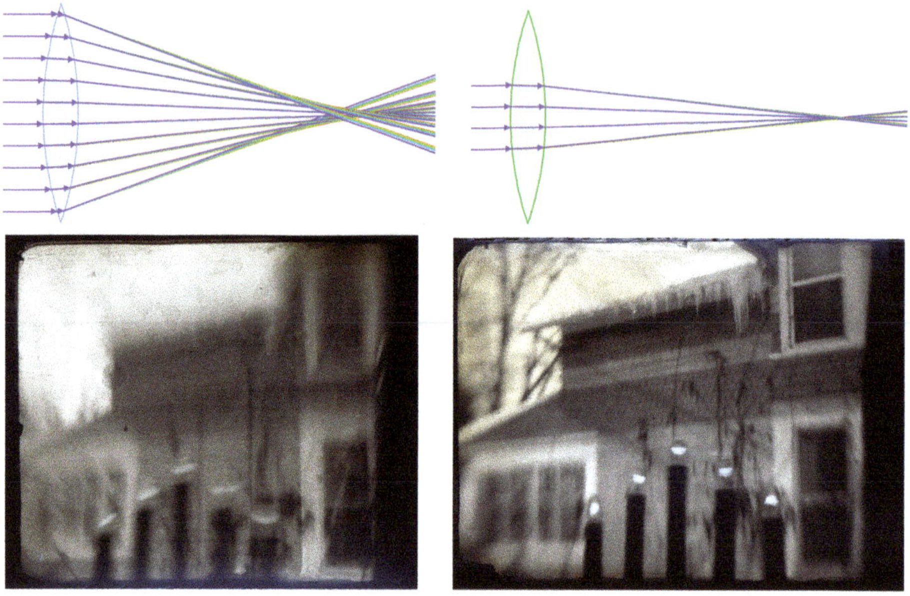

Figure 11.19. If a larger focal ratio (smaller aperture) is used, aberrations are greatly reduced. Both of these images were taken with the same simple double-convex lens, and they show pronounced coma and both chromatic and spherical aberration. But the detail is sharper in the example on the right because an aperture stop was used to allow only rays passing through near the center of the lens to enter the camera.

For any given lens, all aberrations are generally worse when more of the lens is used. And so a common way to get a sharp image with a not-so-great lens is to 'stop it down' to a very small aperture. This is especially true for lower-quality lenses that are not very well corrected for aberrations.

If we take this idea to the extreme, we are back to pinhole photography where the image is formed simply by geometry. Even then, however, the image would still be blurred by diffraction; no amount of lens-design cleverness can sidestep the fact that light is a wave. But also one often wants to use a large aperture because more light is needed, or a shallow depth of focus is desired.

11.8 Lens design

What photographers call a 'lens' is actually a combination of three to as many as a dozen or more individual lenses of different shapes and spacing and even different types of glass. Because of this we usually refer to the combination of many lenses that work as a unit and that one attaches to a camera as a 'lens,' while each individual glass lens inside it is called a lens element. So, a given camera lens may contain many lens elements.

There are really only two ways to design a camera lens with minimal aberrations and distortions—even at small focal ratios. One can use a combination of many, many lens elements, each correcting for the aberrations of the others. One can also use fewer lens elements, but with complex aspherical surfaces. Both choices are expensive. But using a large number of lens elements has the added disadvantage that every lens surface, both front and back, reflects a little bit of light. One problem with this is that the reflected light decreases the brightness of the image because, if it is reflected, it does not pass through. But even worse is the fact that about half of these reflected light rays do eventually make it through the lens and onto the image, but they end up completely in the wrong place.

Most of this unfocused light is spread diffusely over the image, and this adds an overall diffuse light to the image and thus reduces contrast. For color photography, not only is the contrast reduced, but the colors are also less brilliant. And if there is a very small and bright source of light (the Sun for example) within view of the lens (even if it is out of the picture), then a *flare* can be produced—an oddly-shaped streak of light on the image produced solely from these errant reflections.

Long ago, it was nearly impossible to make precision aspherical lens elements, so there was an inevitable trade-off between making a sharper image with fewer aberrations and distortion and using fewer lens elements. Thus one could buy a lens that had a large aperture and produced sharp, low-distortion images, but it would inevitably produce pictures of low contrast due to the large number of lens elements. A high-contrast lens, with only a few lens elements, would have noticeable aberrations at small focal ratios. Many of these old and simple high-contrast lens designs were completely unusable at focal ratios less than about $f/4$, which is very modest by modern standards.

Modern lenses are much improved for three principal reasons. First, it is no longer impossible to make precision aspherical lens elements. It is still expensive but not absurdly so. Secondly, a lens can employ anti-reflection coatings that use wave interference to cancel out most of the reflections. Thus, many more lens elements can be used to cancel aberrations without reducing the contrast to unacceptably low levels. Third, computers can be used to optimize the design of the lens over a wide range of conditions. Thousands of rays can be now be easily traced through a particular complex lens design, so it can be tested before it is even manufactured. A good modern lens is still expensive, but an expensive lens today makes a much more 'perfect' image than did its equivalent-dollar counterpart of 50 years ago.

See the left-hand image in figure 11.20 for an example of a modern camera lens disassembled to show the individual lens elements. Such a lens also has a complex

Figure 11.20. Left: A modern camera lens contains many lens elements and precision moving parts. **Right:** *St. Mary's church*, John Beaver 2003. This picture was taken with only a simple double-convex lens.

mechanical construction; the spacings between many of the lens elements must change in a precise way as the lens is focused to different distances. This particular example also had an adjustable focal length, so both wide-angle and telephoto pictures could be taken with the same lens (called a *zoom lens*). The zoom feature greatly adds to both the optical and mechanical complexity.

And so, are old camera lenses useless today? Not to me. A perfect image does not necessarily make a perfect picture. The right-hand image in figure 11.20 was taken with an intentionally-awful lens—just a single-element double-convex lens. For many years now, the Soho Photo Gallery in Tribeca, NY has hosted an annual Krappy Kamera Competition. The only rule is that the picture must be taken with an awful lens. In 2005 I won second place with this picture, so I guess I can be almost as Krappy as the best.

11.9 Telescopes and viewfinders

We have so far discussed systems of lens elements that are meant to work together as a unit, to converge or diverge light rays toward or away from a particular focus to make a real or virtual image, while correcting for aberrations. Here we consider *compound lens systems*—combinations of lenses that produce more than one image. In particular we discuss some related cases that are of particular interest to both photography and astronomy.

A visual telescope is a system of (at least) two lenses arranged so as to provide an enhanced direct view of relatively-distant objects. Such an instrument does three things.

1. It gathers more light than the unaided eye and thus can allow one to see things that are otherwise too faint.
2. It allows for the gathering of finer detail than what can be seen with the unaided eye.
3. It gives a magnified view.

For astronomers it is the first two of these functions that are the most important, but here we will concentrate on the third.

The most basic visual telescopes employ two lenses (each usually comprised of two or more lens elements, to correct for aberrations). The *objective lens* faces the object to be viewed, while the *eyepiece* or *ocular* faces the viewer's eye. We will consider the two most important ways to arrange these two lenses—both popularized in the early 1600s. The *Galilean telescope* was perfected (although not invented) by Galileo, while the alternative Keplerian telescope was described by his contemporary—Johannes Kepler.

11.9.1 The Galilean telescope

When Galileo pointed his telescopes at the heavens in the early 1600s, he immediately made important discoveries. Although he did not invent the telescope, he apparently reconstructed it from simple descriptions, and his telescopes were the best of his day.

The particular arrangement of lenses used by Galileo consisted of a converging lens facing the object, which gathers and focuses a larger bundle of light rays than the eye. But since the eye does not focus on converging light rays, a negative diverging lens was used to bring the rays parallel again. See figure 11.21 for a diagram.

The vertical green line on the right represents the 'objective' converging lens, while the red line on the left represents the 'ocular' or 'eyepiece' diverging lens. Two blue rays coming from a distant object to the right are traced through the lenses. When the two lenses are spaced at the proper distance, parallel light rays emerge, which can then be readily focused by the eye.

To make this happen for incoming parallel light rays (from a very-distant object), the lenses must be spaced a distance equal to the sum of their focal lengths. But recall that the diverging eyepiece lens has a negative focal length, so the magnitude of its focal length subtracts from the focal length of the positive converging lens. Thus, the eyepiece is located before the converging lens brings the parallel rays to a focus.

Notice that the rays emerge from the eyepiece at a greater angle than they entered the objective lens. This is the heart of the *magnification* of the lens system. The angular size of the distant object (the top of a distant tree, for example) appears larger through the telescope. But also notice that the light rays are above the optical axis both with and without the telescope, and so an erect image results.

11.9.1.1 The reverse Galilean viewfinder

By the principle of the reversibility of light rays, we can switch eyepiece and objective lens. Look through the converging lens and point the diverging lens at the object. Clearly, a magnification smaller than unity would result; light rays entering at a large angle are compressed into a smaller angle for the eye.

This optical arrangement is useful as a *viewfinder*—a (usually small) optical system to allow one to see a prearranged wide field of view. If we attach such a thing to the outside of a camera, we can use it to help us to accurately aim the camera at the subject while the camera's lens focuses its image onto the light detector. The particular example shown here is known, for obvious reasons, as a *reverse Galilean viewfinder*. We will consider camera viewfinders in more detail in chapter 14.

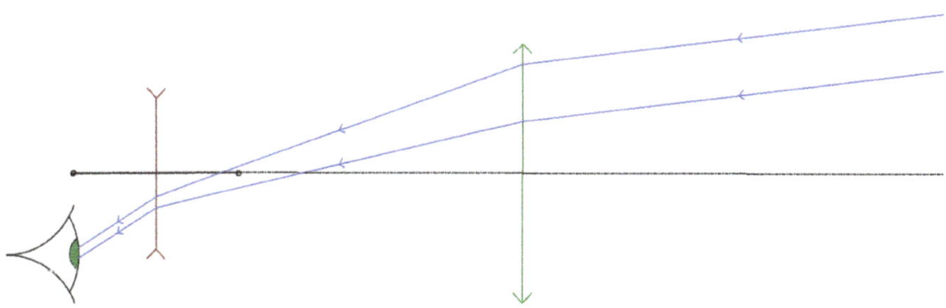

Figure 11.21. In this diagram of a Galilean telescope, parallel light rays from part of a distant object enter a converging lens (vertical red line) from the right and then pass through a diverging lens (vertical green line). If the lenses are separated at the correct distance, the light rays emerge parallel and can thus be focused by the eye.

11.9.2 The Keplerian telescope

The Galilean telescope has two primary flaws: it has a rather limited field of view (like looking through a straw) and only relatively small magnifications are practical. At the same time Galileo was making hay with his telescopes, discovering the satellites of Jupiter and craters on the Moon, his contemporary—Johannes Kepler—proposed a different arrangement of lenses. Instead of a diverging lens to make the image magnified and viewable to the eye, Kepler proposed using a converging lens (see figure 11.22).

As is the case for the Galilean telescope, when a Keplerian telescope is pointed at a very-distant object, the spacing between the two lenses must be roughly equal to the sum of their focal lengths. Since both lenses have positive focal lengths, the eyepiece lens must be placed after the focus of the objective lens rather than before it. An inverted image results; notice that, while the rays arriving at the objective lens point (in this example) downward, the rays emerging from the eyepiece point upward.

My formal academic training is in astronomy, and I say both truthfully and from bitter experience that an astronomer rarely knows up from down or left from right. The Moon is round, and an upside-down circle is a circle. Thus the inverted image of Kepler's lens arrangement has no disadvantage as an astronomical telescope and many advantages; nearly all modern visual telescopes are Keplerian[2].

Kepler's design allows for a much larger apparent field of view than the Galilean telescope, while also allowing for much higher magnification. The magnification of a Keplerian telescope is easily modified, simply by using a shorter focal length eyepiece lens. The magnification (*m*) is related to the focal lengths (*F*) of the two lenses by the simple formula shown in equation (11.9).

$$m = \frac{F_{\text{objective}}}{F_{\text{eyepiece}}} \tag{11.9}$$

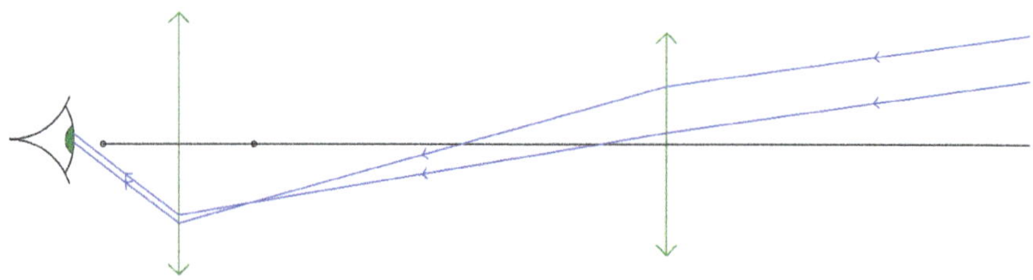

Figure 11.22. As with a Galilean telescope, the Keplerian telescope passes parallel light rays from a distant object through a converging lens. But the light rays are allowed to converge to a real image and then cross over before they pass through a second converging lens of shorter focal length.

---

[2] It is true that most modern telescopes use mirrors instead of lenses, but the principle of the Keplerian telescope translates in a straightforward way to mirror-based telescopes.

The objective lens of the Keplerian telescope produces a real image before the eyepiece. One could then dispense with the eyepiece altogether; instead of looking through the telescope, one could simply place a light detector at the location of the real image. This is indeed how modern research telescopes are used. In that sense, a modern research telescope used for imagery is simply a camera with a giant lens.

11.9.2.1 The Keplerian viewfinder

We can make a camera viewfinder modeled after the Keplerian telescope. Usually we want a viewfinder to show a wider view rather than a magnified one. But equation (11.9) provides the solution: simply choose a shorter focal length objective lens and a longer focal length eyepiece lens, and a magnification less than one results.

The inverted image, however, is a serious disadvantage for a camera viewfinder. It is difficult to point a camera if one's viewfinder shows everything upside down. An easy solution is to incorporate a mirror into the design—as in figure 11.23.

And so one looks downward into the viewfinder in order to see straight ahead. Notice that, if the viewer is leaning down from the left to look into the eyepiece, the light rays now enter the eye from above the axis rather than below. The mirror inverts the image so that it is now right-side up, but it is still reversed left-to-right. This takes a little getting used to, but it works.

Since a real image is produced in a Keplerian viewfinder, a GGS can be placed at the focus; the eyepiece is then a magnifier with which to better view the image focused onto the screen. This arrangement is particularly useful for some

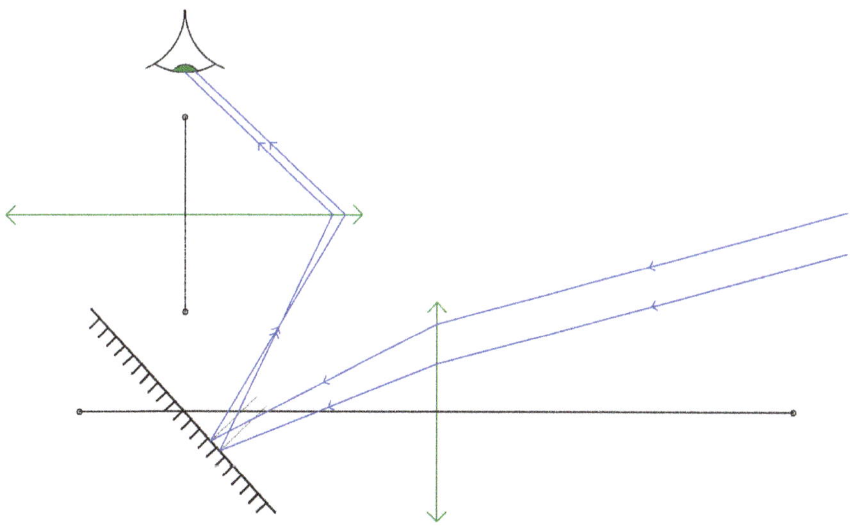

Figure 11.23. When a right-angle mirror is introduced into a Keplerian telescope, the image appears upright if viewed from the proper vantage point. But the mirror leaves the image reversed left-to-right.

kinds of cameras, as the viewfinder can also be used to help focus the camera. We further explore these and other camera viewfinders and focus mechanisms in chapter 14.

Reference

Jenkins F A and White H E 1950 *Fundamentals of Optics* (New York: McGraw-Hill)

IOP Publishing

Photography (Second Edition)

Physics and art in focus

John Beaver

Chapter 12

Symmetry

12.1 Transformations and invariance

Everyone knows that the image in a mirror is reversed left-to-right; it seems too obvious to mention that the image in a mirror would be just what we call a mirror image. With this obvious fact comes an obvious question. Why does a mirror show an image that is reversed left-to-right but not top-to-bottom? A mirror is simply a flat, reflective plane—how does it know the difference between left-and-right on the one hand and up-and-down on the other? The answer is surprising to many: a mirror reverses neither left–right nor top–bottom; we are the ones who do the reversing.

To make the picture on the left side in figure 12.1, I used a permanent marker to write the letters L, R, T and B (for left, right, top and bottom) on a sheet of two-tone origami paper. When I hold the paper in front of the mirror, the image of the letters has the same orientation as my direct view of them. The object on the left (the letter L) is on the left, the object on the right is on the right and the same is true for top and bottom. If an object is to the left of center, its image is also to the left of center. And the analogous rule holds true for objects that are above, below or to the right of center. So why then do the words and letters appear reversed when one views them in a mirror?

The trick is obvious—the mirror images of the letters in the left-hand image in figure 12.1 are of the marker bleed-through, as if the letters were objects floating in space rather than simply written onto a surface. What one usually means by the phrase 'image in the mirror' is not what is represented by the left-hand image in figure 12.1 but rather that of the image on the right—where I have rotated the paper 180° to face the mirror.

The point is that when we look into a mirror, we want to see the image as if we were looking not from our vantage point in front of the mirror but rather from the vantage point of the virtual image behind the mirror. So we imagine ourselves (or the sheet of paper with writing on it) undergoing a *transformation.* When I stand in

doi:10.1088/978-0-7503-3703-8ch12

© IOP Publishing Ltd 2022

Figure 12.1. Left: The images of an object reflected in a mirror appear in the same left, right, top or bottom orientation as they do in real life. **Right:** By 'mirror image,' we usually mean the image in the mirror once we have rotated it (about a vertical axis) to face the mirror. It is our act of rotation, not the mirror, that reverses left and right.

Figure 12.2. We can instead rotate the object about a horizontal axis to face the mirror. In this case, left and right are not reversed while top and bottom are.

front of the mirror, objects behind me—but to my left—also appear to the left in the mirror. But for me to see those objects I would have to turn around, and then they would be to my right.

So back to our original question: why does this work for left–right but not for top–bottom? The answer is that it in fact does work just as well for a top–bottom reversal. It is all about how I transform the paper in order to position it according to the mirror's perspective rather than my own. In figure 12.1, I have rotated the paper 180° around a vertical axis passing through the center of the paper. But I can also rotate it 180° around a horizontal axis, as seen in figure 12.2. In this case the letters instead read correctly left-to-right, but they are reversed top-to-bottom. Note that the vertical stem of the L is still on the left as it should be, but the horizontal bar is now on the top rather than the bottom.

And so text usually appears reversed left-to-right (but not top-to-bottom) when seen in a mirror simply because we most often rotate the text horizontally, about a vertical axis, so that it faces the mirror.

Figure 12.3. Top: The sheet of paper undergoes a rotation about a horizontal axis through the paper. The word OXIDE is invariant under this transformation, but the word WOW is not. **Bottom:** The sheet of paper undergoes a rotation about a vertical axis through the paper. The word WOW is invariant under this transformation, but the word OXIDE is not.

This concept of a transformation—a precisely-defined change—is one half of the important concept of *symmetry*. Figure 12.3 shows the images in a mirror of the words 'WOW OXIDE' both unrotated (left side) and rotated about a horizontal axis (top right) and a vertical axis (bottom right). For the rotation about the horizontal axis, the word 'WOW' has reversed top-to-bottom, while the word 'OXIDE' seems to have not. But for the rotation about the vertical axis, the word OXIDE has reversed left-to-right, while the word WOW has not.

The reason of course is that, for each kind of rotation, one of the two words has a particular kind of symmetry, while the other does not. The concept of symmetry is highly intuitive, but it is important for us nonetheless to define it precisely. An important mathematical way to define symmetry is as follows: a symmetry is an invariance under a transformation.

A transformation is a precisely-defined change, while an invariance is a precisely-defined quantity that remains the same. A symmetry is when there is something that stays the same while some other thing is changed in a particular way. So, for our example of the word 'OXIDE,' the two-dimensional (2D) shape of the word remains the same when it is rotated 180° around a horizontal axis passing through its center. Thus, 'the 2D shape of the word' is the invariance, while 'rotate the word 180° around a horizontal axis passing through its center' is the transformation. The word 'WOW,' of course, has a similar kind of symmetry. But the appropriate transformation is instead 'rotate the word 180° around a vertical axis passing through its center.'

12.2 Symmetry in physics

If all physical laws can be distilled down to a few simple rules, what are they? When I say 'simple,' I mean simple when expressed in the appropriate mathematical language—a language that may be as yet unknown and almost certainly difficult to learn. The search for the answer to this basic question has been one of the primary concerns of theoretical physics over the past 100 years.

And the notion of symmetry—invariance under a transformation—plays a crucial role. It has been known for centuries that there are conserved physical quantities in nature–calculable quantities that stay the same as other measurable quantities change in complex ways. Two important examples are *momentum* and *energy*. The conservation of momentum, as a principle, was at least partially understood even before Isaac Newton; the conservation of energy was established as a principle by the mid-1800s and hinted at much earlier.

The law of conservation of energy and the law of conservation of momentum are two examples of eight known conservation laws. Each conservation law specifies a conserved quantity that can be calculated exactly. As other measured quantities (velocity, position, acceleration, etc) change with time in complex ways, the conserved quantity stays the same. These conservation laws are some of the most powerful tools available to a physicist when she applies physical principles to the real world.

If one thinks of a conservation law as a statement that a particular quantity stays the same as everything else changes, it is not too difficult to see that this idea seems related in some way to the concept of symmetry. Symmetry is, after all, an invariance (something that stays the same) under a particular transformation (something that changes).

The two concepts do seem similar, but they are clearly not the same. For one thing, symmetry involves a particular transformation (change), while a conservation law involves a quantity that stays the same even as all other quantities change.

But despite this difference, symmetries and conservation laws are intimately related, a fact first clearly established by the German mathematician Emmy Noether in 1915 (published in 1918). She proved what is now known as *Noether's First Theorem*, which states that for every conserved quantity, there is an associated symmetry (figure 12.4).

This means, for example, that if the quantity energy is conserved for all physical laws, then there must be some symmetry that all of those physical laws share. Furthermore, Noether showed how to determine precisely what symmetry is associated with each conservation law. For the example of the conservation of energy, the symmetry is this: the very laws of physics themselves are invariant (remain the same) under the transformation of a translation in time.

Another way to say this is that as time passes, the laws of physics themselves do not change. The inevitable result of this symmetry is that a particular quantity (the total energy of a system) is conserved. The reason, then, that all of our physical laws seem to conserve energy is that they all have this basic symmetry.

Figure 12.4. Emmy Noether (1882–1935) first established the mathematical connection between conservation laws and symmetry. (Graphic used from Public Domain.)

Two other important cases are momentum and angular momentum: both are conserved quantities, and each is associated with a different symmetry. The conservation of momentum arises because—so it appears—the laws of physics do not change from one place to another. Angular momentum, on the other hand, is conserved because the laws of physics do not depend on what direction one is pointed in space.

What if, for example, one proposed a physical law that did change in some way as time passed? That law would then violate the conservation of energy. And so the fact that, in practice, the conservation of energy works—it seems to hold up to careful measurement in all circumstances—is evidence that nature itself does have the particular symmetry of invariance under a translation in time.

Whatever the basic rules at work that nature uses, it seems that they do not change as time passes, from one place to another or when pointing in different directions. If they did, then we would not observe energy, momentum and angular momentum, to be conserved quantities.

When we say that physical laws have symmetries, such as invariance under translations in space and time, we are making claims about nature. But how do we test whether or not these claims are true? Noether's first theorem shows us the way. The conserved quantities associated with these symmetries are measurable, so we can measure them in experiments that test whether or not they are conserved. As we build more and more experimental evidence for a particular conservation law, Noether's first theorem demonstrates that we also provide evidence for that particular symmetry of nature.

12.2.1 Symmetry and mirrors, again

I have described two particular types of symmetry that seem to be related to mirrors: rotate a flat object placed parallel to a mirror half way about either a horizontal or vertical axis, and it looks the same as before it was rotated. This means that the reflection in a mirror of the rotated surface also looks the same. But we have also

Figure 12.5. The image in a mirror is a transformation of nearer and farther. If an object is 1 m in front of a mirror, its image is 1 m behind the mirror.

seen that this kind of symmetry really has nothing to do with mirrors, per se. The transformation is our act of rotation such that the object that was facing us now faces the mirror. And so '*mirror symmetry*' is a misleading term for invariance under such a transformation. But clearly, a mirror does perform some kind of transformation. So, what is it? The answer is obvious only if we consider three-dimensional (3D) objects.

We have seen that up, down, left and right are preserved when looking in a mirror. The reflection in the mirror of an object to the left of us also appears to the left, and the same is true for up and down. But what about the other dimension—distance? This is where the reflection in a mirror makes a real transformation. Consider figure 12.5; objects closest to the camera are the most distant in the reflection. This is a real transformation of nearer and farther; notice that the nearest example to the camera of *Castor canadensis* is in focus but its reflection is not. That is because I set the camera focus on the nearby rodent, while its reflection is the most distant part of the picture.

By a mirror transformation we really mean this: a mirror transformation is a reflection about a plane rather than a half rotation around a line. Thus mirror symmetry is invariance under this definition of a mirror transformation.

This gives us another way of thinking about the symmetry of the word 'WOW.' Instead of transforming the word by rotating it about a vertical line through its center, we can accomplish the same end result by putting a vertical plane through the center of the 'O,' and reflecting it about that plane. So WOW does have mirror symmetry, but it is not the mirror of figure 12.3.

We can easily extend this kind of mirror symmetry to 3D objects. Place an imaginary vertical plane through the center of a 3D object. If what is on the right side of the plane is exactly the same as the reflection in the plane of the left side, then we say the object has *bilateral symmetry*.

Mirror transformations are even more important when we consider motion. Imagine that I stand in front of a mirror and a bee flies from my left shoulder to the mirror, and then returns to land on my right shoulder. The bee started from the left side, moving both toward the right and also further away. When it reached the mirror, it then—while still moving toward the right—moved closer and closer to land on my right shoulder. But what about the bee's reflection in the mirror? The reflection of the bee also moved from left to right. But it first moved closer and closer to me and then—after reaching the mirror—moved farther and farther away.

We can consider an even more-dramatic example of a mirror transformation. Let us imagine that the bee instead flies in a horizontal circle in front of the mirror and that, as seen from above, it travels clockwise around its circular path. Clearly, the bee's reflection in the mirror also travels in a circular path. But as seen from the same vantage point, the bee's reflection in the mirror moves *counterclockwise* around its circular path. And so a mirror transformation turns clockwise motion into counterclockwise motion.

12.2.2 Mirror symmetry and *P* invariance

The relation between the motion of a particle and the motion of its reflection in a mirror is called by physicists a *P transformation*, and it points to another symmetry and associated conservation law. Symmetry under a *P* transformation is called *P invariance*, and it leads to what physicists call *the conservation of parity* (Sachs 1987, section 2.3). As is the case with the symmetries associated with the conservation of energy, momentum and angular momentum, *P* invariance applies not to the motions of particles themselves but rather to the physical laws that describe those motions.

Let us now reconsider the bee flying in a clockwise motion in front of the mirror. Let us pretend to use Newton's laws of motion to analyze all of the forces acting on the bee to make it fly in its circular path. To say that Newton's laws obey the conservation of parity means that we can apply those same laws to the bee's reflection, and the reflected motion of the bee will be accurately predicted.

To do this, however, we must also apply the *P* transformation to the forces acting on the bee and to its initial motion. When we do this, Newton's laws do, in fact, predict the reflected motion of the bee. This particular example is simple, for if the bee is traveling in a perfect circle at a constant speed, the forces acting on the bee must add up to a force of constant magnitude that always points toward the center of the circular motion. Clearly, if the force acting on the bee is toward the center of the circular path so too is the reflection of that force in the mirror; the reflection of the force on the bee points toward the center of the reflection of the bee's circular path.

The conservation of parity means that, regarding fundamental physical law, there is nothing special about clockwise versus counterclockwise. It is all simply one's point of view. This seems rather obvious; whether a clock runs clockwise or counterclockwise, for example, is only a matter of from which side one views it. A clock hanging on the wall runs clockwise when seen from the vantage point intended—in front of the clock's face. But if one goes around the wall to the room on the other side and uses one's superpower x-ray vision to view the second hand of the clock through the wall, one will see it move counterclockwise instead.

Considerations of conservation of parity have played an important role in physics, giving meaningful clues to the nature of fundamental physical law. And even though conservation of parity seems, on the surface, to be an obvious truism, it is not always true. We consider this odd fact further in section 12.4.

12.3 Symmetry in art

Physicists have their own operational definition of symmetry, but the word is often used by artists in a somewhat different way—usually to specifically signify what we have here called mirror symmetry (for 2D works) or bilateral symmetry (for 3D works). But there are other forms of symmetry evident in 2D art as well that can be described as an invariance under a transformation. I will first describe some examples of this *formal symmetry*, and then describe the related idea of *balance*.

12.3.1 Formal symmetry in art

Although photography—an example of 2D art—is the main concern of this book, no discussion of symmetry can omit the example of a clay vase formed on a potter's wheel. *Azimuthal symmetry* is defined according to a particular axis in space, called the *symmetry axis*. If an object has azimuthal symmetry it means that one can rotate the object about the symmetry axis by any angle, and it still looks the same. For the potter's vase, it is the very act of forming the wet clay on a spinning platform that imposes this symmetry.

There are many examples of mirror symmetry in 2D art; figure 12.6 shows a couple of my own. The grid in the background of the image on the left has a mirror

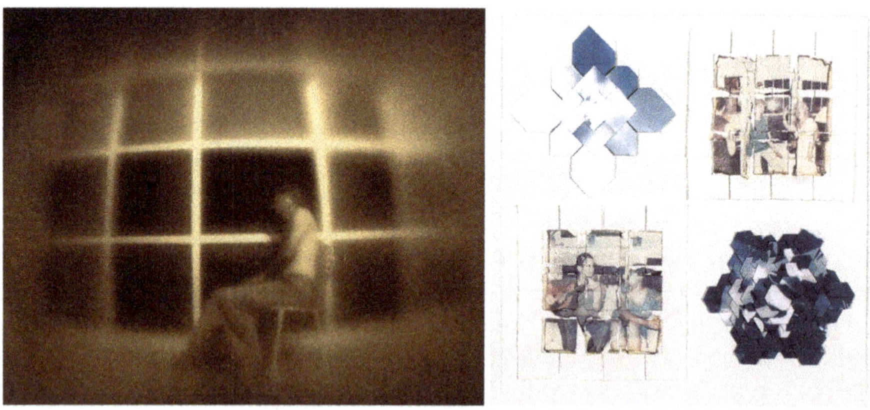

Figure 12.6. Left: *Self-portrait with grid #2*, John Beaver 2004. The background grid pattern in this photograph exhibits several types of symmetries. **Right:** *Celebration*, John Beaver 2015 (folding by Elizabeth Breese and John Beaver). The overall arrangement of the pieces has mirror symmetry about both diagonals. The upper-left piece has four-fold symmetry, while the piece on the lower right has six-fold symmetry.

symmetry when reflected about a vertical plane cutting through the center. Note that the same is not true for a reflection about any horizontal plane that cuts through the image because of the uneven barrel distortion. Yet there is still a sense that it has this type of mirror symmetry as well. Symmetry is so ingrained into our consciousness that we ‘fill in’ the missing pieces with our imagination in order to give it symmetry. And so if one reflects the background of the image about a horizontal plane that cuts through the lowest horizontal line in in the grid, we can easily imagine the symmetrical form that would be created. I believe that part of the power of this particular image is that we see that cut-off part of the picture in our minds, even though it is not really there.

This basic idea that symmetrical forms can be implied or suggested is very important for art. It is one example of how a picture can evoke images that are not even in the picture. A picture is worth not only a thousand words but also a thousand pictures.

There are many forms of symmetry evident in the right-hand image in figure 12.6. First, it is made of four parts of two different types, and they are arranged symmetrically. We can describe this particular symmetry in a couple of ways. Clearly, we could rotate the whole thing in its own plane by 180°, and we would still have the same arrangement of its four parts. But this particular symmetry is better captured by noting that it has mirror symmetry about a plane that cuts through the image across either diagonal.

The grid-like pattern in the origami-folded paper on the upper right and lower left clearly have many types of mirror symmetry, but what about the other two pieces? These patterns are best described by what is often called *N-fold symmetry.* Rotate (in its own plane) the pattern on the upper left by a quarter of a complete circle, and it is still the same pattern. This means we can rotate it four times to come back to its original orientation, and each of those four rotations produces the same shape. This is called four-fold symmetry, and by that same measure it is clear that the figure on the lower right has six-fold symmetry.

There are many connections between the forms found in nature and those used by the artist—a fact well documented in Nathan Cabot Hale’s fascinating study *Abstraction in Art and Nature* (1993). *N*-fold and bilateral symmetries are particularly common in European art of the middle ages perhaps because those same symmetries are so often found in nature. And so a rose window on a medieval cathedral (e.g., see Duby 1992, cover) is like the petals of a flower. And the bilateral symmetry of the arrangement of figures in a thirteenth-century liturgical illustration (e.g., see Duby 1992, pp 31–32) is shared with the arrangement of appendages on vertebrates such as hummingbirds, whales and beavers.

12.3.2 Balance in 2D art

Although it still finds its home in architecture, formal symmetry is much less common in the 2D art of the Renaissance and later. I believe there is a rather obvious reason that art evolved away from formal symmetry as it became more self-aware. Since any particular formal symmetry can be described by a set of simple mathematical rules of transformation and invariance, then where does the art come in? In more recent 2D art, obvious formal symmetries most often appear as the exception that proves the rule. Its obvious presence in a contemporary work is intentional and glaring, and it begs for a specific interpretation.

But there is a related concept that is still very strong—that of balance (Preble and Preble 1994, pp 92–93). The word balance is used in many ways by artists, but a picture has *visual balance* if it appears to be in a sort of equilibrium of weight (Barrett 2011). The idea of *visual weight* is a subjective combination of perceived size and weight that makes some forms in the picture look—well—heavier than others. There is no operationally-defined mathematical way to capture this concept. One must, instead, have a bit of faith that it will be known when it is seen. And so the picture is said to have visual balance if the weights of its elements appear to be in an almost gravitational equilibrium. It is almost as if one asks, 'if a shaft and frictionless bearing were passed through its center, would the painting want to twist one way or the other?'

There is a physical analog to this. The *center of mass* of some object is the average location of its parts, except that one must calculate the average by counting higher-mass parts more in the average than lower-mass parts. For a 2D object, if one allows it to freely rotate about a shaft placed through it, then it will want to rotate such that its center of mass is below that pivot point. If one places the rotating shaft directly through the object's center of mass, it will have no tendency to rotate either one way or the other.

An important case is when the center of mass of the object is below the geometrical center. From one point of view, it is unbalanced; the object is weighted below-center. But there is something different about this type of unbalance because clearly the object would have no tendency to pivot either one way or the other if the pivot point is placed through its center. And if one did give it a push, it would desire to return to its original position. This is called by physicists a *stable equilibrium*, and something like it is probably also a part of the artist's concept of visual balance.

If, on the other hand, the object's center of mass is above a pivot point at its geographical center, then it might possibly be in equilibrium—so long as the center of mass is exactly above center. But all it takes is a gentle nudge from a fly sneezing in the Andromeda Galaxy, and it will flip one way or the other. This is also called an equilibrium, but it is instead an *unstable equilibrium*.

It may be that one could gather these physicists' ideas of center of mass and stable and unstable equilibrium and produce some kind of detailed mathematical model of the artist's idea of visual balance. But I doubt it, and I am skeptical that it would be worth the bother to try. The artist knows it when they see it—whether or not it can be defined in operational terms.

12.4 Asymmetry and broken symmetry

If symmetry is so important, then what about its absence? There are two important types of nonsymmetry.

1. **Asymmetry:** one can identify no type of symmetry, no matter how hard one tries. A perfect asymmetry implies that even if small features are ignored, there is still no identifiable symmetry.
2. **Broken symmetry:** there would be an obvious symmetry, if it were not for some particular small detail.

An amoeba is asymmetrical as is its larger cousin, *The Blob*. The small mole on Marilyn Monroe's left cheek breaks the bilateral symmetry of her face.

Broken symmetries have played an important role in the search for the most fundamental physical laws. A good example is the mirror symmetry of *P* invariance, which seems so obvious and which leads to the conservation of parity. But in 1956 it was discovered, surprisingly, that a certain physical process (called the weak interaction) sometimes violates the conservation of parity (Sachs 1987, p xi).

This is not the only case. Two other well-known almost-symmetries are seen to be sometimes broken—namely the reversal of time (called *T* invariance), and the inversion of electric charge (called *C* invariance). These three most-of-the-time symmetries—*C*, *P* and *T* invariance—are broken in certain cases. This means that the conservation laws associated with these symmetries are, in those same special cases, violated.

The gradual recognition by physicists of these broken symmetries led to much reconsideration of the so-called 'standard model': the gathering together of the fundamental physical laws as currently understood. It was discovered that these broken symmetries led to a different and more subtle symmetry that (so far) does appear to hold as a fundamental property of nature. And so it seems that a particular combination of *C*, *P* and *T* invariance does hold, even though the symmetries of *C*, *P* and *T* are all broken when looked at separately from each other. This new symmetry is called *CPT* invariance and so far, although many questions remain, it does seem to be one of the fundamental symmetries of nature.

But there is another, even more important sense to the physicist's idea of a broken symmetry. *CPT* invariance is an example of a symmetry in nature that is deeper (and apparently more accurate) than the imperfect and sometimes-broken symmetries of *C*, *P* or *T* invariance alone. But physicists began to realize in the 1960s that symmetry at the level of a theory can even describe a physical reality that lacks that symmetry (Weinberg 1992, chapter 8). This may seem to be a contradiction, but it is not, and physicists call it *spontaneous symmetry breaking*. For it seems that some of the laws of physics are such that—at their most fundamental level—the theory itself has a simple but deep symmetry. When that theory is applied in practice to the real world, the symmetry is broken and so produces the complex often-asymmetrical reality we experience. The elementary particle physicist Steven Weinberg (1992, p 195) described the experience.

> It is principles of symmetry that give our theories much of their beauty. That is why it was so exciting when elementary particle physicists started to think about spontaneous symmetry breaking in the early 1960s. It suddenly came home to us that there is much more symmetry in the laws of nature than one

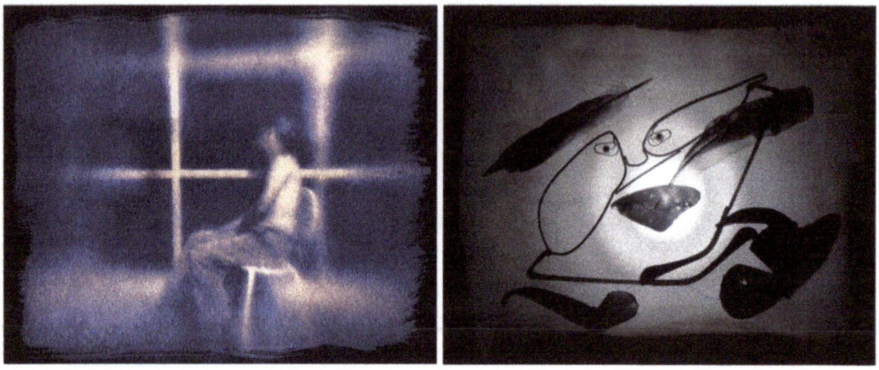

Figure 12.7. Left: *Self-portrait with grid #1*, John Beaver 2004. Compare this to the left-hand image in figure 12.6. Some of the symmetries present in that image are implied here, partly because of our expectations regarding the grid-like structure. But here those symmetries are broken. **Right:** *Revolution of the tear duct*, John Beaver 2018. The background illumination has a very simplistic radial symmetry that contrasts with the asymmetry of the shadow forms and the (literally) broken symmetry of the deformed eyeglasses.

> would guess merely by looking at the properties of elementary particles. Broken symmetry is a very Platonic notion: the reality we observe in our laboratories is only an imperfect reflection of a deeper and more beautiful reality, the reality of the equations that display all the symmetries of the theory.

For the artist, the breaking of an obvious formal symmetry in a picture can be a powerful tool. It evokes, raising questions rather than giving simple answers. Some of my own photography makes use of simple lenses, and their pronounced aberrations impose a simple radial symmetry on the image. The seated subject in figure 12.7 (left side) breaks that symmetry, and our expectations of the symmetries of a grid play off its here-distorted depiction, calling attention to the cross-like form to the left of the subject. The image on the right side of figure 12.7 has an even more pronounced radial symmetry; it was exposed by shining a laser through a smear of red blood cells (kindly donated by a colleague). The symmetry of the diffraction pattern is broken by the shadow forms, and the expected bilateral symmetry of the recognizable eyeglasses is broken by their distorted shape.

References

Barrett T 2011 *Making Art: Form and Meaning.* (New York: McGraw-Hill)

Duby G 1992 *Medieval Art: Europe of the Cathedrals 1140-1280* (Geneva, Switzerland: Bookking Int.)

Hale N C 1993 *Abstraction in Art and Nature* (Dover: New York)

Preble D and Preble S 1994 *Artforms* 5th edn (New York: Harper Collins College Publishers)

Sachs R G 1987 *The Physics of Time Reversal.* (Chicago, IL: University of Chicago Press)

Weinberg S 1992 *Dreams of a Final Theory* (New York: Pantheon Books)

IOP Publishing

Photography (Second Edition)
Physics and art in focus
John Beaver

Chapter 13

Two-dimensional design

A common mistake of the beginning photographer is to take the 'capture the instant' aspect of photography too literally. A picture is not really a capture of a moment of your visual experience. Rather, it is a flat surface with marks on it. And they had better be interesting marks, or no one will want to look at your picture.

I find it useful to think of what I call a *visual event*—a moment of your experience in seeing the world. This can be very powerful, and it is a natural impulse to want to capture that with a camera. But the experience of looking at a photograph is different from the photographer's experience of the visual event that triggered its taking.

You were there before and after the moment. You experienced motion directly of either yourself, the subject or both. You had cues of depth not directly connected to the two-dimensional (2D) image that you saw. This is not to mention all of the nonvisual sensory experience and the context of your particular state of mind when it happened.

The viewer of the picture has none of this; they have only your picture. If you want them to have an experience like the one you had when you took the picture, you will have to trick them into it. And more commonly, you may want them to have a completely different experience, and there is nothing wrong with that.

Thus, there is an important sense in which every photograph is abstract just as is every painting. In the end, the artist uses marks on a flat surface to cause things to happen in the mind of the person looking at their picture. But instead of paint applied directly by hand, the photographer uses straight-line rays of light coming from the three-dimensional (3D) world and redirected by lenses, in an ordered fashion, onto some light-sensitive material.

This is not to say that a photograph is, in essence, the same as a painting. That a photograph is taken at a particular moment, from light coming from the real world often beyond the artist's control, can be a fact that is of the utmost importance. The knowledge of this inevitably changes the way one looks at a photograph, and thus it changes what one sees.

 © IOP Publishing Ltd 2022

Imagine a parallel universe in which Robert Capa had never taken his famous photograph *Death of a Loyalist Soldier* (Cooke and Kinneberg 2014, p 94), and instead the painter Chuck Close had produced a photorealist painting that happened to look exactly like it in every detail. Would it be the same picture? We explore these issues in more detail in chapter 44. Here we discuss some of the properties of a picture that relate only to the formal arrangement of its parts.

13.1 Elements of 2D design

2D design is the intentional arrangements of formal elements in a picture. A picture's 2D design may have a deep connection to its *content*; the interplay between the two may be much of the point. But the design elements can still be considered separately from that content.

First we should agree upon some terminology. There are many different but overlapping ways to choose such a terminology (e.g., see Ocvirk *et al* 2009, Preble and Preble 1994, Barrett 2011, and Brainard 1991), but I list below terms as I will use them in this chapter. We can read a picture, or a part of it, as having some combination of these fundamental elements.

Point: A dot of no particular size or shape.
Line: A line, either curving or straight. This could be simply an edge of a shape (see below), and it may not necessarily be continuous.
Shape: A 2D shape in the geometrical sense, such as a circle or triangle.
Form: A representation of a 3D object, such as a ball or a cone. So a pyramid is a form and a triangle is a shape.
Size: The physical dimensions of a form, line or shape.
Value: The darkness or lightness of a mark.
Texture: The visual effect of a mark. Is it rough or smooth, furry or glassy?
Color: The *hue* and *saturation* of a mark.
Space: The space taken up by objects is called *positive space*, while the space in between objects is called *negative space*.
Depth: The perceived distance from the observer often separated into *foreground*, *background*, and—sometimes—*middle ground*.

13.2 Figure and ground

A particular form is *figure*, while all that is not that form is *ground*. There is a formal sense to this, even when the picture does not represent anything identifiable in the world. So, for example, consider a red triangle on a blue background. Note the 'ground' in 'background.' It is difficult to get around the notion that we seem to see a thing in contrast to all that is not that thing.

It is also well known that one person's figure can be another's ground and vice versa. There are famous optical illusions whereby the figure and ground reverse, depending on the state of mind at the particular moment one sees it. A red circle on a green background can also be a green square with a red hole in it. The most famous example is the 'faces-vase' illusion made popular by the Danish psychologist, Edgar Rubin.

In representational images, and especially in photography, depth is often used to distinguish between figure and ground. Thus a figure can be in front of the background and behind the foreground, although a given photograph may not have all three of these elements. Nevertheless, a common problem in photography is to make a clear distinction between figure, foreground and background. There are many tricks for this, but one we will talk about in detail in chapter 11 is the use of *selective focus*. There are many other depth cues as well, and we discuss some of them in section 13.7.

13.3 Lines

Apparently, our brains will often see a line even when it is not there in any literal sense. Thus, for example, we often speak of a 'dotted line,' which if one thinks about it is something of a contradiction in terms. It is, after all, just a bunch of dots. But if the dots 'lie along a line,' then we see a line that just so happens to be made of dots. Furthermore, a form—such as an arm—may create what one sees as a line in a photograph or painting. Even the edges of a series of unconnected forms can form a line in an image, and the eye–brain will likely see that line (perhaps unconsciously) whether or not that is the artist's intent.

Lines are important because they connect and thus establish a relationship between different elements in a picture. Furthermore, our eyes tend to follow lines when we look at a picture. Where, then, do the lines lead? Do they lead the eye off the picture entirely (probably an unwise choice), or do they lead from one thing to another in a way that establishes important connections between different forms? Finally, lines can form boundaries, signaling our minds to attach separate meaning to different forms

Hale (1993) discusses the meaning of line in art, and how the Impressionists recognized that line is an abstract concept that does not really exist in nature. It is, rather, something we impose upon the natural world.

The history of astronomy provides an interesting historical example of the power of lines. In the late nineteenth and early twentieth centuries, the planet Mars was the subject of much interest to astronomers. Photography had not yet advanced enough to be useful under the dire conditions of recording the telescopic view of the tiny planet, and so it was still the reign of the visual observer.

As seen through a telescope, Mars has barely-discernible surface features that appear with very low contrast. Furthermore, the image constantly wavers and shimmers, going repeatedly in and out of focus due to distortions caused by the Earth's atmosphere. Astronomers of the time squinted for hours, attempting to map out the most subtle of details.

They saw far more than what was actually there—a fact later proved by close-up photographs from orbiting spacecraft. In particular, many astronomers saw an intricate network of lines on the red planet. This fit in with, and added to, an idea

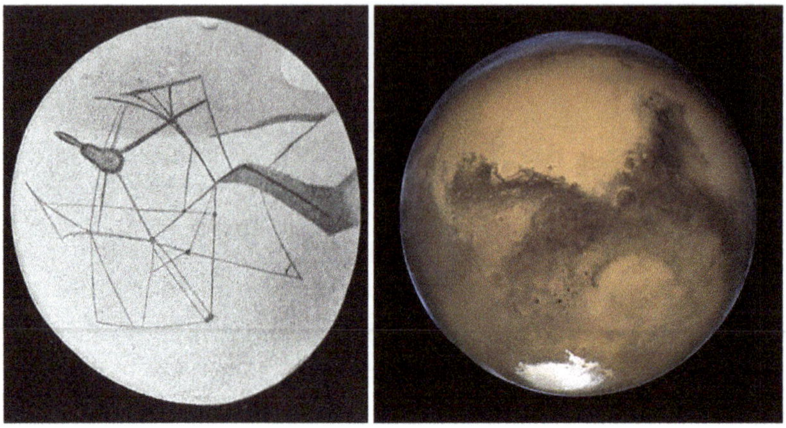

Figure 13.1. **Left:** A drawing of Mars made in the early twentieth century by Percival Lowell (Public Domain). **Right:** A late twentieth-century photograph of Mars from the Hubble Space Telescope (Credit: NASA/ESA, J. Bell (Cornell U.) and M. Wolff (SSI)).

that was popular at the time—that Mars had once been inhabited by an ancient civilization, now perished, that had established a grand network of irrigation canals to bring water from the poles to the more temperate regions. Figure 13.1 shows an early twentieth-century drawing of Mars made by the astronomer Percival Lowell, along with a late twentieth-century image from the Hubble space telescope.

13.4 Geometric shapes

Just as we easily see lines, we also naturally see geometric shapes, especially simple ones such as circles, triangles and rectangles. Often a picture can be broken up into a series of geometric shapes independently of the forms they may represent in the world. These shapes have their own formal qualities in addition to the subject of the picture.

13.5 Value and contrast

Marks (lines, shapes, points, etc) may be distinguished by a difference in value from the surroundings. And so a line may be represented by a lower-value (darker) linear mark on a higher-value (brighter) background. The term *contrast* refers to the difference in value between a mark and its surroundings. A mark of high contrast shows a large difference in value between it and its surroundings, while a mark of very low contrast has a value barely distinguishable from its surroundings.

13.6 Hue and saturation

Hue (for example, is it red or green?) does not tell the whole story of the color of a mark. Even when one allows for differences in value along with hue, there is more to say. That is, one can have two marks that are the same hue and the same value but have different levels of saturation. A pure color has the highest possible saturation.

Figure 13.2. Left: *Road to peace*, John Beaver 2007. Most of the features in this picture are delineated by low contrast marks. The exceptions are the figure at the center and the areas of low-value background peeking through the trees. **Right:** *Waiting for the jaguar*, John Beaver 2018. The forms in this photogram are set off from the background with high contrast in both value and hue.

For the case of pigments, one can lower the saturation by adding grey. If the right shade of grey is chosen, one then obtains a color of the same hue and value but with 'less color,' and this is what we mean by 'less saturation.'

Elements of a photograph can be distinguished by hue and saturation as well as value and contrast. One sees a figure differently if it is against a ground of contrasting hue as opposed to adjacent hues. Furthermore, strongly saturated colors stand in stark contrast to colors of low saturation even if of the same hue (see figure 13.2).

13.7 Depth cues

A sense of depth can be added to a picture by many different cues. Here are just a few.

- *Atmospheric perspective*: Because of scattering of haze in the atmosphere, more-distant forms appear at lower contrast and saturation and may have a more bluish hue.
- The perspective of *converging lines*: Place in the photograph railroad tracks converging into the distance, and a sense of depth is ensured. The photograph may be mocked as cliché, but at least it will not appear 'flat.' See the left-hand picture in figure 13.3.
- Forms may appear more-or-less distant depending on their *vertical placement* in the picture. This is especially true for pictures of landscapes. Objects closer to the top of the picture usually appear as more distant, while those closer to

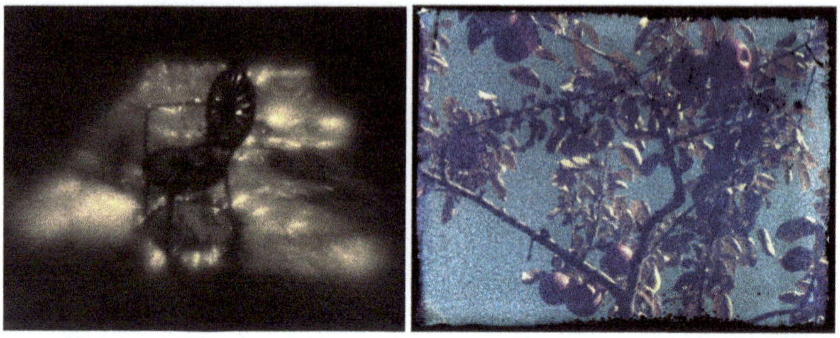

Figure 13.3. Left: *Chair*, John Beaver 2003. The converging lines formed by the edges of the shiny tarp give a sense of depth to the photograph. **Right:** *Apples*, John Beaver 2007. The sense of depth arises from the overlapping of elements in the image.

the bottom of the picture seem to be closer. If there is a horizon line, however, then the opposite is true for the parts of the picture that are above the horizon.

- Nearby objects may *overlap* more-distant objects but not the reverse. This may seem like an obvious statement, but intentionally allowing figures to overlap is one way of showing depth. In the picture on the right side of figure 13.3, the sense of depth arises almost entirely from the multiple overlapping of the apples, branches and leaves.

13.8 Unity and repetition

The way design elements relate to each other is an important part of composition. A set of diagonal lines made by one set of forms can be echoed in another form in a different part of the picture, implicitly linking the two. An image has *unity* if all of the formal elements of design are in harmony with each other.

One way to establish unity is with *repetition*. When one formal element is repeated in a different context elsewhere in the picture, it is a bit like a *canon* in music: one melodic line is altered in some simple way. The altered line is then played at the same time and the two (or more) lines harmonize with each other. Elements in a photograph can do this too. As an example see figure 13.4; diagonal lines are repeated on different scales in the backs of the chairs, the floor decking, the pattern on the shirt and—on a very fine scale—in the hair of the figure.

One can also look for unity in the color palette, and one can attribute other design elements to the picture taken as a whole. Look again at figure 13.4. The placement of the figure makes the picture rather top-heavy, which gives it a sense of uneasiness. The planking in the bottom two frames seems to form 'support' for the figure at the top. But the planks are skewed at odd angles, and so it looks as though the whole structure should fall apart. This effect is partly balanced by, and in contrast to, the very stable base of the chairs directly supporting the figure.

Figure 13.4. *Self-portrait*, John Beaver 2005. The diagonal lines—repeated in the backs of the chairs, the decking, the rail of the deck, the pattern on the shirt, and the hair of the figure—provide a sense of unity to the photograph. But the regular arrangement of four photographs provides a different kind of unity that contrasts with the diagonal lines, and this adds tension.

Like other aspects of 2D design, there can be too much of a good thing. Many bad photographs lack unity, but the presence of perfect unity does not, by itself, make a good picture. In fact, one may intentionally subvert the unity of a picture to add tension to a photograph.

In figure 13.4, there is a second kind of repetition in the uniform rectangular arrangement of the four individual image transfers themselves. This is a very stable arrangement, much at odds with the unstable-looking pattern of skewed lines. It does, however, harmonize with the relaxed expression of the figure. And so one might say this photograph, taken as a whole, lacks unity. Some elements relate to an angular and unstable, almost vibrating picture, while other design elements are of its opposite with a sense of regularity, peace and stability.

There is, however, another way to look at it. Perhaps rather than a lack of unity, this particular photograph has two different kinds of unity. It is saying two things at once, and these two ideas play off each other to make a more interesting photograph that evokes conflicting emotions in the viewer. That complexity of interpretation does not necessarily mean that it is a good picture. The point is that one can intentionally subvert the unity of a photograph; however, to subvert something, one must understand it.

Figure 13.5. Some examples of rhythm. **Left:** The rails of the deck and the planks of the floor have regular rhythm. **Center:** The lines made by the whitecaps have flowing rhythm. **Right:** The birds show progressive rhythm. They are separate birds, but they look like multiple images of the same bird, moving from lower-left to upper-right as time passes.

Figure 13.6. *Celeste with cat*, John Beaver 2008. The tree on the right side of the picture can either be a vertical form or it can be part of a frame, depending on how the photograph is cropped.

13.9 Rhythm

A regular, repeated pattern that moves from one part of a picture to another is called *rhythm*. It may be simple and *regular*, such as the boards of a fence or a pattern of bricks on a wall. But rhythm may also be more organic and *flowing*, like a pattern of choppy waves in the ocean. Or, instead, rhythm can be *progressive*, showing a migration—not necessarily in a straight line—of forms from one place to another. See figure 13.5 for examples of all three.

13.10 Framing

At some point, one must decide what is part of the picture and what is not. And the overall shape of the picture can have a large impact. Is it a long, skinny rectangle or square? If rectangular, is it oriented vertically (portrait) or horizontally (landscape)?

The inclusion, or not, of a single element can drastically affect the composition of a picture. Does that tree on the right side have space to its right and so look like a vertical form in the picture? Or, instead, does the edge of the picture cut through the middle of the tree so that the tree acts as part of a frame (see figure 13.6)?

The example on the right is simpler, with a clear center of interest. In the more mysterious version on the left, the eye sometimes wanders away from the center of interest to travel down the path at the right edge of the frame. Which is better?

13.11 Composition: some useful rules of thumb

Are there rules of design that should not be broken? Probably no one would make such a strong claim. Many rules, however, do have reasons, and it is probably wise to understand them. It is often said that a rule cannot be successfully broken until it is first understood.

I would not argue with that, but I would put it differently. There are fundamental principles—facts if you will—as best we understand them. If one does such and such in a picture, then particular things are likely to happen when someone looks at it. Does one want those things to happen or not? We are not talking about rules from an authority, which we either obey or disobey. Rather, we are trying to understand the consequences of our actions so we can act accordingly.

Note that this list is neither exhaustive nor definitive. Some would include others but not include some of these or use different names for similar ideas.

13.11.1 The rule of thirds

A common way to make a boring picture is to either center a single form or to bisect the frame with forms either vertically or horizontally (or both). Place a flat horizon line right in the middle of a picture where the only strong forms are the bright sky and the dark foreground, and one likely has a dull picture indeed. The rule of thirds proposes dividing the frame into thirds, both horizontally and vertically, with imaginary lines. Then place the forms at the intersections of these lines, thus avoiding bisecting the image.

13.11.2 The rule of odds

An odd number of similar forms in an image is easier to compose than an even number of the same forms. This is because the two outermost forms provide a frame for the symmetric arrangement of inner forms. Put two artichokes in a picture, and the brain says, 'Which one do I look at? What am I supposed to do? What's the right answer?' Put three artichokes in the picture, and the brain sees it as one symmetrical form comprised of three elements; the outer two artichokes frame the one between.

13.11.3 The rule of space

For a form that is moving, leave some space in front of it. Otherwise, it will 'move' right out of the picture when one looks at it. If a figure is looking in a particular direction, leave some space toward where they are looking. Otherwise the viewer will follow the figure's eyes right off the picture. Intentionally violating this rule adds an element of tension.

Figure 13.7. **Left:** *Sumac*, John Beaver 2007. This photograph has no particular center of interest, and so it violates the so-called rule of simplicity. The photograph is more challenging than those that have an obvious center of interest. **Right:** *The ties that bind*, John Beaver 2013. Diagonal lines move the viewer's eye around the picture more than do horizontal and vertical lines.

13.11.4 The rule of simplicity

Keep it simple. There should be a *center of interest* in the photograph—a subject if you will. A corollary of this rule is that one should avoid clutter. A common reason camera-phone pictures are often uninteresting is that the subject is lost in a confusing background clutter. What is one supposed to look at?

On the other hand, there are many successful images for which there is no identifiable subject at any particular location in the picture. In the left-hand image of figure 13.7, it is the geometrical patterns within the textures, combined with larger shapes arising from differences in hue and value, that form the composition.

13.11.5 The rule of diagonals

Diagonal lines are more interesting than horizontal lines. They add a dynamic element to the picture; in extreme cases they are unsettling. See the right-hand image in figure 13.7.

13.11.6 The rule of triangles

Triangular forms with the base at the bottom add a peaceful and stable quality to an image, while inverted triangular forms can result in a sense of tension and uneasiness in the viewer.

13.11.7 The golden rectangle and the rule of the golden mean

There is a particular proportion of rectangle that has held fascination for many over many centuries. The ratio of the lengths of the sides of the *golden rectangle* are such that, if one sections off a square, the remaining section forms a rectangle of the same proportions (see figure 13.8). The small rectangle has the same proportions as the larger rectangle it has been sectioned from by the square. With a little algebra (see appendix B), it is easy to show that the *golden ratio*—the ratio of the sides of the golden rectangle—must be approximately 1:1.618.

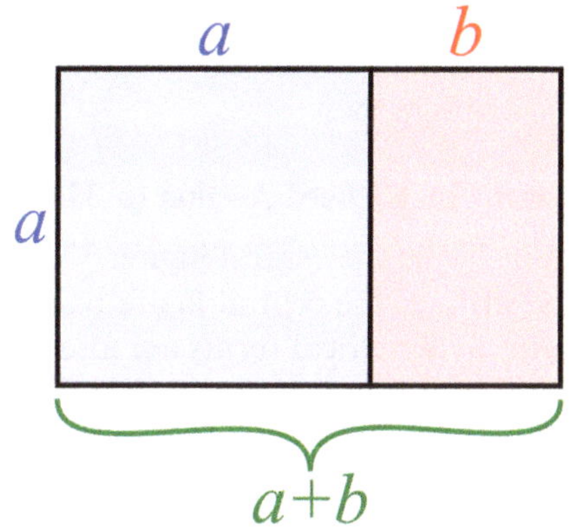

Figure 13.8. A golden rectangle has the dimensions such that, if one sections off a square, another golden rectangle remains.Copyright © Phecht (Original); Pbroks13 (Derivative work) – Own work, Public Domain, https://commons.wikimedia.org/wiki/File:SimilarGoldenRectangles.svg, SimilarGoldenRectangles.svg.

The mathematical elegance of the golden rectangle has inspired some artists to prefer, over the rule of thirds, a *rule of the golden mean*. Instead of thirds, we split up the picture plane into segments of 3/8 and 5/8, which approximates the golden ratio: $\frac{5}{8}\colon\frac{3}{8} = 1.67 \approx 1.618$. The portion of $\frac{5}{8}$ itself, compared to the whole, also approximates the golden ratio: $1\colon\frac{5}{8} = 1.60 \approx 1.618$. Since one of the purposes of the rule of thirds is to avoid bisecting the image, the rule of the golden mean accomplishes this as well but with a different feel.

13.12 Some examples of 2D design in photography

Let us, by way of example, consider five famous photographs. Each of these can be easily seen by way of an internet search, and they are all included in only the first few pages of *The Photography Book* (Cooke and Kinneberg 2014). Here I point out only some of the most obvious elements of composition in these photographs; it is a useful exercise to search for additional examples.

13.12.1 *Child with toy hand grenade* by Diane Arbus (p 27)

Converging lines from the tree shadows provide depth. The two vertical forms of the twin tree trunks are echoed, almost, in the single suspender strap, calling attention to the fact that one strap is missing. The two sets of out-of-focus figures in the distance frame the picture. The placement of the subject neither obeys the rule of thirds, nor is the picture bisected.

13.12.2 *Marilyn Monroe, Hollywood* by Eve Arnold (p 31)

Forms in the chair back are echoed canonically in the figure. The rule of space is used, giving her room to see in the direction her head is pointing (although her eyes are pointed at us). The line of the arm leads one right to the central figure, as do the triangular shape of the exposed foot and leg brace on the chair. The picture is framed to allow a small amount of space beyond the backdrop, emphasizing the posed nature of the event being photographed.

13.12.3 *Dovina with elephants* by Richard Avedon (p 34)

The rule of threes is used to frame the central figure. Violating a standard 'rule' of composition, the body and arms of the central figure bisect the picture plane with a strident, high-contrast form. Symmetrical forms are also used; note the bend in the front legs of the two elephants framing the figure. The line of the white scarf leads right to the neck and head of the figure, which then leads they eye directly to the eye of the elephant. The right-hand elephant is framed so as to add tension to the picture, as it leads off the edge. The chain on the foot forms a line, also leading the eye off the picture. The central figure itself forms an inverted triangle, adding instability to that form, but it is made stable by the anchors of the two elephants.

13.12.4 *Andean boy, Cuzco* by Werner Bischof (p 61)

One can see use of the rule of space and an avoidance of bisecting the image (although the rule of thirds is not exactly followed). The position of the legs turns the figure into a stable triangle, and this form is echoed by the triangle made by the flute, hands and back edge of the sack. Differences in contrast are used both to differentiate between figure and ground (both foreground and background) and also to add depth (through atmospheric perspective). The curved pattern of lines in the cloth is repeated in the agricultural terraces in the background.

13.12.5 *The lambeth walk* by Bill Brandt (p 75)

Notice the converging lines of the apartment building, which provide a strong depth cue. But also, they are echoed in the converging lines formed by the line of faces and the row of arms. And so one cannot look at the picture without connecting the girls to the row of working-class tenement houses. This repeated form appears once again in the dancer's blouse.

References

Barrett T 2011 *Making Art: Form and Meaning.* (New York: McGraw-Hill)

Brainard S and Design A 1991 *Manual* (Englewood Cliffs, NJ: Prentice-Hall)

Cooke T and Kinneberg C (ed) 2014 *The Photography Book* 2nd edn (London: Phaidon Press Ltd.)

Hale N C 1993 *Abstraction in Art and Nature* (Dover: New York)

Ocvirk O G, Stinson R E, Wigg P R, Bone R O and Cayton D L 2009 *Art Fundamentals: Theory and Practice* (New York: McGraw-Hill)

Preble D and Preble S 1994 *Artforms* 5th edn (New York: Harper Collins College Publishers)

IOP Publishing

Photography (Second Edition)
Physics and art in focus
John Beaver

Chapter 14

Camera design

14.1 Introduction

Here we consider the many factors that go into the design of a camera. The goal is to somewhat demystify the camera, such that the *camera obscura*—the 'dark box' that arranges an image onto a detector of light—is a bit less of a 'black box.'

We first consider the different types of cameras that use relatively high-sensitivity photochemical detectors, most typically what is called 'film.' This brings up most of the important issues, and we then apply these principles to the specific cases of digital cameras and certain kinds of homemade cameras, which raise their own issues.

Much of the content of this chapter is about cameras that are no longer manufactured. This may seem odd, but there are several reasons for the choice.

- Most of the important principles of camera design—contemporary digital cameras included—are literally embodied in the history of predigital cameras. The bodies of these cameras are opaque, but the working principles behind their mechanisms are often far-more transparent to understand than those of contemporary cameras.
- Mechanical, predigital cameras have limitations, and such limits—though a hindrance to the professional photographer who needs to satisfy a client—can actually be helpful to the art photographer.
- Working vintage cameras are easily available in online markets often at a tiny fraction of their original cost. Many such cameras are amazing machines that are fascinating in and of themselves.

14.2 Photochemical detector formats

The *format* of a particular light detector refers to the physical dimensions of its image area. Throughout the history of photography, photochemical light detectors have been produced in a myriad of different formats, many going in and out of commercial production over the decades. And with each of these formats comes a

doi:10.1088/978-0-7503-3703-8ch14 © IOP Publishing Ltd 2022

different type of camera to accommodate it. Photographers usually organize this chaos into four rough categories.

1. **Large format:** The light-sensitive material measures approximately 4 × 5 inches (or the slightly-smaller 9 × 12 cm) or larger. Formats larger than 8 × 10 inches are sometimes called *ultra large format*. Since the early twentieth century, large-format detectors have been used most often in the form of individual sheets of sensitized plastic film, but sensitized sheets of glass, metal or even paper have also been used.
2. **Medium format:** The light-sensitive material is smaller than 9 × 12 cm, but larger than 35 mm (see below). Medium format detectors are most commonly in the form of long rolls of film meant to accommodate many separate exposures.
3. **Miniature:** This term was coined for one of the key innovations in film photography—35 mm roll film, first popularized in the 1930s by Leica. Still one of the most ubiquitous film formats, it is usually referred to simply as '35 mm.' The film has an image area that is 24 mm across, and usually 36 mm long. A variation is *half-frame 35 mm*. The same roll film is used, but the half-fame camera produces an image area that is instead 24 × 18 mm.
4. **Subminiature:** Roll-film formats that are smaller than 35 mm. The most common are 16 mm and 8 mm width—two popular cinema formats that are sometimes also used for still photography.

The boundaries of these categories are not set in stone. The (rather uncommon) sheet-film format $3\frac{1}{4} \times 4\frac{1}{4}$ inch, for example, is called medium format by many but large format by some. Only a little smaller than 4 × 5, it is used most often in cameras that are otherwise identical to a typical 4 × 5 camera. Most medium formats, on the other hand, are roll films and so are used in cameras very different from a typical large-format sheet- or plate-film camera.

A photochemical detector is, obviously, sensitive to light, and so it must be kept in the dark prior to exposure. After exposure in the camera, it must somehow be removed from the camera while kept in darkness until chemical processing. Two basic strategies have evolved over the years to accomplish this task. The earliest approach was to load individual sheets of film or plates made of metal or glass into a special *holder* designed for a particular type of camera. This strategy is still used for large-format photography, but most medium and smaller formats use *rolls* of long strips of flexible film. In the sections that follow we explore the different strategies that have been employed to keep all of this in the dark, except for the brief moment of exposure to light focused by the camera lens.

14.3 Plates and sheet film

One of the earliest and simplest strategies for keeping the light-sensitive material in the dark is to place an individual sheet in a special *sheet film* or *plate holder*. The procedure works as follows.

1. An individual sheet of film (or a sensitized sheet of glass or metal) is loaded—in the dark—into a special light-tight holder. A thin, removable dark slide covers the detector.
2. The loaded film holder is placed so the detector inside is located exactly at the camera's focal plane.
3. The dark slide is removed.
4. The camera shutter is tripped to make the exposure.
5. The dark slide is replaced.
6. The film holder is removed.
7. The light-sensitive sheet is removed and chemically processed in the dark.

See figure 14.1 for an assortment of sheet-film and glass plate holders. Clearly, the design of the camera, detector format, and holder must all work together. The most significant challenge is that the detector inside the holder is located exactly at the camera's focal plane prior to exposure. Many early cameras had their own system, and so film holders were often not interchangeable between different makes, even if they were meant for the same detector format.

Most film and plate holders are two-sided, so that a second exposure can be made simply by flipping the holder over. There were also clever mechanisms for stacking several sheets of film—usually held in separate metal *septums*. The individual sheets could be shuffled from the outside with the most recent exposure slid to the back of the stack, readying for the next exposure. The holder on the lower left in figure 14.1 holds 12 sheets of film. The most recent exposure is pulled off the top of the stack with a lever and into the light-tight leather bag. The next unexposed sheet of film is thus situated for an exposure while the exposed sheet is pushed to the back of the 'deck' by hand.

Most detectors up until the late 1800s were for sensitized glass or metal plates. When thin plastic sheet film became practical in the early 1900s, the first film holders were modifications of existing glass plate holders. A thin metal insert held the sheet film and was inserted in place of the glass plate. See figure 14.2 for a 4 × 5 inch format example.

Figure 14.1. An assortment of film and glass plate holders with formats arranging from 6 × 9 cm (upper **left**) to 8 × 10 inches (**right**). A sheet of film and a dark slide can be seen in the top row.

Figure 14.2. A glass plate holder with a septum for sheet film and the dark slide.

Since there must be room for the insert, the corresponding sheet-film format was chosen slightly smaller than the glass-plate format it largely replaced. Thus the so-called 4 × 5 format is, literally, 4 × 5 inches when glass plates are used, but the sheet-film format is only 3.9 × 4.9 inches even if the holder is meant to accommodate only sheet film.

From the dozens of film, plate and holder formats that arose over the past 150 years, by far the most common today are for 4 × 5 inch and 8 × 10 inch sheet film. These two formats have for many years had standard sizes by international agreement. A modern 4 × 5 film holder should work with any brand of 4 × 5 film in any modern 4 × 5 camera. Quite popular is the *4 × 5 international back system*—a common mounting for focus screens and assorted film holders, including instant film and medium-format roll-film holders. Figure 14.3 shows an example.

14.4 Roll film

The development of a durable plastic, flexible base for silver-gelatin emulsions allowed for *roll film*. Many exposures-worth of film are coated onto a long strip and then rolled up. The roll is placed in the camera to one side of the image frame with a *take-up spool* on the opposite side. After an exposure is made, a mechanical *film advance mechanism* winds just enough film from the roll to the take-up spool in order to position a new piece of unexposed film in the image frame. The process is repeated until the end of the roll is reached.

There are two different approaches to removing the exposed film from the camera. Either the take-up spool itself, with its roll of exposed film, is removed from

Figure 14.3. Bottom center: A 4 × 5 international back with spring-back focus screen, and a partially-inserted film holder with assorted instant film and roll-film holders that work with the same back.

the camera for processing. Or the camera has a *rewind mechanism*—the film advance process is reversed, and the exposed film is rewound onto the original spool and is then removed from the camera for processing.

Of course, all of this must be carried out in the dark, including the crucial steps of getting the unexposed roll into the camera and the exposed roll out. Two basic approaches are used—film cannisters and paper-backed rolls. We consider each, and their variations, in turn.

14.4.1 Cannister film

For many smaller formats—35 mm is the most familiar example—the roll of unexposed film is contained within a light-proof cannister. The film exits the cannister through a narrow slit in one edge of the cannister. The slit is usually lined on both sides with a layer of velvety cloth to help prevent light from entering through the slit.

A roll of cannister film is loaded so that a short length of film—a *leader*—is left emerging from the cannister. The camera is opened, the film cannister put in one side, and the leader is then connected to the camera's take-up spool. In many cameras the film advance mechanism must be employed, and a little bit of film must be wound around the take-up spool in order to ensure that the film is well connected to it. The camera is then closed.

Obviously, this process ensures that a short length at the beginning of the film roll is wasted; whatever length of film is outside of the cannister before the camera is closed has been hopelessly exposed to light. Before the camera is closed there is at least one frame already stretched across the image area, plus whatever must be initially wound around the take-up spool, plus an additional small amount that is out of the film cannister but not yet onto the image area. This means after the camera is loaded and closed, and before the first picture is taken, the film advance must be engaged to bring unexposed film out of the cannister and onto the image area. For most cannister-film cameras, the equivalent of two to four frames are wasted.

Most cannister-film cameras use a rewind mechanism. The last-exposure end of the strip of film is fixed to the spool inside the cannister. When the end of the roll is reached, the film will no longer advance. The mechanism is then reversed and the film is wound—with an external knob or crank or an internal motor—back onto the spool inside its light-proof cannister. The light-proof cannister with exposed film wound up inside can then be safely removed from the camera. Most photographers rewind the film completely into the film cannister; the presence or absence of the film leader outside the cannister reminds the photographer whether or not the film has been exposed.

14.4.2 Paper-backed roll film

The earliest roll films used a simple but effective strategy for keeping it all in the dark. The strip of film is layered with a somewhat-longer strip of opaque paper, and both are wound together onto a spool with the paper as the outer layer. Since the paper strip is longer, the light-sensitive film is protected by several extra turns of opaque paper.

When loaded into the camera, the excess paper leader is connected to the take-up spool and the camera is closed. The take-up spool is wound until the first frame of the film comes into the image area, and a picture can then be taken. Once the end of the film is reached, the roll is advanced so that excess paper exits the film spool, winds onto the take-up spool and covers the exposed film. The exposed roll can thus be safely removed from the camera. The end of the protective paper is then usually adhered to the roll so it does not accidentally unwind and leave the exposed film unprotected. The remaining empty film spool is then moved to the other side of the camera to serve as the take-up spool for the next roll of film.

The elegance and simplicity of this scheme resulted in great popularity—many different widths and formats went in and out of production over the 130-year history of such film. Figure 14.4 shows the most venerable—type 120 film. It is still produced today for medium format cameras. The empty film spool on the left is for 120 film, while the two on the right are for different roll-film formats.

As an example of the great historical diversity of roll films and the cameras designed for their use, the middle spool in figure 14.4 is for type 620 film. Although the film stock itself is identical to 120 film, the spool is different. Most cameras designed for 620 film will not accept 120 film and vice versa. Figure 14.5 shows three different cameras, all of which use 120 film. The width of the frame is the same for each, but they have different lengths, varying from the square 6 × 6 cm (also called the $2\frac{1}{4} \times 2\frac{1}{4}$ inch) to the rather oblong 6 × 9 cm ($2\frac{1}{4} \times 3\frac{1}{4}$ inch) on the right. The format of the center camera is 6 × 7 cm, and many 120 film cameras were also made for a $4\frac{1}{2} \times 6$ cm format.

There is an important variation on the paper-backed roll-film concept. Instead of the entire strip of film being layered with a longer continuous strip of paper, paper strips are attached only to the ends of the film. This still allows for both the

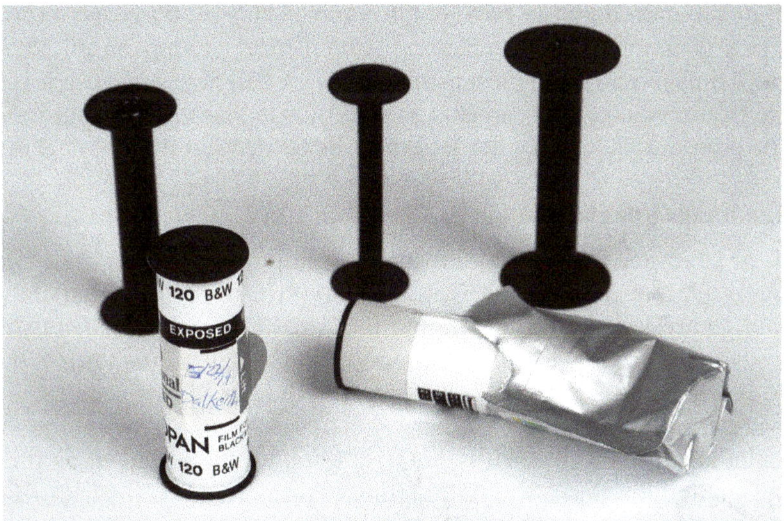

Figure 14.4. **Left** and **front:** Type 120 roll film. **Rear:** Take-up spools for two other formats of roll film.

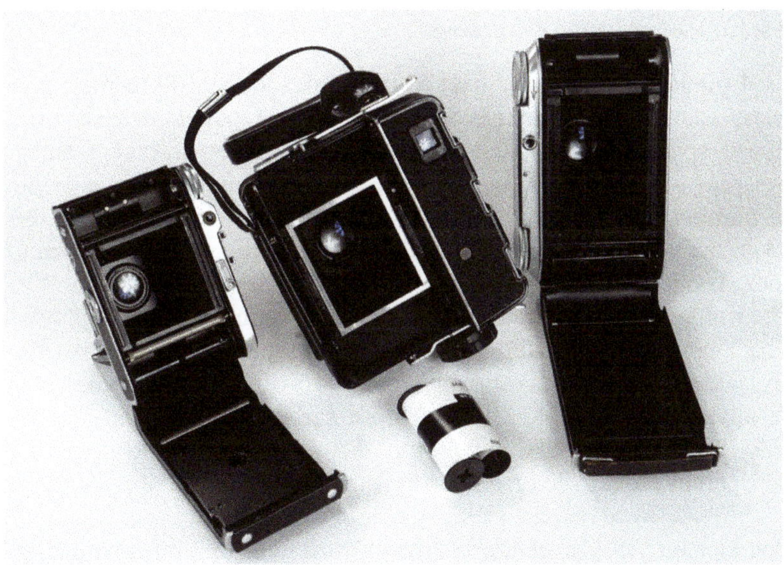

Figure 14.5. Three cameras with different image formats, all of which use type 120 film.

unexposed and exposed film roll to be protected by layers of opaque paper. But a film strip twice as long (producing twice as many exposures) can then fit onto the same roll. As we shall see in section 14.4.3, this obvious advantage comes with a cost; it requires a significant increase in the complexity of the film advance mechanism. Such a twice-as-long roll of type 120 film is referred to as 220 film format.

Type 120 film, introduced in 1901, is still commercially produced as of this writing; it works in a plethora of vintage cameras. Commercial production of 220 film was discontinued only recently, but most cameras that use 220 film can also use 120 film. It is possible to use vintage 620 cameras as well by winding 120 film onto a 620 film spool. The procedure takes practice and must be carried out in total darkness.

14.4.3 The framing mechanism

Any roll- or cannister-film camera must solve the problem of framing. Given that everything inside the camera is in the dark, how does one know that the film has been advanced precisely to the next frame? If it is advanced too far, then there will be large gaps between the images, and thus wasted film and fewer exposures than what should be possible on the same roll. Worse yet, if the film is advanced too little then images will partially overlap each other.

If the film advance mechanism is simply a knob or crank that turns the take-up spool, one might think that a simple counting of turns could do the trick. But with a moment's thought, it is clear that will not work. As more and more of the film winds onto the take-up spool, the circumference of the spool increases. And so a given rotation of the take-up spool near the end of the roll advances significantly more film than at the beginning of the roll. There are four basic strategies by which roll-film cameras solve the problem of framing.

The red window: Fully paper-backed roll film, such as 120 format, provides the simplest solution to framing. The back of the paper strip is printed with numbers, spaced so as to correspond to the center of each frame. A small transparent window in the back of the camera then allows one to directly see these numbers on the backside of the paper. One simply rotates the take-up spool with a knob on the outside of the camera while watching the window until the next frame number appears.

The window must be small enough and press tightly enough against the paper film backing so as to not allow a significant amount of light to enter the camera. Many cameras employed a movable cover, so the window would only be open while advancing to the next frame. Since early black-and-white films were not sensitive in the red part of the spectrum, the film advance window is usually made of transparent red plastic or glass—thus the name 'red window' (see figure 14.6).

One elegant feature of the red-window system of film advance is that the framing is controlled simply by the spacing of the numbers printed on the back of the paper. Even better, different frame-spacing numbers can be printed on the paper back of the same roll of film as parallel rows of numbers. Which is visible depends upon the location of the red window. Thus 120 film, for example, has rows of numbers for 6×4.5, 6×6, and 6×9 spacing. Thus the same roll of film can be used in cameras with three different formats.

Red-window framing has the disadvantage that it is slow. One must carefully look through the little window and slowly turn the advance knob so as not to accidentally miss the next frame number or advance past it (there is usually no

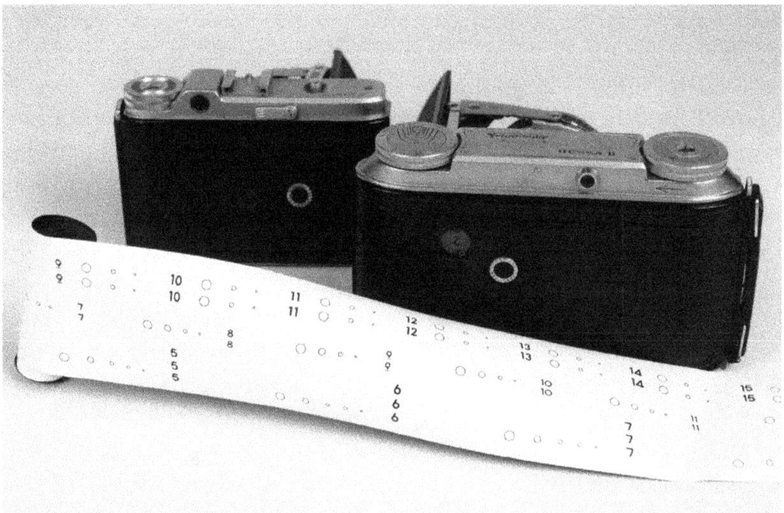

Figure 14.6. Two 120 roll-film cameras that use a red window system for framing pictures. The location of the red window reveals sets of numbers for different image formats.

way to back up). Most brands of 120 film have warning marks to help with that, but it can still be a slow process, especially in dim lighting when the frame numbers are difficult to see.

Clearly the red-window method will not work with 220 film, which only has the paper at the ends and so is bare film once loaded into the camera and advanced to the first frame.

Sprocket holes: Sprocket holes are a series of regularly-spaced small holes (usually rectangular in shape) that run along the edges of the film, outside the image frame area. Thus the film must be significantly wider than the image width. The most common sprocket films are 35 mm and various cinema films—8 mm, 16 mm and 70 mm (along with 35 mm which is also used in cinema).

The sprocket holes mean that the film can be advanced by a sprocket—a pointy gear that pokes through the holes and pushes the film along as it rotates. Even if the film is advanced by the simple winding of a take-up spool—so that the film advances the sprocket instead of vice versa—the rotating sprocket can still be used to precisely mark the position of the film.

Either way, the position of the film, compared to the image frame, is directly connected to the rotation of the sprocket regardless of how much film has been wound onto the take-up spool. Thus cameras that use sprocket-hole film, such as 35 mm, simply count rotations of the sprocket to determine where the next frame is. In most such cameras, when the next frame is reached the film advance mechanism locks. After the picture has been taken, the mechanism is released, and it can be advanced again until the sprocket rotates by the correct amount to bring in the next frame.

Most better-quality 35 mm cameras use the sprocket to advance the film itself. This means the same mechanical action is required at both the beginning and end of the roll of film and allows for a more convenient single-action lever mechanism. But there is a complication—the sprocket always rotates the same amount for each frame, but the take-up spool must rotate more at the beginning of the roll and less at the end of the roll. If the take-up spool turns too little, the sprocket will try to push more film onto the spool than what is being wound, and the film will bunch up and jam. If the take-up spool rotates too much compared to the sprocket, then it will try to tear the film or its sprocket holes.

This problem is usually solved by putting the take-up spool on a *friction clutch*. The take-up spool slips with friction on a geared mechanism that tries to wind it at a speed that is faster than what the sprocket will allow, even at the beginning of the roll. The spool simply slips and turns at whatever speed the film —advanced by the sprocket—allows thus also keeping tension on the film and ensuring it winds tightly.

This basic mechanism goes back to (for example) Leicas of the 1930s and is still used today in most 35 mm film cameras. See figure 14.7 for a comparison of the mechanisms of a Fed 2 from the 1950s (with a mechanism copied from the Leica II) to a Nikon FM3a from the early 2000s.

Friction wheel: A sprocket cannot be used to help frame the pictures with roll films that do not have sprocket holes—120 and 220 medium-format film for

Figure 14.7. **Left:** Nikon FM3a. **Right:** Fed 2. Although manufactured 50 years apart (and the older Fed is a copy of the Leicas of 20 years earlier), the two 35 mm cameras employ similar mechanisms for film advance. From **top** to **bottom:** Film cannister compartment with rewind clutch and crank, focal plane shutter, sprocket and slip-clutch take-up spool.

example. But a *friction wheel* can serve some of the same functions. A small wheel presses up against the film or its paper backing and turns as the film moves past. Like a sprocket, the rotation of the friction wheel is directly connected to the position of the film, regardless of how much film has been wound onto the take-up spool.

For completely paper-backed film such as 120, the friction wheel usually presses against the paper backing of the film at one of the two spools. The wheel must be on some type of spring-loaded pivot so that it always stays in contact with the film backing. Alternatively, the friction wheel is sometimes located just to the outside of the image frame edge; a plate presses the film against the spring-loaded wheel. This latter approach is necessary for 220 film; the friction wheel would mark the film if it turned against the image area directly.

Film without sprocket holes must still be wound by turning the take-up spool with either a knob, crank, or ratcheted lever. The friction wheel simply measures the position of the film and, through a series of gears, locks the film advance once the next frame is in position.

The simplest such mechanisms still use a red window to position the first frame. The friction wheel and gear mechanism is then used for the remaining frames. More complex mechanisms have the ability to wind the film the correct amount to also center the first frame. They still must be manually loaded with the paper leader starting in the correct position. For 120 and 200 films, this has fortunately been made standard between different manufacturers of cameras and films; a 'start' mark printed on the paper leader is lined up with markings on the camera.

Miscellaneous gear mechanisms: Some film advance mechanisms use neither sprocket holes, a friction wheel, nor a red window. Instead they use gears to rotate the take-up spool by just the right amount to pull the correct amount of film to the next frame—taking into account the fact that as more film is wound onto the take-up spool, the same rotation advances more film. This requires a more-complex gear mechanism that acts differently for each frame. Such a mechanism uses the rotation of the take-up spool to only indirectly control the amount of film that passes the image frame. If the mechanism fails to maintain careful control of how tightly the film winds onto the spool, then uneven framing results.

An unusual example is the Koni-Omega Rapid camera (see left image in figure 14.8) that advances the roll film with a ratchet mechanism. The ratchet is pulled out and then quickly pushed back in to rotate the take-up spool a fixed amount. The internal gear mechanism causes the ratchet to move less for each successive frame in order to correct for the larger circumference of the fuller spool.

The right-hand image in figure 14.8 shows three lever-wind roll-film holders, each of a different image format. The right-hand example is disassembled to show the loading mechanism. They have standard mounting mechanisms to allow them to fit onto the backs of many different modular-design cameras.

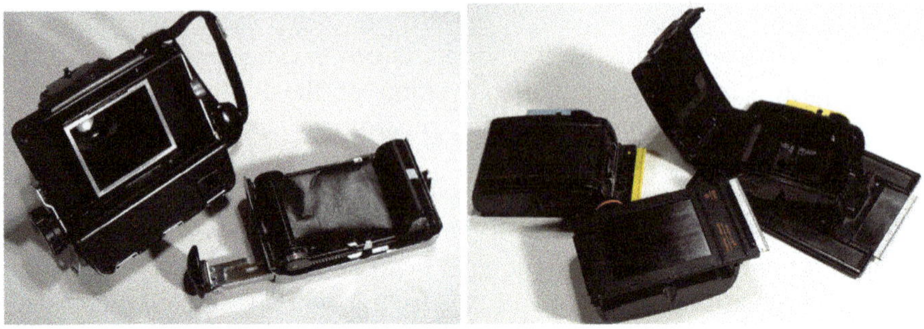

Figure 14.8. Left: Koni-Omega rapid camera and its unique ratchet-based roll-film advance mechanism. **Right:** Roll-film lever-wind advance mechanisms, of different formats, adapted to fit onto the back of different cameras.

14.5 Pointing and focusing

Figure 14.9 shows a photograph I took on a narrow mountain trail. The modified camera uses a simple lens of fixed aperture and focus with a moderately wide-angle field of view. It employs the simplest possible mechanism for pointing at the subject and composing the picture—it has no mechanism at all.

To take the picture I simply pointed the camera, making a rather wild guess as to what would be in the image frame. I held the camera at waist level, tripping the shutter carefully as I edged past the mountain goat, hoping it would refrain from butting me over the side. I call this method 'point and pray,' and sometimes it works. In the sections that follow we consider strategies by which the photographer can have some more detailed control over what is in the picture and what is not. We consider several common approaches for making a camera.

1. **Viewfinder cameras:** A separate device attached to the camera, through which the photographer looks, shows a view that is (hopefully) similar to what the camera sees. Focus is achieved by either a numerical *focus scale*, some kind of *autofocus* or a *rangefinder* mechanism.
2. **View cameras:** The image from the lens is focused directly onto a ground-glass screen (GGS). To take the picture, the screen is removed and the light-sensitive material put in its place.
3. **Twin-lens reflex (TLR) cameras:** Two lenses of the same focal length are used—one to focus onto the film and the other to focus onto a GGS in a right-angle Keplerian viewfinder. Both lenses move together to focus.
4. **Single-lens reflex (SLR) cameras:** The camera lens focuses either to the light detector or to a GGS in a right-angle viewfinder for composing and focusing. A pivoting mirror is used to flip the image from one to the other.

14.5.1 Viewfinder cameras

The simplest strategy for pointing a camera and framing the image is to couple a *viewfinder* to the camera body. The viewfinder is an arrangement of lenses or masks

Figure 14.9. The lucky photograph was 'shot from the hip' with a fixed-focus camera with no mechanism for framing the image.

that, when looked through, shows the same angle of view as does the camera lens. As the camera points in different directions and at different subjects, so too does the viewfinder.

For many cameras, the viewfinder is incorporated directly into the camera body. But for cameras with interchangeable lenses, the viewfinder may be a separate unit that can be attached to the camera by some clip mechanism. This allows for different viewfinders—each with their own angle of view—to be interchanged to accommodate the switching of lenses of different focal lengths.

14.5.1.1 Parallax error

The separate viewfinder optical system must necessarily occupy a separate space than the optics of the camera itself, and it must thus see the world from a vantage point slightly different from that of the camera lens. This results in a deviation in framing and perspective, called *parallax error*. If the viewfinder is above the camera lens, it will point slightly high.

This framing error is most prominent at the nearest focus. Some viewfinders optically superimpose a set of *frame lines* that show the approximate borders of the picture for both near and far focus (e.g., see figure 14.10). More sophisticated viewfinders have *moving frame lines* that automatically adjust as one focuses. Or in some cases the viewfinder tilts toward the lens to frame for different distances, either manually or automatically with focus.

Even if the viewfinder has adjustments to adjust the framing for distance, a viewfinder still shows a somewhat different view than the camera, especially if the scene contains objects at many distances. For example, a distant object may be visible in the viewfinder, peeking just around the edge of some nearby object. But that does not mean it is visible from the camera lens; from that perspective the distant object may be hidden.

Figure 14.10. The optically superimposed frame lines in the viewfinder of a compact 35 mm camera. The two dashes at top are to mark the location of the top edge of the frame when pointed at objects in its nearest focus.

14.5.1.2 Frame viewfinders

The simplest type of viewfinder is a replica of the geometry of the camera itself. One looks through a hole at a rectangular frame held some distance away. If the frame width and height, compared to the frame distance from the eye hole, are in the same ratio as the film format as compared to the camera focus distance, then the same angle of view is reproduced.

A frame viewfinder may be a small unit—a box with a small hole for the eye on one side and a rectangular opening on the other. In many very old cameras, the rectangular frame is simply a piece of wire, and both it and the eye hole fold down out of the way when not in use.

One of the most versatile wire-frame viewfinders can be found on old press cameras such as the Graflex Speed Graphic and Crown Graphic. A wire frame can be extended upward from the front standard, where the lens is located; the frame is the same height and width as the film format. The eye hole is a small unit located at the film plane. Thus the geometry of the wire frame and eye hole is an exact—but reversed—duplicate of that of the lens and film (see figure 14.11). This has the advantage that the same finder works no matter the focal length of the lens mounted in the camera. A longer focal length produces a smaller angle of view. However, since the rectangular frame moves with the lens, the viewfinder shows a correspondingly smaller angle of view.

This type of arrangement is sometimes called a *sports finder*; it can be used quickly for rapidly-moving subjects, since the eye is used naturally, without having to accommodate for the focus of a lens-based viewfinder. The Speed Graphic sports finder also demonstrates a simple method for adjusting the framing for nearby subjects because of parallax. The viewfinder is mounted higher than the film and lens, so it would point a few inches above the subject focused by the lens. For a nearby subject, one simply adjusts the eye hole upward—above the center of the wire frame. Thus the viewfinder points somewhat downward to correct for the framing error due to parallax. See the right-hand image in figure 14.11.

14.5.1.3 The reverse Galilean viewfinder

One of the simplest and most common lens arrangements for a viewfinder is the *reverse Galilean* system, described optically in section 11.9.2. It is popular because it is inexpensive, and even a physically-small one can produce a relatively-large angle of view. One looks through a small convex lens, through a larger concave lens. This arrangement works well for moderate angles of view, as would be used with a 'normal' or somewhat wide-angle camera. Figure 14.12 shows a Fed 2 camera with a reverse Galilean viewfinder mounted on top. Since this camera can accommodate lenses of different focal lengths, the external viewfinder fits into an *accessory shoe*;

Figure 14.11. The wire-frame sports finder of a 4 × 5 Graflex Speed Graphic camera.

Figure 14.12. A Fed 2 camera with wide-angle lens to the right and a matched reverse Galilean viewfinder mounted on top.

the viewfinder is designed to show the same angle of view as the moderately wide-angle lens to the right.

14.5.1.4 Right-angle viewfinders

As described in section 11.9.2, a converging lens produces an inverted real image. But if a 45° mirror is placed before the focus, the image will appear right-side-up, although it is still reversed left-to-right. This real image can then be focused onto a GGS and viewed with a magnifying glass (or directly if a comfortable viewing distance can be arranged). As an alternative, a large short-focus lens can be placed at the focal plane to form a *brilliant viewfinder.*

The right-angle view lends itself to a *waist-level* as opposed to an *eye-level* system of pointing the camera. Small waist-level brilliant and GGS viewfinders are common on many old, simple 'box' cameras. Figure 14.13 shows two examples. The Kodak 2A Brownie Model B at right has two tiny GGS waste-level viewfinders—one for portrait and the other for landscape mode. The square-format Brownie Hawkeye at the right has a brilliant viewfinder at the top.

Variations on the GGS right-angle viewfinder are used in the TLR and SLR cameras we discuss in sections 12.5.3 and 12.5.4. For these cameras, the real image of this arrangement is used not only to point the camera but also to precisely focus its lens.

14.5.1.5 Complex optical viewfinders

More sophisticated viewfinder cameras may have a complex arrangement of lenses and beamsplitters to produce a bright, distortion free image for framing, and with frame lines that both correct for parallax and can be selected for different focal length lenses. This is most common on expensive rangefinder-focus cameras (see below), such as the Bessa R shown in figure 14.14 or the far-more-expensive modern Leica cameras.

The viewfinder window is built into the camera body (the rectangular window on the upper right as seen from the front). A lever on top of the camera switches

Figure 14.13. Two simple medium-format cameras with Keplerian viewfinders. **Left:** Kodak No. 2A Brownie Model B, circa 1920–24. **Right:** Kodak Brownie Hawkeye, circa 1950–61.

Figure 14.14. A modern rangefinder camera with a sophisticated viewfinder with frame lines that can be switched for lenses of different focal length.

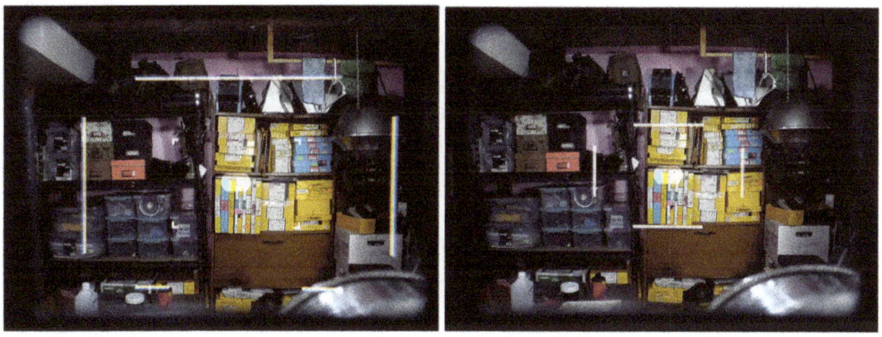

Figure 14.15. Views through the viewfinder of a modern 35 mm rangefinder camera with frame lines set for wide-angle (**left**) and telephoto (**right**) lenses. The bright spot at center is the coupled rangefinder for focusing.

between a choice of frame lines to duplicate the angles of view of lenses of three different focal lengths. The translucent white window at top center allows light to enter to illuminate the frame lines. The frame lines move to adjust for parallax as one focuses the lens. The camera is shown here mounted with a super-wide-angle lens beyond the capabilities of the built-in viewfinder; a separate viewfinder made for this lens is attached at top. Two views through its viewfinder can be seen in figure 14.15. For the left view, frame lines for a wide-angle lens have been selected (by a switch on the top of the camera); for the right-hand image, the frames lines are selected for a telephoto lens.

14.5.1.6 Focusing a viewfinder camera

A viewfinder camera is two parts—the viewfinder and the camera. The photographer does not literally see what the camera sees, and there must be some separate arrangement for insuring that the image in the camera is in focus. This is usually done in one of four ways.

1. **Fixed focus:** The optical arrangement is such that the camera will produce images that are acceptably in focus over a range of conditions, and the focus is simply fixed as such. This arrangement is most common on very inexpensive cameras of rather large focal ratio.
2. **Scale focus:** There is a numerical scale that one adjusts, corresponding to the subject distance. One must thus either first measure the distance to the subject or else make an adequate estimate.
3. **Rangefinder focus:** An arrangement of mirrors, prisms and beamsplitters allows one to compare the view from two separate windows, spaced a couple of inches apart on the camera. A mechanical coupling pivots the view of the subject through one window, as compared to the other, until the two views coincide. From this the distance to the subject can be determined to set the camera focus.
4. **Autofocus:** The distance to the subject is measured through either optical parallax or by sonar, and the focus is adjusted automatically.

Figure 14.16 shows the scale-focus mechanism of an inexpensive plastic Smena 8M camera from the 1980s. The lens is on a helical screw; rotating it moves it closer or farther from the film plane in order to focus. On the outside is a (hopefully!) calibrated distance scale, allowing one to dial in approximate distances between one meter and effective infinity. There are also symbols that correspond to a typical scene that would fill the frame at the focused distance. For example, if three full-bodied persons standing side-by-side would just fit in the frame in landscape mode, then the distance must be a bit less than 4 m.

Figure 14.17 shows the rangefinder-focus mechanism of a Graflex Speed Graphic camera. The rangefinder in figure 14.17 is *coupled*: both the rangefinder and focus are adjusted simultaneously. When working properly[1], bringing the double image of the rangefinder into alignment on some particular object ensures that object will be in perfect focus.

Figure 14.16. The scale focus adjustment of an inexpensive plastic Smena 8M.

[1] The mechanism must be well-maintained and carefully-adjusted.

Figure 14.17. The rangefinder of a speed graphic camera.

Figure 14.18. A scale-focus Perkeo II camera from the 1950s; the square window is its reverse Galilean viewfinder. Mounted on top is a separate mechanical rangefinder to measure the subject distance. The rear side of the rangefinder is shown at **right**.

The rangefinder shown in figure 14.18 is *uncoupled.* It is a separate unit that (in this case) clips onto the top of a scale-focus viewfinder camera. A knob on the rangefinder is turned until the double image is aligned; the distance is then read off a scale marked on the rangefinder. The camera lens is then set manually for that distance.

Figure 14.19 shows two roll-film coupled rangefinder cameras. The camera at right has a moving prism arrangement. When one looks through the rangefinder, the top of the image is from the left-hand window, while the bottom of the image is from the right-hand window. A vertical line appears split in the middle; the focus is adjusted until the two halves merge. A separate wide-angle reverse Galilean viewfinder is used for framing and composing; the rangefinder window, on the other hand, has a more restricted view. One must thus focus and frame the image in two steps.

Figure 14.19. Two different coupled rangefinder systems. **Left:** Bessa II from the 1950s. **Right:** Rangefinder Bessa from the late 1930s. For the older model, one must look through separate windows to compose and to focus.

Figure 14.20. The double image means the bird feeder is out-of-focus in this view through a camera that has a combined viewfinder–rangefinder.

The left-hand cameras has a viewfinder and rangefinder that is integrated into the same window. A beamsplitter superimposes a small double image for focusing in the center of the viewfinder window. A picture taken through the viewfinder of such a camera can be seen in figure 14.20. Notice the double image in the bright spot centered on the out-of-focus bird feeder.

14.5.2 View cameras

A view camera is in some ways the simplest approach to framing and focusing. The image from the lens is focused directly onto a GGS, and the upside-down image is viewed from the other side. A sheet-film holder is put in place of the GGS, and the recorded image is then of exactly what was viewed on the GGS. Thus there is no issue of parallax error, and if the GGS is focused, so too will be the image recorded on the sheet film. Because of their importance for technical and art photography, we consider view cameras separately and in detail in chapter 15.

14.5.3 TLR cameras

A TLR camera has a real-image, right-angle viewfinder that is of the same format and focal length as the picture-taking part of the camera. Imagine two cameras, each with its own lens, one on top of the other. The actual camera is in the bottom. The image focused by the upper lens is reflected by a 45° mirror to a viewfinder that is viewed from above. Since both the *viewing lens* and *taking lens* are of the same focal length, the image in the viewfinder is the same as what is focused onto the detector.

If the viewfinder uses a GGS, then the view is only in focus if the viewing lens is exactly at the focus distance. In a TLR camera, the viewfinder lens is movable; it can be moved closer or farther from the 45° mirror and GGS. The trick is that the viewing lens and taking lens move together; as the viewing lens is focused, so too is the taking lens.

Thus one holds the camera at waist level and looks down into the top of the camera to see the lens's view in the forward direction. Both lenses are adjusted until the image on the viewfinder's GGS is in focus. The *waist-level viewfinder* has both advantages and disadvantages compared to the more-familiar *eye-level viewfinder*. It gives a somewhat more natural perspective for a full-body portrait of a nearby person. It is much easier to use if one wants a picture from a perspective that looks upward at the subject from a low vantage point. Of course, one is seeing the viewfinder image as reflected in a mirror; this flips the upside-down focused image to right-side up. However, it leaves the viewfinder image flipped left-to-right, which can be a bit confusing until one gets used to it.

The TLR approach is most commonly used for medium-format roll film. The space behind the angled mirror in the viewfinder provides a convenient location for the take-up spool. Figure 14.21 shows three examples, all using 120 film in a square $6cm \times 6cm$ format. For the Yashica (left), both the viewing and taking lenses move back and forth on the same plate as one focuses. The less-sophisticated Voigtländer (center) and Lubitel (right) focus the viewing and taking lenses together with a clever geared mechanism. Each lens is on a helical mount; rotating the lens' housing moves it closer or farther from the film or GGS. But they work opposite each other and are simply geared together by meshed teeth on the outside of the lens' housing.

All three cameras in figure 14.22 have pop-up lens hoods to block outside light while looking into the viewfinder. The hoods incorporate flip-out magnifiers to allow for a more detailed viewfinder image. The Yashica at left has a GGS in its viewfinder, and so critical focus of the viewfinder automatically focuses the taking lens onto the film. The center and right-hand cameras instead have brilliant viewfinders—a large short-focus lens at the focal plane redirects the image to the eye. This does not allow for critical focus since the viewer's eye is as much a part of the optical system as the other lenses. But a small GGS spot on the bottom of the

Figure 14.21. Three vintage 120 film TLR cameras. **Left:** Yashica EM; **Center:** Voigtländer Brillant; and **Right:** Lomo Lubitel 166 Olympic.

Figure 14.22. A look into the waist-level viewfinders of two TLR cameras. **Left:** The GGS viewfinder of the Yashica EM. **Right:** The brilliant viewfinder, with center GGS focusing spot, of the Voigtländer Brillant.

brilliant viewfinder lens allows for critical focus in that region. See figure 14.22 for views through (left) the GGS viewfinder of the Yashica and (right) the brilliant viewfinder of the Voigtländer.

A TLR camera can have a significant parallax error. The distance between the viewing lens and the taking lens must be at least the full height of the film format—twice the minimum distance of a simple viewfinder camera. Some TLR cameras have a near-focus framing mark in the viewfinder; in more elaborate versions the mark may be coupled to the focus mechanism. In at least one example (see figure 14.23), the entire viewfinder tilts by just the right amount as one focuses so as to always point at the subject.

There is a clever arrangement for some TLR cameras that truly eliminates parallax—but only for motionless subjects. A mechanism is placed between the camera and tripod that allows one to, after framing but just before the exposure, flip

Figure 14.23. To partially correct for parallax, the entire reflex viewfinder assembly of this 1930s era Voigtländer Superb TLR camera tilts downward for closer subjects. **Left:** The camera is focused for infinity, and the viewing lens is parallel to the camera. **Right:** The camera is set for its closest focus. Notice that the lenses not only protrude farther, but the viewing lens points noticeably downward.

a lever that raises the camera such that the taking lens is now in the same position as the viewing lens was.

Although the TLR arrangement naturally leads to a waist-level viewfinder, most such cameras provide other options. The Voigtländer, at center in figure 14.21, incorporates a reverse Galilean viewfinder into the pop-up finder hood. In the Lubitel at the right, a hinged door (surrounding the Lomo logo) can be folded down to reveal a frame viewfinder. Thus, both cameras can be pointed at eye level as an alternative to the waist-level TLR viewfinder.

14.5.4 SLR cameras

An SLR camera is in some ways similar to a TLR camera. A 45° mirror is used to reflect the image from the lens by a right angle to a viewfinder onto a GGS. But there is only one lens—the taking lens. The light detector is placed directly behind the mirror. Immediately before the picture, the mirror flips upward, closing off the viewfinder and allowing the image to focus upon the detector instead. This means that the viewfinder image goes dark just before the exposure, but it also means that what one sees in the viewfinder is exactly what is focused onto the detector. An SLR camera has several advantages over a TLR camera regarding pointing and framing.

- There is no parallax error.
- Only one lens is needed rather than the two lenses for a TLR camera.
- The overall bulk is potentially reduced as compared to a TLR camera because the viewing and photographing portions are combined into one unit.
- Both TLR and SLR designs lend themselves to cameras with *interchangeable lenses*. But this is far simpler with an SLR camera, since there is only the one

lens to replace compared to the matched viewing–taking lens pair of a TLR camera.

- A metering system that works with interchangeable lenses can be incorporated directly into the viewfinder since it measures the very light that will expose the film.

Because of the 45° mirror it shares with TLR cameras, an SLR is at its most basic a waist-level viewfinder system. The first SLR systems were large-format sheet-film cameras. Probably the last such camera to be made commercially was the Graflex RB Super D, which ceased manufacture decades ago. An example can be seen in figure 14.24. This particular example uses an $3\frac{1}{4} \times 4\frac{1}{4}$ inch sheet film, but the larger 4 × 5 inch version is more famous, used notably by Dorothea Lange, Margaret Bourke-White and others.

Note the roughly cubic shape (left) of the box part of the camera. This contains the reflex mirror, which in its down position directs the image from the lens to a GGS at the top of the camera body. The pop-up hood is used as a shade through which to comfortably view it. The camera lens (at the right) has its own pop-up sun shade. The knob at lower right is geared to a rack that moves the lens assembly (connected with a light-tight bellows) back and forth to focus. Awkward contortions would be required to look through the viewing hood if the entire camera were tipped on its side, and so this model has a *rotating back*. The sheet-film holder at the back of the camera can rotate to choose between portrait and landscape formats.

The film holder attaches at the back, behind a *focal plane shutter*. Located as close as possible to the film itself, a horizontal slit in an opaque sheet of cloth moves rapidly across the film, briefly exposing each part to the image from the lens. The effective time of exposure is controlled by the width of the slit and the speed at which it moves. Focal plane shutters in smaller-format SLR cameras usually use two

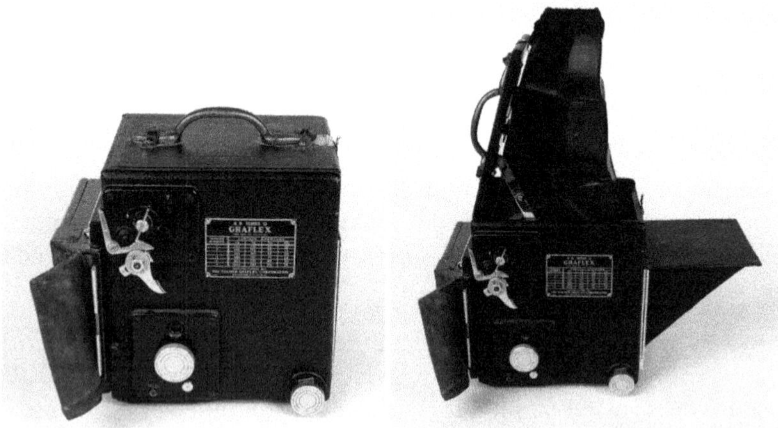

Figure 14.24. A Graflex RB Super D camera, for an $3\frac{1}{4} \times 4\frac{1}{4}$ inch sheet film. It is shown at **right** with its top-mounted pop-up viewing hood extended and its lens hood open.

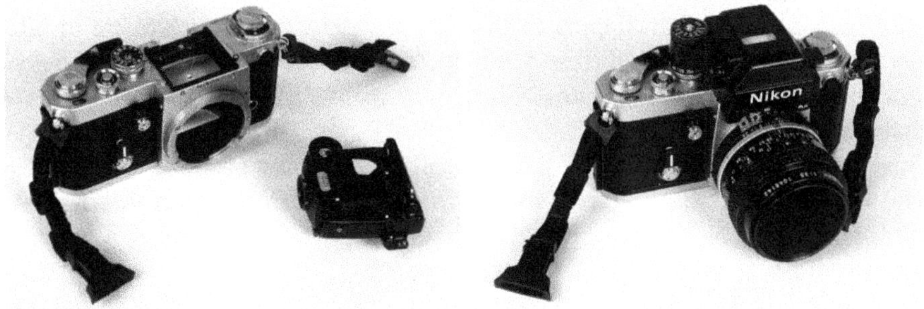

Figure 14.25. A Nikon F2 SLR camera, seen at **left** with its lens and pentaprism eye-level viewfinder removed.

separate curtains that move across the film plane—one leads to expose the film and the other follows to halt the exposure.

Waist-level viewfinders for small film formats, whether SLR or TLR, require additional lenses. The view screen is simply too tiny for the photographer to see any detail without the aid of what is essentially a magnifying glass. Usually, this viewing lens is incorporated into the pop-up hood.

This becomes more problematic at the smallest formats—35 mm roll film, for example, or even-smaller digital detectors. Nonetheless, most SLR cameras in the world are indeed of these tiny formats. The problem is solved by turning the waist-level viewfinder into an eye-level viewfinder with a special *pentaprism*. The overall effect of the special shape of the glass prism is to add an extra 90° bend to the light path. But the additional internal reflections have the added desirable consequence of correcting the left-right image reversal produced by the reflex mirror.

The finder lens of the pentaprism SLR viewfinder allows the photographer to place their eye right against the camera and focus directly onto the GGS focus, just as the lens focuses the image onto the other side of the ground glass. See figure 14.25 for an example of a 35 mm SLR camera, the Nikon F2. With the pentaprism assembly removed, one can see the resemblance of its central reflex assembly—minus the film rewind and advance mechanism—to the much larger Graflex in figure 14.24.

14.6 Digital cameras

Digital cameras use a photoelectronic detector (chapter 38). The exposure to light produces an electrical signal that is measured and stored digitally and from which a picture can be reconstructed by a computer. One key difference is that, as compared to photochemical detectors such as silver gelatin film, the detector is reusable. Thus, one of the key technical challenges of designing a film camera is automatically solved: there is no need to have some clever mechanism by which to get the light detector in and out of the camera while keeping it in the dark.

Many digital detectors are able not only to record an image from a brief exposure to light, they can also output a live video signal. Thus an electronic display screen can show the photographer, in real time, the image focused onto the detector. We can then design a digital camera that has no viewfinder mechanism at all; one simply watches the video feed from the detector on its on-board display screen. It is also possible for a digital camera to couple the live video feed to an autofocus mechanism.

Thus it is possible for a digital camera to be simply a dark box with a lens that focuses directly onto the detector and a live-view electronic display to show the image for framing. No separate lenses, mirrors or prisms are needed. But it is also possible to make a digital camera in the same way that a film camera is made but with the film mechanism replaced by a permanently-mounted internal digital detector. Arguably the most versatile (and often the most expensive) digital cameras use the SLR approach—the so-called *digital SLR* camera or *DSLR* camera. There are also digital versions of viewfinder and rangefinder cameras. The increasingly-popular *mirrorless* system incorporates an eye-level viewfinder that uses lenses to focus onto a tiny high-resolution video screen for a live view of the image focused onto the detector by the camera lens.

14.7 Homemade cameras

Most film camera designs require somewhat sophisticated mechanics, often beyond the means of a home camera builder. One of the most problematic parts to make from scratch is the *shutter*—the mechanical device that allows the light to pass through the lens to the film for only a tiny, but controlled, amount of time. There are two important situations, however, for which no elaborate timed shutter mechanism is needed because the exposure times are measured in seconds or minutes rather than tiny fractions of a second. Thus one needs only a removable cover and a timer to control the exposure.

- **Pinhole photography:** The aperture is so small that, even on a bright sunny day, exposure times are at least many seconds long.
- **Very-low sensitivity (VLS) photography:** The photochemical light detector is of such low sensitivity that, even though it may be a sunny day and a large-aperture lens is used, exposure times are still minutes long.

For pinhole photography there remains the problem of how to get the light-sensitive material into the camera before the exposure, and out of the camera for processing after the exposure. A common and simple solution is to load the camera in a darkroom, make the exposure and then bring it back into the darkroom for unloading and processing. Inconvenient to be sure, but it is the way many pinhole photographs are made. Such a camera can easily be (and often is) made from a simple cardboard box.

As we will describe in more detail in chapter 24, a VLS camera is in one way simpler than a pinhole camera. The low sensitivity of the detector means that it can be transferred back and forth between the camera and a dark bag while protected from light by only a dark cloth—even in broad daylight. But the low sensitivity also means that a lens is required rather than a tiny pinole. Thus there must also be some kind of simple focus mechanism. Detailed design considerations for both pinhole and VLS photography can be found in Beaver 2023.

Reference

Beaver J 2023 *LoFi Photography: Art from Do-It-Yourself Chemistry and Physics* (Cambridge: Royal Society of Chemistry) (forthcoming book)

IOP Publishing

Photography (Second Edition)
Physics and art in focus

John Beaver

Chapter 15

The view camera

The lens of the most familiar type of camera has an axis that is centered on, and perpendicular to, the plane of the light detector. One can adjust the distance between the lens and the detector in order to focus. But that is the only way in which the lens and detector can move relative to each other. A *view camera*, on the other hand, has movements—adjustments other than focus for changing the relation between the lens and the detector.

A common way to accomplish this is to dispense with the dark box altogether. Instead, we mount the lens and detector separately, each on its own *standard*: a *front standard* for the lens and a *rear standard* for the detector. If we mount both standards separately on a common rail, this allows each to separately shift and swivel while allowing us to move the two standards closer or farther apart in order to focus. We then use a flexible bellows to connect the two and keep out the light.

What I have described is a special type of view camera called a *monorail*, as shown in figure 15.1. Although relatively heavy and bulky, it allows for the greatest

Figure 15.1. A monorail view camera. **Left:** Front standard. **Right:** Rear standard with an image focused on the ground-glass screen.

Figure 15.2. A vintage German-made Linhof Technika III technical camera. Although it is capable of a wide range of movements (**left**), it can fold into a compact box (**right**).

flexibility in movements. There are more compact versions, called *field cameras* or *technical cameras*, that fold up for portability, the trade-off being somewhat greater limitations in the movements (see figure 15.2).

The name 'view camera' comes not so much from the movements themselves but from the way focus is achieved. The lens is allowed to project onto a ground-glass screen (GGS) mounted in the rear standard. This screen can be examined directly from the outside in order to compose and focus the image (see the right-hand image in figure 15.1). The screen is then replaced by the light detector, and the picture is taken.

Nearly all view cameras have removable lenses, so different focal lengths can be used. Although some have used a single large shutter at the focal plane itself (right in front of the film), most have the shutter mounted right on the lens. This means that each lens has its own separate, built-in shutter, and also its own adjustable iris for changing the lens' aperture.

Since the addition of mechanical movements requires a certain amount of physical space to accomplish, a view camera is usually also a *large-format camera*: a camera with a detector that is 9 × 12 cm or larger. A single frame of 35 mm film, on the other hand, is only 2.4 × 3.6 cm. Since it is very difficult and expensive to make a digital detector even as large as 35 mm film, large-format cameras almost always use traditional film, most typically in the form of individual sheets (rather than rolls).

A large-format view camera with movements opens up new possibilities and raises new issues.

- Since the picture must be composed on the view screen before the film is even put in place, a view camera is impractical for photographing moving subjects.
- Since sheet film is used instead of roll film, each individual exposure can be processed differently to better control for contrast.
- Since the view screen shows exactly what will be on the film (the screen is removed and the film put in its place), one can very carefully control every detail of the composition of the photograph.

- The camera movements allow one to alter the geometrical perspective of the image. One can control what lines are parallel and what lines converge.
- The camera movements allow one to alter the orientation of the focal plane. The plane of focus no longer must be parallel to the film. It can now be, for example, at an angle stretching from the nearby left to the distant right.
- The elaborate setup and the careful focusing, arranging of movements and composing ensures that one takes fewer photographs but spends much more time on each one. Thus use of a view camera emphasizes quality over quantity.
- The larger-format view cameras (18 × 24 cm for example) are so big and heavy that one often finds the most interesting subjects within only a few meters from the car.

15.1 Description of movements

A view camera with *full movements* allows for the following adjustments of both the front and rear standard (except for *rotation*, which applies only to the back).

- **Focus:** This is the basic movement that most cameras possess, so ubiquitous that is sometimes not called a movement at all. But for a view camera, focus means that the distance between the front and rear standard can be changed. On some view cameras the rear standard is fixed while the front standard can move forward and back on either a rail or a set of tracks. On others (monorails especially), both standards can be moved independently of each other. Sometimes the tripod mount can be moved independently as well. The focus movement is usually limited in several ways.
  - The monorail or focus tracks are only physically so long. Many monorail view cameras have rails that can be extended.
  - There is usually also a minimum distance at which one can physically place the front and rear standard before their mechanisms touch each other.
  - The bellows will only stretch so far and will only compress so tightly. For this reason many view cameras have interchangeable bellows to make it adaptable for both very-short focus or very-long focus lenses.

In addition, the other movements listed below both affect and are affected by the focus range.

- **Rise:** The standard moves vertically upward.
- **Fall:** The standard moves vertically downward.
- **Shift:** The standard moves to the left (left shift) or right (right shift), as seen from behind the camera facing the subject.
- **Swing:** The standard pivots about a vertical axis. As seen from above the camera, a clockwise rotation is called *right swing*, while a counter-clockwise rotation is called *left swing*. Some view cameras are not equipped with swing, but in my opinion 'it don't mean a thing if it ain't got that swing.'
- **Tilt:** The standard pivots about a horizontal axis. If the top of the standard is tilted forward (toward the subject) it is called *forward tilt*. If the top of the standard is tipped rearward, it is called *backward tilt*.

Figure 15.3. Front movements on a 9 × 12 cm monorail view camera. **Left:** Front rise. **Center:** Front right swing. **Right:** Front forward tilt.

Table 15.1. The possible movements for a view camera. Most view cameras have only some of these movements.

Front Rise	Front Fall	Front Left Shift	Front Right Shift
Rear rise	Rear fall	Rear left shift	Rear right shift
Front right swing	Front left swing	Front forward tilt	Front backward tilt
Rear right swing	Rear left swing	Rear forward tilt	Rear backward tilt
Front focus	Rear focus	Tripod mount movement	Back rotation

- **Rotation:** The view screen and film holder rotates so as to change the orientation of the rectangular piece of film. Cameras that have this feature are said to have a *rotating back*. More often, the back does not fully rotate, but the view screen and film holder can be quickly removed and repositioned at a 90° angle.

See figure 15.3 for illustrations of some common examples of front movements as demonstrated with a 9 × 12 cm monorail camera.

The full list of movements can be found in table 15.1. Most view cameras have only a subset of these, but it is often the case that two movements can be combined to give the same effect as a third movement. For example, one inch of front rise while lowering the tripod by one inch has the same effect as one inch of rear fall. If one uses both front and rear forward tilt of 10° while also using the tripod to point the entire camera upward by 10° it has almost the same overall effect as front rise. Thus many view cameras save on complexity (and thus expense, weight and bulkiness) by choosing a strategic base of movements with the intention that the rest can be 'faked.'

15.2 Movements and the image circle

The movements of a view camera are necessarily limited by the physical mechanism. The front and rear standards will shift only so many inches or swing by only so many degrees. But the lens itself provides another constraint. Recall the concept of *vignetting* from chapter 9 in the context of a pinhole camera[1]. Vignetting is a feature of lenses too, though it can be mitigated somewhat by clever lens design. Even so, every lens has a maximum angle of view outside of which no light is able to pass through the lens.

Thus the image made by a given lens lies within a circle formed by the intersection of the maximum angle of view and the image plane at the light detector. This is called the *image circle*, and it is an important feature of any lens used for large-format photography with a view camera.

For one thing, the diameter of the image circle must be larger than the diagonal of the film format. The diagonal of 9×12 cm inch film, for example, is (by the Pythagorean theorem) $\sqrt{9^2 + 12^2} = 15$ cm, and so a lens for this format must have an image circle at least that large. The lens is then said to cover the format.

See figure 15.4 for an example of a picture taken with a lens that does not cover the film format. Notice the clear presence of an image circle, outside of which there is no image at all. Near the edge of the image circle there is an image, but it is darker. So, in this case, there is an image circle that does not cover the format, and it shows partial vignetting at the edges of the image circle.

Movements make the issues of coverage and vignetting even more important. A lens that covers a particular piece of film without movements may fail to do so when movements are employed. As an obvious example, consider the case of a lens that has an image circle only slightly larger than the diagonal dimension of the 9×12 cm film being used. With no movements, all is well. But clearly, if one employs two inches of rear shift, part of the detector will be moved right out of the image circle.

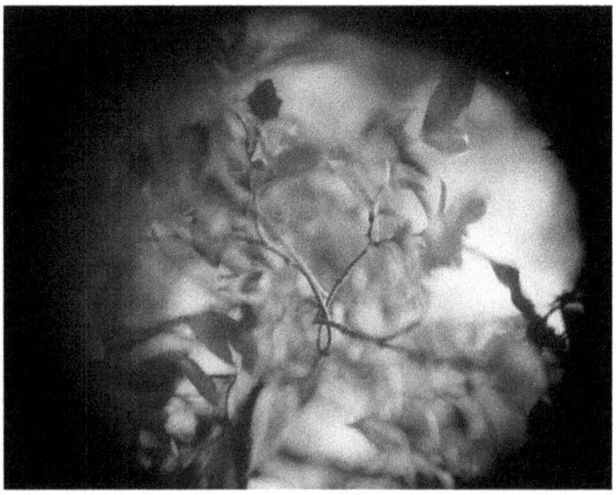

Figure 15.4. An example of a picture taken with a lens that does not cover the particular film format used. The image circle is smaller than the film, and the picture shows pronounced vignetting.

[1] We discuss vignetting further in chapter 20.

This is most obvious for rear rise, fall and shift, but all movements (even focus) can move part of the film out of the image circle—thus causing vignetting. For this reason most view cameras have a view screen with *clipped corners*: the corners of the GGS are literally cut off. The focus screen in the right-hand image of figure 15.1 shows an example. When composing the picture the photographer looks through those holes at the lens. If the full aperture of the lens is visible from all four corners, then vignetting is likely to be negligible.

This can be a more-or-less thing rather than an either-or thing. For it is possible that, as seen from a particular corner of the film, some but not all of the lens' aperture is visible. Figure 15.5 shows two views through the clipped corner of a view camera screen, showing partial vignetting and no vignetting.

Partial vignetting most often occurs with large-lens' apertures, and this can lead to a fading out of the image at the edges. For a very small aperture, the lens is pretty much either blocked or not (as seen from the film). And so vignetting at small apertures tends to give a sharp-edged circular cutoff of the image.

Clearly, in order for a view camera lens to allow for movements without vignetting, its image circle must be significantly larger than the film being used. For a given focal length it is usually more expensive, sometimes considerably more, to produce a lens with a large image circle than with a small image circle. This is especially true for a wide-angle lens because a wide-angle lens already, by definition, has a large angle of view; to allow for movements, the angle of view must be even larger still. Figure 15.6 shows for comparison two lenses with very different image circles and angles of view. The lens on the left—although it is physically much smaller—has a larger aperture by which light enters the camera (it has both a longer focal length and a smaller focal ratio). The lens on the right is physically so much larger because it is better corrected for aberrations and has a much larger image circle and angle of view, allowing for its use as a wide-angle lens for large format with movements. The lens on the left, on the other hand, was intended for only medium format and with no movements. It takes a lot of glass to get all of those light rays, coming from such a wide range of angles, to their proper places on the film.

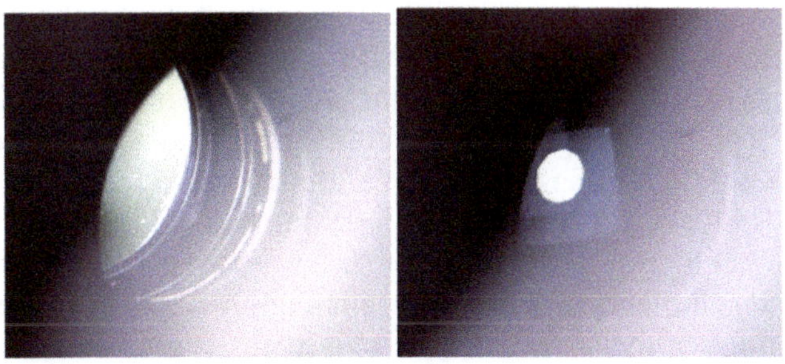

Figure 15.5. Views from the clipped corners of a view camera focus screen. **Left:** At large aperture part of the lens opening is obscured, and so partial vignetting would occur. **Right:** At a smaller lens opening all of the aperture is visible, and so little vignetting will occur at this corner of the image.

Figure 15.6. Left: 114 mm *f*/4.5 Voigtländer Anastigmat Skopar. **Right:** Schneider 90 mm *f*/5.6 Super Angulon. Even though the Skopar on the left has an aperture that is physically over 50% larger, the Skopar is much smaller and lighter because it is less corrected for aberrations and has a much smaller image circle.

15.3 Selective focus

In section 11.6 we considered the use of a large lens opening (small focal ratio) to achieve a shallow depth of focus. This can be useful to ensure that only the subject is in focus while rendering other parts of the picture blurry. In a traditional camera, this *plane of best focus* is perpendicular to the film and lies all at the same distance from the camera.

But the use of movements on a view camera introduces new possibilities. Look carefully at the images in figure 15.7. These two photographs were taken with large apertures, and so for each there is a well-defined plane of best focus—an imaginary flat plane out in the world for which objects are rendered in good focus, while objects on either side of this imaginary plane are rendered in poor focus.

The difference is that the plane of focus for these images is not all at the same distance, and it is not parallel to the film. In the left-hand image, the plane of best focus is vertical, but it runs from nearby on the left (putting the plant in the window in focus, while the distant background is blurry) to the distant right (putting the right side of the window out of focus, while the distant trees on the right side are sharp).

A similar technique was used for the right-hand image in figure 15.7, which shows what seems to be an in-focus path going almost straight away from the camera. The left-hand image in figure 15.8 is more subtle. But close inspection shows that, while the table cloth is blurry, the top of the candle and the top of the pomegranate and spoon are in perfect focus. Although the picture was taken at a downward angle, the plane of best focus lies almost parallel to, but slightly above, the table.

These alterations of focus can be done only with a view camera. The trick is to angle the lens relative to the film. I used front swing for the images in figure 15.7 and front forward tilt for the left-hand image in figure 15.8. The geometry of these movements is shown in figure 15.9, and it demonstrates that the plane of the lens lies

Figure 15.7. Left: *Out the window*, John Beaver 2008. Front swing allows one to put the foreground in focus on one side of the picture while the background is in focus on the other side. **Right:** *Creative nonviolence (for Kathy Kelly)*, John Beaver 2008. I used front swing to make an in-focus path running from lower left to upper right.

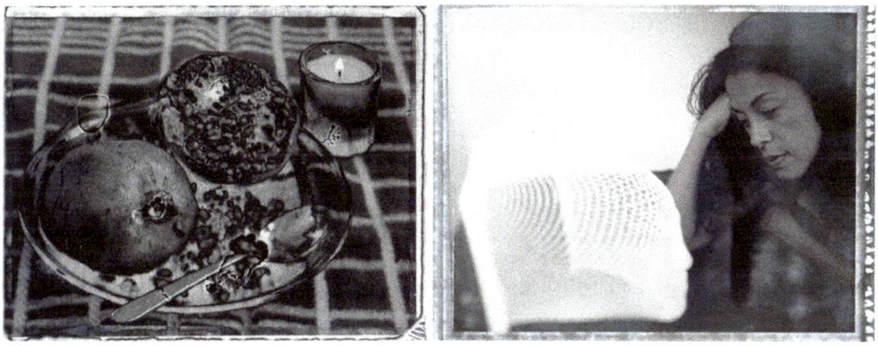

Figure 15.8. Left: *Pomegranate*, John Beaver 2006. Front forward tilt was used to place the plane of best focus parallel to, but above, the table. **Right:** *Mask*, John Beaver 2011. Front swing was used to connect the forehead of the subject to the corresponding part of her angiography mask.

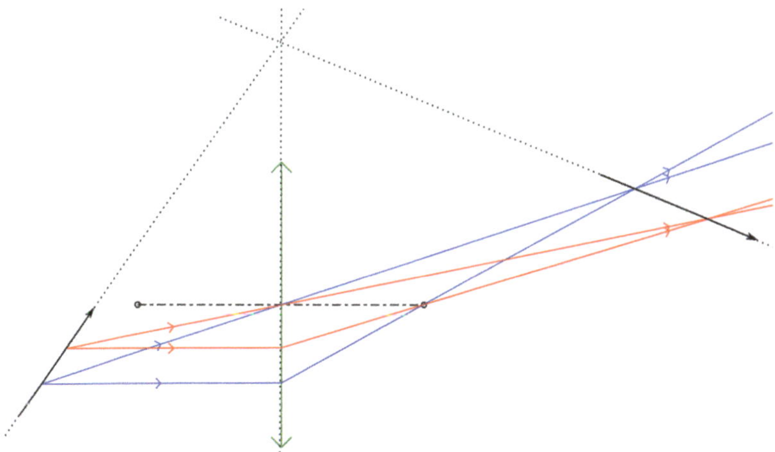

Figure 15.9. If a view camera is used to angle the lens (green vertical line) with respect to the image plane (black arrow on right), then the plane of best focus out in the world (black arrow on left) is also at an angle. The image, object and lens planes converge when extended.

at an angle somewhere in between the planes of the film and the plane of best focus; when extended, the three planes intersect each other.

Control of the angle of the plane of best focus opens up many exciting possibilities and can solve many technical problems. Say, for example, that one wants to photograph several people in a line among a lot of clutter. It would be nice to use selective focus to isolate the people from the clutter around them. But without camera movements, the only way to do this would be to photograph them from directly in front. A large aperture could then put foreground and background objects out of focus.

But what if one wants to compose the picture for dramatic effect, perhaps from an angle instead? From an angle, without movements, one cannot use selective focus to isolate only the people in focus, since they are all at different distances. With a view camera, however, one simply swings the lens until the plane of best focus lies along the line of people.

Large amounts of front swing or tilt (or both) can sometimes be cleverly used in conjunction with large apertures for shallow depth of focus to draw the eye in such a way as to focus on two widely-separated elements in the photograph, forcing comparison. For the right-hand image in figure 15.8, I used this technique to draw the viewer's eye to both the forehead of the subject and the in-focus geometric pattern on the part of her angiography mask (for treatment of a brain aneurysm), corresponding to her forehead.

15.4 Controlling perspective

A rectilinear lens takes straight lines in the world and projects them onto the image also as straight lines. But this does not mean that parallel lines in the world also appear parallel in the picture. For example, look upward from near a tall building. The sides of the building are (presumably) parallel, but they appear to the viewer as converging off into the distance.

This is just an inevitable consequence of *linear perspective*—the mapping of lines in the three-dimensional world onto a two-dimensional surface. But without access to all of the depth cues that one unconsciously employs while looking at that building in the real world, a photograph of the same scene can give the impression that the building is falling over backwards. This is yet another example of how the experience of looking at a photograph can be very different from seeing the same scene in real life. Sometimes an 'unaltered' photograph is less realistic than an altered one.

For both of the wide-angle images in figure 15.10, I used ordinary cameras that were incapable of movements. For the image on the left I centered the building, and it looks like it is falling over backwards. For the image on the right, I centered the horizon line instead, so the perspective appears normal. The trade-off is, however, that half of the picture is foreground. That works for this particular composition but for many pictures it would not.

Thus it is often desirable to alter the linear perspective as seen by the camera, and this can only be done with movements. While front swing and tilt are the preferred ways to alter the angle of the plane of best focus, perspective is best altered by the use of rear-standard movements.

Figure 15.10. Left: *San Francisco*, John Beaver 2012. **Right:** *Power*, John Beaver 2006. Both of these wide-angle images were taken with cameras incapable of movements. If the tall building is centered (left), it appears as if it is falling over backwards. If, on the other hand, the horizon line is centered (right) then the building appears normal, but half of the picture is foreground. Employing rear fall (or front rise) would have allowed me to center the buildings while correcting the perspective.

The example of photographing a tall building from close up (with a wide-angle lens) is a textbook case. If one points the camera exactly horizontal, then the parallel vertical sides of the building will also appear parallel on the film. But then the building will only be in the top half of the picture; the bottom half will be foreground (and perhaps the top of the building will be out of the frame). If one tilts the camera upward, to precisely frame the building, then normal linear perspective will force the parallel sides of the building to converge, and the picture will look like a building falling over backwards.

The solution is simple. Point the camera horizontal, so the sides of the building appear parallel, and then simply move the rear standard (and thus the film) vertically until the building is centered in the picture. Since the image is upside down, one must use rear fall to recenter the building. If the camera lacks rear movements, then front rise will accomplish the same thing as rear fall. If it matters that the lens is then positioned a couple of inches higher than it was before, so it is now seeing from a slightly different vantage point, then simply use the tripod to lower the entire camera by the same amount. Thus front rise, one of the mechanically simplest movements to incorporate into a camera, can be used to solve one of the most common technical problems. As such, if a view camera has only one movement, it is likely to be front rise.

IOP Publishing

Photography (Second Edition)
Physics and art in focus
John Beaver

Chapter 16

Perspective and pinhole photography

A pinhole camera is an excellent tool for photographic experiments that alter geometrical perspective in unusual ways. It is the simplest of cameras—a dark box with a tiny hole. And since the exposure times are measured in seconds or minutes, a complex mechanical shutter is unnecessary. But also, the most commonly-used light detector for a pinhole camera is black-and-white enlarging paper meant for producing prints in the darkroom from black-and-white negatives. Since the light detector is a piece of paper, it can be angled, folded or curved at will, as it is placed in the pinhole camera. Thus one can do all that a view camera can do, and more, regarding alterations of perspective.

Recall that a pinhole camera does not have a particular best-focus distance; whatever the distance between the detector and pinhole, the focus is essentially the same. This means that one can twist, bend and tilt the light-sensitive paper to the heart's content, and all of the image will be in the same focus. For a pinhole camera this is, admittedly, an imperfect focus—but it does mean that one can explore perspective independent of the effects of selective focus.

Whereas the typical use of view-camera movements is to correct perspective to a 'natural' look, the pinhole photographer is more likely to take advantage of the everything-goes nature of a pinhole camera to produce an image with a strikingly unnatural perspective.

16.1 Curved paper

The most common way that pinhole photographers distort perspective is to curve the detector inside the camera. Figure 16.1 shows examples; the same scene was photographed with three different pinhole cameras. The upper left image was made with the light-sensitive paper flat; it is an ordinary wide-angle rectilinear photograph. For the upper right image, I curved the paper, forming it into a half cylinder. For the lower left image, I curved the paper into nearly a full cylinder with the left and right edges meeting nearly at the pinhole. For comparison, the image at lower right was made with a fisheye lens on a digital camera.

doi:10.1088/978-0-7503-3703-8ch16

© IOP Publishing Ltd 2022

Figure 16.1. Four images of the same scene recorded with (**upper left**) a pinhole camera with a flat detector, (**upper right**) a detector curved into half a cylinder, (**lower left**) the detector curved into nearly a full cylinder, and (**lower right**) a fisheye lens on a digital camera.

Notice that for all three pinhole images, the center of the image is essentially the same. Since the paper was curved in the horizontal direction, it is that dimension that is most distorted. When the detector is curved in this manner, the edges of the picture are closer to the pinhole compared to a flat detector. This means the part of the image at the edges is at a smaller scale than it would be with an ordinary camera —thus objects in the center seem overly large. If the camera is pointed horizontally, horizontal lines curve around the center, while vertical lines remain straight. For the fisheye lens image at lower right in figure 16.1, on the other hand, both horizontal and vertical lines curve around the center.

There is also a pronounced difference in the degree of vignetting between the pinhole images in figure 16.1—a topic we take up in more detail in section 16.8. The images made with curved paper are more evenly illuminated at large angles from the camera.

16.2 Folded paper

One can fold the light-sensitive paper before placing it in the pinhole camera. When unfolded and processed, different sections of the paper will have been exposed from a different perspective. Areas that were hidden inside the folds remain unexposed. One can thus combine origami with photography.

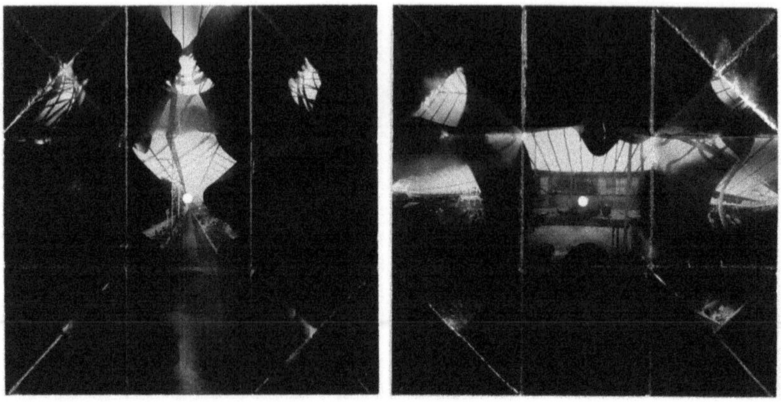

Figure 16.2. *The greenhouse*, John Beaver 2015. These two pinhole photographs were made simultaneously in opposite directions. Two sheets of light-sensitive paper were folded together, using origami techniques, into a cube. A pinhole was placed on each side of the cube and placed inside an identical origami cube made from opaque black paper. Once developed and flattened, the images show a perspective that is not possible to see from any particular vantage point.

For figure 16.2, I attempted the most extreme example of which I could think. I took two square sheets of light-sensitive enlarging paper, punched a hole in the center of each sheet and taped over each hole a small piece of sheet metal with the proper-size pinhole for a pinhole camera. I then used origami techniques to fold the two sheets of paper into a cube, thus forming a pinhole camera[1]. But it had two pinholes, each forming an image on the opposite side of the camera; the camera was formed from the light-sensitive paper itself. To make the exposure, I uncovered both pinholes simultaneously and each made an image on the opposite side of the cube. But since the light-sensitive paper was folded in a complex way, each part of the paper formed a different perspective of the same subject once it was unfolded and flattened.

16.3 Anamorphic pinhole cameras

An *anamorphic lens* uses cylindrical elements to produce a larger image scale in one dimension as compared to the other. Such lenses are widely use in cinema to compress a panoramic view onto film stock that has a more square format. The same lens is then used the other way when the image is projected in order to expand the nearly-square image on the film into a wide format on the screen.

Anamorphic lenses are expensive, but we can easily construct a pinhole equivalent that exaggerates sizes in either the horizontal or vertical direction. The trick is to note that a hole, albeit a square one, is formed by the intersection of two

[1] I placed this origami pinhole camera inside a similar origami cube made with opaque black paper so that light could enter only through the pinholes and not through the backside of the not-entirely-opaque light-sensitive paper.

perpendicular slits—one horizontal and the other vertical. The vertical slit then produces the horizontal part of the image, while the horizontal slit is responsible for the vertical dimension of the image.

If the vertical slit, for example, is farther from the detector than the horizontal slit; it will be as if the camera has a longer focus distance in the horizontal direction than in the vertical direction. Thus the image scale would be greater in the horizontal dimension than in the vertical dimension. If such a camera used a square detector, it would image a greater angle of view vertically than horizontally.

The left-hand image in figure 16.3 was made with a anamorphic pinhole camera with the vertical slit twice the distance from the detector as the horizontal slit. This produces an image scale in the horizontal dimension that is twice that in the vertical dimension. Compare it to the image on the right, made with an ordinary digital camera.

The odd vertical lines in the left-hand image in figure 16.3 result from the crude nature of my anamorphic 'pinhole.' The two slits were made quickly from strips of black paper placed next to each other with a small gap between. Each part of the horizontal slit, for example, is responsible for a particular vertical strip of the image. Thus even the slightest variation in width of the slit leads to different exposures on different parts of the picture.

My particular anamorphic camera has a rectangular format. If it were an ordinary camera, rotating it by 90° would produce an image that is identical except for how the image is cropped. But a fundamentally-different image results when one rotates an anamorphic camera; it changes which axis has the largest image scale. Figure 16.4 shows the same scene photographed with the same camera but rotated 90° and 45°, respectively.

It is easy to make a crude anamorphic 'pinhole' camera. But to construct one that exposes all parts of the image evenly is difficult. It is far easier to form a round pinhole than a perfectly-even long slit.

Figure 16.3. Left: Image made with a crossed-slit anamorphic 'pinhole' camera. The vertical slit was twice the distance from the detector as the horizontal slit. **Right:** The same scene imaged with an ordinary camera.

Figure 16.4. Images made with the same anamorphic camera as used for figure 16.3, but with the camera rotated 90° (**left**) and 45° (**right**).

Figure 16.5. The subtle character of out-of-focus parts of an image made by a lens is called bokeh.

16.4 Pinhole bokeh

The term *bokeh* usually refers to the often-subtle character of out-of-focus spots of light in a photograph, as imaged by a particular lens (see figure 16.5). The word would seem to have no relevance for pinhole photography; there is no lens and no particular focal distance. Although the causes are different, a pinhole photograph may also have a unique character to its blurriness.

Different distances do 'focus' somewhat differently in a pinhole camera. Recall that there is an optimal-size pinhole that produces the sharpest image, and that this optimal size depends upon the focus distance. For most pinhole cameras, the focus distance varies across the picture, being smallest in the center if the camera is a simple box. Thus if the pinhole is the best possible size for the center of the image, it will be somewhat too small for the edges. Equally important, however, is that at significant angles off axis the pinhole becomes more and more like a thin line—wider in one dimension than the other.

Let us consider, by example, the case of a circular pinhole that is the optimal size for the center of the image in a camera with a center focal length of $F = 100$ mm and a full angle of view of 120°. How does the image sharpness at the edges compare to that at the center? For optimal sharpness (see appendix C), the pinhole should have a diameter of $D_0 = \sqrt{2.44F\lambda} = 0.35$ mm.

In appendix C, I make a crude estimate of the size (x) of the image of a point source in our pinhole camera, as a function of the pinhole diameter (D).

$$x(D) = D + 2.44\frac{F\lambda}{D} \tag{16.1}$$

$$x_c = x(D_0) = 2D_0, \tag{16.2}$$

Here, x_c is the size of the image at the center of the picture, using a pinhole of optimal size for the focal length. This means that for light of a wavelength of $\lambda = 500$ nm, an imaged point at the center of the picture would have a width of about $x_c = 0.70$ mm if an optimal-size pinhole is used for that wavelength and focal length.

What about the edge of the image? From that vantage point the pinhole appears elliptical with a width equal to D_0 in one dimension, but smaller in the other by a factor of $\cos\theta$, where θ is half the full angle of view ($\theta = 60°$ in our example). Furthermore, the focal length is larger by a factor of the reciprocal of $\cos\theta$. And this elliptical image is intercepted by an angled piece of paper, further stretching out its image. Let us use equation (16.1) to estimate the sizes of the long axis (x_a) and the short axis (x_b) for this elliptical image. We can then compare them to x_c and to each other. For the short axis (x_b), we use the full width of the pinhole (D_0) but the increased focus distance $F/\cos\theta$.

$$x_b = D_0 + \frac{2.44F\lambda}{D_0\cos\theta} \tag{16.3}$$

$$= D_0 + \frac{D_0^2}{D_0\cos\theta} \tag{16.4}$$

$$x_b = D_0\left(1 + \frac{1}{\cos\theta}\right) \tag{16.5}$$

For $\theta = 60°$, this gives us $x_b = 3.0\ D_0 = 1.5x_c$.

For the long axis (x_a), we use the pinhole diameter reduced by a factor $\cos\theta$ and the focus distance increased by the reciprocal of $\cos\theta$. We divide by an additional factor $\cos\theta$ to account for the fact that the elliptical image, coming from the angled pinhole, intercepts an angled detector surface.

$$x_a = \left[D_0\cos\theta + \frac{2.44F\lambda}{D_0\cos^2\theta}\right] / \cos\theta \tag{16.6}$$

$$= D_0 + \frac{D_0^2}{D_0\cos^3\theta} \tag{16.7}$$

$$x_a = D_0\left(1 + \frac{1}{\cos^3\theta}\right) \tag{16.8}$$

For $\theta = 60°$, this gives us $x_a = 9.0\ D_0 = 4.5x_c$. We can also see that the ratio of the long axis to the short axis of this elliptical image is given by the following equation.

$$\frac{x_a}{x_b} = \frac{1 + \cos^3\theta}{\cos^2\theta(1 + \cos\theta)} \tag{16.9}$$

Figure 16.6 shows a graph of equation (16.9). Notice that it climbs very steeply as the angle approaches 90°. Between 60° and 70° off axis, the elongation of the image more than doubles.

Figure 16.7 shows images made by a laser (the collimated beam simulates a distant point source) passing through a pinhole toward the center of the image (left) and at an angle of 60° (right). Both images are shown to the same scale.

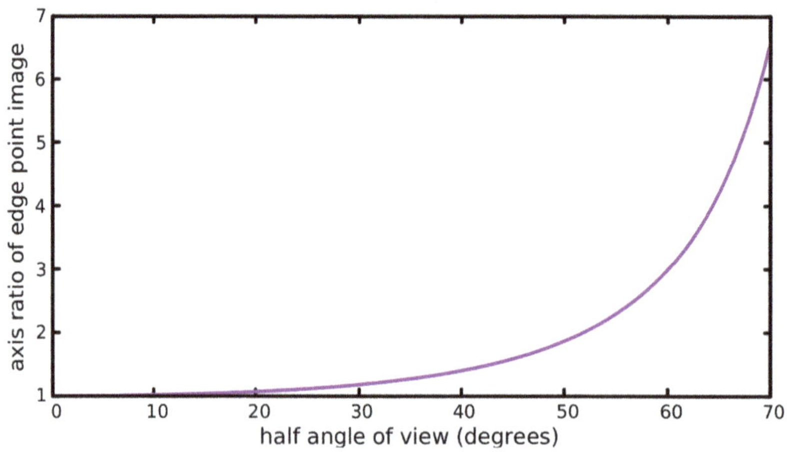

Figure 16.6. Ratio of the long axis to short axis of the elliptical image of a point source, formed by a pinhole when the light enters at an angle.

Figure 16.7. Images of a laser passing through a pinhole at an angle of (left) 0 degrees and (right) 60 degrees.

Precision measurement with a scanner shows that both the size of the left-hand image and the narrow dimension of the right-hand image are in good agreement with, respectively, x_c and x_b from equations (16.2) and (16.5) for the wavelength of the light and the size of the pinhole. But even a casual inspection of the right-hand image indicates that the long dimension of the image is much larger than x_a as given by equation (16.8). From equation (16.9), the axis ratio should be 3.0, and it is clearly more than that.

The result is unsurprising if we consider a typical real pinhole instead of the imaginary perfect one we have used in our calculations. In particular, the pinhole is in a brass sheet, and—although thin—its thickness becomes increasingly relevant at steeper and steeper angles. It serves to narrow the apparent width, and so it also stretches—and thus blurs—each part of the image outward by even greater amounts than one would expect from a pinhole in a sheet of zero thickness.

It is easy to show from simple geometry that the width of the apparent ellipse when the pinhole is seen from angle θ is given not by $D_0 \cos\theta$ but rather $(D_0 - T\tan\theta)\cos\theta$, where T is the thickness of the metal sheet. This changes equation (16.8) to the following equation.

$$x_a = D_0\left[1 - \frac{T}{D_0}\tan\theta + \frac{1}{\left(1 - \frac{T}{D_0}\tan\theta\right)\cos^3\theta}\right] \tag{16.10}$$

Measuring my pinhole and the thickness of the sheet metal, I have $D_0 = 0.4$ mm and $T \approx 0.1$ mm. Such a pinhole would have zero width as seen from an angle greater than 76° ($\tan^{-1}\frac{D_0}{T}$). This would imply x_a is really closer to 15 times the diameter of the pinhole rather than only nine, giving an axis ratio of the image at 60° of about five instead of three, which is more in line with the right-hand image in figure 16.7.

This illustrates the importance of using the thinnest sheeting possible when making a pinhole for a pinhole camera, especially if wide-angle images are desired. Of course, if one is looking for exactly the effect seen in figure 16.7, then a thicker sheet helps exaggerate it.

Thus point sources would make images at the edge of our wide-angle 120° picture that are overall blurrier than at the center, and elliptical in shape with the long axes, three times the short axes. The long axis of the elliptical image is oriented away from the center of the picture.

For all of this analysis I have used the crude approximation for the image size laid out in appendix A. It is based on the assumption that we can find the approximate size of the image by simply adding that given by geometry to that given by simple Fraunhofer diffraction theory. A proper analysis using full diffraction theory would be far more complex, but it would likely yield results that do not differ markedly from this simple approach.

In particular I have used the naive idea that the diffraction by a slit when light is incident upon it at an angle is simply the same diffraction pattern in the new direction but calculated with a slit width decreased by its projection—a factor of $\cos\theta$. This turns out to give results that are surprisingly close to the correct calculation.

We can see this most easily for single-slit diffraction, but the same argument holds for the circular aperture as well. The correct result is that the first minimum (and thus the edge of the central bright region) appears at an angle θ given by the following equation (Biswas *et al* 2021).

$$\theta = \sin^{-1}\left(\frac{\lambda}{D} + \sin\theta_i\right) \tag{16.11}$$

Here, θ_i is the angle of incidence of the light. Our naive calculation, on the other hand, is that θ is given by the following equation.

$$\theta = \theta_i + \sin^{-1}\left(\frac{\lambda}{D\cos\theta_i}\right) \tag{16.12}$$

This is simply the single-slit diffraction formula solved for the angle of the first minimum but recognizing that all of this is now centered around our new direction—the angle of incidence θ_i. The angle θ represents half the width of the diffraction pattern, and it is larger than it would be because of the $\cos\theta$ factor, recognizing that coming from an angle the light 'sees' a narrower slit.

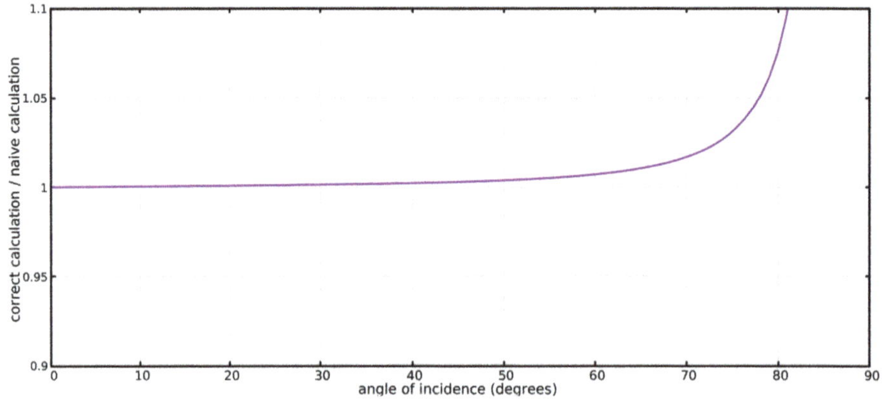

Figure 16.8. A plot of the ratio of the correctly-calculated spreading of the pinhole image with angle compared to the simple and naive calculation derived here. The naive calculation is surprisingly accurate for angles of incidence less than 80°.

This is not really the correct way to do the problem; we should be adding all of those waves coming from different parts of the slit. Furthermore, the incident light encounters one side of the slit before the other, and this too should be taken into account. But when it is done correctly, the answer is surprisingly close to the naive simple calculation unless the angle is very close to 90°. See figure 16.8 for a comparison; I have plotted the ratio of the correct and naive calculations as a function of angle of incidence (with an assumed wavelength of 400 nm and a slit width of 0.1 mm). It is to within 10% of unity out to even an extremely oblique 80°. This is of course, for a single slit illuminated from an angle, not a circular pinhole. But the comparison turns out to be the same, and figure 16.8 would apply to the long axis of our pinhole diffraction image as well (Biswas *et al* 2021, equations 8 and 10).

Reference

Biswas S, Bhattacharyya R and Bhattacharyya S 2021 Explicit derivation of the fraunhofer diffraction formula for oblique incidence *Eur. J. Phys.* **43** 015301

Part IV

Energy and photography

IOP Publishing

Photography (Second Edition)
Physics and art in focus
John Beaver

Chapter 17

The nature of energy

Energy has many forms. It appears that although energy can transform from one form to another, the total amount remains constant as stuff happens. This is called the *conservation of energy*, and it is one of the most important principles in physics.

There is no good one-sentence definition of energy that does not raise more questions than it answers. As with other physical quantities, to be thorough we must use an *operational definition* that describes all of its properties in different circumstances. To do so in detail would require a book of its own; here I give only a brief overview.

There are many types of energy, some of which have familiar names: electrical energy, solar energy, chemical energy and nuclear energy are examples. These different types of energy manifest themselves in different ways and under different circumstances (and so they have different names). But when it comes down to it, all types of energy can be described as some variety of four basic types, of which I give only brief descriptions.

- **Kinetic energy:** This is the energy due to an object's mass and motion when compared to some other frame of reference.
- **Gravitational energy:** This is energy associated with the force of gravity. A glass is nudged off the table, falls to the floor and shatters. The energy to make the glass shatter came at the expense of a decrease in gravitational energy.
- **Electromagnetic energy:** The energy associated with electric and magnetic forces is the basis for solar energy and chemical energy among many other forms. Since light carries electromagnetic energy, it is of particular importance to photography.
- **Nuclear energy:** This is the energy associated with the enormous forces at work in the nuclei of atoms.

All but the first of these are forms of energy associated with the three fundamental *forces* of nature, and I have listed them in order of the relative strength of these

doi:10.1088/978-0-7503-3703-8ch17

© IOP Publishing Ltd 2022

forces. This likely fits with one's prior knowledge; compare the power[1] output from an old-fashioned water wheel (gravitational energy of the falling water) to typical sources of electrical power. And then compare these to nuclear energy.

I say 'three' forces, but the number of fundamental physical forces in nature depends upon how one counts. There are, for example, really two fundamental types of nuclear forces and thus two fundamental types of nuclear energy. I have lumped these together as simply 'nuclear energy.' Furthermore, most physicists believe that all of these forces are (in ways that are, at present, only partially understood) different manifestations of a single, even more fundamental, force.

In addition to these explicit forms of energy, there is a direct correspondence between energy and *mass*. Albert Einstein discovered, as part of the theory of Special Relativity, that there is a direct equivalence between mass and energy that is embodied in the most famous of all equations.

$$E = mc^2 \tag{17.1}$$

Here, c is the speed of light, and equation (17.1) says that a tiny bit of mass is equivalent to an enormous amount of energy. And so matter can manifest itself as energy and vice versa. This is one of the key discoveries of modern physics, and it is the foundation for many important phenomena, including the fact that the Sun shines. But it has little direct relevance for photography, and so we will not consider it further in *Photography: Physics and Art in Focus*. In most circumstances of interest to photographers, mass acts like mass, and energy acts like energy.

Energy can take many forms, and it can change from one form to another (or from one form to several others), but it does not disappear or appear from nothing. From a fundamental standpoint, we now know that there is a connection between this odd principle of the conservation of energy and a basic *symmetry* of the Universe itself. This symmetry is called the symmetry of time translation, and it means that whatever are the fundamental laws of physics, they are the same today as they were last Wednesday and as they will be next Thursday. This, and other symmetry principles, are now recognized as some of the most important ways for a physicist to look at the Universe. Indeed, physicists often look to symmetry principles in their search for new and undiscovered physical laws, most strikingly in the ongoing search for a *unified field theory* of physics. I point out this somewhat arcane fact just to illustrate that one can find many metaphors for art in science itself; symmetry is equally important in art as it is in physics.

But in any event, light has energy, and as light travels from place to place it carries its energy with it. When light does stuff, it is in part because some transfer of energy has taken place. When a dark surface absorbs sunlight, for example, the energy carried by the light is transformed to other forms of energy. The light disappears, but the energy it carried does not; it changes to an equal sum of other forms of energy.

The joule (J) is the official SI unit for energy. How big is a joule? Here is an example: to lift 500 ml of Irn-Bru (or any other aqueous liquid) from the floor to the kitchen counter requires an increase of gravitational energy of about five joules.

[1] We will define power more precisely in section 17.1.

17.1 Energy transfer

Whenever some type of energy is seen to increase or decrease, it is because energy was *transferred* from one kind to another. Let us consider some examples of this process. Although there are four basic forms of energy, there are many other names to describe common cases of energy transfer that are often complex combinations of the four basic types.

1. **Thermal energy:** The individual atoms and molecules in a gas, liquid or solid are constantly in motion, and so these molecules—each individually—have kinetic energy. Thermal energy is related to the total kinetic energy of these individual motions all added up. The transfer of thermal energy is called *heat*, and the average kinetic energy of the individual atoms or molecules is directly related to an object's *temperature*.
2. **Solar energy:** Light carries electromagnetic energy, and when it arrives at some object, it can transfer some of that energy. In many circumstances, whatever energy is not redirected (by reflection, for example) is transferred to thermal energy.
3. **Electrical energy:** In a conducting material some of the electrons are free to move easily from atom to atom. It is then possible to use electric forces to move these electrons from one part of the conductor to another in a continuous fashion. Electrons have an electric charge, and the motion of this charge under the influence of electrical forces results in a change in electromagnetic energy.
4. **Hydroelectric energy:** A given volume of water has mass, and so gravity acts upon it. If one allows water to flow downhill, the mass of the water moves closer to the center of the Earth. This is a decrease in gravitational energy, and so other forms of energy must increase. In the normal flow of a stream downhill, this gravitational energy is turned mostly to thermal energy. But if one builds a dam and allows the water to undergo a sudden decrease in height, a turbine-generator can be used to transfer this decrease in gravitational energy to an increase in electrical energy (a form of electromagnetic energy).

   If one looks at the big picture, however, there is more happening. How does the water get back up to higher elevations so that this process can continue? If water from lower altitudes is to somehow arrive at the higher-elevation source of a river, its gravitational energy must increase, and that means other forms of energy must decrease.

   It is, of course, the natural process of evaporation and precipitation that carries the water to higher elevations. But evaporation requires an increase in thermal energy, and this is provided by the absorption of sunlight. Thus, solar energy (electromagnetic energy from the Sun) is transferred to thermal energy, which is ultimately transferred to gravitational energy, which is then transferred by the dam's turbine-generator to electrical energy—another form of electromagnetic energy.
5. **Chemical energy:** Atoms are comprised of positively-charged nuclei and negatively-charged electrons, and it is the electrical force between opposite

charges that holds an atom together. But when atoms are place near each other the situation is far more complex, and some of the electrons may feel significant electrical attraction to more than one atomic nucleus at a time as well as repulsion by the electrons of more than one atom.

And so the rearrangement of the electrons in neighboring atoms results in a change in electromagnetic energy. Whenever atoms rearrange from one configuration of electrons to another, the resulting changes in electromagnetic energy are known as *chemical energy*.

6. **Elastic energy:** Certain solid materials can be stretched or compressed, but energy is required to do so. If the material is elastic, that energy can be efficiently transferred to the electromagnetic energy of the interatomic and intermolecular bonds. When the material goes back to its unstretched (or uncompressed state), that electromagnetic energy stored in those bonds can be transferred to other types of energy.
7. **Potential energy:** Potential energy is a category, rather than a specific type of energy. As an example, consider gravitational energy. If I lift a particular weight so it is one meter higher, its gravitational energy increases. If I then lower it by one meter, its gravitational energy decreases by the same amount. In this particular case, the gravitational energy depends only on the weight and its height. And so a particular change in height (of a particular weight) results in a particular change in gravitational energy no matter how that change in height came to be.

   For example I could raise the weight 12 m, then lower it by 27 m, and then raise it again by 16 m. Since the end result is to raise the weight by one meter higher than its original position ($12 - 27 + 16 = 1$), the gravitational energy increases by exactly the same amount as if I had simply raised the weight by one meter in the first place.

   For any case in which a transfer of energy depends on only the initial and final orientation—and not how we got there—we can define a *potential energy*. The particular definition is mathematical, and it depends on the details of the physical system being considered. Besides gravitational energy, elastic energy can also be described as a form of potential energy.

Each of these types of energy can be explained in terms of the basic energies described earlier. And these are not the only kinds of energy. Energy is partly a matter of context; one's definition of the total energy of a given situation—what a physicist calls a *system*—depends on the precise definition of that system. This does not, however, mean that energy is subjective or that 'anything goes.'

Let us consider, as an example, hydroelectric energy. A dam holds back water so that a reservoir forms, the top of which is much higher than the stream below it. Thus, as water goes through the dam, it decreases in height (h) by so many meters. Since the water moves closer to the center of Earth its gravitational energy decreases, and the law of conservation of energy says that this must be accompanied by an equivalent increase in other forms of energy. The point is that it is the transfer of energy that is meaningful rather than the quantity of energy. An energy of 137 J could mean anything, or it could mean nothing. But the statement, '137 J of electrical energy was transferred to thermal energy,' means that something physical has happened.

17.2 Energy, power, force and momentum

A word such as *energy* has an operational definition in physics; its meaning is defined by the very experiments used to measure it. The same is true for other physical terms: *momentum, power* and *force* are examples. To learn the operational definitions of these and other such terms is, in effect, to learn how to do physics.

These same words also exist in everyday English usage, and their meanings there are vague and overlapping. One might say, 'I have little momentum today,' or instead one might say, 'I have little energy today,' and mean very much the same thing. In the context of the practice of physics however, such terms do not overlap. In a given sentence, if 'power' is the correct term then 'energy,' 'momentum' and 'force' are not—full stop. Thus, let us take a little time to delineate the meanings of these four important and related—but still distinct—physical terms. Table 17.1 provides some hints regarding the differences and similarities of energy, power, force and momentum and their relationships to more-basic concepts of length, time and mass. One important distinction is whether the quantity is a *vector* or a *scalar.* Leaving aside the precise mathematical definitions of these terms, there is one simple distinction between them: a vector quantity has both an amount (a *magnitude*) and a *direction* in space. A scalar quantity, on the other hand, has only a magnitude.

Mass, for example, is a scalar quantity. A liter of water has a mass of about one kilogram; it would make no sense to say its mass is 'one kilogram in a direction 33° South of East.' Energy and power are also scalar quantities. Force and momentum, on the other hand, are vector quantities. The direction of a force is just as important as its magnitude, measured in the SI unit of newtons (see table 17.1).

We can also see from the third column of table 17.1 that energy, power, force and momentum all have different physical *dimensions*—the basic mathematical combination of length, time and mass. Representing these physical dimensions in terms of their corresponding SI units, we can see, for example, that energy has dimensions of

Table 17.1. Comparison of the physical terms energy, power, momentum and force, and the base quantities of length, time and mass.

Term	Symbol	SI Unit	Vector?	Notes
Length	Varies	Meter (m)	Sometimes	Base quantity
Time	t	Second (s)	No	Base quantity
Mass	m	Kilogram (kg)	No	Base quantity
Energy	E	Joule (J), (Nm), (kg m^2 s^{-2})	No	
Power	P	Watt (W), (J s^{-1}), (kg m^2 s^{-3})	No	$P = \Delta E/\Delta t$
Momentum	$\vec{\mathbf{p}}$	(N s), (kg m s^{-1})	Yes	
Force	$\vec{\mathbf{F}}$	Newton (N), (kg m s^{-2})	Yes	$\vec{\mathbf{F}} = \Delta\vec{\mathbf{p}}/\Delta t$

mass times length squared over time squared, while momentum has dimensions of mass times length per time. This means they are fundamentally different kinds of quantities—apples and oranges if you will—defined by very different experimental measurements.

Despite the fundamental differences between these physical quantities, there are important relationships between them, as can be seen by the many equivalent ways to represent their units listed in the third column of table 17.1. As an example, the SI unit of energy—the joule—is equivalent to a newton-meter, and this means there is a relationship between energy on the one hand and force times length on the other. This book is not the place to go into those details, but the concept of *mechanical work* (familiar from any introductory physics text) defines that physical relationship.

More important for our purposes are the relationships between energy and momentum on the one hand and time on the other. Energy and momentum are related to conservation laws. For a closed system, both the total energy and momentum remain constant as other aspects of the system change with time. Since, for example, the total energy remains constant, changes in energy are always transfers; an increase in energy for one part of the system is countered by a corresponding decrease in other parts. The same is true for momentum, which has the added complication that not only the amount but also the direction must remain constant.

This means that whenever anything happens, there are always transfers of energy and momentum. Often this is a more-or-less continuous process, and we are interested more in the *rate of transfer* than the total amount transferred. Thus we have the two important physical quantities of power and force. Power is the rate of transfer of energy, while force is the rate of transfer of momentum. We will be especially concerned in this book with energy and power, as these quantities most directly relate to important interactions between light and matter that are the basis for photographic processes.

IOP Publishing

Photography (Second Edition)
Physics and art in focus
John Beaver

Chapter 18

Energy and exposure

18.1 Defining our terms

When light falls upon a light *detector*, such as traditional photographic film, a digital detector or the retina of the human eye, energy must transfer for anything to happen; some physical change must occur in the detector, and this requires a transfer of energy. The amount of energy per unit area that arrives at the detector is related to what photographers call *exposure*. The amount of physical change that the detector undergoes as a result of exposure to light is called *density*. Density is the effect, it is what makes the picture, and exposure is the cause.

We must untangle several interrelated concepts; I have listed them in table 18.1 along with their SI units, the symbols I will use in this book to represent them, and some of the relations between them. We will consider each in detail and how they relate to each other in the following sections. Here I give a brief summary of each.

Before one immerses oneself too deeply in table 18.1, I must give a warning. There are several ways to lay out the relevant definitions, and the exact terminology that is used in practice depends, unfortunately, upon the context and the particular field of study. In particular there is a distinction between the physics of the light itself and our human perception of that light. Also, we often define quantities in a practical way that takes into account the particular technical methods by which we measure them. For now, I concentrate on the physics of the light itself with no special regard to our perception of it or the technical details of the devices we use to measure it. I choose a terminology that seems to me the most straightforward for that purpose.

Power, *P*: A rate of transfer of energy per unit of time, and it is measured in watts or joules of energy transferred per second. The total emitted brightness of a source of light is an example of power, and that is why a 100 W lightbulb is brighter than a 40 W lightbulb[1].

[1] I am assuming the two lightbulbs are of the same type. We will later consider some of the complexities of the relation between the power of a light source and its perceived brightness.

doi:10.1088/978-0-7503-3703-8ch18

© IOP Publishing Ltd 2022

Table 18.1. Summary of physical terms and their SI units—relating to energy, power and flux.

Term	Symbol	SI Units	Equivalent to	Relation
Energy	E	Joule (J)	—	—
Time	t	Second (s)	—	—
Area	A	Square meter (m^2)	—	—
Power	P	Watt (W)	J/s	E/t
Intensity	B	W/m^2	$J/s/m^2$	P/A
Illuminance	I	W/m^2	$J/s/m^2$	P/A
Specific intensity	I_0	W/m^2/solid angle	$J/s/m^2$/solid angle	B/solid angle

Intensity, *B*: The brightness of the light itself as it travels through space. We know that if direct sunlight passes through a tiny window, it has less ability to light a room than if the same light passes through a large window. And so the intensity of the light itself (rather than what makes it through the window) is not power, but power per unit area crosswise to the direction the light travels. A more physical name for this basic concept is *energy flux*.

Illuminance, *I*: The power per area arriving at a surface. This has the same physical units as intensity, but there is a subtle difference. Intensity is the power per area crosswise to the direction the light is traveling. Illuminance, on the other hand, is power per area incident upon the surface the light arrives at. These two areas may not be the same because the surface may be tilted with respect to the direction of the light. Illuminance thus often contains a separate factor that includes this angle.

Specific intensity, I_0: The intensity of light traveling along an individual ray. It is an important but subtle (and often confusing) concept. We will use the basic idea in this book, but we will sidestep its mathematical definition by finding other routes to the answers for which we are looking.

18.2 Power, *P*

Although it is the total energy absorbed per surface area that determines exposure, it is the rate at which energy is transferred that relates to the brightness of light. To complicate matters, we could mean a few different things when we use the term *brightness*.

- The rate at which energy is emitted by a source of light.
- The rate at which electromagnetic energy of light is carried through a given area (an open window, for example), as it travels through space.
- The rate at which energy, carried by light, is transferred to some surface—a solar panel, for example.

And so instead of ‘brightness,’ I will use other terms instead that distinguish between these cases. But for all of these, it is the rate of energy transfer, rather than the total energy transferred, that is the important concept.

The rate of transfer of energy is so important that it has its own name—*power*. In SI units *power is the amount of energy that is transferred per second.* Since power is energy per time, the SI units for power are joules per second (J s^{-1}). This has its own name—the watt (W). And so 1 J s^{-1} = 1 W. When a lightbulb is labeled '100 W,' it means that every second 100 J of electrical energy are transferred to other forms of energy (thermal energy and electromagnetic energy in this case).

There is an equation hidden in here. When we say that power (P) is energy (E) per time (t) it means the following equation.

$$P = \frac{E}{t} \tag{18.1}$$

It also means that

$$E = Pt, \tag{18.2}$$

and that

$$t = \frac{E}{P}. \tag{18.3}$$

These three relations are also true for our official units for these three quantities. A watt is a joule per second. A joule is a watt times one second, and a second is a joule per watt.

Since the brightness of the light relates to power, a given total amount of energy is transferred only when the light of a particular brightness is absorbed for a particular length of time. And the product of the two, the power multiplied by the time, tells us the energy transferred and thus the total physical effect on a given area of the light detector.

18.3 Intensity, *B*

The total power emitted by a source of light can be measured directly in watts. The power emitted by the Sun, for example, is 6.67×10^{28} W. But what about the *intensity* of sunlight passing through a window on a sunny day here on planet Earth, nearly 150 million kilometer away?

The intensity of light passing through space and the power of a source of light are related, but somewhat different, concepts. Emitted power is energy per time. But the intensity of the light itself is an 'energy flux.' A flux is a flow of something through a surface. And so the energy flux of light is the rate at which energy travels crosswise to some (maybe imaginary) surface, and it is measured per surface area.

As sunlight streams through an open window, for example, one may calculate how many joules of energy are carried by light through the window per second, and that would be a calculation of power in watts. But clearly, the answer one gets would depend upon the size of the window, even if it is the same sunlight. And so the intensity (B) of the light itself is not just energy per time, it is energy per time per area, and is thus measured in watts per square meter (W m^{-2}).

A photovoltaic solar panel provides an excellent example; it converts the energy carried by sunlight directly into electrical energy. The sunlight itself on a particular day carries a certain energy flux—a certain number of joules per second that it can transfer per square meter. And so the power that is intercepted by the solar panel depends directly on the number of square meters of sunlight that it intercepts. If the sunlight has an intensity of 1000 W m^{-2}, then at least one-square meter of solar panel is needed to transfer energy at a rate of 1000 W.[2]

Go to Calacalí, Ecuador on the first day of Spring, at local noon, and the Sun is directly overhead. On a clear day the intensity of this sunlight is about 1000 W m^{-2}. Place a one-meter-square piece of cardboard horizontally on the ground, and it will intercept 1000 W of power from the sunlight. Place a bigger piece of cardboard, say two meters on each side in the same place, and it will intercept proportionally more power. Since this new piece of cardboard would have an area of $2 \times 2 = 4\ m^2$, it would intercept $1000 \times 4 = 4000$ W of power. There is thus a simple relation between the intensity (B) of the light and the power (P) it carries through a given area (A). Equation (18.4) shows this relationship.

$$P = B \times A \tag{18.4}$$

18.4 Illuminance, *I*

Light has a direction, and the number of watts of power that arrive at a given surface depends not only on the intensity of the light but also at what angle the light strikes that surface. And so there is yet another distinction we must make: there is a difference between the intensity of the light itself—how many watts per square meter that are carried by the light in whatever direction it is traveling—and the number of watts per square meter that actually strike a particular surface upon which the light is incident. We call the former intensity (B), while we call the latter *illuminance* (I).

A fact of key importance for the numeric example of intensity given in section 18.3 is that the experiment is performed on the equator, and so the noon-time Sun is directly overhead, and the sunlight thus arrives normal to the piece of cardboard. Let us, by way of comparison, perform this same experiment in Suring, Wisconsin, at a latitude of 45°. At local noon on the first day of Spring, the Sun is not straight overhead; rather, it makes a 45° angle with the vertical. It is, however, still the same sunlight (assuming the air is just as clear as in Calacali), so it still has an energy flux of 1000 W m^{-2}.

Nevertheless, a one-meter-square piece of cardboard placed horizontally on the ground in Suring will intercept less than 1000 W of power from that sunlight. The sunlight carries 1000 W of power only through every 1 m^2 surface that is normal to the direction the light is traveling. When striking a surface at an angle, this 1000 W is spread out over an area greater than 1 m^2. And so the energy flux striking the piece of cardboard is less than 1000 W m^{-2} simply because it is 1000 W divided by an area greater than 1 m^2 (see figure 18.1).

[2] In practice, not all of the energy is transferred to electrical energy. Some energy is always transferred to other forms as well.

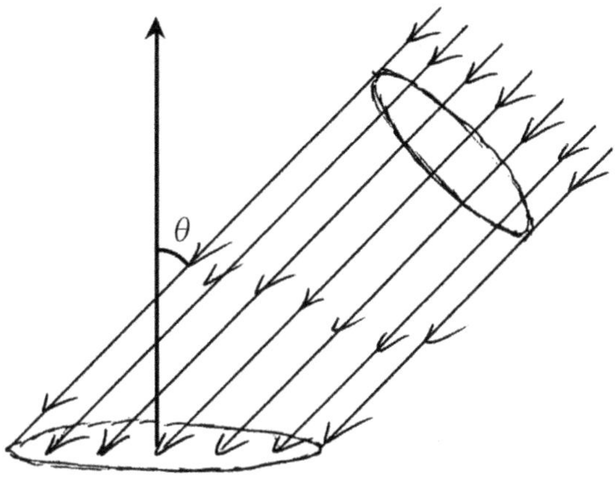

Figure 18.1. An example of the relation between the transfer of power and angle. Light with an intensity of 1000 W m^{-2} strikes a surface at an angle. This means the light carries 1000 W of power through any one-square-meter surface oriented normal to its direction of travel. However, this 1000 W is then spread out over a greater area when it strikes a surface at an angle. Thus the power transferred is less than 1000 W for every square meter of surface.

Simple trigonometry shows that the relationship between the energy flux of the light itself and the energy flux intercepted by a surface, oriented with its normal at an angle (θ) to the direction of the light, is given by a simple cosine factor. This fact, combined with equation (18.4), gives us equation (18.5) for the power actually intercepted by a surface even if it is not oriented perpendicular to the direction the light is traveling.

$$I = B\cos\theta \tag{18.5}$$

$$\text{P(transferred)} = I \times A \tag{18.6}$$

$$\text{P(transferred)} = BA\cos\theta \tag{18.7}$$

And so the 1 m^2 piece of cardboard in Suring would intercept only 1000 W $m^{-2} \times 1$ m^2 $\times \cos 45° = 707$ W.

18.5 Specific intensity, I_0

With the exception of parts of our image that may be unresolved (such as stars in the night time sky), the brightness of the light along a particular ray from the surface of something in the world is what it is. And so the surface of the Sun is blindingly bright, while the surface of black asphalt in the moonlight is very dim. Moving closer to the blacktop will not make its surface appear brighter, and going to Mars so one is farther from the Sun will not make the Sun's visible surface appear any less bright. We refer to the inherent brightness of a ray of light coming from a particular surface as the *specific intensity* of the light ray.

Technically, specific intensity is an *energy flux per solid angle*[3]. In *Photography: Physics and Art in Focus* we will not consider in detail how to calculate specific intensity numerically. We will, however, use the concept; it is the appropriate term to describe the 'brightness' of an individual ray of light coming from a particular point on the visible surface of some object.

Intensity decreases with distance as rays of light spread out from some source. But the specific intensity along an individual ray stays the same. Similarly, the apparent total brightness of a source of light decreases as one moves farther away from it, but its surface brightness remains the same. It is this surface brightness of the object that is recorded by the camera, so long as the object is not so small or so distant that it appears as only a point of light.

There is an important exception to the statement that the specific intensity of a ray of light is constant—whatever the distance. If the ray of light passes through a diffuse medium that absorbs light—a cloud of smoke, for example—then the medium itself can remove energy from the light ray as it passes through, deflecting it off to other directions or converting it to thermal energy. If that happens, then the specific intensity of the ray becomes smaller and smaller as it passes through the absorbing medium.

18.6 The meaning of exposure

A light-sensitive material is affected by light that falls upon it and is thus changed. By 'exposure' we mean the portion of this process that is about the light—an issue separated from however the particular material may respond. The illuminance of light upon a surface is a measure of power per surface area, but it is the total transfer of energy that affects the material in the end. And so it is clear that time, as well as power, is involved.

But the two-dimensional surface of the light-sensitive material is—well—a surface. The relevant measure of exposure for the light incident upon the photographic medium is thus not simply total energy transferred but rather the total energy transferred per surface area— $\mathrm{J\ m^{-2}}$ in our SI units.

We have seen in section 18.5 that the illuminance of the light can be characterized as a transfer of power per time, or energy per time per surface area: $\mathrm{J\ s^{-1}\ m^{-2}}$ in our SI units. These facts taken together imply the following most basic of relationships regarding the exposure of a photographic material to light:

$$\boxed{\text{Exposure} = \text{Illuminance} \times \text{Time}} \tag{18.8}$$

We will use this relation often in the chapters that follow. But I must here make a confession—I have pulled a 'fast one' so far, in this exposition. I stated that we define exposure in terms of the illuminance of the light alone, independent of the response to it of the particular photographic material. But this is true only within a

[3] A solid angle is related to area at the end of a cone, whereas a 'normal' angle is related to the length at the end of a triangle.

certain agreed-upon set of assumptions regarding the range of wavelengths of light to which the photographic material is sensitive.

To give an example, light of 600 nm wavelength light has no effect whatsoever on a light-sensitive material that happens to have no sensitivity at that particular wavelength. Thus any energy per surface area incident at that wavelength contributes nothing to the photographic effect—in this particular case—and thus should not be considered as part of the exposure. We will consider these complications in more detail in later chapters. In short, a particular photographic material responds either more-or-less to each wavelength, so this too must be taken into account regarding our measure of exposure.

Chapter 19

Tracing the energy from source to camera

Let us consider the terms: *energy*, *power*, *intensity* and *illuminance* and their relationships to size, surface area, distance and time as described in chapter 18. Here we consider a concrete example. We trace the light that makes a photograph—all the way from its source through a lens and to its interaction with the detector in a camera. Let us consider in detail the following numerical example.

> Light emitted by the Sun arrives at Earth and strikes a 2.5 cm diameter flat gray disk at an angle of 37° to its normal, with 18% of the light reflecting off the disk—some of which travels toward the camera, which is 3.5 m directly in front of the disk. This light is then focused by the camera onto the detector with a 135 mm, *f*/5.6 lens, for an exposure of 60 seconds. How many joules of energy per unit area strike the detector?

The answer to this question is important because it is the physical basis of *exposure*. We will determine this answer by going step-by-step, following the light and using our basic relationships between energy, power, area, time, intensity and illuminance. Along the way, we will have to think some more about how light reflects off objects.

19.1 The power of light emitted by the Sun

The Sun emits light with a power (P) of 3.85×10^{26} watts (W) or 3.85×10^{26} joules (J) of energy every second.

19.2 The intensity of sunlight at Earth

The 3.85×10^{26} W of light emitted by the Sun spreads out equally in all directions. And so by the time it arrives here at Earth, it has spread out over a sphere with a huge radius equal to the Sun–Earth distance (d) of 1.50×10^{11} m. The intensity or

doi:10.1088/978-0-7503-3703-8ch19 © IOP Publishing Ltd 2022

energy flux (B) of the light that arrives here is thus equal to that enormous power but divided by an equally-enormous area. And so we are left with a modest intensity of sunlight—here at Earth.

$$B = P/A \tag{19.1}$$

$$A = 4\pi d^2 \tag{19.2}$$

$$B = \frac{P}{4\pi d^2} \tag{19.3}$$

$$= \frac{3.85\times10^{26}\text{ W}}{4 \times 3.141\,59 \times (1.5 \times 10^{11}\text{ m})^2} \tag{19.4}$$

$$= 1362\text{ W m}^{-2} \tag{19.5}$$

Here we have used the fact that the surface area of a sphere is 4π times the square of its radius. Thus the intensity of light arriving at Earth, and so also at our gray metal disk, is 1362 W m^{-2}.

19.3 The inverse square law

We derived equation (19.2) for our special case of sunlight reaching Earth. It should be clear that this equation would apply equally to any case in which a single source of power (P) emits light *isotropically*—uniformly in all directions. Equation (19.2) gives the intensity of light at a distance (d) from any such *isotropic emitter.*

This is called the *inverse square law*, and it is easy to see why: the intensity varies with the inverse of the square of the distance from the source of light. If one were to move three times farther away, the intensity would not reduce to 1/3 as bright but rather to only $1/3^2 = 1/9$ as bright.

19.4 Illuminance of the light on the subject

Since sunlight of intensity $B = 1362$ W m^{-2} strikes the disk at angle of $\theta = 35°$ to its normal, then the illuminance on the disk is given by the following equations.

$$I = B \times \cos\theta \tag{19.6}$$

$$I = 1362\text{ W m}^{-2} \times \cos(35°) \tag{19.7}$$

$$I = 1115\text{ W m}^{-2} \tag{19.8}$$

And so the illuminance of sunlight on our metal disk is only 1115 Wm^{-2}.

19.5 The power of the light intercepted by the metal disk

The power (P) of light intercepted by the metal disk is given by the following.

$$P = I \times A\,, \tag{19.9}$$

Here I is the illuminance on the disk, and A is the surface area of the 2.5 cm diameter disk. Therefore, we have the following equations.

$$= I \times A \tag{19.10}$$

$$= I \times \pi r^2 \tag{19.11}$$

$$= I \times \pi \left(\frac{D}{2}\right)^2 \tag{19.12}$$

$$= 1115 \text{ W m}^{-2} \times 3.141596 \times \left(\frac{0.025 \text{ m}}{2}\right)^2 \tag{19.13}$$

$$= 0.5475 \text{ W} \tag{19.14}$$

The metal disk intercepts 0.5475 W of sunlight.

19.6 The power of the light reflected by the metal disk

The disk does not reflect all of the light that it intercepts; a certain percentage is absorbed. In this example, we have chosen a disk that reflects only 18% of the light that falls upon it; instead of reflecting 0.5475 W, it only reflects 18% of that, or 0.09855 W. Incidentally, a reflectivity of 18% is considered by photographers to be a mid-tone gray.

19.7 The intensity of the reflected light when it reaches the camera

Here is the complicated part. If the disk reflects a total power of 0.09855 W, how much of that is directed towards the camera, and what is its intensity by the time it reaches the camera 3.50 m away?

To answer this question we must make some assumptions about how the partially-reflective surface of the disk spreads out the reflected light in different directions. Obviously one can imagine different possibilities. If it were a mirror it would send all of the light in the same direction according to the law of reflection for mirrors. We instead assume here that it is a *diffuse reflection*: each ray of light striking the disk is spread out into rays that reflect in all directions.

The simplest possibility to calculate is if the disk were to reflect light with an intensity that is equal in all directions—what we could call an *isotropic reflector*. In that case, the 0.09855 W reflected by the disk would, by the time it reaches the camera a distance d = 3.50 m away, spread out uniformly over one half of a sphere of that radius. Since intensity is power per area, we would have the following equations.

$$B = \frac{P}{A} \tag{19.15}$$

$$= \frac{P}{2\pi d^2} \tag{19.16}$$

$$= \frac{0.09855 \text{ W}}{2 \times 3.141\,59 \times (3.50 \text{ m})^2} \tag{19.17}$$

$$B = 1.280 \times 10^{-3} \text{ W m}^{-2} \tag{19.18}$$

Here I have used the fact that the area of a hemisphere of radius (d) is given by $2\pi d^2$.

Thus we have the result that if our metal disk were to produce a diffuse reflection in this particular way (isotropic) the intensity of the reflected light, by the time it reached the camera, would be 1.280×10^{-3} W m^{-2}.

The problem is that this assumption is very unrealistic, and it flies in the face of everyday experience. Consider that as seen from a sharp angle, the flat disk would appear as only a thin sliver. And so for it to reflect the same amount of light at a sharp angle as it does straight forward, it would have to appear much brighter per surface area as seen from that angle. Experience shows that this is not true for most objects that have a matte finish—flat gray paint, for example. Instead the *surface brightness* appears approximately the same from all angles. This type of reflecting surface is called a *Lambertian reflector* after Johann Heinrich Lambert who, in the eighteenth century, first described its properties mathematically.

Since the more-realistic Lambertian reflector appears the same brightness per surface area in all directions, it must send more light directly forward than it does at an angle. This means that, by making an assumption of an isotropic reflector, we must have underestimated the amount of light that is sent directly forward. And so our calculation of 1.280×10^{-3} W m^{-2} must be too small.

The question is—too small by what factor? Calculus is required to determine that answer, and I show how to calculate it correctly for a Lambertian reflector in appendix D. The answer turns out to be simple: the Lambertian reflector sends twice as much light as the isotropic reflector in the forward direction. To correct our calculation, we need simply double it. Thus the light arrives at our camera with an intensity of $B = 2.561 \times 10^{-3}$ W m^{-2}.

Another way to distinguish between the (very-unrealistic) isotropic reflector and the (much-more-realistic) Lambertian reflector is as follows. The isotropic reflector reflects light with the same intensity in all directions. But in order to do this, it would have to send light off at angles with a greater specific intensity. The Lambertian reflector sends off light at all angles with the same specific intensity. The consequence is that it sends light off at an angle with a smaller intensity.

19.8 The power of the light that enters the camera lens

The light arriving at the camera passes through the aperture of the lens. The power (P) that makes it through the aperture is simply given by the following equation.

$$P = B \times A \tag{19.19}$$

Here B is the intensity of the light at the camera lens, and A is the area of the circular lens opening. This is a simple calculation if we know A, which we have not been given. We know only that it is a lens of focal length $F = 135$ mm with a focal ratio of

$f/5.6$. But the focal ratio (f) of a lens is simply its focal length (F) divided by its diameter (D), and so we have $D = F/f$. Thus, the following must be true.

$$A = \pi\left(\frac{D}{2}\right)^2 \tag{19.20}$$

$$A = \pi\left(\frac{F}{2f}\right)^2 \tag{19.21}$$

$$A = 3.1416\left(\frac{0.135\ \text{m}}{2 \times 5.6}\right)^2 \tag{19.22}$$

$$A = 4.564 \times 10^{-4}\ \text{m}^2 \tag{19.23}$$

The power that makes it through our camera aperture is then represented by the following equations.

$$P = B \times A \tag{19.24}$$

$$P = 2.561 \times 10^{-3}\ \text{W m}^{-2} \times 4.564 \times 10^{-4}\ \text{m}^2 \tag{19.25}$$

$$P = 1.169 \times 10^{-6}\ \text{W} \tag{19.26}$$

Thus 1.169×10^{-6} W of power makes it through our camera lens.

19.9 The illuminance of the light on the camera detector

The light that reflected off our gray metal disk and then reached the camera and passed through the aperture of its lens has a power of 1.169×10^{-6} W. But to determine the exposure we need the illuminance of this light as it is focused onto the detector. This light does not spread out uniformly; rather, it is focused onto the detector by the lens (or so we assume).

The lens will focus this light to a circular image, and if we know the area (A) of that image, we can easily calculate the illuminance. If the lens is focused properly, then all of that $P = 1.169 \times 10^{-6}$ W spreads out over only the area (A) of the image. And so we have the following equation.

$$I = \frac{P}{A} \tag{19.27}$$

But what is the area of the circular image on the detector? To determine this we must use the relationship between the size of an image focused by a lens, the distance to the object and the focal length of the lens. If we let D_O and D_I represent, respectively, the distances of the object and the image from the lens and S_O and S_I represent, respectively, the sizes of the object and image, then simple geometry gives the following equations (see figure 11.4 in chapter 11).

$$\frac{S_I}{S_O} = \frac{D_I}{D_O} \tag{19.28}$$

$$S_I = S_O \frac{D_I}{D_O} \quad (19.29)$$

We have already said that the camera is 3.5 m from the metal disk, so we know that $D_O = 3.50$ m. But what about D_I? How far is the image from the lens? Equation (11.1) in chapter 11 allows us to calculate the image distance exactly, but we do not really need that. If the object is located much farther away than the focal length of the lens, then the image distance very closely equals the focal length of the lens (F) (135 mm in our example). And so we have the following equations.

$$S_I \approx S_O \frac{F}{D_O} \quad (19.30)$$

$$S_I \approx 0.025\ \text{m} \frac{0.135\ \text{m}}{3.50\ \text{m}} \quad (19.31)$$

$$S_I \approx 0.009643\ \text{m} \quad (19.32)$$

$$A = \pi \left(\frac{S_I}{2} \right)^2 \quad (19.33)$$

$$A = 3.141\,6 \left(\frac{0.009\,643\ \text{m}}{2} \right)^2 \quad (19.34)$$

$$A = 7.303 \times 10^{-7}\ \text{m}^2 \quad (19.35)$$

Now that we know the area of the image on the detector, we can calculate the illuminance.

$$I = \frac{P}{A} \quad (19.36)$$

$$I = \frac{1.169 \times 10^{-6}\ \text{W}}{7.303 \times 10^{-7}\ \text{m}^2} \quad (19.37)$$

$$I = 1.600\ \text{W m}^{-2} \quad (19.38)$$

And so the image on the detector has an illuminance of 1.600 W m^{-2}.

19.10 The exposure imparted to the detector

We now know the illuminance of the light that makes the image on the detector. To calculate the exposure, however, it is not the power per square meter, but rather the total energy, in J, of light that strikes the image per square meter. Since $E = Pt$, we simply have the following basic relation between exposure, illuminance and time.

$$\text{exposure} = \text{illuminance} \times \text{time} \quad (19.39)$$

Since we stipulated that the light is exposed to the camera detector for 60 seconds, we have an exposure of the following.

$$\text{exposure} = 1.600\ \text{W m}^{-2} \times 60\ \text{s} \tag{19.40}$$

$$=1.600\ \text{J s}^{-1}\ \text{m}^{-2} \times 60\ \text{s} \tag{19.41}$$

$$= 96.0\ \text{J m}^{-2} \tag{19.42}$$

And so, the image on the detector receives 96.0 J m^{-2} of energy from the exposure to light from the subject. This is typical for the real case of taking a picture in the daytime with a real camera.

It is important to note that this is a tiny amount of energy. A square meter is a lot of area, and the image on the detector in our example is very tiny in comparison—only 7.303×10^{-7} m^2. Thus a total of only 96.0 J m^{-2} $\times$ 7.303×10^{-7} m^2 = 7.01×10^{-5} J of light strikes the image area during the exposure in our example. This is roughly the amount of energy required to lift a honey bee to a height of seven centimeters.

19.11 Summary of steps

Now let us retrace these steps, from the Sun to the camera, and list them in order.

1. The Sun emits light with a power, $P = 3.85 \times 10^{26}$ W, uniformly in all directions.
2. This light spreads out over a huge area by the time it reaches Earth, and so the Earth—and the subject—is illuminated by light with an intensity of $B = 1362$ W m^{-2}.
3. Our subject, inclined at an angle of 37° to the incoming sunlight, receives an illuminance of $I = 1115$ W m^{-2}.
4. Our subject, only 2.50 cm in diameter, intercepts a power of $P = 0.5475$ W.
5. Our subject reflects only 18%, or $P = 0.09855$ W, of the total that it intercepts.
6. The subject reflects the 0.098 55 W in all forward directions as a Lambertian reflector. By the time this light reaches the camera, directly in front of the subject and 3.50 m away, it has an intensity of $B = 2.56 \times 10^{-3}$ W m^{-2}.
7. The particular lens on our camera intercepts some of this light, and a power of $P = 1.169 \times 10^{-6}$ W enters the camera.
8. This light is focused by the lens as an image of a particular size (and thus area) on the detector, giving the image an illuminance on the detector of 1.600 W m^{-2}.
9. Multiplying the illuminance by the exposure time (set by the camera shutter) of 60 s gives an exposure of 96.0 J m^{-2} .

Notice that the illuminance of the image inside the camera, focused onto the light detector, is much less than the illuminance of light that fell upon the subject. It is only 1.6 W m^{-2}, compared to 1115 W m^{-2} for the sunlight—nearly 700 times dimmer. This is typical, and as we shall see in chapter 20, the in-camera illuminance is always at least several times dimmer than the subject illumination for even the fastest of camera lenses pointed at the most reflective of subjects.

19.12 What about focus?

We have assumed here that the image is perfectly focused onto the detector. That is to say, we assumed that all of the intensity (B) of light reflected off the subject and that reaches the lens, is all focused onto a specific area of the detector. What if that were not the case?

The case of an out-of-focus image is complex, but we can understand some of the basic principles. One way to look at it this: the power that would have arrived at area (A) on the detector, had it been perfectly focused, instead arrives at a greater unfocused area: $A' > A$. Since exposure is energy transferred per surface area, the exposure delivered in a particular time is thus reduced by a factor of A/A'.

For a real out-of-focus image of the world, this reduction in exposure may be rendered somewhat moot. If the image of some particular object is out of focus, then each part of the detector receives less exposure from that object. But light from nearby out-of-focus objects is just as likely to add exposure to those same locations on the light-sensitive material. This may lead to an image that, on average, has the same exposure as if it had been in focus, but with greatly reduced contrast.

Chapter 20

The Jones–Condit equation

In chapter 19 we worked through a specific example of light reflecting off a subject and focused by a lens onto a detector in a camera. We can also work through this problem in a general sense to find the mathematical relation between the illumination of light upon a subject to the exposure inside a camera. In so doing, we derive a version of what is known as the *Jones–Condit equation* (Ware 2019, p 211ff). In its traditional form, it uses photometric variables that are useful in a practical sense but that obscure, somewhat, the basic physical principles involved.

In section 20.1 (with some of the more technical mathematical details left for appendix D) I derive a simple form of the Jones–Condit equation that better illustrates issues important for this book. In particular, I derive an equation that relates the illuminance (I_D) of the image on the detector inside the camera and the illuminance (I_{SUB}) of the light shining upon the subject. In section 20.2 we consider the factors that alter the in-camera image illumination for rays of light that enter the camera at an angle to the normal and develop some simple relations for *vignetting*.

20.1 The Jones–Condit equation

The Jones–Condit equation in its simplest form may be written as equation (20.1).

$$\boxed{I_{SUB} = \frac{4f^2}{R} I_D}, \tag{20.1}$$

Here f is the focal ratio of the camera lens, and R is the fraction of light reflected by the subject. Equation (20.1) ignores absorption of light by the glass in the lenses, lens distortion and vignetting, and it assumes the object is much farther away than the focal length of the camera lens.

To derive equation (20.1), let us imagine a flat, matte-finished reflective circular disk of radius: (S_O [for 'object size']). And let us say that our subject is illuminated by light with an illuminance (I_{SUB}) in W m^{-2}, and that the disk reflects a fraction (R) of the light that falls upon it. At a distance (d_O [for 'object distance']) directly in front of

doi:10.1088/978-0-7503-3703-8ch20

© IOP Publishing Ltd 2022

the disk, we place a camera, the lens of which forms an image of the disk on the camera's light detector. See figure 20.1, which also defines the size of the image as S_I, the image distance inside the camera as d_I, and the diameter of the lens aperture as D.

The light reflected from the disk will spread out over a hemisphere in the forward direction with area equal to $2\pi d_O{}^2$,[1] and it will have an intensity before entering the camera lens of B_C. The camera lens will then focus this light to a circular image of radius (S_I) and the light in this image that falls on the detector will have an illuminance (I_D). Our question is, then, what is the mathematical relation between I_{SUB} and I_D?

First, we must recognize that the total power (P_R) reflected by the disk is the illuminance of the light falling on the disk multiplied by its reflectivity (R) and by its surface area ($\pi S_O{}^2$); recall that the product of illuminance (W m^{-2}) and area (m^2) is power in watts (W). The reflectivity is simply the fraction of this that is reflected rather than absorbed.

$$P_R = I_{SUB}R(\pi S_O{}^2) \qquad (20.2)$$

These watts of power then spread out over the forward-facing hemisphere of radius (d) before reaching the camera. And so, over that hemisphere, that includes the camera; it has an average intensity, power per area, of simply $P_R/(2\pi d_O{}^2)$. But is the intensity the same over different parts of the hemisphere? In particular, is the intensity the same directly in front of the disk as it is at a large angle?

If we do assume that the disk reflects light in such a way that the intensity, at distance (d) from the disk, is the same in every direction, we could call the disk an *isotropic reflector*. The intensity at the camera (B_C) would then be represented in the following equations.

$$B_C = \frac{P_R}{2\pi d_O^2} \qquad (20.3)$$

$$B_C = \frac{I_{SUB} R \pi S_O^2}{2\pi d_O^2} \qquad (20.4)$$

$$B_C = \frac{R S_O^2}{2 d_O^2} I_{SUB} \qquad (20.5)$$

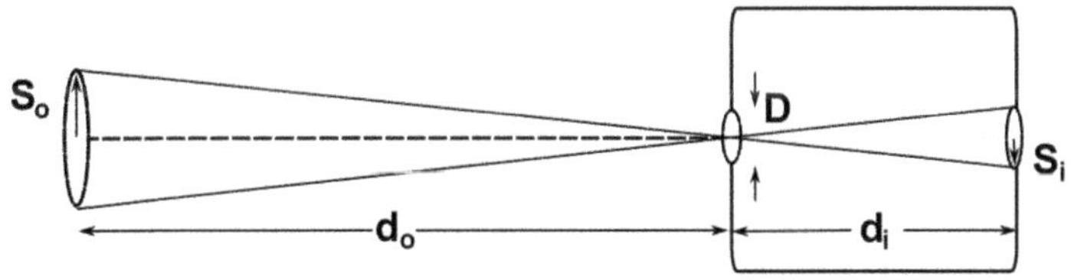

Figure 20.1. Definitions of image distance (d_I), object distance (d_O), image size (S_I), object size (S_O) and lens diameter (D) for our derivation of the Jones–Condit equation.

[1] The surface area of a full sphere of radius (r) is $4\pi r^2$.

But we know that, when viewed from an angle, the round disk appears as an ellipse—not a circle. And so the visible surface area is less as seen from an angle than when viewed directly from the front. If this smaller visible surface as viewed from an angle is to reflect the same power as when viewed from the front, the disk's apparent brightness (its specific intensity) would have to be proportionally greater. Thus equation (20.4) implies that the surface would appear brighter per surface area as seen from an angle than as seen from straight-on in order to make up for the fact that less area is seen from an angle.

A more realistic assumption is the *Lambertian reflector*, which reflects with the same specific intensity in every direction. But to do so, a Lambertian reflector must reflect relatively greater power per area normally than at an angle.

The Lambertian reflector emits a specific intensity (W m^{-2} per solid angle) that is constant with direction but an intensity (W m^{-2}) that varies with angle (at a given distance). Thus by assuming an isotropic emitter, we must have underestimated the energy flux emitted in the forward direction. It requires calculus to calculate the face-on reflected energy flux from a Lambertian reflector. But the correct result, derived fully in appendix D, turns out to be simply two times the equal-energy-flux assumption used to derive equation (20.4). Thus equation (20.4) for the intensity of light (B_C) arriving at the camera lens should be as follows.

$$B_C = \frac{RS_{O^2}}{d_O^2} I_{SUB} \tag{20.6}$$

Let us now assume that this reflected light is incident normally upon a lens of diameter (D) and is focused onto the detector a distance (d_I) from the lens. Thus the power (P_L [in W)]) that makes it through the lens is given by the product of B_C (in W m^{-2}) and the area (in m^2) of the circular lens aperture.

$$P_L = \pi\left(\frac{D}{2}\right)^2 B_C \tag{20.7}$$

If we assume the lens focuses this light to a circle of radius (S_I) on the detector, the illuminance in W m^{-2} (I_D) on the detector would then be as follows.

$$I_D = \frac{P_L}{\pi S_I^2} \tag{20.8}$$

Combining equations (20.6), (20.7) and (20.8) gives us equation (20.11)

$$I_D = \frac{\pi\left(\frac{D}{2}\right)^2 B_C}{\pi S_I^2} \tag{20.9}$$

$$I_D = \frac{\pi\left(\frac{D}{2}\right)^2 \frac{RS_O^2}{d^2} I_{SUB}}{\pi S_I^2} \tag{20.10}$$

$$I_{\mathrm{D}} = \frac{D^2 R S_{\mathrm{O}}^2}{4 d_{\mathrm{O}}{}^2 S_{\mathrm{I}}^2} I_{\mathrm{SUB}} \tag{20.11}$$

From figure 20.1, it is clear (from the law of similar triangles) that the ratio of image size to object size is equal to the ratio of image distance to object distance.

$$d_{\mathrm{I}} = \frac{S_{\mathrm{I}}}{S_{\mathrm{O}}} d_{\mathrm{O}} \tag{20.12}$$

Combining equations (20.11), and (20.12), we have the following.

$$I_{\mathrm{D}} = \frac{D^2 R}{4 d_{\mathrm{I}}^2} I_{\mathrm{SUB}} \tag{20.13}$$

Or, we have equation (20.14).

$$\boxed{I_{\mathrm{SUB}} = \frac{4f^2}{R} I_{\mathrm{D}}} \tag{20.14}$$

Here f, the focal ratio of the lens, is defined by $f \equiv F/D$, and I have made the approximation $d_{\mathrm{I}} \approx F$, the focal length of the lens. This approximation holds so long as $d_{\mathrm{O}} \gg F$. Thus equation (20.14) is accurate only when the subject is significantly more distant than the focal length of the lens.

For our numerical example in chapter 19, the illuminance on the subject was 1115 W m^{-2}, the illuminance on the detector was 1.600 W m^{-2}, the reflectivity was $R = 0.18$, and the focal ratio of the lens was $f = 5.6$. Given the reflectivity and focal ratio in our example, equation (20.14) says that the illuminance of the image on the subject should be $4 \times 5.6^2/0.18 = 697$ times larger than the illuminance on the detector in agreement with our step-by-step calculations (1115/1.600 = 697) in chapter 19.

We can use this result to make an interesting comparison. Suppose I were to place a particular light detector not in the camera but rather in place of the subject, illuminated by the same light (I_{SUB}) and exposed it for a time (t_{SUB}). What length of exposure time (t_{D}) would be required to give the detector the same exposure inside the camera? We note that exposure is the product of illuminance and time, and so for the in-camera and at-subject exposures to be the same, we must have the following equations.

$$I_{\mathrm{D}} t_{\mathrm{D}} = I_{\mathrm{SUB}} t_{\mathrm{SUB}} \tag{20.15}$$

$$t_{\mathrm{D}} = \frac{I_{\mathrm{SUB}}}{I_{\mathrm{D}}} t_{\mathrm{SUB}} \tag{20.16}$$

Thus, combining equations (20.14) and (20.16) renders equation (20.17).

$$\boxed{t_{\mathrm{D}} = \frac{4f^2}{R} t_{\mathrm{SUB}}} \tag{20.17}$$

Equation (20.17) means that however much time is required to expose our detector when illuminated directly by the light on the subject, when placed in a camera a much longer exposure time is required—by the factor $4f^2/R$. For a subject with a mid-tone gray reflectivity ($R \approx 0.2$), this factor is thus about $20f^2$. For larger film formats, an $f/4$ lens is fast indeed, and so we should expect that in-camera exposure times are likely to be at least dozens of times longer than what would be required if the detector were directly exposed to the light falling upon the subject. In section 25.4 we use ephemeral-process photography to test this relationship experimentally.

Notice that, when all is said and done, the illuminance of the image on the detector (and thus the exposure time needed to achieve the correct density) does not depend at all on the size of the subject or on its distance. We did use those values in our step-by-step calculations in chapter 19; however, when we instead carry them all the way through the algebra, we find that both of those quantities cancel out in the end. Instead, for a subject reflecting light as a Lambertian reflector, the illuminance on the detector depends on only three factors.

1. The illuminance on the subject
2. The reflectivity of the subject
3. The focal ratio of the camera lens

For equation 20.14, I have made the additional assumption that the camera is far from the subject, compared to the focal length of the lens. This is usually true, but it is decidedly untrue for close-up macro photography. For that case the distance to the subject does affect the image illuminance and sometimes greatly so. We consider this complication more fully in section 22.3.1. I have also ignored vignetting and the reduction of the transmitted light through the lens by reflection off the lens' surfaces and absorption by the glass. The full Jones–Condit equation takes all of these factors into account—see Ware (2016, pp 104–106) for a more-complex example; however, equation (20.14) is a good approximation for the simple case reviewed here.

I have, here, laid out the foundations for the relationships between light and exposure in terms of physical quantities such as illuminance and power. But photographers take a very different approach to these same goals, as we shall see in what follows. From a physical standpoint, the photographer's approach is roundabout and obscures somewhat the physical principles involved. But from the practical standpoint of taking pictures with real cameras in the real world, it makes a lot of sense.

20.2 Vignetting

Our results in section 20.1 are based on the assumption that the light is entering the camera normally. But light entering the lens aperture at an angle to the normal produces a smaller intensity on the detector, making the image darker at the edges. This vignetting of the image has many separate causes, some of which may be related to the particular lens' design. Here we consider only *natural vignetting*, which is caused by the simple geometry of light entering a circular aperture and imaged onto a surface. A well-designed lens can mitigate some of these factors, producing less vignetting than described here.

Figure 20.2 lays out the basic geometry for two important cases: a flat detector normal to the axis of the lens aperture and located a distance F from it, and a curved detector with radius equal to F. Let us imagine some distant object, emitting or

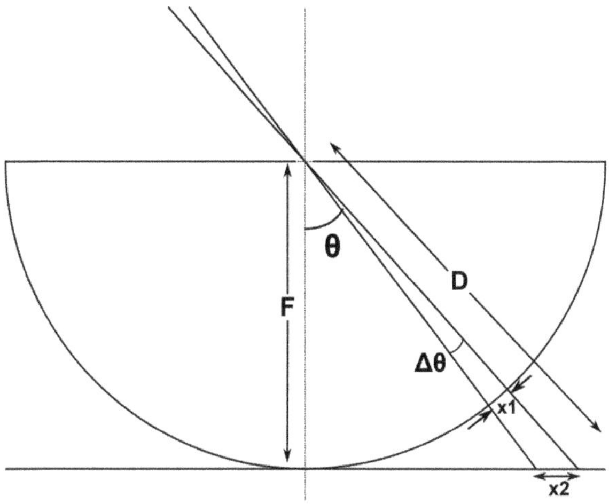

Figure 20.2. The geometry of a bundle of rays passing through a pinhole and intersecting both a curved and a flat detector.

reflecting light, and the extent of which makes an angle ($\Delta\theta$). This is imaged onto one of our two detectors, making an image width of either $x1$ or $x2$, respectively. For the flat detector, the illuminance is smaller compared to $\theta = 0$ for three reasons.

1. The light rays enter through an aperture that no longer has a circular profile. The rays, instead, see an ellipse with a smaller projected area.
2. The light rays travel a greater distance to the detector and so are more greatly diminished by the inverse square law.
3. The bundle of light rays intersects the detector at an angle and so spreads out over a greater area.

A quick glance at figure 20.2 shows that, for the case of the semicircular curved detector, only the first of these reasons would apply.

We recall that the illuminance at the detector is proportional to the power transmitted through the aperture, and this is proportional to the area of the aperture normal to the path of the light. But at angle θ this is diminished by a factor $\cos\theta$. Thus, for this reason alone, our image at our detector is dimmer by a factor of $\cos\theta$.

The light that does make it through the aperture diminishes in intensity by the inverse square law; the cone of light spreads out over a larger area, proportional to the square of the distance. Since the illuminance is the ratio of power to area, a larger area means a proportionally smaller image illuminance.

But the light spreads out over a given area on the detector for another reason as well: the circular cone of light makes a larger-area ellipse, as it intersects the flat image plane at an angle. For our flat detector, the light travels a distance (D) to the detector, given by the following equation.

$$D = \frac{F}{\cos\theta} \tag{20.18}$$

For $\theta = 0$, $D = F$, the focal length of the camera. The diameter of the cone of light at the image plane is then given by $(F/\cos\theta)\Delta\theta$.

As the cone of light intersects the image plane, it makes an ellipse with a width of $(F/\cos\theta)\Delta\theta$. But the geometry extends the length, $x2$, of this ellipse by another factor of $1/\cos\theta$.

$$x2 = \frac{F\Delta\theta}{\cos^2\theta} \tag{20.19}$$

The image thus occupies an area (A), given by the following equations.

$$A = \pi\left(\frac{F\Delta\theta}{2\cos\theta}\right)\left(\frac{F\Delta\theta}{2\cos^2\theta}\right) \tag{20.20}$$

$$= \frac{\pi F^2\Delta\theta^2}{4\cos^3\theta} \tag{20.21}$$

Here I have used the fact that the area of an ellipse is the product of π, half the long axis and half the short axis.

The area of an image, focused at the center, is given by putting $\theta = 0$ into equation (20.21).

$$A_0 = \pi\left(\frac{x1}{2}\right)^2 = \frac{\pi F^2\Delta\theta^2}{4} \tag{20.22}$$

Here I note that $x1 = F\Delta\theta$. Thus our image off-axis at angle θ would have an area larger on our flat detector by a factor of $A/A_0 = 1/\cos^3\theta$ when compared to the center. This diminishes the image illuminance by that same factor.

Our flat detector would thus experience a total lessening of the image illuminance due to the combined effects of the smaller projected area of the aperture, the greater distance to the detector and the increased area the bundle of rays spreads over on the detector. The first of these diminishes the image illuminance by a factor $1/\cos\theta$, while the combination of the second and third decreases it by an additional factor of $1/\cos^3\theta$. We thus have an off-axis image that is $1/\cos^4\theta$ times dimmer than that at the center. We can also write it as the following equation.

$$B_{\text{off-axis}} = B_{\text{center}}\cos^4\theta \tag{20.23}$$

For the semicircular curved detector, on the other hand, only the first factor applies, and so we have the following equation.

$$B_{\text{off-axis}} = B_{\text{center}}\cos\theta \tag{20.24}$$

Figure 20.3 shows the results for both cases. The red curve shows the image brightness as a function of angle for the curved detector, while the green curve is for the flat detector.

As a benchmark, a quite-large 120° angle of view means $\theta = 60°$. Since $\cos 60° = 0.5$, the semicircular detector would be half as bright at the edges when compared to the center. But the edges of the flat detector would be only 1/16 as bright.

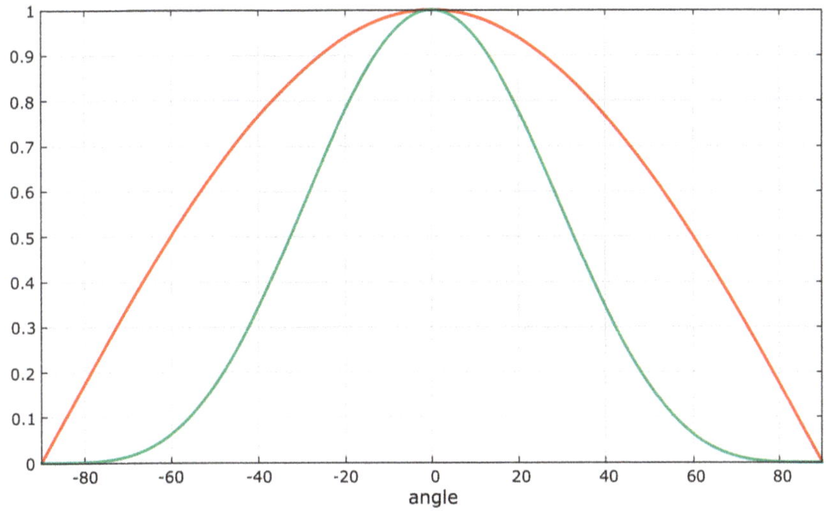

Figure 20.3. A comparison of simple $\cos\theta$ (red curve) vignetting and $\cos^4\theta$ vignetting (green curve).

Mitigating vignetting is much of the challenge of designing a good wide-angle lens. If we recall our pinhole fisheye camera of chapter 10, it is clear that we can use refraction to greatly reduce our first factor—the diminished projected area of the aperture. Light rays arrived at the camera from the world at angles close to 90°, but after refracting at the surface of the water they passed through the pinhole at angles less than the critical angle of 49°.

If a wide range of incoming angles is compressed by our lens, so the rays pass through the aperture at much smaller angles, then barrel distortion is produced (section 10.1). We can, however, undo this distortion by placing a similar arrangement of lenses, in reverse, on the other side of the lens' aperture. So long as the rays both enter and exit at the same angle, little distortion is introduced. For this reason, wide-angle rectilinear lenses typically have a shape as in figure 20.4 with front and rear lens groupings cone-shaped and physically much larger in size than the actual lens' aperture. The small circles in front of each of the lenses in figure 20.4 show the actual sizes of their respective apertures.

The two lenses in figure 20.4 have an angle of view of 105° or $\theta = 52.5°$. If it were a mere pinhole, the light at the edges would be only $\cos^4 52.5° = 0.137$ as bright as the center of the image—smaller by a factor of 7.3. These two lens do show some vignetting but not nearly that much; the price is over 450 g of glass for the larger-aperture version.

To reduce vignetting in this way, but to also maintain a relatively large effective lens aperture, significantly larger front and rear lens' groupings are required. Figure 20.4 shows two versions of the same wide-angle lens' design. Both lenses have a focal length of 90 mm and similar angles of view. The right-hand version has an effective aperture that is only 43% larger, but it weighs nearly three times as much.

Figure 20.4. Left: A 90 mm *f*/8 Super Angulon wide-angle lens. **Right:** A faster *f*/5.6 lens of the same focal length and design. The circles in front of each lens show the actual sizes of their respective maximum apertures. The disk at the center of each is the aperture and shutter mechanism.

If one tries to get the largest image circle possible out of any real camera lens, it will likely vignette significantly at its largest apertures for a very different reason than what we have laid out in this section. The combination of many lens elements spaced out along the optical path means the lens' assembly is something like a tube. And so the edges of this assembly, set by the limited diameters of the individual lens elements, block some of these rays. This type of vignetting is discussed and illustrated in section 15.2. Both of the lenses, shown in figure 20.4, have this additional vignetting when used at their maximum apertures.

References

Ware M 2016 Cyanomicon II–History, Science and Art of Cyanotype: Photographic Printing in Prussian Blue https://www.mikeware.co.uk/downloads/Cyanomicon.pdf

Ware M 2019 Argyronomicon: Silver Photographs on Paper: Chemical History of Their Invention, Deterioration, and Conservation https://www.mikeware.co.uk/downloads/Argyronomicon.pdf

IOP Publishing

Photography (Second Edition)

Physics and art in focus

John Beaver

Chapter 21

Illumination and photograms

Here we consider the relationship between illuminance at the subject and the nature of the source of light. We start with the simple case of a spherical, isotropically-radiating blackbody, and then see what is different for more typical sources of light.

21.1 Illumination from blackbodies

For a spherical source of light radiating isotropically, we have the inverse-square law for the intensity (B) of the light.

$$B = \frac{P}{4\pi d^2} \tag{21.1}$$

Here P is the power (in watts) emitted by the source, and d is the distance to the center of the sphere.

Since we have specified that the light source is a spherical *blackbody* (section 5.2), we can relate the power (P) emitted to the blackbody's radius (R) and its temperature (T).

$$P = 4\pi R^2 \sigma T^4 \tag{21.2}$$

This is simply the power emitted per surface area of a blackbody (σT^4) multiplied by the total surface area of the sphere ($4\pi R^2$). The symbol σ—the Greek letter sigma—represents a scale factor known as the Stefan-Boltzmann constant. In SI units of its value are 5.67×10^{-8} W m^{-2} K^{-4}.

We can combine equations 21.1 and 21.2 into the following equation.

$$B = \sigma T^4 \left(\frac{R^2}{d^2} \right) \tag{21.3}$$

The factor in parentheses in equation (21.3) is closely related to the square of the angular diameter (θ) of the light source as seen from the surface it is illuminating.

 © IOP Publishing Ltd 2022

Taking note that the diameter of the light source is twice its radius, we have the following equation.

$$\theta = \frac{2R}{d} \tag{21.4}$$

We can, thus, rewrite equation (21.3) in terms of the light-source angular diameter.

$$B = \frac{\sigma T^4 \theta^2}{4} \tag{21.5}$$

We can also solve this for the angular diameter.

$$\theta = \frac{2}{T^2}\sqrt{\frac{B}{\sigma}} \tag{21.6}$$

A surprising feature of equation (21.5) or (21.6) is the absence of the distance to the light source as a factor. Of course all else being equal, the distance to the light source clearly does matter. If we lived on one of Jupiter's moons, five times farther from the Sun than at our present earthly distance, our photograms would need an exposure that is 25 times longer.

But when we think in terms of the *angular* size of the source—instead of its true size in centimeters or meters—we see that it is not the distance alone that matters. If we lived on Ganymede, the Sun would not only appear dimmer, but its angular diameter would be smaller too. It is the comparison of the light source's size to its distance that matters, and that is precisely what angular diameter means.

As we have already seen in chapter 8, the concept of light-source angular diameter is an important one for the making of photograms. If we want, for example, relatively sharp shadows from a thick, three-dimensional (3D) object, we must illuminate our photogram with a light source that has a small value of θ. We can thus use equation (21.6) to make an important comparison. What illumination can we expect from an artificial light source that is (at least approximately) a spherical blackbody—the small filament from an incandescent quartz-halogen lamp, for example?

We can compare the intensity from sunlight ($B_\odot$) to the intensity (B) from our artificial light source, given their respective angular diameters ($\theta_\odot$ and θ) and temperatures ($T_\odot$ and T). From equation (21.5) we have the following result.

$$\frac{B}{B_\odot} = \left(\frac{T}{T_\odot}\right)^4 \left(\frac{\theta}{\theta_\odot}\right)^2 \tag{21.7}$$

Let us say we are making photograms, and we want a source of artificial illumination that has both the intensity of sunlight and the same, small angular size as the Sun so as to produce relatively sharp shadows. Equation (21.7) reveals an important fact. Within our assumptions, the source of light must have a temperature as hot as the surface of the Sun—5770 K. No incandescent light source is that hot; even tungsten is vaporized at that temperature.

And so if we use incandescent lighting, with a filament temperature significantly lower than that of the Sun, the only way to achieve a small angular size is to settle for a much dimmer illumination. A typical quartz-halogen flood light, for example, has a filament temperature of about 3400 K. From equation (21.7), if we keep the angular diameter the same but reduce the temperature by a factor of 1.7, the illumination will be smaller by a factor of 1.7^4 or 8.4 times. Thus a more than eight-fold increase in exposure time would be required to compensate. This is one reason why direct sunlight is such a useful source of light for photograms.

I stress again that the equations of this section are valid only under our starting assumptions—that the source of light can be approximated as a blackbody, is spherical, radiates isotropically and has a relatively small angular diameter. We have also ignored the fact that our photographic light detector inevitably responds differently to different wavelengths of light. We consider these assumptions in turn in the sections that follow.

21.2 Nearby sources of light

In deriving equation (21.3) we assumed that the source of light is spherical and emits isotropically. The decrease of brightness with distance along a particular line-of-sight would also hold approximately even for a nonisotropic source if it is significantly farther away than its radius—that is, if its angular diameter is relatively small. But what about a nearby, very spread-out flat source of light—an array of fluorescent tubes in a light box, for example? This situation is more complex.

Let us, by way of example, consider as a source a flat, circular light box that is evenly illuminated across its surface, which has a radius of R = 10 cm and emits in a Lambertian fashion rather than isotropically. At distances much larger than R ($d \gg R$), the illumination provided by the box roughly obeys the inverse-square law. Move the light box twice as far away, and we receive only $1/2^2 = 1/4$ as much light from it. The intensity at a distance of 40 m, for example, is four times greater than at a distance of 80 m.

When very-much closer to the surface of the light source than its radius of 20 cm ($d \ll R$), on the other hand, the situation is radically different. At this extreme, the intensity of the light remains the same as one changes the distance. And so, for example, the intensity is very nearly the same at one millimeter from the surface of the light box as at five millimeters—five times farther away.

The intensity at these close distances is simply equal to the intensity of the light source itself—its total emitted power (P) divided by the surface area ($A = \pi R^2$) of the source of light.

$$B_{\text{near}} = \frac{P}{\pi R^2} \tag{21.8}$$

For our particular example, the area of the light box is about 0.126 m^2. For an emitted power of P = 100 W, for example, the intensity very near the surface would be about 796 W m^{-2}.

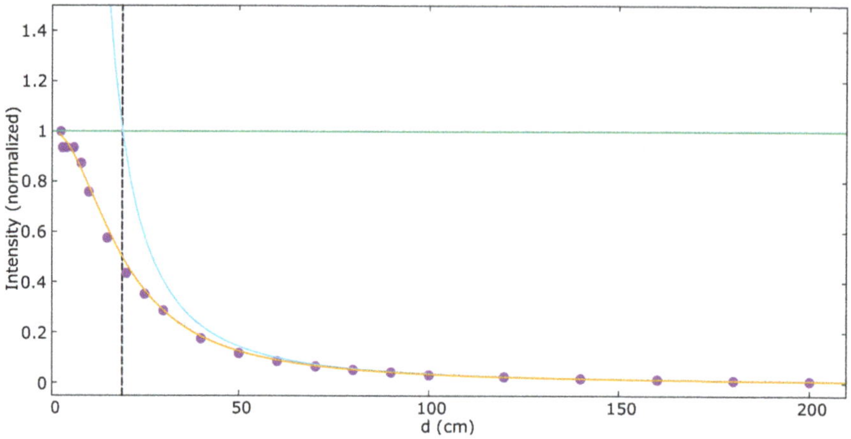

Figure 21.1. Relative intensity versus distance (d) from a circular Lambertian emitter of radius ($R = 19$ cm), marked as the black vertical line. The purple dots are (rather crudely) measured data from a homemade circular light box. The orange curve is the theoretical function derived in appendix D for a Lambertian emitter of this radius. The green and blue lines are the approximations for the cases of, respectively, $R \ll d$ and $R \gg d$.

We know the inverse-square law should hold at large distances, as one moves along a line directly away from the light box. But what would be the scale factor? We can use our discussion of Lambertian reflectors in chapter 20 and appendix D to help answer this question. We can, for simplicity, assume the light box emits with angle in the same way as does a Lambertian reflector—it appears to have the same surface brightness as seen from any angle. From equation (D.29) in appendix D, we would then have the following equation.

$$B_{\text{far}} = \frac{P}{\pi d^2} \tag{21.9}$$

Here, we have distance (d) along a line directly in front of (normal to) the surface of the light box. Figure 21.1 shows graphs of both B_{near} (green line) and B_{far} (blue curve) versus d plotted for the case of $P = 100$ W and $R = 19$ cm. Note that our approximation for B_{near} is clearly too big for large values of d, while our approximation for B_{far} is too big for very small values of d, approaching infinity at $d = 0$.

The vertical dashed line in figure 21.1 marks the radius (R) of the light box. Notice that the near and far approximations are of the same order at distances near the dimensions of the light source. In this range of distances, to calculate the intensity of light is complex.

The detailed calculation of the intensity versus distance for this case can be found in appendix D. In figure 21.1 the results of these calculations are plotted as the orange curve along with some measured data. Note that the correct calculation has the features we have already described regarding the near and far limits. The circle used had a radius of 19.05 cm, and the curve shown is for that radius. The best fit curve has a radius of 18.28 cm, only 4% smaller, and so the crude light box I built to test the theory must have been very close to Lambertian.

The inverse-square law would imply that for any given light source, we could achieve whatever illumination desired simply by moving the source of light closer. The reciprocal of the square of zero distance is, after all, infinity. But figure 21.1 shows that for a real light source, the maximum intensity one can achieve is the surface brightness of the light source itself. For the geometry of our example, we can see from figure 21.1 that for distances greater than about $3R$, the inverse-square law is a fair approximation.

For a thermal blackbody source of light, the surface brightness is given by the Stefan Boltzmann law: $B = \sigma T^4$. In the next section we consider nonthermal light sources.

21.3 Thermal versus nonthermal light sources

Let us reconsider the nature of the blackbody-emitting source of light we have so far considered. Let us suppose the surface of my experimental light box from section 21.2 were to emit with a surface-brightness intensity of $B = 1000 \text{ W m}^{-2}$—the same order of magnitude as sunlight at Earth. If its surface were a blackbody, how hot would it need to be? The answer is given by the Stefan-Boltzmann equation as solved for temperature.

$$1000 \text{ W m}^{-2} = \sigma T^4 \tag{21.10}$$

$$T = \sqrt[4]{\frac{1000 \text{ W m}^{-2}}{5.670 \times 10^{8} \text{ W m}^{-2} \text{ K}^{-4}}} \tag{21.11}$$

$$T = 364.4 \text{ K} \tag{21.12}$$

Thus a blackbody-emitting surface need only be 364.4 K (91.25 °C) to radiate 1000 W m^{-2}. But 91.25 °C will not even boil water! How do we reconcile this with the fact that the intensity of sunlight at the surface of Earth is only a little brighter (about 1300 W m^{-2})?

For one thing, the enormous intensity at the surface of the nearly 6000 K Sun has been diluted over the vast distance to Earth, by the inverse-square law, to only 1300 W m^{-2}. So there should be no surprise that this intensity corresponds to the surface emission of a far-cooler blackbody.

But we must also take into account *Wein's Law*: the wavelength that is most-strongly emitted by a blackbody is inversely proportional to temperature. A hotter blackbody emits shorter-wavelength light. Thus the 1300 W m^{-2} that reaches us from the Sun, emitted at a much-higher temperature than boiling water, means this light is of much shorter wavelengths. In particular, the sunlight arrives at visible wavelengths. A blackbody emitting at boiling-water temperatures, on the other hand, emits mostly thermal-infrared light that is invisible to both our eyes and our photographic emulsions.

This leads to another question—how in practice would we make our hypothetical visible light, 1000 W m^{-2} blackbody-emitting light box? We could, for example, put a 113 W incandescent tungsten-filament lamp inside a closed, internally-reflective box with a diffusing glass surface on one side for the light to emit in all forward directions. If the internal surfaces reflect diffusely and absorb no light, and if the diffusing glass effectively scatters the light but absorbs none, then one could reason that all 113 W must emerge uniformly spread out through the $\pi R^2 = 0.113\ m^2$ surface of our $R = 0.19$ m circular box. Thus an emitting-surface intensity of 1000 W m^{-2} would result.

Of course, this would have the spectrum of a tungsten filament, not sunlight; it would emit proportionately less visible and ultraviolet (UV) and more near-infrared. Thus, the bulb would need to emit considerably more than 100 W if we wanted it to emit proportionally as much visible light as the equivalent intensity of sunlight.

In practice, it is even worse; the inside-reflective surfaces and the diffusing glass would absorb a significant amount of the power emitted by the tungsten lamp, and the inside of the box would heat up markedly over time unless considered efforts were taken to keep it cool. Furthermore, it would be impossible for the tungsten filament to both emit light and also be thermally isolated from its surroundings. Thus the hot filament would transfer heat through conduction and convection as well as radiation, heating up the light box even more. And the absorption means the lamp would need to be all that much brighter still in order to make up for the light lost. This means even more power would be absorbed, necessitating even greater cooling measures. This could be accomplished by using a fan, for example, to circulate air through the box from the outside. It would, however, be a challenge to keep it all from getting dangerously hot.

My real-world light box (used for the data in figure 21.1) has a surface brightness, over the range of visible wavelengths, that is only about 1/6 that of sunlight. This would correspond to roughly 200 W m^{-2} if the spectrum of the light emitted by my light box was the same as that of sunlight. Even that lower intensity would require significant external cooling to keep it all from overheating if I had used a blackbody-emitting source of light such as an incandescent lightbulb.

Although my light box does have holes in it to allow outside air to flow through, it stays safely cool even with no cooling fan. The difference is that my light box does not use a blackbody-emitting incandescent light source. Instead it uses light-emitting diodes (LEDs), a decidedly nonthermal source of light.

Part of the trick is that the LEDs emit nearly all of their electromagnetic energy in the visible part of the spectrum. Even though they emit less light per surface area than a blackbody of the same temperature would, they emit far more visible light for their temperature. Even sunlight, the spectrum of which peaks in the middle of the visible spectrum, emits nearly 60% of its energy as invisible infrared and UV. The situation for an incandescent bulb is far worse, with at most about 20% falling in the visible spectrum for the most efficient quartz-halogen bulbs.

Most importantly, the LEDs are a nonthermal source of light, and neither Stefan-Boltzman nor Wein's law apply. The light emitted is simply unconnected to the temperature of the bulb in any straightforward way. Since the light is not emitted from the natural thermal motion of the molecules, as in a blackbody source, it is possible for an LED to emit a high intensity of short-wavelength light while remaining cool.

21.4 Laser photograms

We have so far considered sources of light that spread out markedly—either isotropically or as a Lambertian emitter. But what about a *collimated* light source—one that emits all of its power over only a small range of angles? A laser is a good example; it can be as close to unidirectional as the physics of waves allows. The angular spread of even an inexpensive laser is so tiny that very-high intensities can be achieved even with a power of only a few milliwatts. But of course, the area exposed is tiny indeed, perhaps less than a millimeter across. Over that minuscule area, however, the illuminance can be significantly greater than that of direct sunlight.

The beam of a laser does spread out naturally by a tiny amount given by diffraction theory. The angle (θ) at which the beam spreads is related to the diameter (D) of the laser's exit aperture.

$$\theta_{\text{min}} = 2.44\frac{\lambda}{D} \tag{21.13}$$

This is simply the size of the diffraction pattern from a circular aperture, as described in section 3.7; however, it uses the approximation $\sin\theta \approx \theta$, appropriate for the very tiny values of θ in this context.

Let us consider the example of an inexpensive 405 nm battery-powered, deep-violet pen-style laser shining on a piece of paper 1.0 m away. My laser pen has a labeled power of $P = 10$ m W, and it appears to have an exit aperture of roughly $D = 2$ mm. From equation 21.13, and noting the relationships between power, illuminance (B), area (A), angle and distance (d), we have the following equations.

$$\theta = 2.44\frac{405 \times 10^{-9}\text{ m}}{2 \times 10^{-3}\text{ m}} \tag{21.14}$$

$$= 4.94 \times 10^{-4}\text{ rad} \tag{21.15}$$

$$A = \pi(d\theta/2)^2 \tag{21.16}$$

$$= 1.91 \times 10^{-7}\text{ m}^2 \tag{21.17}$$

$$B = \frac{P}{A} \tag{21.18}$$

$$= 5.2 \times 10^4\text{ W m}^{-2} \tag{21.19}$$

The result demonstrates that in theory even a low-power laser can produce an illuminance many times greater than direct sunlight. Of course my real laser pen does not produce such a high illuminance. The labeled power, for example, is not its measured power but rather the maximum allowed for its class. And the imperfect

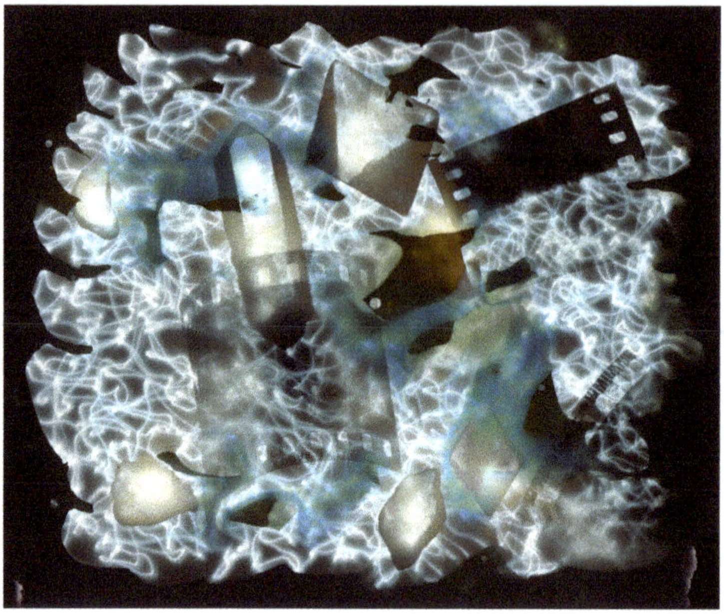

Figure 21.2. *The six cornered snowflake*, John Beaver 2016. A laser swiveling about a fixed point was used to expose a photogram of 3D crystals. Since the beam always came from the same fixed location, it was as if the source of light were a point source, and so the shadows are perfectly sharp.

optics produces a spot that is likely many times larger than the theoretical smallest. In practice, however, this small laser does have the ability to expose light-sensitive paper in significantly less time than direct sunlight.

Another feature of using a laser to expose a photogram is that it is a light source with a tiny angular size. Thus, one can achieve extremely sharp shadows. There is, however, a nearly-fatal flaw: it exposes a circle on the paper (in our example) that is only 0.2 mm across! That is a small photogram indeed, even for an alternative-process-photographer ant.

There is, nonetheless, a way to use a laser to make a photogram of large objects with shadows much sharper than possible with sunlight. Attach the laser to a fixed mount that can swivel in different directions. Then slowly scribble over the light-sensitive paper, exposing it with a pattern of random lines. Each position on the paper receives an exposure of only a tiny duration, so different parts of the photogram are built up over time. See figure 21.2 for an example, where I used ephemeral-process photography.

To produce such a sharp image in this way, the laser beam must come from exactly the same location the entire time. I mounted the laser on a tripod, but in such a way that it could pivot somewhat about the center of its light-emitting part. I could point the beam in different directions (within limits) and expose any portion of the light-sensitive paper. But from the point of view of the light-sensitive paper, the beam always originated from exactly the same location. Thus the effect, regarding the sharpness of the shadows, was the same as if the photogram had been illuminated with a tiny point source of light—much smaller even than the angular size of the Sun.

The scribble pattern is obvious, of course; it is impossible to produce an even illumination in this way. But the shadows of the semitransparent 3D crystals have a near photographic quality, and the scribbled background can be interesting in the proper context. Consider it yet-another tool in the artistic toolbox.

IOP Publishing

Photography (Second Edition)
Physics and art in focus
John Beaver

Chapter 22

The elements of exposure

22.1 Shutter speed and aperture

We have seen that exposure = illuminance × time. Given the amount of light coming from the subject, the photographer makes two different adjustments to the camera itself in order to alter exposure. Time is the most directly-controlled factor; a physical *shutter* allows the light to enter the camera for only a fixed time. This adjustable *shutter speed* allows the photographer to choose the time over which the image interacts with the detector.

There is no camera setting, however, that directly sets illuminance—the other factor of exposure. The camera, of course, has no control over the amount of light from the subject that arrives at the lens. It can only control the fraction of that incident light that enters, by controlling the size of the *aperture*. We have already seen in chapter 20 that, given the brightness of the light arriving from the subject, it is the *focal ratio* (f) that controls the illuminance at the detector.

Thus time is controlled by a shutter and illuminance is controlled, albeit only in part, by the focal ratio of the aperture. There is another important feature about the way photographers choose to change aperture and shutter: adjustments are made in *steps*, and the difference between steps is a *factor of two* in exposure.

Photographers measure and set exposure quite differently than how a physicist might; they do so in *relative* terms, without the use of physical units such as $W\ m^{-2}$. The idea of the exposure step is key. If all factors are counted in steps that affect exposure by a factor of two, then all we need is some agreed-upon starting point for each of these factors. After that, it is just counting the number of steps. We lay out this procedure in detail in section 22.4.

22.2 Power and shutter speed

In most cameras the shutter is a precise timer that controls a mechanical cover that opens, exposing the detector to light, and then closes, covering it up. In some cases

 © IOP Publishing Ltd 2022

Figure 22.1. An old mechanical leaf shutter for a large-format camera. It uses quickly-moving leaves of thin metal to briefly uncover the lens opening. The lens' elements have been removed for this picture, and I am holding the shutter leaves open halfway. The timing mechanism—a complex arrangement of gears, levers and springs—is inside the outer housing. Note the shutter-speed settings, listed in reciprocal seconds, on the right-hand side of the outer ring.

(called a *leaf shutter*) this mechanism is located between the elements of the lens itself, while in others (known as a *focal plane shutter*) it covers the entire detector and is located directly in front.

Since the most commonly-used exposure times are fractions of a second, they are usually labeled with their reciprocals. And so 1/60 s is labeled '60,' 1/30 s is labeled '30,' etc. Furthermore, photographers generally refer to this not as 'exposure time,' which is literally what it is, but rather as 'shutter speed.' This makes sense if one thinks about it; '60' means the shutter is 'moving faster' than does '30.' If the exposure time is greater than 1 s, then 's' is added to clarify: '4 s' means four seconds. An unusually-long exposure time measured in many seconds, minutes or even hours is known as a *time exposure*. Because of the use of reciprocals instead of the actual exposure times, a larger shutter-speed number results in less exposure.

These are the shutter-speed settings available on most cameras: 1, 2, 4, 8, 15, 30, 60, 125, 250, 500 and 1000. In the order I have written them, each step would give half the exposure of the previous step, all else being equal. In addition, many modern cameras with electronically-controlled shutters have settings of 2 s, 4 s, 8 s, 15 s and 30 s, and additional fast speeds of 2000, 4000 or even 8000. Some also have 'half-step' settings in between the standard values. Furthermore, shutters may have settings of 'B' or 'T' that are not controlled by a timer and so allow the photographer to take time exposures of whatever length desired. See figure 22.1 for an example of an old mechanical leaf-shutter mechanism.

Notice that the standard list of shutter-speed numbers does not exactly represent steps of factors of two. For example, $60 \times 2 = 120$ and not 125. The standard numbers are tweaked slightly for convenience. So we start with 1 s and end with 1/1000 s rather than 1/1024 s. In all but the most technical photography, the difference is so slight as to be unimportant[1].

22.3 Aperture and focal ratio

To control the power of the light entering the camera, we use an *aperture stop*—a variable-sized hole of some sort. It is sometimes also called an *iris* after the part of the human eye that accomplishes the same task. A camera iris is usually made of several overlapping thin metal leaves, and it is almost always located inside the camera lens between the different lens elements (e.g., see figure 22.2). Our measure of aperture is the focal ratio (f): the ratio of the focal length of the lens to the diameter of the lens opening.

Equation (20.1) in chapter 20 relates the illuminance of light on the subject to the illuminance of light on the detector from the image focused by the lens. We can solve this equation for the image illuminance as follows.

$$I_{\mathrm{D}} = \frac{R}{4f^2} I_{\mathrm{SUB}} \tag{22.1}$$

Equation (22.1) means that, for a given amount of light on the subject (I_{SUB}) and a given subject reflectivity (R), the illuminance of the image on the detector is proportional to $1/f^2$. We can thus adjust the image illuminance simply by adjusting f for a given amount of light arriving from the subject. Since this illuminance scales as the *reciprocal* of the square of the focal ratio, it means that increasing the focal ratio decreases the illuminance of the image.

As is the case for shutter speed, photographers choose focal-ratio settings in steps, the difference of which results in a factor of two in exposure. Since the exposure is proportional to the illuminance, and the illuminance is proportional to $1/f^2$, then to increase the exposure by one step—a factor of two—we must decrease the focal ratio not by a factor of two but rather by a factor of $\sqrt{2}$.

Figure 22.2. An iris-aperture stop in the lens of a medium-format camera. The front lens elements have been removed to make the aperture blades more easily visible. The aperture is set, from left to right, at $f/4.5$, $f/8$ and $f/32$. The marked f numbers can be seen at the top outer edge of the lens assembly.

[1] Many very-old shutters have a slightly different set of shutter numbers: 1, 2, 5, 10, 20, 50, 100, 200 and 400.

To adjust f according to the photographers' exposure steps, we need a list of settings that are the square roots of successive factors of two: $\sqrt{1}=1$, $\sqrt{2}=1.414$, $\sqrt{4}=2$, $\sqrt{8}=2.828$, $\sqrt{16}=4$, etc. Here is a complete list that covers the possible f settings on nearly any camera lens one might come across. By tradition, the settings are rounded off to at most two significant digits.

$$1,\ 1.4,\ 2,\ 2.8,\ 4,\ 5.6,\ 8,\ 11,\ 16,\ 22,\ 32,\ 45,\ 64,\ 90$$

No real camera lens has all of these settings. For a given lens the minimum value of f is determined by the glass of the lens itself, while the maximum value is set by the mechanical limitations of the iris inside the lens. Since the minimum f of the lens itself may not match up exactly to one of the standard exposure steps, the smallest f setting is often in between the standard steps. For example, the f settings on the lens in figure 22.2 are 4.5, 5.6, 8, 11, 16, 22 and 32.

I must point out a particular quirk of how photographers refer in print to focal ratio. To emphasize that it is a ratio, $f = 5.6$ (for example) is often referred to as '$f/5.6$.' To the mathematically-inclined this looks odd, as if one means to say, 'f divided by 5.6,' which is of course simply one if $f = 5.6$. There is more quirky terminology: focal ratio is often referred to as 'f stop.' The term arises from the earliest days of photography, before the advent of continuously-adjustable mechanical irises, when the focal ratio of a lens was controlled by inserting an 'aperture stop' into the lens—one of a set of metal slides with different size holes in them. As such, a photographer will often refer to increasing the focal ratio of a lens (decreasing the diameter of the aperture) as 'stopping it down.'

Since $f = F/D$, where F and D are, respectively, the focal length and diameter of the lens, a smaller f means a larger aperture diameter. And so the minimum f of a lens is when the aperture stop is 'wide open' and not covering the lens at all. Since it is far more difficult to design and manufacture a lens of small f than large f, this *maximum aperture* (minimum f) is one of the key features of any lens. As such, lenses with a focal ratio less than 1.4 are very rare. For larger detector formats, everything is scaled up and even more expensive. And so the vast majority of lenses for such cameras have a minimum focal ratio that is greater than $f/4$.

Because of the way focal ratio is defined, a large focal ratio number results in less exposure. Recall that this is also true for the shutter-speed numbers. The settings for both aperture and shutter speed are marked with numbers such that larger numbers result in less exposure.

22.3.1 The effect of focus on exposure

If an object is very far away from the lens, the image distance is roughly equal to the focal length of the lens. But when a camera is focused on a nearby object, the detector is at a distance from the lens greater than the focal length. This larger image distance means the same light is spread out over a greater area on the detector, and thus a smaller exposure results.

This means that exposure is really proportional, not to the inverse square of the ratio of the *focal length* (F) to the lens' diameter (D), but rather to the inverse square

Figure 22.3. A camera lens focused at infinity (**left**) is only half the distance from the film as one focused at 1:1 macro (**right**). The *f* settings on the lens correspond to infinity focus. For close-up work, the effective *f* is larger than this. For this example of 1:1 macro, a lens marked as $f/5.6$ would actually expose the detector at $f/11$ (two exposure steps less exposure).

of the ratio of the *image distance* (d_i) to the diameter (D). Recall from chapter 20 that we used an approximation in deriving our relationship between focal ratio and illuminance—that the image is much farther away than the focal length of the lens. A lens is marked with *f* numbers that correspond to $f = F/D$, not d_i/D, and so the photographer may have to apply an exposure correction when focusing on a very nearby subject. This is especially relevant for macro photography, where the image distance may be as much as twice the focal length, leading to an effective focal ratio twice as great as what is marked on the lens. This results in only 1/4 the exposure intended—a difference of two exposure steps—if the photographer (or the camera) does not apply an appropriate correction (see figure 22.3).

In this discussion of focal ratio and exposure, we have assumed that doubling the image distance doubles the diameter of the image on the detector (and so increases its area four times). This is true the vast majority of time, but it may not always be true. For what if the object is so distant that no lens can resolve it as anything but a perfect point? In that case the image on the detector would be only a point as well, no matter what the focal length of the lens.

The most obvious examples are the stars in the night sky—most appear as essentially perfect points in even the world's largest telescopes. There is the issue of diffraction; a perfect point-like object forms a diffraction pattern rather than a truly-perfect point-like image. But for most cameras this diffraction pattern is smaller than the smallest detail that can be resolved by the light detector, so that is usually not a factor. Thus images of stars are *unresolved*: they appear as points no matter what focal length is used.

If one sharply focuses a camera on the stars in the night sky, a larger aperture lets in more light (and thus increases the exposure), but a longer focal length has essentially no effect on the size of the focused image of the stars. This means that exposure for an unresolved image is simply proportional to the square of the diameter of the lens rather than the inverse of the square of the focal ratio.

When taking real pictures of the night sky it is usually more complicated than this, as the stars are often not all that is in the picture. Besides the unresolved, point-like stars, there may be resolved, diffuse elements in the photograph—objects in the foreground or perhaps the glow of the Milky Way. The exposure of the resolved elements are related to the focal ratio of the lens, while the exposure of the unresolved stars are related only to the lens diameter.

22.4 Density and the elements of exposure

When all is said and done, there are four factors that determine how dark or light a particular part of the picture will be. The photographer's name for the degree of physical effect on the detector is *density*. For traditional negative film, for example, density is how dark the film gets when it is developed. For a digital detector, density is related to the amount of electric charge that builds upon a part of the detector due to its exposure to light.

The density that is produced is a result of a combination of the exposure from the light and the properties of the detector itself. We have already considered exposure, and it has three elements, which are listed below. The fourth factor that determines density is a measurement of the *sensitivity* of the detector to exposure. These four factors are summarized below.

Shutter: This determines the length of time for which light enters the camera and thus the total energy transferred. All else being equal, *a smaller shutter-speed number (slower shutter speed) results in greater density.*

Aperture: This varies the power of the light entering the camera compared to how much it spreads out on its way to the detector, by adjusting the relative size of the opening in the lens compared to the focal length. All else being equal, *a smaller f/ number (larger aperture) results in greater density.*

Light: While the first two factors are settings on the camera itself, the third factor that determines exposure is, of course, the intensity of the light arriving at the camera. *Metering* is the process of measuring the amount of light. All else being equal, *more light results in greater density.*

ISO: ISO is the official measure of the sensitivity to light of the light detector (whether it is film or digital). We will discuss the physics of many kinds of light detectors in part VI of *Photography: Physics and Art in Focus*. The same exposure on a more-sensitive detector produces a greater density and thus a brighter picture. All else being equal, *a larger ISO number means a more-sensitive detector and results in greater density*. Typical ISO numbers are 50, 100, 200, 400, 800 and 1600. Each represents a doubling of the previous and thus a doubling of the detector sensitivity. With twice the ISO, half the exposure is needed to produce the same density.

As a warning, we will consider factors that complicate this a bit. For example, the specific intensity of light entering the camera is different in different directions, obviously, or else our picture would be one uniform shade of gray. That is why some parts of our picture are dark and some are light. So what do we do if we are faced with a choice of either making some parts of the picture too dark or some parts too light? What if there is no single exposure that gets both the dark parts and the bright parts right at the same time? Life is complicated and so is photography. We will have more to say about this important issue, but let us put it aside for now.

Another complication is color. Although aperture and shutter speed for the most part have the same effect on exposure whatever the color of light, the same may not be true for ISO. And so this fact, that light is comprised of a range of wavelengths, will have significant consequences for exposure and for metering.

22.5 The definition of ISO speed

Aperture and shutter speed are defined in essentially physical terms, but there is something arbitrary about the definition of the ISO speed of a detector. Why, for example, does ISO 100 film have the particular sensitivity to light that it does? The answer is simple: we have all agreed to say that it does. It is an international standard of measurement. Much can be said about the gritty details of the ISO speed standard, but its very-rough outline can be summarized by the useful *sunny 16 rule*.

> A 'proper' density results for a typical landscape scene on a bright sunny day when the exposure is made at $f/16$, and the reciprocal shutter number is the same as the ISO speed of the detector.

Thus a shutter speed of 1/100 s with an aperture of $f/16$ gives the 'correct' exposure for a sunny-day landscape when using a detector with an ISO speed of 100.

22.6 Reciprocity and exposure

Exposure is the total energy per area that strikes the detector. Since the illuminance of the light falling on the light detector is energy per time per area, it is easy to see that exposure is given by the following equation.

$$\text{Exposure} = \text{Illuminance} \times \text{Time} \tag{22.2}$$

Notice that it is neither illuminance nor time alone that matters but rather the product of the two. If, for example, one is quadrupled while the other is reduced by a factor of four the same exposure results: $4 \times \frac{1}{4} = 1$.

The density that results from an exposure depends on the detector ISO, clearly. But there is another important question. Does it depend only on the exposure—the product of illuminance and time? Or does density depend on both illuminance and time separately in some more complicated way?

The short answer is that most of the time, density is a function of exposure alone and not illuminance or time specifically. When this is true, it is called the *law of reciprocity*, and it means that any two exposures that are the same—but have different combinations of illuminance and time—will produce the same density. And so one can reduce the illuminance by a factor of five while increasing the time by a factor of five, and the detector will produce the same density.

There are occasions, however, when the law of reciprocity does not hold, and this is called *reciprocity failure*. The most common case is that some detectors have, in effect, a lower ISO value at very low levels of illuminance. We will revisit reciprocity failure specifically in chapter 37, but for what follows I will assume that the law of reciprocity holds.

22.7 Camera settings

We have seen that exposure is the product of illuminance and time. Exposure is the cause, but density—what actually happens to the detector—is the effect. It is the detector sensitivity, ISO, that connects density to exposure, and so we can say the following is true.

$$\text{ISO} + \text{Exposure} \Longrightarrow \text{Density}$$

Taken together, these facts mean that density is determined by a combination of three measurements: detector ISO, exposure time and illuminance of the light falling on the detector. The catch is that one of these three measurements—illuminance of the image focused on the detector—is usually not directly measured or controlled by the photographer. Instead, image-plane illuminance is determined by a combination of aperture and light. And so the photographer has the following four factors: ISO, exposure time (shutter speed), aperture (focal ratio) and intensity of light arriving at the camera.

$$\begin{aligned}
\text{Light} + \text{Aperture} &\Longrightarrow \text{Illuminance}\\
\text{Shutter} + \text{Illuminance} &\Longrightarrow \text{Exposure}\\
&\text{and so:}\\
\text{Light} + \text{Aperture} + \text{Shutter} &\Longrightarrow \text{Exposure}\\
\text{ISO} + \text{Exposure} &\Longrightarrow \text{Density}\\
&\text{and so:}\\
\text{Light} + \text{Aperture} + \text{Shutter} + \text{ISO} &\Longrightarrow \text{Density}
\end{aligned}$$

Aperture, shutter and ISO are settings on the camera. These three factors, combined with the light, determine the density. All three of our camera settings use steps that, taken individually, double or halve the density from one step to the next. Table 22.1 shows typical settings, each representing one step in density, for shutter, aperture and ISO. Taking each of the three settings individually, the top of the table represents increased density while the bottom of the table represents lower densities. Most cameras have only a subset of these possible settings, and it is possible to extend the table beyond what I have reproduced here.

ISO is a special case. Traditionally, photochemical detectors were used in cameras (ordinary film is an example), and a given piece of film has a particular ISO. Thus once the camera is loaded with film, the ISO is already chosen. Most cameras have, on the other hand, many possible settings for aperture and shutter. Digital cameras, however, are different. It is possible to internally change the sensitivity (and thus the

Table 22.1. Possible camera settings to affect the density of the picture, given the light reaching the camera. A greater effect on density is toward the top of the table. Each number in a given column represents a doubling of the physical exposure as compared to the number below it or a halving of exposure as compared to the number above it. We call such a doubling (or halving) a single step.

f/ Number	Shutter speed	ISO
1.4	1	6400
2	2	3200
2.8	4	1600
4	8	800
5.6	15	400
8	30	200
11	60	100
16	125	50
22	250	25
32	500	12
45	1000	6
64	2000	—
90	4000	—

ISO) of the light detector in a digital camera. And so ISO can be 'set' from picture to picture just like aperture and shutter.

Below I list three examples, each of which, taken alone, increases the density by one step.

1. Change the shutter speed from 250 to 125,
2. Change the aperture from *f*/5.6 to *f*/4, or
3. Change the ISO from 200 to 400.

What if I did all three? What if I changed the shutter speed from 250 to 125, changed the aperture from *f*/5.6 to *f*/4 and changed the ISO from 200 to 400? Well, since each individually would increase the density by one step, doing all three increases the density by three steps. The same logic applies if decreasing density. A decrease in any of the three decreases the density. If some increase and some decrease, then one must both add and subtract to determine the overall change. Armed with this knowledge, one should now be able to explain why this next example results in an increase in density of three steps.

1. Change the shutter speed from 250 to 15,
2. Change the aperture from *f*/5.6 to *f*/16, and
3. Change the ISO from 200 to 800.

If one is having trouble with this exercise, it can be thought of this way: The change in shutter speed from 250 to 15 moves up the table by four steps. And so let us record that as '+4.' The change in aperture from $f/5.6$ to $f/16$ moves down the table by three steps; so let us record that as '−3.' And finally, the change in ISO from 200 to 800 moves up the table by two steps or '+2.' To find the overall result, simply add the numbers $+4 - 3 + 2 = +3$, indicating an increase in density of three steps.

22.8 Choosing between equivalent settings

It should be evident that there are many different camera settings that achieve exactly the same density for a given amount of light entering the camera. For example, one could do any of the following actions.

1. Increase *f* by three steps while decreasing the shutter number by three steps,
2. Decrease the ISO number by two steps while decreasing the shutter number by two steps, or
3. Increase the ISO number by three steps while increasing the shutter number by two steps and increasing *f* by one step.

The exposure would remain exactly the same. So why choose one combination over another? The reason is, of course, that density is not the only issue. Each of these three settings has other consequences for the picture.

22.8.1 Aperture and depth of focus

Apart from exposure, focal ratio alters the *depth of focus* of the picture—the range of subject distances that are in acceptable focus. A smaller focal ratio leads to a more narrow range of distances that are in good focus, while a large focal ratio makes both nearby and distant objects in focus at once. For example, one might want to use a combination of shutter speed and ISO that achieves the desired exposure, while using a large aperture (small *f*) because this results in a shallow depth of focus and so isolates the subject from the background. Or perhaps one wants the opposite—to use a small aperture to get both the foreground and background in focus simultaneously. One can then pick, for the amount of light present, a combination of ISO and shutter speed that gives the proper exposure for the aperture desired (e.g., see figure 22.4).

22.8.2 Shutter speed and motion blur

It should be unsurprising that shutter speed is important in relation to time and motion. A fast shutter speed (large shutter-speed number) freezes motion both for the subject and the camera. And thus there is the rule of thumb that when holding a camera by hand, a shutter-speed number lower than 30 will likely produce a blurred picture, as it will be too difficult to hold the camera steady enough to ensure it does not move appreciably during the picture[2].

[2] This rule of thumb assumes that a lens with a 'normal' angle of view is being used. There are modified versions of this rule for wide-angle or telephoto lenses.

Figure 22.4. The **left**-hand image was taken at *f*/1.4 with a shutter speed of 1/2000 s, while the **right**-hand image was taken at *f*/16 with a shutter speed of 1/15 s. Both combinations produce the same exposure, but the smaller focal ratio results in a greatly-reduced depth of focus in the left-hand image.

Figure 22.5. Left: The camera was panned with one of the dancers, and the shutter triggered when she was relatively motionless. A moderately-slow (1/8 s) shutter speed was used to produce motion blur in the other dancers. **Right:** A very-fast shutter speed (1/2000 s) was used to freeze the motion of the falling water so that it would look like a crystal sculpture.

As is the case for selective focus, careful control of *motion blur* can be an essential part of a composition. Take two pictures of a waterfall, one at a shutter speed of 1/4000 s and the other with a shutter speed of 1/2 s (using a tripod). The former will freeze the motion of every drop of water, and the waterfall will look like a crystal sculpture. The latter will make the water look like a dreamy cloud flowing over the sharply-defined rocks. Which do you want?

Typical examples of using shutter speed as part of the composition can be seen in figure 22.5. For the photograph on the left I used a shutter speed fast enough that I could successfully pan the hand-held camera with one of the dancers, triggering the shutter at a moment when she was relatively motionless. But the shutter speed was slow enough to produce motion blur in the rapidly-moving dancers surrounding her. This had the effect of isolating the subject, while reducing the other dancers to abstract swirling patterns that emphasize the motion of the dance.

For the photograph on the right, I did the opposite. I used a very fast shutter speed (1/2000 s) to freeze the motion of the falling water so that the waterfall would look like a crystal sculpture.

Figure 22.6. From the series *Motion in scotland*, John Beaver 2017. I panned the camera from a moving bus, using a slow shutter speed.

For a much better example, consider Carlo Bevilacqua's intriguing use of motion blur in his photograph *Catari* (see Cooke and Kinneberg [2014, p 59]). A large depth of focus (from a small aperture) leaves background detail distinct so as to ground the figures in a real place. The animated face of the lead figure is seemingly frozen in time while her clothes are blurred by motion due to the slow shutter speed. The girl's skirt seems to have moved mostly in the vertical direction, making its outline indistinct but leaving intact its vertical stripes.

Successful use of motion blur often involves an element of luck combined with the experience to know how to make the most of the situation. In a real scene some parts usually move faster than others, while other parts do not move at all. A slow shutter speed will cause the faster-moving parts of the scene to blur the most. Panning the camera with the moving object has the opposite effect: the stationary parts of the scene blur the most. Finally, the blurring is always in the direction of the relative motion between the subject and camera. In a real situation one must often react quickly, and so it is crucial to think ahead about what might happen and to have already chosen a shutter speed that is appropriate for the desired effect.

The photographs in figure 22.6 are a complex example of the use of motion blur. I took the photographs out the window of a quickly-moving bus while panning the camera with particular features as we passed them. The effect of the panning depends strongly on distance, and I now had two motions—the motion of the bus and the panning of the camera. The complexities of which features were sharp, which were blurred and in what direction they were blurred often defies after-the-fact explanation. Luck was a big part of the result. As such, I took over 200 pictures but exhibited only 15 of them.

22.8.3 ISO and noise

There are, at times, competing reasons for wanting to set aperture and shutter speed. For example, suppose one wants to freeze motion but also show a large depth of focus. Both choices will result in a decrease in exposure. And so it is nice that there is the third factor of ISO. If increasing both the shutter-speed number and the *f* to the desired values gives too little exposure for the amount of light, one may be able to simply increase the ISO number to compensate.

But this, too, has its consequences. We take up this issue more fully in chapter 40, but I point out here that an increase in ISO inevitably results in a loss of detail in the picture. This can show up as 'graininess' and a lessening of the range of subtle values and colors that can be portrayed (e.g., see figure 22.7).

22.8.4 Changing the light

Finally, it may be possible to change the light itself. In a studio, one can shine more light on the subject with flashes or flood lights. In the field, one can use a portable flash or a reflector to add some extra light on the subject. To reduce the light, one can use a neutral-density filter.

22.8.5 Navigating the trade-offs

Here I summarize the consequences, apart from exposure, of a particular choice of aperture, shutter speed and ISO.

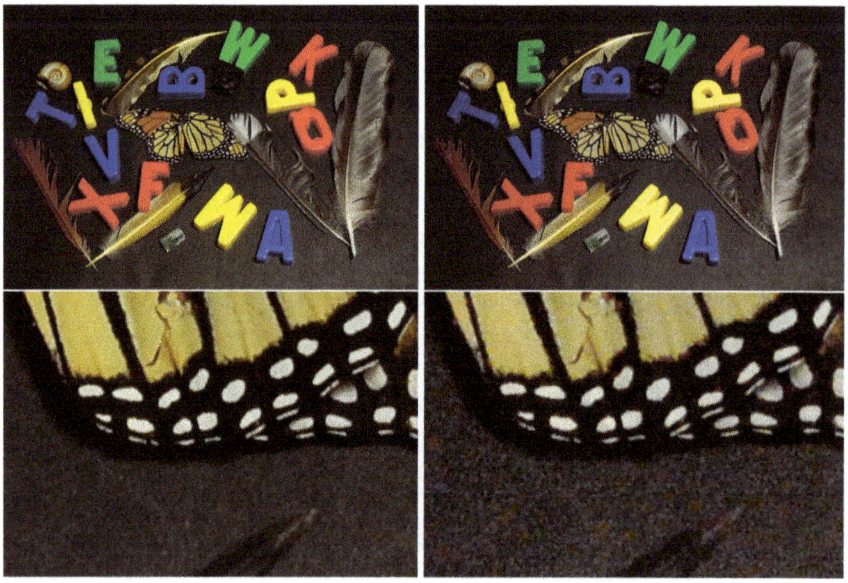

Figure 22.7. **Left:** ISO 200, *f*/5.6, 1/15 s. **Right:** ISO 3200, *f*/5.6, 1/250 s. Both images have the same density, but the right-hand version, taken at higher ISO and correspondingly lesser exposure has more noise. This is apparent in the loss of fine detail and the grainy nature of the blacks in the close-up view. But the loss of richness in the colors is apparent even in the unmagnified view.

1. A smaller aperture (larger *f*) results in a greater depth of focus. This can be either a good thing (if one wants the viewer to compare nearby objects to far-away objects) or a bad thing (if one wants the in-focus subject to be isolated by selective focus from its surroundings.)
2. A faster shutter speed (larger shutter-speed number) results in a greater freezing of motion of the subject and camera as well as less motion blur.
3. A more light-sensitive detector (larger ISO number) results in a picture that is more grainy and less subtle in its range of values and hues.

If we add to this the constraints of getting the proper exposure, we also have the following consequences.

1. Shallow depth of focus, all else being equal, means one will need a faster shutter speed (because *f* is smaller) and thus a greater freezing of motion. On the other hand, if one wants a large depth of focus (large *f*), the consequence is a slower shutter speed and thus more motion blur.
2. Greater motion blur (slower shutter speed) means that, all else being equal, a larger *f* is required and thus a more shallow depth of focus. If one wants to freeze the motion (faster shutter speed) then the consequence, all else being equal, is a smaller *f* and thus a more shallow depth of focus.
3. If we do not want motion blur from the vibrations of a hand-held camera, then we must use a tripod for shutter numbers smaller than 30 or so.

Thus there are trade-offs, imposed by the constraints of exposure, on our compositional choices regarding motion blur on the one hand and depth of focus on the other. What if one wanted to freeze motion with a large depth of focus in dim light and with relatively low amounts of noise and graininess? It should be clear that all of these desires are in conflict with each other. Sometimes we cannot have it all; we have to make tough choices.

22.9 Exposure value (EV)

For traditional film cameras, the ISO is a pretty-much fixed property of the film loaded into the camera. Particularly when using roll film, once the camera is loaded one is stuck with whatever ISO of film is in the camera. It is possible to alter the ISO in the process of chemical development of the film, but one must make the choice of ISO for the entire roll and not for each individual picture.

Before the advent of digital photography exposure was thus changed from picture to picture primarily through shutter speed and aperture alone. For this reason, it has long been convenient to talk of an *exposure value* (EV) that represents the effect of only aperture and shutter speed on exposure.

> A particular EV value represents all of the combinations of aperture and shutter speed that give the same exposure.

Table 22.2. EV numbers for some different combinations of aperture and shutter speed. For a particular ISO, these EV values represent correct exposures for different lighting conditions. The fourth column gives lighting conditions for EV_{100} (scene descriptions adapted from Exposure value (2018), table 2 and references cited therein).

EV	*f*/	Shutter Speed (s)	Scene for ISO 100
−1	1.4	4	—
0	1.0	1	—
+1	1.4	1	—
+2	1.4	2	Distant view of lighted buildings at night
+3	1.4	4	Floodlit building at night
+4	1.4	8	Floodlit building at night
+5	1.4	15	Home interiors
+6	1.4	30	Home interiors
+7	2	30	Home interiors
+8	2.8	30	Office or work area
+9	4	30	Night sports
+10	5.6	30	Skyline just after sunset (clear)
+11	8	30	Skyline just after sunset (clear)
+12	11	30	Heavy overcast or open shade on sunny day
+13	16	30	Cloudy bright (no shadows)
+14	16	60	Hazy Sun (soft shadows)
+15	16	125	Sunny bright (distinct shadows)
+16	16	250	Distinct shadows, sand or snow

Like aperture, shutter speed and ISO, EV numbers represent steps that correspond to a doubling of exposure. As is the case for shutter and aperture numbers, a smaller EV number represents greater exposure. And so EV = +1 represents, all else being equal, one step higher (greater exposure) than EV = +2. I say '+2' because EV numbers can be negative. And so EV = −3 represents an increase of five exposure steps over EV = +2 (because '−3' is five less than '+2'). And yes, there is EV = 0; it represents an increase of exposure of one step over EV = +1 and a decrease in exposure of one step from EV = −1.

An EV of zero is defined to be any combination of aperture and shutter speed equivalent to a 1 s exposure at *f*/1.0. Very, very few lenses have an *f*/1.0 setting, but it represents one aperture number step smaller than *f*/1.4 (which is not so uncommon). From this definition and the fact that EV numbers represent our normal exposure steps, we can make a table of EV values for some combinations of aperture and shutter speed (see table 22.2).

We can also look at this a different way, as seen in table 22.3. Here we represent different combinations of aperture and shutter speed, all of which represent the same EV ('+15' in this example). Thus, for a given amount of light, these all represent the same exposure. Notice that if we read down the table, aperture numbers get bigger while shutter-speed numbers get smaller. This leads to the *one up, the other down rule*.

Table 22.3. Shutter speed and aperture combinations for EV = +15.

EV	*f*/	Shutter Speed (s)
+15	2	8000
+15	2.8	4000
+15	4	2000
+15	5.6	1000
+15	8	500
+15	11	250
+15	16	125
+15	22	60
+15	32	30
+15	45	15
+15	64	8
+15	90	4

Increase the aperture number by a set number of steps while decreasing the shutter-speed number by the same number of steps, and the exposure remains unchanged. Similarly, increase the shutter-speed number by a set number of steps while decreasing *f* by the same number of steps, and the exposure is unaffected.

When EV is combined with a particular ISO, it represents all of the camera settings that determine a particular density given the amount of light. Thus, for a particular ISO setting, EV values can be used to represent the proper exposure for different lighting conditions. This is often given in a chart of EV_{100}, which represents EV settings combined with ISO set to 100. The fourth column in table 22.2 gives lighting conditions that correspond to EV_{100} exposures.

Looking at table 22.2, notice that entries at the top of the table (smaller EV values) represent camera settings that would result in an increase of exposure over choices from lower in the table (larger EV numbers). But the opposite is true for the lighting conditions listed in the fourth column. A moment's thought should render this unsurprising; a darker scene requires a camera setting that corresponds to a greater exposure compared to the camera setting for a brighter scene.

The particular example I have chosen for table 22.3, $EV_{100} = +15$, is an important one; it represents the normal exposure for taking pictures outdoors on a clear, bright sunny day, assuming a 'normal' landscape. It is often referred to as 'sunny bright,' and one can see, in the fourth column of table 22.2, names for related outdoor scenes that require either more or less exposure.

Notice that one of the combinations in table 22.3 is 16 for the aperture and 125 for the shutter. In this case the shutter number (125) and the ISO (100) are almost the same. If we made, instead, a similar EV chart for ISO 400 (two steps greater density for the same exposure), all of the shutter-speed numbers would have to be adjusted to two steps lesser exposure (larger shutter numbers) for any given *f*, in order to give the same density, and thus would still be appropriate for the same lighting condition. And thus for that EV_{400} chart the same density would result at EV = +17 instead of EV = +15. This would correspond to 16 for the aperture and 500 for the shutter, and we would again find that our ISO number (400) is about the same as our shutter-speed number (500) for an aperture choice of $f/16$.

This demonstrates the 'sunny 16 rule' described in section 22.5. When taking pictures outdoors on a bright sunny day, set the aperture for $f/16$ and then set the shutter number as close as possible to the ISO of the film, and the exposure will be approximately correct. To use an aperture other than $f/16$, simply use the 'one up, the other down' rule to pick another combination of aperture and shutter speed that has the same EV.

References

Cooke T and Kinneberg C (ed) 2014 *The Photography Book* 2nd edn (London: Phaidon Press Limited)

Exposure value 2018 https://en.wikipedia.org/wiki/Exposure_value (collected September 20, 2018)

Chapter 23

Metering

Metering is the process of measuring how much light from the subject will arrive at the camera. To meter properly we must make the distinction between forms in the picture that are emitting their own light (the sky, for example) and forms that we see because of light reflected from some other source. There is also another special case, and that involves the use of flash. Furthermore, some objects reflect more light than others. These considerations lead to different approaches to metering, and we will consider each in turn.

1. **Ambient-light metering:** The scene is lit by ordinary light sources that deliver luminous energy gradually over time. And thus the *intensity*—proportional to the luminous energy per second, or power—is measured.
2. **Flash metering:** The scene is lit by a *flash*, which delivers all of its luminous energy at once in a short burst. And thus the total luminous energy, not the power, is measured.
3. **Incident-light metering:** The light falling on the subject is measured.
4. **Reflected-light metering:** The light arriving at the camera is measured.

And so a light meter may measure either ambient light or flash. Whether it does one or the other (or both), it may be either a reflected-light or incident-light meter.

In all of these cases the starting point is to render on the picture dark forms dark and bright forms bright. That may not be, in every case, the final goal; one may want to make a picture that is very different from how the forms in the real world appeared to the human eye. But even then, the starting point is usually to determine what would be the proper exposure to render the forms in the picture as they appear in real life. One may then adjust the exposure accordingly, to make either an *overexposure* (to brighten the picture) or an *underexposure* (to darken the picture).

doi:10.1088/978-0-7503-3703-8ch23 © IOP Publishing Ltd 2022

In what follows we first consider meters that measure ambient light, leaving the special case of flash photography for section 23.4. Furthermore, many cameras use a built-in meter to set the exposure automatically, according to some prearranged scheme; we will explore these automatic systems further in section 23.2.2. We first consider meters and cameras that are used in a manual mode.

Photographic light meters do not record the intensity of light in physical units, such as $W\ m^{-2}$. Instead they display, for the amount of light measured, the required combination of camera settings in order to achieve a 'normal' range of densities on the detector. Most often ISO is set first, and then different combinations of aperture and shutter (or exposure value [EV]) can be selected that are appropriate for the measured amount of light.

23.1 Direct-read versus null meters

Whether the light meter is a separate hand-held device or is built into a camera, there are two basic approaches to how the meter communicates with the photographer. In a *direct-read meter*, there is some kind of indicator that directly shows the proper exposure for the measured light. In older meters, this was usually some kind of moving needle. A more modern direct-read meter typically uses a digital display to show the properly-exposed combination of shutter and aperture numbers, usually with some method for displaying alternative combinations with the same EV or with a different ISO. As an alternative, some direct-read meters indicate only the EV number (for a particular ISO setting) for proper exposure, leaving it up to the photographer to pick a combination of aperture and shutter with that EV. A direct-read meter built into a camera may point to shutter numbers for a particular aperture or to a focal ratio for a particular shutter speed, depending on the design of the camera.

A *null meter*, on the other hand, does not directly communicate the meter reading. Instead one adjusts aperture, shutter and ISO (or EV and ISO), and the meter tells the photographer whether that combination would be under or overexposed. Typically, there is some system of arrows or other marks to either side of a 'perfect exposure' marking. And so one adjusts the exposure by playing the higher-lower game.

Null meters are very popular for manual exposure modes on in-camera meters because they are easy to understand. They have the disadvantage, however, that the photographer only knows the camera setting once the meter is 'nulled' to a perfect exposure (by adjusting the shutter speed, aperture or both). With a direct-read meter, on the other hand, it is possible for the photographer to point the meter at different parts of the scene and quickly determine a range of readings.

23.2 Reflected-light metering

Reflected-light metering occurs at the camera. It measures the amount of light from the subject that actually makes it to the camera. As such, reflected-light meters are often located inside the camera itself. If a reflected-light meter measures the amount

Figure 23.1. TTL metering systems in two vintage 35 mm cameras. At **top left** is the view through a Nikon F2, which uses a null meter. The centered needle indicates correct exposure, in this case it is set at aperture *f*/8 and shutter 1/125 s. At **top right** is the view through a Nikon FM3a, which uses a direct-read meter. The blue indicator shows the shutter is set for 1/125 s while the aperture of *f*/8 is shown by the number at the top of the frame. A black needle points to the shutter speed needed for correct exposure. One then matches the blue indicator and the black needle (or leaves them mismatched if under or overexposure is desired). The **bottom** image shows a Nikon FM3a set for *f*/8, 1/125 s at ISO 200.

of light that actually makes it through the camera lens, it is called a through-the-lens, or *TTL*, meter.

Reflected-light metering, especially of the TTL variety, is very convenient since it measures just what the camera sees. And so it can easily be used to guarantee that the detector will be exposed by an amount that puts the recorded light and dark values in the middle range of what the detector is able to record (see figure 23.1).

There is, however, a serious disadvantage to reflected-light metering: the light meter does not know what the picture is supposed to look like. Maybe there is a lot of light entering the camera because one is taking a picture of a snowy field on a sunny day. In that case, the picture is supposed to be pretty much all white, but the light meter does not know this. Left to its own devices, the light meter will decrease the exposure to make that bright, snowy field look a mid-tone gray instead. Similarly, if one takes a picture of a bunch of dark rocks, that are supposed to appear dark on the picture, the light meter in the camera will increase the exposure so as to make the black rocks appear a mid-tone gray.

The point is, things in the world appear dark or light partly due to the light that is falling on them but also due to how they reflect light. The black rock appears dark to us because it reflects only a small percentage of the light that falls on it, while the snow appears bright white because it reflects a very high percentage. Shine a lot of light on an object of low reflectivity, and then shine only a little light on something of high reflectivity: both situations may result in exactly the same amount of light arriving at the camera, and the meter cannot tell the difference between the two. The meter, in other words, must make some assumptions.

By convention, the photographic metering system is designed with a reflectivity of 18% as the standard[1]. Thus any object that reflects 18% of the light that falls on it will be exposed correctly when metered with a reflected-light meter. If one uses a reflected-light meter to measure the light reflected off—for example—white snow, which reflects close to 80% of the light falling on it, the meter will nonetheless assume the snow is reflecting only 18% of the light. And so it will suggest an exposure that would make the snow look grey and not bright white. An increase in exposure over the meter's suggestion would be required, in this case, to make the snow look bright white.

One solution is to read the meter only while it is pointed at something that is illuminated by the same light as the subject, but that reflects about 18%. Then one sets the exposure on the camera manually, based on that 18% reflectivity meter reading. Many photographers carry around a *gray card*: a piece of cardboard coated with a dull surface that reflects 18%. They then put the gray card in the same position as the subject (or with the same light falling on it as the subject), read the meter while it is pointed at the gray card, and then set the camera exposure manually before taking the picture. It is even possible to buy a t-shirt that is 18% gray, so one can always have at least a 'gray shirt' handy.

A variation on this technique is to simply memorize what 18% gray looks like, and find something in the scene that looks like that and has the same light falling upon it, and so use that in place of a gray card. It is interesting that when we humans look at things in the world, we see them not in terms of the light they reflect, even though that is what literally enters our eyes. Instead we see things in the world according to our unconscious assumptions about the light that is falling upon them. We know, instinctively, that a white object in very dim light reflects very little light. And so, in context, we still see it as white. The camera, on the other hand, does not know this; therefore, we must tell it.

23.2.1 Spot, center-weighted and matrix metering

When one points a light meter at a subject, just what light does it record? Does it measure only the light in one precise direction? Or does it record an average of all of the light in the general direction it is pointed?

[1] This is approximately true, but it is more complicated than this when other factors, such as the various wavelength responses of different detectors, are taken into account.

The answer is that it depends. First off, recall that if we are talking about a TTL meter built into a camera, then clearly it is only measuring the light coming through the lens. Even in this case, however, we still must ask a question: it is measuring the brightness of only one small spot on the picture, or is it measuring all of the light the picture will record? So let us make a distinction between the three most-common types of reflected-light meters.

1. **Spot:** The meter responds only to the light coming from a small region in a very precise direction.
2. **Center-weighted:** The meter responds to a large angle of view, but the light coming from a small region in the center counts for the majority of the meter's reading.
3. **Matrix:** The meter makes many small individual measurements of different small parts of the scene. It then combines these into one reading according to some program.

Many modern cameras have a built-in meter that can be switched between spot, center-weighted and matrix modes.

A *spot meter* measures only a small spot on the picture, usually at the direct center of the field of view. Thus if one points the meter at someone wearing a white shirt and black pants, it will read a low value when pointed at the person's pants and a high value when pointed at their shirt. To use this kind of meter, one takes many readings on different parts of the image in order to determine the range of EVs. Then the exposure decision is made based on how one wants the picture to appear. Needless to say, this is a slow and meticulous way to determine the exposure, but it is also the most versatile.

Many photographers carry a separate dedicated spot meter, such as the one shown in figure 23.2. It shows a magnified view, somewhat like a small telescope, and the meter needle (visible from the inside) responds only to the light in a small circle (5° across, for this particular meter) at the center of the field of view. The needle on the inside reads EV_{100} values, and the photographer reads off aperture,

Figure 23.2. A typical hand-held spot meter. This model shows a magnified view of the subject, like a small telescope. The meter responds only to the small 5° circle at the center of the field of view. The EV value is read from inside the meter, and the aperture and shutter speed can then be calculated by the set of wheels on the side of the meter.

shutter speed and ISO by dialing in the meter reading to the rotary dial on the outside of the meter.

One of the most-common types of built-in TTL light meters is the *center-weighted meter*. A small percentage of the meter reading corresponds to the entire image (in a TTL meter) or field of view (in a hand-held meter). Most of the reading, however comes from a usually medium-sized circle in the center. For most center-weighted meters, it is about 60%–70% for the center and 30%–40% for the rest. This works pretty well on its own for many common situations, but it still allows the photographer to move the camera around and point at different objects to crudely sample the possible range of meter readings.

The most sophisticated type of meter is the *matrix meter*. It records light readings at many small places scattered (usually in some kind of grid pattern) over the image. It then makes a decision for the best exposure based upon some kind of program. Cameras with matrix meters often have many different programs from which the photographer can choose. Each program makes different assumptions about just what is a 'typical' scene, and thus what the light and dark values may represent. The idea is that, once one chooses the appropriate program, the photographer can just trust the meter to do the right thing. This actually works pretty well much of the time, and it is usually the best type of metering for taking pictures in a fully-automatic mode.

Matrix metering gives the photographer far less control over individual pictures since, due to its complex program, it is often difficult to know just what the meter is doing and why. Spot metering gives by far the most control, but it is also the most difficult and time consuming to use, and it is pretty much hopeless when combined with automatic settings. In many situations, center-weighted metering provides a nice balance for the photographer between speed and control.

23.2.2 Manual, automatic, semiautomatic and program exposure modes

Many cameras have a prominent dial with, among other symbols, four enigmatic letters: 'M,' 'A,' 'S' and 'P.' They represent, respectively, manual, aperture-preferred, shutter-preferred and program.

In most cases, the ISO is set separately from the information provided by the meter, and we can think of that as the starting point for metering. It is the first thing one chooses, based on whether there is, in general, a lot of light (use a low ISO number) or very little light (use a high ISO number). Many digital point-and-shoot cameras, as well as most cell-phone cameras, do this automatically and may not even allow the photographer to adjust the ISO at all. Here we consider cameras that at least allow for the possibility of manual metering.

The *manual exposure mode*, usually symbolized on a camera by 'M,' means just what it says. The photographer adjusts both aperture and shutter speed according to the reading of the meter. This means, of course, that one does not have to do what the meter recommends. For example, if the meter says the camera should be set for a shutter speed of 60 and an aperture of $f/$ 5.6, one is perfectly free to use an aperture of $f/4$ instead.

Reflected sunlight flickering off a flowing stream is a good example of a situation that can be difficult to meter; in this case the light (and thus the meter reading) changes rapidly with time. If one were to use any form of automatic metering, the result would be essentially random. But with a manual exposure mode, one can watch the meter for a few seconds, get a feel for the range of values and then set it accordingly. Or one could point the camera at something else entirely to read the meter because the processes of reading the meter and setting the camera are separate from each other.

Aperture-preferred metering ('A') allows the photographer to set the aperture manually, and the meter then automatically adjusts the shutter speed for the chosen aperture, according to the meter reading. *Shutter-preferred* metering ('S') is also an automatic exposure mode, but it works in the opposite sense; the photographer adjusts the shutter speed, and the camera adjusts the aperture according to the meter reading. These are called *semiautomatic metering* modes, and one or the other is more appropriate depending on whether the photographer wants more control over aperture or shutter speed. For example, if one is using a tripod to photograph a nonmoving subject, then a particular choice of shutter speed is unimportant, and an aperture-preferred mode is more appropriate to allow for control of depth of focus. If, on the other hand, one wants a particular amount of motion blur in a moving subject, then a shutter-preferred mode is more useful.

For either semiautomatic mode, once one picks the aperture or shutter speed, the possible range of the other settings limits the EVs that can be photographed. For example, if a camera has shutter numbers of only 1, 2, 4, 8, 15, 30, 60, 125, 250, 500 and 1000, then once one picks the aperture, the aperture-preferred mode will only have a range of 11 EVs to choose from. It may turn out, for example, that there is too much light for the chosen aperture even for the highest available shutter speed. Most cameras would give some kind of warning in that case. The moral is that one must still pay attention even when using an automatic mode.

A *program exposure mode* ('P') adjusts both aperture and shutter speed automatically according to a prearranged scheme (a program). Most modern cameras actually have many different program modes. The 'P' setting is usually reserved for the simplest and least-sophisticated program that makes the fewest assumptions about the scene. It is thus, of all the program modes, the easiest for the photographer to understand.

Understanding the camera's autoexposure modes is important, as many cameras still allow the photographer some choice even with fully-automatic metering. This is usually done in one (or both) of two ways, the most common of which is called *exposure compensation*—the exposure is altered by a selected number of steps above or below that which the meter indicates. Less common, but even more useful in my opinion, is some sort of easily-accessible button that serves as an *autoexposure lock*. One presses and holds the lock button, and the camera remains at the same exposure setting even as the camera is repositioned and fired.

With a little practice, a combination of semiautomatic exposure, spot or center-weighted metering with a direct-read meter, and use of an autoexposure lock button can give the photographer nearly all of the flexibility of a fully-manual setting but at much greater speed. Even so, manual metering is still necessary for conditions of randomly-changing lighting conditions.

Finally, I point out that in certain cases manual meeting is, surprisingly, the fastest method for determining the exposure. Sometimes the situation is such that the lighting conditions are consistent from one picture to the next over a significant period of time. In that case one need only use the light meter once to determine the proper camera settings. For subsequent pictures, the camera can be preset with no metering at all.

23.3 Incident-light metering

A completely different approach to metering is to directly measure the light falling upon the subject rather than the light reflecting off the subject and arriving at the camera. An *incident-light meter* is almost always hand held, separate from the camera, and it is placed at the subject so that the light falling on the subject falls on the meter. This avoids altogether the issue of the light meter not knowing what the subject is supposed to look like, and so no gray card is needed. Snow will look like snow; asphalt will look like asphalt.

An incident-light meter is, of course, helpless in the face of a subject that emits its own light. It is useless, for example, for metering a sunset. And if the subject is distant, it may be difficult to get the meter physically in the same position as the subject. It does work, however, to place the meter in a different (and more convenient) location so long as the same light falling on the subject also falls upon the meter.

A hand-held meter, whether it is reflected-light or incident-light (some allow for both), has all of the adjustments present on a camera—ISO, shutter speed and aperture. Some are null meters, some have a needle that reads EVs, and some read shutter speed, aperture and ISO directly with a digital display. For all of these methods of communicating with the photographer, it is possible to easily choose any of the combinations of aperture and shutter speed that give the same EV.

23.4 Flash

The use of flash dates back to the earliest days of photography. It can be quite convenient to carry around one's own source of light, especially when it is mounted directly on the camera itself. See, for example, *The Endurance by Night* by Frank Hurley (Cooke and Kinneberg 2014, p 240). The use of flash, however, requires a completely different approach to metering.

For most modern flash units, the energy is delivered in a burst of light that lasts for only about 1/1000 s. For this reason it is ineffective to use the shutter speed to limit the exposure when the subject is lit by a flash. Obviously if the source of light (the flash) is only giving off light for 1/1000 s, it hardly matters whether the shutter speed is 1/30 s or 1/250 s. Both shutter speeds would allow all of the energy of the flash pulse to enter the camera. And so when using a flash alone to illuminate the subject, the exposure is controlled entirely by the flash, ISO setting and aperture; shutter speed does not enter into the calculation.

This is not to say that shutter speed is completely unimportant for all flash photography. For example, there may be ambient light falling on the subject from sources other than the flash. Also, most cameras have a maximum shutter speed for which flash is usable. This is called the *sync speed* of the camera. If one attempts to use flash at a higher shutter speed, the camera will not be able to synchronize the shutter with the flash, or the shutter may not expose all of the image frame during the flash exposure. For a digital or 35 mm single-lens reflex (SLR) type camera, the maximum sync speed is typically between 1/90–1/250 s, depending on the camera.

Recall that the exposure is determined by the total energy per surface area delivered to the light detector. For nonflash sources of light, we achieve this by using a shutter to limit the time of the exposure by a light source of a certain power. And so energy equals power multiplied by time. For flash, however, we cannot do this; instead it is a pulse of light with a particular *total energy*. In effect, the short duration of the flash itself becomes our camera shutter. Thus for a given brightness of light from the flash on the subject, the exposure is set at the camera solely by the ISO and aperture. Most flash units have the ability to vary the power and duration (and thus the total energy) of their flash pulse. There may be a direct setting for this on the flash unit, but it may also be done automatically as part of the flash-metering process.

23.4.1 Distance and flash

There is another complication for metering that arises whenever the flash is mounted on the camera. Recall that the specific intensity of light is independent of distance. And so for a particular amount of light incident on a particular subject, one sets the camera exposure the same whether the camera is close to the subject or far away. Not so for a camera-mounted flash!

Remember the inverse square law? The light coming from a small source of light decreases with the inverse square of the distance from the source. It is true that once the amount of light falling on the subject is fixed, the exposure setting of the camera does not depend on distance to the subject. But if the flash is mounted on the camera, there is less light from the flash on the subject if the subject is farther away from the camera because the subject is also farther away from the flash. Thus a given setting of flash brightness, ISO and aperture will expose subjects at different distances by different amounts.

Since the light from the flash varies as the inverse of the distance squared, this is a huge effect. Twice the distance means only 1/4 the light, corresponding to two whole exposure steps. A common flash fail arises when someone uses flash to take a picture of a football game. The flash exposes the backs of the nearby people in the next row a lot and the distant game field not at all. The result is a couple of bright, overexposed backs framing a completely dark football field.

23.4.2 Flash metering

As is the case for metering in general, one can take either an incident-light or a reflected-light approach to metering flash. But an ordinary light meter cannot be used for flash; a special *flash meter* is needed. The reason is simple; a flash meter measures energy, while an ordinary light meter measures power, which is energy per time.

An incident-light flash meter is a hand-held meter placed at the subject. The flash is triggered in the same way as it would be for the picture, and the meter tells the photographer the appropriate camera settings for ISO and aperture (remember that shutter speed does not affect the flash exposure). Some flash meters can be set in a 'ready mode,' whereby the flash itself triggers the meter. This is convenient, but usually slightly less accurate, as some of the energy of the flash is needed to trigger the metering process and so goes unmeasured (the flash unit corrects for this bias as best it can). Sometimes a wire is connected between the flash unit itself and the meter, and the meter itself triggers the flash. The most sophisticated (and expensive) process uses a small radio transmitter mounted in the meter and a radio receiver mounted in the flash unit. This saves one from tripping over wires.

In a portrait studio there may be many flash units used at once, lighting the subject from different angles, and there may be no flash mounted directly on the camera at all. In this case use of a separate, incident-light flash meter is essential. The separate flash units can be easily synchronized with each other with an inexpensive device known, most unfortunately, as a *slave unit*. A slave unit, mounted on a flash, triggers the flash whenever it 'sees' the light from another flash. Thus only one of the flash units need be triggered by the camera or flash meter, and all others will trigger themselves when the 'master unit'[2] flashes.

Since flash exposure depends so critically on the distance between the flash and the subject, built-in, reflected-light meters are commonly used when a flash is mounted directly on a camera, and this allows for fully-automatic flash exposure. But this arrangement can also be accomplished in a fully manual mode. If the energy output of the flash is known, it is possible to calculate the proper camera ISO and aperture settings given the distance to the subject. Once the ISO and flash power are set, there is then a particular aperture setting for a given subject distance[3].

The brightness of the flash can be represented in physical units related to joules, but photographers use a more convenient measure known as a flash guide number (GN). The GN of a flash is defined by equation 23.1.

$$\mathrm{GN} = \text{distance} \times f \tag{23.1}$$

Since f is simply a unitless ratio of two lengths, the GN of a flash has units of distance. One important complication is that GN depends on ISO. That is, the GN for a flash is always expressed assuming a particular camera ISO setting (usually ISO 100). If one is using a different ISO setting, then the exposure must be adjusted accordingly. But given the ISO number, it is easy to use the flash GN to calculate the exposure.

---

[2] Another very unfortunate term.

[3] There is another important factor that we will consider later, and that is how well-focused the light coming from the flash is. Does the flash emit all of its energy in a narrow beam, or does the flash instead allow energy to spread out over a wide area?

Figure 23.3. A modern digital incident-light meter. This model works with either ambient light or flash.

For example, if our flash has a GN of 40 feet for ISO 100, then a subject at a distance of 20 feet will require an aperture of $f/\,2$. This is because equation 23.1 can be rearranged as follows, and 40/20 = 2.

$$f/\ \text{number} = \frac{\text{GN}}{\text{distance}} \tag{23.2}$$

Most on-camera flash units, however, employ some kind of automatic exposure using a built-in reflected-light flash meter called a *thyristor*. The thyristor, usually located on the flash unit itself, measures the accumulated energy flux of the light returning to the camera as time passes. Thus the flash can begin the flash pulse, and the thyristor can automatically turn off the flash when the correct amount of light has returned to the camera from the subject. This all happens in only a tiny fraction of a second.

Of course, the correct amount of light returning to the camera depends on the camera's ISO and aperture settings. If the flash is a separate unit, attached to the camera via a cable or *hot shoe* (see figure 23.4), then the camera and flash must communicate with each other somehow. First, the camera shutter must tell the flash to fire only when the shutter has opened. And the flash must somehow know the ISO and aperture settings of the camera in order for the thyristor to know how much light must return to the camera. In some cases the camera and flash communicate with each other electronically, but it may involve manually turning a dial on the flash unit to 'dial in' the camera aperture and ISO settings.

Some cameras allow for TTL flash; the thyristor is mounted inside the camera itself, and it measures only the light returning from the subject directly through the lens. This may seem impossible; remember that the thyristor measures the light entering the lens as the picture is being taken. It then turns off the flash (in a tiny fraction of a second) once enough light has returned for a proper exposure. But if one is taking a picture, the light coming through the lens is going directly to the

Figure 23.4. The small metal bracket on top of this SLR camera is a hot shoe. The flash unit shown to the right can be mounted on the camera, and they communicate with each other by means of the electrical contacts in the hot shoe that connect to those on the bottom of the flash unit (shown upside down). The large center contact synchronizes the flash with the camera shutter while the other contacts convey information specific to the manufacturer of the camera and flash. For this reason a flash from one manufacturer may not work on the camera of a different manufacturer.

detector, whether it be film or a light-sensitive digital chip. So how then can that light also go to the thyristor? Would the thyristor just block the light and not allow it to strike the detector?

If the thyristor were placed directly in the light path to the detector, it would indeed interfere with the picture. Instead the thyristor for a TTL flash meter is located inside the camera off to the side of the detector, just outside the light path from the lens. A real-light detector only detects a fraction of the light that falls upon it; some light is inevitably reflected off the detector surface. It is this little bit of reflected light that the TTL flash thyristor measures. And so a TTL flash meter in a film camera sees the little bit of light reflected off the shiny film surface as the exposure is made.

TTL flash metering is more complicated in that the thyristor inside the camera must communicate with the flash. This is not much of a problem if the flash is built into the camera itself—as with most point-and-shoot cameras. But if it is a separate flash unit mounted on the camera via a hot shoe, the flash and camera must be compatible with each other. TTL flash has a big advantage for cameras with interchangeable lenses because one can switch between wide-angle and telephoto modes, and only the part of the returning light that corresponds to the actual picture is measured.

23.4.3 Fill flash

The most complex use of flash is the mixing of light from flash and nonflash (ambient) sources. A common example is what is known as *fill flash*. Consider, for example, the case of taking a close-up picture of a person's face with a beautiful sunset in the background. If the camera is facing the sunset, then inevitably there is very little daylight shining on the face. And so if one makes the correct exposure for the face, the sunset will be overexposed. If one sets the exposure for the sunset, then the face will be nothing but a dark silhouette.

Fill flash to the rescue! The key point is that the flash exposure is set (for a given ISO) only by the aperture, while the exposure for the nonflash light is set by a combination of both aperture and shutter speed. And so one can choose an aperture setting that will give the proper flash exposure of the nearby face and then select a shutter speed that, in combination with the aperture already chosen, gives the proper exposure for the sunset.

Reference

Cooke T and Kinneberg C (ed) 2014 *The Photography Book* 2nd edn (London: Phaidon Press Limited)

IOP Publishing

Photography (Second Edition)

Physics and art in focus

John Beaver

Chapter 24

Low-sensitivity detectors in photography

24.1 Regimes of photographic exposure

When thinking about the different ways in which exposure is determined in photography, it is interesting to consider eight basic categories of in-camera photography.

1. A shutter speed of a fraction of a second results from a combination of a sensitive detector (ISO 6–1600) and a relatively small ($<f/64$) focal ratio, combined with ordinary lighting. 'Sunny 16' at ISO 100 is a good example; this is the exposure regime for most photography.
2. A medium-length exposure, typically of several seconds, results from a small focal ratio combined with significant light and a lower-sensitivity detector. This is the lens-cap-as-shutter photography of processes preceding modern silver-gelatin emulsions. Examples are tintype, ambrotype, daguerreotype and wet-collodion photography.
3. A long time exposure of many minutes results from a combination of very-dim light with a sensitive detector and small focal ratio. A night-time exposure with an ordinary digital or film camera is a good example.
4. A long time exposure of many minutes results from a combination of a very large focal ratio ($>f/100$) with a moderately sensitive detector (ISO > 1) and bright daytime lighting. This is pinhole photography.
5. A long time exposure (many minutes to possibly hours) results even with the combination of significant lighting and a small focal ratio because a detector of *very-low sensitivity* (VLS) is used.
6. A VLS detector is used in conjunction with a very large focal ratio (such as a pinhole) under bright light conditions. This may result in a months-long exposure time. The practice of *solargraphy* is an example (Beaver 2023).

 © IOP Publishing Ltd 2022

7. An *extremely-low sensitivity* (ELS) detector is used in conjunction with small focal ratios and bright light in an exposure that takes many hours or even several days or weeks. The very first in-camera photograph is an example.
8. A VLS or ELS detector is used in conjunction with a very large focal ratio (such as a pinhole) under very-dim light (at night for example). To my knowledge, this has never been attempted.

The first and third categories involve traditional cameras, lenses and detectors (ordinary film or digital detectors). The second category is of historical interest, but as we shall see in part VII of *Photography: Physics and Art in Focus*, it is of contemporary interest as well for art photographers. We have already discussed the fourth category, pinhole photography (chapter 9); it has enduring popularity because the camera construction is so simple. The eighth category seems to be a nonstarter, but it is the rather unfamiliar fifth category we consider here—what I call VLS photography.

We intentionally choose a detector that is of such low sensitivity that time exposures of at least several minutes are required for an exposure even with ordinary outdoor lighting conditions and relatively small focal ratios. As is the case for pinhole photography, no shutter is needed because of the long time exposures. But there is the added feature that the low-sensitivity light detector is more easily handled, and the camera construction is, in some ways, even simpler than for pinhole photography.

What we want is a detector that is just barely sensitive enough that it can be used to take outdoor photographs with a moderately-fast lens but has a sensitivity low enough that it can be handled in dim ambient light for brief periods of time without risk of significant exposure during handling. This allows for a greatly simplified camera design. In particular, no mechanism is needed for keeping the detector in total darkness before and after the exposure. And for brief handling of the VLS detector outside of the camera, no photographic darkroom is needed—only a 'dimroom.' This discussion would be simply academic, except that there really are detectors with these properties, and we will see how to use them in chapters 35, 36 and 48.

24.2 A benchmark for VLS photography

To set up a practical benchmark for exposure with a VLS detector, it is convenient to state the following assumptions.

1. We must handle the detector in subdued (but not totally dark) lighting for 30–60 s. This should provide plenty of time to prepare the detector and load it into the camera before the exposure and to transfer it from the camera to a dark bag after the exposure.
2. While handling the detector, we can shield it from light well enough by working in a dimroom, under a black sheet, or in a 'dim tent' to reduce the lighting by ten exposure steps dimmer than the lighting used to take the picture. A difference of ten steps in lighting is equivalent to the difference

between bright sunlight (exposure value $[EV]_{100} = +15$) and very-dim home interior lighting ($EV_{100} = +5$). One can accomplish this difference in the field on a sunny day simply by going into the shade and working under a dark sheet.

3. We want the unwanted exposure of the detector during handling to be at least five exposure steps less than the correct in-camera, mid-tone exposure of the subject. Whatever exposure it takes to bring the subject up to a mid-tone density in the camera, we want to expose the detector during handling by no more than five exposure steps less than this. This is to ensure that the dim-light handling has little detectable effect on the final picture.

Let us consider the time (t_H) to expose the detector, while handling it in dim light, to some particular barely-acceptable background density. And let us compare this to the time (t_C) to expose the detector in the camera from reflected light off an 18% reflectivity subject and achieve a mid-tone density on the detector for our exposure.

Let us also assume we are using a camera with a focal ratio (f), and let us define a couple of new variables. Let Z be the number of exposure steps less than the subject exposure that we can accept as a barely-acceptable level of exposure during handling. And so, for example, if $Z = 8$ it would mean that whatever exposure achieves the desired mid-tone density of the subject, we would need to keep the unwanted exposure resulting from handling the detector to eight exposure steps less. This translates to $2 \times 2 \times 2 \times 2 \times 2 \times 2 \times 2 \times 2 = 2^8 = 256$ times less exposure. For our benchmark we will choose $Z = 5$.

Let us also define N to be the number of steps by which we can dim the light while handling the detector by—for example—putting a black sheet over our head. I have already indicated that $N = 10$ is a reasonable assumption, but let us keep this as a variable so we could plug in other numbers at the end if we so choose. Since N refers to the number of steps in illuminance, then $N = 10$ means that the illuminance on the detector while handling is 2^{10} times less than the illuminance on the subject during the in-camera exposure.

Note the distinction between Z and N. They both refer to the difference between the case of handling the detector in dim light and exposing the detector in the camera to an image of the subject. But Z is in regards to the difference in exposure that results from those two cases, while N refers to the difference in illuminance on the detector while handling and on the subject while its picture is taken.

We now have everything we need to derive an equation that relates t_C to t_H. What this equation tells us in practice is the following: if I can successfully limit the dim-light handling of the detector to t_H seconds, then how long would the exposure time (t_C) be in order to correctly expose our subject?

One other crucial assumption we must make is regarding the subject. Does it reflect a lot of light or only a little? Since I have specified that our exposure of the subject should be whatever is needed to yield a mid-tone density in its image on the detector, then we should assume a subject that reflects light as some mid-tone level of gray. The photographic standard for this turns out to be a subject reflectivity of $R = 0.18$ or 18% reflectivity.

We have already seen (equation 20.1) that the illuminance (I_D) on the detector inside the camera is related to the illuminance (I_{SUB}) on the subject.

$$I_{SUB} = \frac{4f^2}{R} I_D \tag{24.1}$$

Here R is the reflectivity of the subject (18% in our example). And we also have the most-basic of facts about exposure.

$$\text{exposure} = \text{illuminance} \times \text{time} \tag{24.2}$$

We can write this equation for three cases: inside the camera, handling the detector and exposure by the illumination directly shining on the subject. I will represent these with subscripts 'D', 'H' and 'SUB,' respectively. And so we have the following equations.

$$E_{SUB} = I_{SUB} \times t_{SUB} \tag{24.3}$$

$$E_H = I_H \times t_H \tag{24.4}$$

$$E_D = I_D \times t_D \tag{24.5}$$

Here E here represents exposure (not energy!), I represents illuminance and t represents exposure time.

We also have our assumptions that the illuminance on the detector while it is being handled is less than the illuminance on the subject by a factor of 2^N and that the exposure due to handling the detector in our subdued light is less than the exposure in the camera by a factor of 2^Z. And so we have the following equations.

$$I_H = \frac{I_{SUB}}{2^N} \tag{24.6}$$

$$E_H = \frac{E_D}{2^Z} \tag{24.7}$$

We can combine these relationships as follows. Equations (24.6) and (24.1) give us equation (24.8).

$$I_H = \frac{4f^2}{R2^N} I_D \tag{24.8}$$

Combining this with equations (24.4), (24.5) and (24.7) gives us the following equations.

$$\frac{E_H}{t_H} = \frac{4f^2}{R2^N} \frac{E_D}{t_D} \tag{24.9}$$

$$\frac{\cancel{E_H}}{t_H} = \frac{4f^2}{R2^N} \frac{\cancel{E_H} 2^Z}{t_D} \tag{24.10}$$

$$t_D = \frac{4f^2}{R}\frac{2^Z}{2^N}t_H \tag{24.11}$$

or:

$$\boxed{t_D = \frac{4f^2 2^{(Z-N)}}{R}t_H} \tag{24.12}$$

We now have an equation that relates the time that we handle the detector in subdued light compared to the time that the detector is properly exposed in the camera. We assume that we handle the detector in light subdued by a factor 2^N, and that this dim-light handling results in a (hopefully insignificant) exposure that is Z exposure steps less than the proper in-camera exposure of the subject.

For reasons I will explain in the next section, let us choose a focal ratio of $f/5.6$ as our benchmark for VLS photography. Given our benchmark values of $N = 10$, $Z = 5$, $R = 0.18$ and $f = 5.6$ we have the following relations.

$$t_D = \frac{4f^2 2^{(Z-N)}}{R}t_H \tag{24.13}$$

$$t_D = \frac{4 \times 5.6^2 \times 2^{(5-10)}}{0.18}t_H \tag{24.14}$$

$$t_D = 21.7\ t_H \tag{24.15}$$

This means that the in-camera exposure would need to be at least 22 times longer than the exposure while handling the detector in dim light. This equation is true for any detector; we have, so far, made no assumptions about the sensitivity of the detector. But it is only for a VLS detector that it would be of any practical use. For a typical film detector with ISO = 100, for example, the 'sunny-16 rule' indicates that on a sunny day we would have, using an $f/5.6$ lens, an exposure time in the camera of $t_D = 1/1000$ s (−3 steps from 1/125 s, and +3 steps from $f/16$). Under our assumptions equation (24.15) states that we would only be able to handle the detector for 1/22 of that already-tiny amount of time. Clearly that would be impractical.

But for a VLS detector, the exposure time under the same conditions would be much longer. For example, what if the sensitivity of our detector means we need an exposure time on a sunny day at $f/5.6$ of not 1/1000 s but instead much, much longer than that—say 15 minutes (900 s). Then equation (24.15) indicates that we could handle the detector in our subdued light for 900 s/21.7 = 41.5 s without adversely affecting the exposure.

If the detector is even less sensitive than this, if we do a better job of shading the detector while transferring it to the camera, or if there is less than full sunlight, then handling is even less problematic. On the other hand, if the sensitivity is much less than this, then exposure times begin to get uncomfortably long, and the range of conditions that can be photographed in practice diminishes greatly. There are photographic processes that have sensitivities near that of our benchmark, and we will see shortly how to use them to take pictures.

24.3 VLS photography in context

Given our benchmark sensitivity for VLS photography of a sunny day exposure in about 15 minutes at *f* /5.6, the 900 s of our VLS exposure time is 9×10^5 times longer than the corresponding exposure at ISO 100 (1/1000 s). It is easy to verify that this is roughly a difference of 20 exposure steps ($2^{20} = 1.05 \times 10^6$). And so our benchmark VLS detector has a sensitivity of about one million times less than ISO 100, corresponding to an ISO of approximately 0.0001.

To illustrate this in a different way, consider table 24.1. Here I have calculated exposure times in a camera at *f* /5.6 and also incident-light exposure times for detectors of different ISO sensitivities—both for 'sunny-16' ($EV_{100} = 15$) and dim-room interior lighting ($EV_{100} = 5$) conditions. The third column gives the exposure

Table 24.1. Effect of detector ISO on exposure times for both 'sunny-16' ($EV_{100} = 15$) and dimroom interior lighting conditions ($EV_{100} = 5$). The third column shows exposure times assuming an in-camera reflected-light exposure at *f* /5.6 of an 18% reflectivity subject. The fifth column gives the exposure time to achieve a similar density if the detector is exposed directly to the same incident light. Reciprocity failure has been ignored in these calculations. The rows marked 'VLS' have outdoor, in-camera exposure times at *f* /5.6 that are typical of traditional pinhole photography. In this case, however, the detector itself can be safely handled for a brief period in dimroom lighting conditions.

EV_{100}	*f*	*t* (In-Camera)	ISO	*t* (Incident)	
15	5.6	1/8000 s	1000	1.8×10^{-7} s	Film
5	5.6	1/8 s	1000	1/6000 s	Film
15	5.6	1/800 s	100	1.8×10^{-6} s	Film
5	5.6	1.25 s	100	1/600 s	Film
15	5.6	1/80 s	10	1.8×10^{-5} s	Paper
5	5.6	12.5 s	10	1/60 s	Paper
15	5.6	1/8 s	1	1/6000 s	Paper
5	5.6	125 s	1	1/6 s	Paper
15	5.6	1.23 s	0.1	1/600 s	—
5	5.6	21 m	0.1	1.8 s	—
15	5.6	12.3 s	0.01	1/60 s	—
5	5.6	3.5 h	0.01	18 s	—
15	5.6	123 s	0.001	1/6 s	VLS
5	5.6	1.5 d	0.001	3 m	VLS
15	5.6	20.4 m	0.000 1	1.8 s	VLS
5	5.6	14.5 d	0.000 1	30 m	VLS
15	5.6	3.4 h	0.000 01	18 s	VLS
5	5.6	145 d	0.000 01	5 h	VLS
15	5.6	1.42 d	0.000 001	3 m	—
5	5.6	1450 d	0.000 001	2 d	—

time in a camera at $f/5.6$, while the fifth column gives the equivalent exposure time if the light-sensitive material were to be placed directly in the same incident light as the subject.

If we pick a range of sensitivities near our ISO of 1×10^{-4} benchmark, then a VLS detector would need to have an effective ISO in the range between 1×10^{-3} and 1×10^{-5}. At the upper end of this sensitivity, it would take about two minutes to expose the detector in the camera on a sunny day, but it would take about three minutes to give the detector the same exposure outside of the camera in dimroom lighting.

Of course, in handling the light-sensitive material outside of the camera, we want to give it much less exposure than this so as not to ruin our picture; I have already suggested that five exposure steps (32×) less is a safe level of out-of-camera exposure. And so at an ISO of 1×10^{-3} we would have only 180/32 = 5.6 s to handle our detector outside of the camera. This may be possible, but it would likely be difficult.

At ISO 1×10^{-4}, close to our original VLS benchmark, we would have an exposure time of 20.4 minutes in the camera, and it would require 30 minutes outside of the camera to give the detector the same exposure in dim light. Thus we could safely handle the detector for 1800/32 = 56 seconds in our dimroom. At 1/10 this ISO (1×10^{-5}), on other hand, handling the detector in a dimroom would be quite easy, but the exposure time outside the camera would then be measured in hours rather than minutes.

What about the other parts of table 24.1? The top of the table, with ISO between 100 and 1000, is the range for the detectors used in most film and digital cameras, while ISO sensitivities in the single digits are typical of the enlarging papers used in the darkroom to make prints from negatives. Lower sensitivities, in the range of 0.1–0.01, are more typical of some of the nineteenth-century photographic processes used before the advent of faster silver-gelatin emulsions. The seconds-long exposure times were often made simply by uncovering and then covering the lens by hand with a lens cap. The very bottom of the chart, on the other hand, could be called ELS photography. We will consider one example in chapter 36—the so-called anthotype. With rare exceptions, ELS detectors are used only for photograms where they are illuminated directly by the light of the Sun.

In chapter 25, we consider two practical VLS detectors. What I call *ephemeral-process* (EP) photography has an effective ISO right in the middle of the practical range of sensitivities for a VLS detector, very near that of our benchmark. What I call *cyanonegative* photography, on the other hand, is at the low-sensitivity end of that range.

I note that the enlarging paper used in the darkroom, with ISO in the range of one to ten, is exactly the same paper as used for EP photography, with an ISO of 1×10^{-4}. How can the same light-sensitive paper be used for both? We explore this question in detail in chapter 35, but the short answer is that in its normal use in the darkroom, its sensitivity is chemically amplified by a *developer*. For VLS photography we use only its native sensitivity to light, without chemical development.

I have chosen for our benchmark example of VLS photography, a focal ratio of $f/5.6$ on a sunny day—but why $f/5.6$? This may seem odd given the well-known 'sunny-16' rule, which uses $f/16$ as a starting point. But if we remember the purpose of VLS photography—to take pictures with a camera—then it should be clear why I have chosen a smaller focal ratio around which to center the discussion. Even at $f/5.6$, the exposure times are still minutes long, and if a 15 minute exposure is required at $f/5.6$, then it would take two hours to expose the detector at $f/16$. And so the VLS photographer is far more likely to choose a small focal ratio over a large one. With the high sensitivities of ordinary film, on the other hand, a large focal ratio is often needed on a sunny day, or a shutter speed faster than the camera's highest setting might be required.

The ISO standard for describing the sensitivity of a photographic detector is clearly rather awkward for VLS photography with resulting ISOs that are tiny fractions. In appendix E, I instead propose describing such detectors with a *slowness* rather than a speed. The slowness is defined inversely to ISO speed; a greater slowness means a less-sensitive detector. The ISO speed is roughly equal to the reciprocal shutter number for a sunny day exposure at $f/16$ (this is the 'sunny-16' rule). And so an ISO 1000 detector is such that one sets the shutter at 1000 (1/1000 s) to properly expose the picture at $f/16$ on a sunny day. My slowness measure takes the opposite approach for the (in many ways opposite) case of VLS photography.

> The slowness of a VLS detector is equal to the proper exposure time (not its reciprocal) in minutes (not seconds) on a sunny day at $f/5.6$.

And so our benchmark VLS detector, for which a 15 minute exposure would be required at $f/5.6$ on a sunny day, would simply have a slowness of 15. And if I am using a VLS detector with a slowness of 30, then I know it will take me one half hour to take a picture at $f/5.6$ on a sunny day.

Reference

Beaver J 2023 *LoFi Photography: Art from Do-It-Yourself Chemistry and Physics* (Cambridge: Royal Society of Chemistry) (forthcoming book)

IOP Publishing

Photography (Second Edition)
Physics and art in focus
John Beaver

Chapter 25

Ephemeral-process and cyanonegative photography

We here consider two examples of very-low sensitivity (VLS) detectors used to take pictures with a camera. Both of these processes have some of the elegance and simplicity of pinhole photography, but the end result is quite different; moreover, in some ways they are even simpler than pinhole photography. Both are printing-out processes: the image becomes visible during the exposure to light with no chemical development needed. I call the two techniques by names of my own coinage; neither process has been part of traditional photographic practice.

- *Cyanonegative photography* uses the antiquarian process of cyanotype as the detector in a camera. Traditionally cyanotype has been used only for making direct contact prints. A cyanotype makes a blue-tone negative on paper.
- *Ephemeral-process (EP) photography* uses ordinary silver-gelatin enlarging paper for the in-camera detector. This is the paper that is used in a darkroom to make black-and-white prints directly from black-and-white negatives. But for EP photography, the paper is not used as intended. No chemical developer or fixer is used, but a nondeveloping accelerator is applied immediately before exposure, greatly (but temporarily) increasing its speed. Once exposed, it is washed and dried, and the paper returns to its extremely-low sensitivity state.

Both of these processes involve time exposures of at least several minutes, and this gives these processes a kinship with pinhole photography. No shutter is needed, and all motion disappears from the picture. Since a fast lens is used and the detector is in the camera only during the length of the exposure, small light leaks cannot compete with the large lens. Thus the camera need not be perfectly light-tight. The very-low sensitivity means no darkroom is needed—only a ‘dimroom.’ This greatly simplifies storage, loading, unloading and processing.

doi:10.1088/978-0-7503-3703-8ch25 © IOP Publishing Ltd 2022

Both of these processes are inexpensive and require no specialized facilities. But there is a catch—for both of these negative processes, the most practical way to make a positive print from the negative is via the intermediate step of digital image capture. With some limited exceptions, these negative must be scanned and then printed (or not) digitally.

An appealing feature of both processes is that neither uses hazardous wet-chemistry processing. Cyanotype does use a chemical sensitizer that is brushed onto the paper before exposure. This wet sensitizer is toxic but can be purchased premixed, and there is also a usable (but inferior) presensitized cyanotype paper available. The accelerator solution used for EP photography can be made entirely from nontoxic ingredients. In fact, it can be made with edible ingredients: water, vitamin C and xanthan gum (used in gluten-free baking). Both cyanotype and EP negatives are simply washed in plain water after the exposure.

25.1 Cyanonegative and EP wavelength response

Before describing how to take pictures with cyanotype and EP, it is helpful to understand to what range of wavelengths the two processes are sensitive. We can use a simple *spectrometer* to determine the wavelength response of cyanotype and EP paper when exposed to sunlight. We use a *diffraction grating* to disperse the light according to wavelength. A diffraction grating is a series of many evenly-spaced parallel lines, usually on some transparent medium. If the spacing between the lines is only several times that of the wavelength of the light itself, then different wavelengths are deflected by different angles. And so a narrow beam comprised of many wavelengths at once is spread out such that each wavelength can be detected and measured separately. Figure 25.1 shows my (rather-crude) spectrometer. It is simply a light-tight box with a place to put a small piece of the light-sensitive paper. Light entering the box passes through the diffraction grating and is deflected such that different wavelengths are sent off in different directions. When this light arrives at the light-sensitive paper, each wavelength is measured separately.

Figure 25.1. A simple spectrometer for recording the wavelength response of cyanotype and EP paper as exposed by sunlight. **Left:** The sunlight is collimated by two narrow slits at either end of a tube. A small sighting telescope is used (by projection, not direct viewing!) to keep the spectrometer pointed at the Sun. **Right:** A diffraction grating disperses the collimated beam according to wavelength, and the light-sensitive paper is exposed inside the dark box (here shown open).

For this to work, the entering beam must be narrow and *collimated*—all pointing in one direction. This is usually accomplished with lenses. But lenses are made of glass, and glass absorbs the very short-wavelength light we want to detect. And so my collimator is simply two slits, one at each end of a tube; geometry provides the collimation with no absorption in lenses to worry about. This is a crude way to make a spectrometer, but we can still use it to get at least a rough measurement of the overall wavelength response of our VLS light detectors.

When a beam of light of a particular wavelength (λ) passes through a diffraction grating, the light is deflected by an angle (θ) given by the following equation.

$$\sin\theta = m\frac{\lambda}{d} \tag{25.1}$$

Here d is the (very tiny!) distance between the parallel lines on the grating and m is called an *order number*.

The order number (m) can take on any of the values: 0, 1, 2, 3, etc. This means there is not just one spectrum produced—there are many. For $m = 0$, the *zero-order spectrum*, the angle is $\theta = 0$ for all wavelengths. And so the light of the zero-order spectrum simply passes through undeflected. Some of the light instead goes to the *first-order spectrum*, given by $m = 1$ or simply the following equation.

$$\sin\theta = \frac{\lambda}{d} \tag{25.2}$$

Thus every wavelength gets deflected by a different angle. If we know d for our grating, and we measure the angle θ with our spectrometer, we can then determine the wavelength.

$$\lambda = d\sin\theta \tag{25.3}$$

A value of $m = 2$ gives the second-order spectrum ($\sin\theta = 2\lambda/d$), which deflects wavelengths by larger angles than does the first-order spectrum. In most cases, less and less light goes into the spectra of higher orders, and so we will use only the first-order spectrum. Figure 25.2 shows four spectra recorded with the spectrometer of

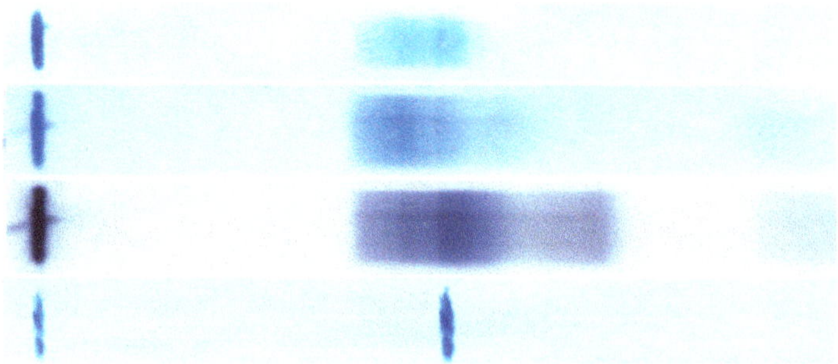

Figure 25.2. Spectra recorded with the spectrometer of figure 25.1. **From top to bottom:** Presensitized cyanotype paper (blueprint paper), hand-sensitized cyanotype paper, EP paper and blueprint paper. The top three were exposed to sunlight, while the bottom was exposed to the single wavelength of a 403 nm laser, thus helping to calibrate the spectrometer.

figure 25.1. The top is presensitized cyanotype paper—blueprint paper essentially. The second was made by hand brushing 'new cyanotype' sensitizer (James 2016, p 212) onto paper, and the third is EP paper. The bottom spectrum is also blueprint paper, but it was exposed not to sunlight but rather to a violet laser pointer that emitted only a single wavelength of about 405 nm right at the short-wavelength edge of the visible spectrum.

The vertical line on the left of each spectrum is the zero-order spectrum; it was exposed by the portion of light that passed through the collimator tube undeflected and directly exposed the paper. The smear to the right in the top three spectra is the first-order spectrum with the shortest wavelengths on the left and the longest wavelengths on the right. Clearly, all three types of paper are sensitive to wavelengths both longer and shorter than our 405 nm laser, and so they are sensitive to ultraviolet (UV) light as well as visible light.

The beginning of the second-order spectrum is just barely visible at the far right edge of the third spectrum. It corresponds to the same wavelengths as the first-order spectrum but repeated at larger angles (so most is off the paper) and much fainter.

25.1.1 Wavelength calibration

The top image in figure 25.3 shows the geometry of the spectrum for the case of a single wavelength from a laser entering the spectrometer. If we know the wavelength of the laser, then equation (25.3) and the measured angle (θ) can be used to find the spacing (d) of the lines in the spectrometer's diffraction grating. I used a nominally 405 nm laser pointer for these tests, but the precise wavelength of this type of laser pointer varies somewhat from one unit to the next.

For the bottom of figure 25.3 the same laser shines through a precision diffraction grating of 600 lines per mm. Thus d, the spacing between the lines, is $1/600\ \text{mm} = 1.667 \times 10^{-6}\ \text{m} = 1667\ \text{nm}$. We can turn equation (25.2) around and use the fact that the geometry in figure 25.3 is that of a right triangle, for which the sine of the angle is the ratio of the side opposite to the hypotenuse.

$$\sin\theta = \frac{\lambda}{d} = \frac{x}{\sqrt{x^2 + y^2}} \tag{25.4}$$

$$\Longrightarrow d = \frac{\lambda\sqrt{x^2 + y^2}}{x} \tag{25.5}$$

$$\Longrightarrow \lambda = \frac{xd}{\sqrt{x^2 + y^2}} \tag{25.6}$$

Here y is the perpendicular distance between the grating and the spectrum, and x is the distance along the first-order spectrum as measured from the straight-through zero-order spectrum.

And so we can use the precision grating with known d to determine the wavelength of the laser and then use that same laser to determine d for the different grating in our

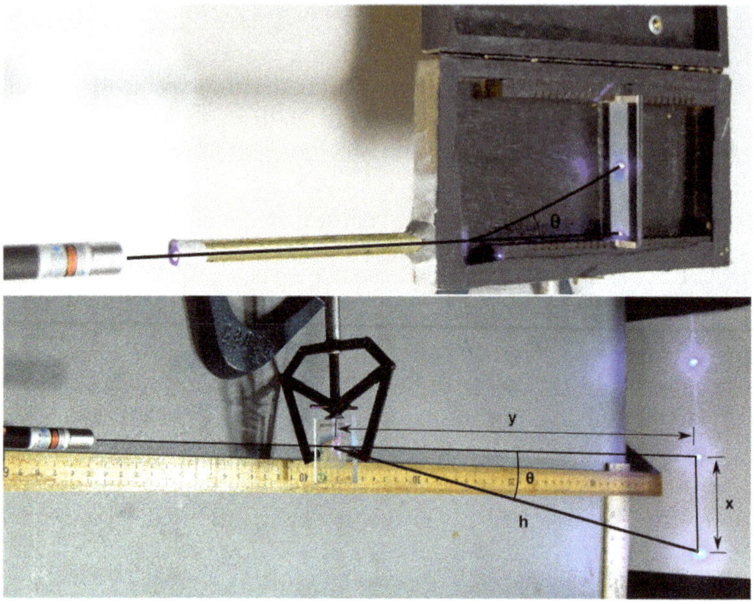

Figure 25.3. **Top:** A violet laser pointer of one single wavelength is projected through the spectrometer, and the light is deflected by an angle (θ). If the wavelength of the laser is known, the measured angle can be used to determine the spacing in the lines of the diffraction grating. The spectrometer can then be used to measure unknown wavelengths. **Bottom:** The same laser shines through a precision diffraction grating of known spacing of 600 lines per mm. By measuring the dimensions (x and y) the angle (θ) can be determined, and the precise wavelength of the laser pointer can then be calculated.

spectrometer. Once d is known we can use equation (25.5) to determine λ for any measured position, x, on our spectrum. For my particular example, the laser turns out to be 403 nm—very close to its marked wavelength of 405 nm. Figure 25.4 shows the result for both the new cyanotype formula and one particular EP paper. I digitized the paper spectra by scanning them and then averaged across many rows to output intensity of the image versus columns in the scan.

The calibration can easily be performed with a spreadsheet. I first noted at what column the peak of the zero-order spectrum occurred, subtracting that value off so that the x coordinate would start at the proper zero. I then applied a correction factor to convert a column number to a physical measurement in units of length, given the resolution at which I scanned the spectra. Finally, I applied a calibration to the data, converting x to wavelength in nanometers with equation (25.5).

The crude wavelength calibration I have described is at the appropriate level of sophistication for the crude spectrometer of figure 25.1. But how are real, noncrude spectra—taken with noncrude spectrometers—calibrated in real life? The answer is at once both simpler and more complex.

It is more complex because there are many other factors than what I have described that affect the calibration in significant ways if one needs high precision. Many of these factors are subtle and specific to the particular construction of the spectrometer in question. But even for our crude version, we can think of some examples.

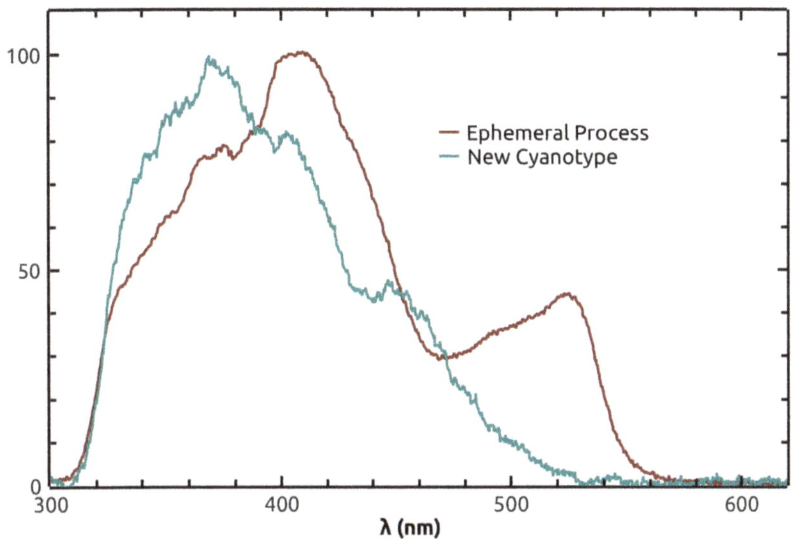

Figure 25.4. The wavelength-calibrated detected spectra for new cyanotype and a particular EP paper, exposed by sunlight. The cyanotype paper responds to a smaller range of wavelength than the EP paper. Blueprint paper (a variation on the cyanotype process) has an even narrower wavelength response. Both detected spectra cut off at about 300 nm because Earth's atmosphere blocks sunlight at shorter wavelengths.

For example, the paper was processed with water and then dried. But does wetting paper and then drying it not alter its size? And so are the dimensions on the processed paper really the same as they were when it was exposed in the spectrometer? And what if the paper wrinkled during exposure? What if there are imprecisions in the construction of my spectrometer so that—for example—the supposedly 90° angles are slightly off? Most important of all, what about other factors I have failed to think of? This is just a small sample of the many reasons that wavelength calibration is complex.

There is also a reason that wavelength calibration is, in practice, simple. A *gas discharge tube* emits light by exciting atoms in a gas with a high voltage. The key is that each gas emits its own set of precisely known wavelengths. And so such a tube can be used to make a *calibration source spectrum*. Iron, for example, emits many well-known wavelengths.

Simply record the spectrum of iron with the uncalibrated spectrometer. Create a table that lists, for each known wavelength of iron, the particular uncalibrated response recorded by the spectrometer. Then find a mathematical function that relates the two and allows one to determine wavelengths in between the iron wavelengths. Thus, the iron spectrum provides a reality check. It allows one to interpolate between those known wavelengths to accurately determine unknown wavelengths.

The crude calibration I described for my spectrometer used, in effect, only one known wavelength—the 403 nm of the laser pointer. I had to assume the relationship was simply proportional and did not change from one part of the spectrum to another. But a scientist performing a precise calibration of a real spectrometer would use many known wavelengths for comparison, thus ensuring that the wavelength calibration is accurate at every part of the spectrum.

25.1.2 EP versus cyanotype

It is clear from figure 25.4 that the EP paper responds to a much broader range of wavelengths than does even the new cyanotype formula (the toy blueprint paper has an even more narrow range (see figure 25.2)). Most of the response of cyanotype is in the near-UV part of the spectrum, while the sensitivity of EP paper extends well into the green part of the visible spectrum.

It appears that both spectra cut off sharply at the same wavelength of roughly 300 nm, which might seem odd since cyanotype and EP are very different physical processes. This is not a coincidence, however, and it is due to the one factor that was shared by both in making these spectra—the source of light. The spectrum of the Sun extends to wavelengths much shorter than 300 nm, but Earth's atmosphere (ozone in particular) absorbs much of that UV light. And so the steep left-hand side of the these spectra is well known to astronomers; they call it *atmospheric cutoff*.

The spectra in figure 25.4 tell us something about the wavelength response of cyanotype and EP paper. They also tell us something about the spectrum of the sunlight that entered the spectrometer. But the curves in figure 25.4 depict neither the response curves of cyanotype and EP paper, nor do they show the solar spectrum. Rather, they are the *detected spectra*—a combination of both as described in chapter 26. But for what follows, that is exactly what we want. For we will use cyanotype and EP paper to take pictures, and it is sunlight that we will use for our source of light. The spectra in figure 25.4 show the overall effect of using those detectors to take pictures with that light.

25.2 Cyanonegative photography

Cyanonegative photography (Beaver 2017, Piper 2021, Ware 2016) uses cyanotype as a light detector. Because of its low sensitivity to (only UV) light, cyanotype has traditionally been used only as a printing process. But it is possible to use it as a detector in a camera for taking a picture with a long time exposure on a bright sunny day. I first began making cyanonegative photographs around 2000 and have since taught the process to students and other photographers; it is easy and inexpensive. There are, however, a few tricks.

1. Use the modern and much-faster 'new cyanotype' process, developed by contemporary photographer and chemist Mike Ware (2016 and James 2016, p 212). Ready-made blueprint paper is also available from toy stores and has a high enough sensitivity, although the tonal range is often poor.
2. Use a simple magnifying glass instead of an ordinary camera lens to focus the image onto the sensitized cyanotype paper. A simple magnifier typically has

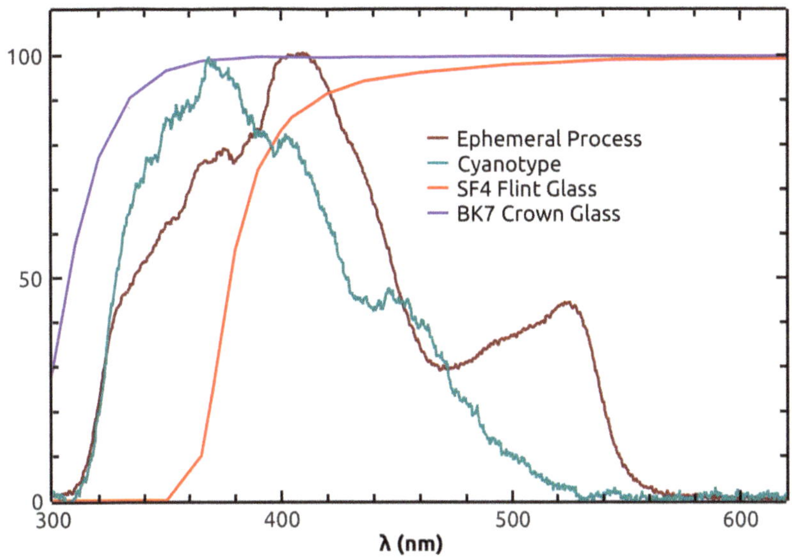

Figure 25.5. The transmission curves (relative transmission vs. wavelength) for two common types of glass are superimposed on the cyanotype and EP response curves. Both types of glass transmit visible light ($\lambda > 400$ nm) well. But the flint glass is opaque over much of the wavelength range of which cyanotype is sensitive (glass data from SCHOTT (2017), pp 13, 93).

a very small focal ration ($\approx f/3$) and absorbs much less UV light than an ordinary camera lens.

3. Offset the focus slightly to correct for the fact that while we focus with visible light, the cyanotype 'sees' at a significantly shorter wavelength.
4. Given that cyanotype only responds to short wavelengths, the daytime sky—by far—is likely to be the brightest part of the picture. And thus the easiest scene to photograph is a silhouette against the sky.

Exposure times in bright sunlight at $f/2.5$, using Mike Ware's cyanotype formula, are about ten minutes to record a silhouette against the sky. Up to several hours may be required to adequately record bright sunlight reflected from a subject.

Cyanotype is only sensitive in the violet and near-UV part of the spectrum, and typical camera lenses block a significant fraction of the light at these wavelengths. Real lenses are affected by *chromatic aberration*; different wavelengths of light focus at different distances because the refractive property of glass is different for different wavelengths. This effect is corrected by using many lenses in combination, all correcting for the chromatic aberrations of each other. But the lens' elements must be made of different types of glass, some of which are denser, high-dispersion glasses (often called *flint glass*). These higher-dispersion flint glasses typically block more UV light than lower-dispersion *crown glass*.

In figure 25.5 (SCHOTT 2017, pp 13, 93) I plot the transmission curves of two typical glasses used for lenses: one a high-dispersion flint glass and the other a

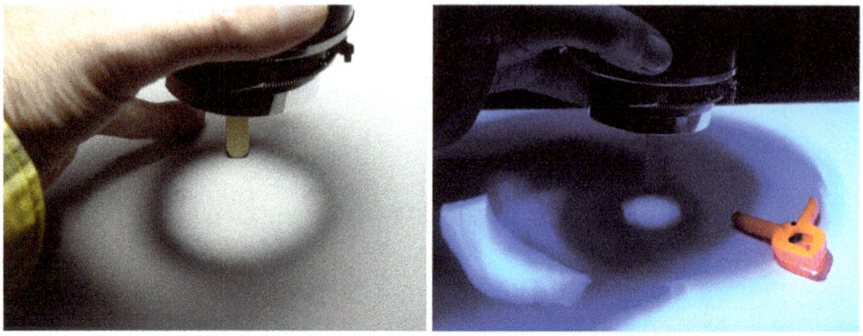

Figure 25.6. Projecting visible light (**left**) and near-UV light (**right**) through a fast medium-format lens. A sheet of fluorescent paper is used to monitor the amount of UV light. Clearly, the lens transmits little light at the UV wavelength.

low-dispersion crown glass. For comparison I superimpose the detected spectra from figure 25.4 of both new cyanotype and EP paper. It is clear that the flint-glass lens would block a significant fraction of the light to which cyanotype is sensitive. Using the ready-made blueprint paper from figure 25.2 would be even worse.

Figure 25.6 shows an example. I project light through a fast lens designed to cover medium format—a Zeiss 100 mm, $f/2.8$ Planar. The left-hand image shows light from an ordinary lightbulb projected through the lens onto white paper. For the right-hand image, the distances are the same, but the light source is a 365 nm light-emitting diode lamp. This is in the near UV, so nothing would be visible to either the eye or my camera. But the white copy paper exhibits strong fluorescence at that wavelength. Clearly, the lens transmits relatively little light in the near UV as compared to visible light.

The simplest way to solve this problem is to use, instead of a normal camera lens, a simple, uncoated, single-element, inexpensive crown or soda-lime double-convex glass magnifier as a lens. Such a lens can often be found in the sale bin at the hardware store as a hand magnifier. They come in many sizes and usually have a focal ratio of about $f/3$. Many of my early cyanonegative pictures were made with simple lenses of this type, mounted in a fixed manner onto a box of some sort. Getting the sensitized cyanotype paper in and out of the camera is easy; its very-low sensitivity means one must simply be quick about it and avoid exposure to direct sunlight.

Use of a simple lens would ordinarily result in a lot of chromatic aberration, but for cyanonegative photography this is not so bad as one might expect. Chromatic aberration means that different wavelengths focus at different distances, so a blurry picture results when many different wavelengths are present at once in a picture. For an ordinary picture taken with visible light, there are wavelengths across the visible spectrum—from 400 nm to 700 nm—a range of 300 nm. For cyanonegative photography, figure 25.4 shows that the lens only needs to focus wavelengths over a range about half that for new cyanotype, even less for blueprint paper. And so the focus error is less for cyanonegative photography than it would be if the same lens were used with an ordinary light detector, sensitive across the full visible range of

Figure 25.7. Left: *Ash tree no. 1*, John Beaver 2000. When cyanotype-sensitized paper is exposed in a special camera, it produces a blue-tone negative. When digitally scanned and reversed to a positive, pleasant amber and sepia tones result. **Right:** A simple box camera for cyanotype negatives, using a 50 mm double-convex lens. The original negative for the image at left can be seen to the right of the camera. Also shown are the negatives for two of the cyanonegative photographs in figure 25.9, which were taken with cameras of larger format.

wavelengths. The simple lens, of course, has other aberrations and distortions as well, and these still remain and give cyanonegative photography much of its look. Figure 25.7 shows one of my earliest cyanonegative photographs, made with toy blueprint paper exposed in the 5 cm × 8 cm box camera shown at the right. The box has a translucent lid that acts like the focus screen in a view camera. It is simply placed on a tripod, and the picture is composed on the lid of the box. Then the lid is removed, and the sensitized cyanotype paper is placed on the inside of the lid—facing the lens. I attach the paper to a precut piece of sheet metal, which can then be held fast to the inside of the lid by placing a strong magnet on the outside. The lid, with cyanotype paper, is then placed back on the camera for the duration of the exposure.

The cyanonegative picture in figure 25.7 is mostly just a silhouette of the tree against the sky. The ground would be completely dark with this 10 m exposure, but it was covered with snow (except for the bare spot under the tree), which reflected the UV light from the sky. The simple lens shows pronounced barrel distortion, spherical aberration, coma and vignetting, and the texture of the grain of the blueprint paper is evident.

Before the advent of digital scanning and printing, a cyanotype negative would have been essentially useless. It will not contact print very well (Ware 2016, p 104), and for similar reasons it would give poor results if used in an ordinary darkroom enlarger. But a scanned and digitally reversed cyanotype paper negative has many intriguing features.

1. The use of a fast, simple magnifier for a lens gives an unusual image that naturally blurs toward the edge—one way in which this process is decidedly unlike pinhole photography.

Figure 25.8. Left: *Self-portrait*, John Beaver 2001. **Right:** *Dandelion*, John Beaver 2004.

2. The grain of the paper adds interesting texture to the image. The overall effect is reminiscent of a lithograph. If the cyanotype sensitizer is hand brushed onto the paper negative, the brush marks can become an interesting element of the composition.
3. The cyanotype Prussian blue color of the negative, upon reversal to a positive, becomes a pleasing sepia.
4. A camera for cyanotype negatives is easy to make, and the lens need not cost more than several dollars. The cost of the cyanotype 'film' can be as low as a few pennies per exposure.

Since a long exposure is needed for a cyanonegative photograph, the relationship between motion and time for the subject is an important factor. For the left-hand image in figure 25.8, I used a system of rods to keep my head in position during the hour-long exposure. For the last several minutes, I got up out of the chair and stood behind it, making a faint ghost in the picture. I was able to take a self-portrait unassisted, composing the picture before I was in it. After loading the camera with cyanotype and beginning the exposure, I simply walked to the chair and took my place; the several seconds required to do so were insignificant compared to the hour's time needed to make the picture.

The cyanonegative progress can also be used—quite easily—to make photograms (shadow exposures). As a blue-tone negative print, this has been one of the most common historical uses of cyanotype. But we can accomplish something very different by scanning the blue-tone negative and inverting it to a positive (see the right-hand image in figure 25.8).

The top two examples in figure 25.9 reveal how the process of hand brushing the sensitizer can be incorporated into the composition. Since this is a negative process,

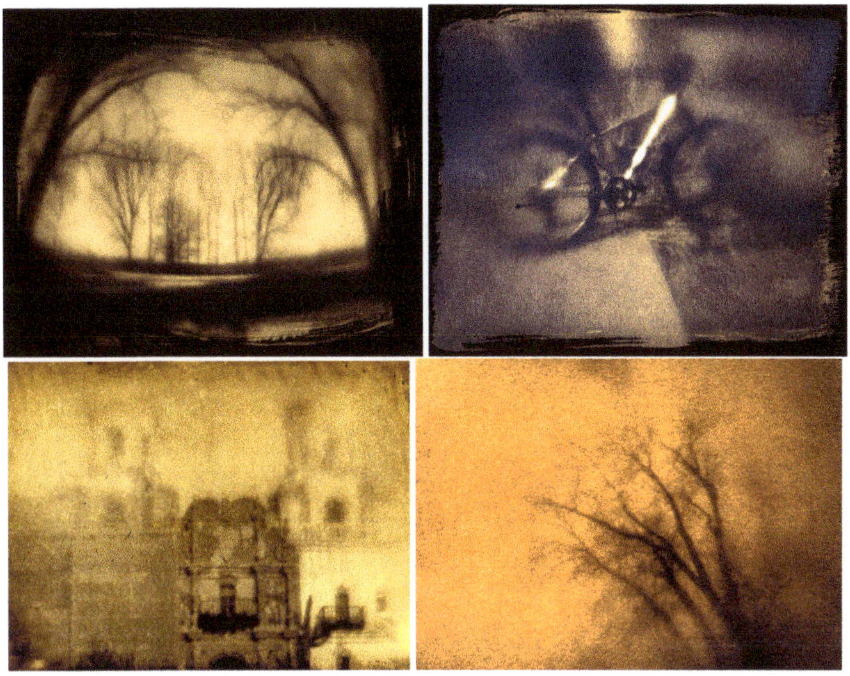

Figure 25.9. Clockwise from upper left: *Trees and ice no. 2*, John Beaver 2004; *Bicycle*, John Beaver 2003; *Take me with you*, John Beaver 2002; *San Xavier Mission no. 2*, John Beaver 2004. The upper two images demonstrate how the brush strokes from brushing the sensitizer onto the paper can be incorporated into the composition of a cyanonegative photograph. Since this is a negative process, it is as if light has been brushed onto a dark background.

to brush the sensitizer onto white paper is, ultimately, to brush light onto darkness—exactly the opposite of the effect achieved with a hand-sensitized print.

Since everything looks very different with this lens–detector combination, one often finds that the most ordinary of photographic subjects can be interesting. It also means that a scene of great power to the eye in visible light may be rendered confusing—perhaps in an uninteresting way—by the cyanonegative process. The bottom two images in figure 25.9 are not at all representative of those scenes in real life, and that is likely much of their power.

25.2.1 Cyanonegative focus offset

Even if a simple lens does pass enough UV light to be useful for cyanonegative photography, it will inevitably have a different focal length for UV light (which the cyanotype paper sees) than visible light (which one sees as one focuses the camera). It is possible to correct for this by first focusing with the eye, which responds to light centered at about 550 nm. We then apply a focus correction, bringing the lens slightly closer to the light-sensitive paper, since the same lens will focus the shorter wavelength light to a slightly smaller distance.

We can easily estimate this focus offset as follows. The focal length of a simple lens is given by the *lensmaker's equation* (section 11.3). If we assume the lens is thin compared to its focal length, the lensmaker's equation reduces to a simple form.

$$\frac{1}{F} = (n - 1)\left(\frac{1}{R_1} - \frac{1}{R_2}\right) \tag{25.7}$$

Here F is the focal length of the lens, n is the index of refraction of the glass, and R_1 and R_2 are the radii of the curvature of the two sides of the lens. Thus for any particular lens we have the following equations.

$$\frac{1}{F} \propto (n - 1) \tag{25.8}$$

$$d_i \propto \frac{1}{n - 1} \tag{25.9}$$

Here $\propto$ means 'is proportional to,' and I have noted that the image distance (d_i) is approximately equal to the focal length of the lens so long as the subject is not too close. From equation (25.8) we can see that a higher index of refraction results in a shorter image distance.

We can use equation (25.8) to compare the image distance for one index of refraction to the image distance with the same lens but for a different index of refraction. That is the point of dispersion; the same piece of glass has a different index of refraction when used to image a different wavelength of light. And so let us use n to represent the index of refraction of the glass in the middle of the visible spectrum (550 nm), and d_i to represent the image distance for that particular n. Let us use n' and d_i' to represent the index of refraction and associated image distance for the shorter wavelength to which cyanotype is most sensitive (350 nm).

When we use our eye to focus, we set the image distance according to the index of refraction at 550 nm. But the index of refraction—and thus the correct image distance—is different for the 350 nm of which cyanotype is sensitive. And so let us calculate the approximate percent error in our setting of the focus.

$$\%\ \text{shift} \approx 100\left(\frac{d_i - d_i'}{d_i'}\right) \tag{25.10}$$

$$= 100\left(\frac{\dfrac{1}{n - 1} - \dfrac{1}{n' - 1}}{\dfrac{1}{n' - 1}}\right) \tag{25.11}$$

$$= \frac{n' - n}{n - 1} \tag{25.12}$$

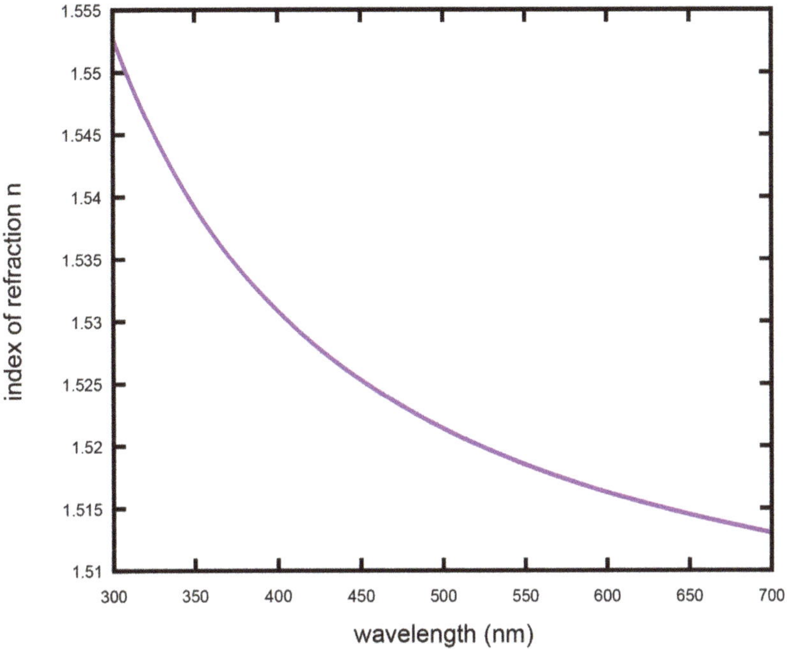

Figure 25.10. The dispersion curve (index of refraction vs. wavelength) for BK7 glass, a common glass used for simple lenses. If a cyanonegative exposure is focused by eye (550 nm) the lens will not be in focus for the cyanotype detector (350 nm) (data from SCHOTT (2017), p 13).

As an example, let us assume we use a simple thin lens made of BK7 crown glass. The dispersion curve for this glass can be seen in figure 25.10 (data from SCHOTT (2017), p 13). For such glass the index of refraction at 550 nm is 1.519, but for 350 nm the index of refraction is 1.539. And so if we assign $n = 1.519$ and $n' = 1.539$ we have the following from equation (25.11).

$$\%\text{ shift} \approx \frac{1.539 - 1.519}{1.519 - 1} = 3.85 \tag{25.13}$$

The positive answer means that d_i', the correct image distance for cyanotype, is less than d_i, the image distance we set by focusing with our eyes. If our lens is made of this type of glass, after focusing with our eye we need to reposition the lens about 3.9% closer in order for the image recorded by the cyanotype to be in focus. Using numbers for a simple soda-lime glass (the type used for bottles, windows and sometimes for inexpensive magnifying lenses) gives a similar result. In practice, trial and error is important.

It should be unsurprising that the image distance would be smaller at shorter wavelengths. Figure 25.10 shows that the index of refraction is larger at shorter wavelengths. A larger index of refraction means that the lens bends the incoming light at a greater angle inward, so the incoming light rays need less distance to converge to a focus.

25.3 EP photography

Since the earliest days of photography, one of the central problems was how to make some material sensitive to light for the desired exposure—but then insensitive to light once it is hanging on the wall. A picture is not much good if the act of looking at it makes it disappear. It was John Herschel, the inventor of cyanotype, who came up with the first really-successful fixer for silver-based photography.

The technology of digital image capture, however, puts a new twist on this. Photographic processes that produce only an ephemeral image—that would have been useless in the predigital era—can now be scanned to produce a permanent digital image. The still light-sensitive material need only have a sensitivity low enough such that it can be scanned without too much degradation from the light of the scanner.

It is well known that ordinary silver-gelatin paper—intended for use in the darkroom to produce a black-and-white print form a film negative—will, with no developer, print out like cyanotype if exposed to enough light. The sensitivity is sometimes about the same as cyanotype but usually less (sometimes much less) depending on the particular paper chosen and other factors. One could use it to take photographs in the same way as cyanonegative photography. The sensitivity is so low that the image from the long exposure in the camera can be scanned to capture a digital negative.

This is a rather obvious idea, so I quite naturally tried it back in the early 2000s, placing enlarging paper in the same homemade cameras I was using then for cyanonegative photography. I found the simplicity and chemical-free nature of the process intriguing, but I was not impressed with the results. Much of the appeal (for me) of the cyanonegative process is that the sensitizer is uneven and applied by hand, and thus introduces nonphotographic detail that can result in happy accidents—a topic we explore in chapter 44. Unlike cyanotype sensitizer brushed by hand onto watercolor paper, stock silver-gelatin papers are manufactured with as much uniformity as possible. And so we are simply back to the traditional silver-gelatin process—albeit without the chemicals—but with none of the high-sensitivity advantages of developing out[1].

This all changed in late 2015 when I realized that it is easy to accelerate the printing out of silver-gelatin enlarging paper by applying a simple liquid *accelerator* immediately before exposure. When printing out, the accelerated sensitivity can be many hundreds of times greater than the dry paper out of the box (see figure 25.11). After exposure the paper is then washed and dried, and it returns to its VLS state, making it easy to scan in order to digitally capture the negative image. One of my first EP photos can be seen in figure 25.12.

The accelerator turns out to be very simple—water alone has a large effect. An oxygen scavenger such as ascorbic acid (vitamin C) or sodium sulfite and a binder

[1] In more recent years, others have discovered the same process, and it sometimes goes by the name *lumenbox*. Since lumenbox practitioners also typically use simple lenses, it often has a similar look to cyanonegative photography but without the hand-brushed framing and interesting paper texture.

Figure 25.11. Left: A piece of dry enlarging paper after a 500 s exposure to sunlight. **Right:** The same paper but with an accelerator brushed onto it after exposure to the same sunlight for only five seconds.

Figure 25.12. *Front swing*, John Beaver 2016. Wet enlarging paper was placed in an 8 × 10 inch format camera and exposed for many minutes with a simple lens. After exposure, the paper was dried, scanned, and the digital image was inverted from a negative to a positive.

(such as xanthan gum) to make the solution brush more readily are all the rest that is needed. More information on the choice of paper and recipes for the accelerator can be found in chapter 48 of this volume. The physical mechanism behind the action of the accelerator is discussed in chapter 35.

And so we have the best of both worlds. We brush the accelerator onto the paper immediately before exposure in order to render it sensitive—far more sensitive than cyanotype. In fact the wet sensitivity brings us up to near our benchmark of the maximum sensitivity that is practical if one wants to handle the negative in a dimroom instead of a darkroom and to transfer the detector from dark bag to camera (and back again) in shaded ambient light. But after exposure we can desensitize the negative simply by washing and drying it, thus rendering it back to a sensitivity low enough for permanent image capture by scanning. Of even more importance, since we effectively sensitize the paper by brushing our accelerator solution onto it, we are back to what is essentially a hand-brushed negative. Only the parts that receive the accelerator are sensitive enough for image making in the camera.

With the right choice of paper and accelerator, EP photography can achieve a usable negative of a standard scene on a sunny day at $f/4$ in only five minutes. The wavelength response of EP negatives is broader than that of cyanotype, extending well into the green region of the visible spectrum. Thus, in contrast to cyanonegative photography, ordinary camera lenses can be used. There is little need for a focus offset, and reflected light from the subject is not nearly so drowned out by the sky.

Cyanonegative photography is barely worth the bother unless it is a bright sunny day, and even then a very long exposure with a fast simple lens is required to capture more than a silhouette against the sky. EP photography, on the other hand, is still possible with a (relatively inexpensive) $f/4.7$ standard lens from an old speed graphic camera even on an overcast day. Even room interiors can be captured if illuminated by daylight through a large window.

See figure 25.13 for two examples of the kind of look that is easily achieved with EP photography. In the example on the left, the patterns in the sky resulted from wrinkles

Figure 25.13. A selection of EP photographs. Fiber-based paper was used, sensitized with water thickened with xanthan gum and with a small amount of ascorbic acid (vitamin C) added. Exposure times were from 10–15 minutes, at focal ratios from $f/4.7 - f/5.6$. The paper negatives of 3×4 inch format were scanned and digitally inverted to positives with adjustments only to overall levels and contrast and minor adjustments of tone.

Figure 25.14. Examples of EP photography. All of the exposures were made with a 3 × 4 inch format camera with an old, but inexpensive, *f*/4.7 lens. Exposure times ranged from 10–15 minutes. For the two images on the right the accelerator was applied to the paper onsite, just before exposure, and under a dark cloth.

in the paper negative and air bubbles in the accelerator. And of course, as is the case with all VLS photography, the exposures are long enough that motion is spread out in time or erased. In the example on the right, taken on a windy day, the ends of the tree branches moved, but the trunk was motionless during the 15 minute exposure. In the two examples on the right in figure 25.14, detail has been added to the sky by the physical process itself. In real life, the sky was hazy and uninteresting for both of those exposures; an ordinary photograph would not likely have been as successful.

As is the case with cyanonegative photography, EP photography can be accomplished with a simple lens attached to a box. But since an ordinary camera lens can be used, it is practical to use a simple large-format camera with adjustable focus. Also, the much-higher sensitivity to the EP paper, once it has been accelerated, requires more care in handling. Still this can be accomplished in a dimly-lit room or even in the field under a dark cloth. In chapter 48, I describe some ways to build or adapt inexpensive cameras for EP (and cyanonegative) photography and some further details on the process.

25.4 Using EP photography to test the Jones–Condit equation

EP photography provides a simple method to test the validity of the Jones–Condit equation from chapter 20. We can use the relation between incident and in-camera exposure time given by equation (20.17).

Figure 25.15. Left: A gray card of 18% reflectivity was photographed with EP photography. **Right:** A scan comparing a two second direct exposure to the incident light (left), and the negative from a 1394 s exposure inside the camera (right).

$$t_{\mathrm{D}} = \frac{4f^2}{R} t_{\mathrm{SUB}} \tag{25.14}$$

Here f is the focal ratio of the lens, and R is the reflectivity of the subject. Moreover t_{SUB} is the time we expose the detector directly to the light incident on the subject in order to bring the detector up to some particular level of density, and t_{D} is the time we expose the detector in the camera in order to bring the image of the subject up to that same level of density.

If equation (25.14) is correct, we should be able to place a subject of known reflectivity in sunlight and expose our detector directly to that same sunlight for time t_{SUB}. We then set up a camera, point it at the subject and take an exposure with the same type of detector, using equation (25.14) to calculate the exposure time (t_{D}) in the camera. In the end, both the direct and in-camera exposures should achieve the same density.

For figure 25.15, I set up a gray card—a piece of cardboard designed to diffusely reflect (as a Lambertian reflector) 18% of the light ($R = 0.18$) that falls upon it. I exposed EP paper for two seconds to the same sunlight (and incident at the same angle) that was falling upon the gray card. I then photographed the gray card at $f/5.6$, using the same type of EP paper in the camera. I calculated the in-camera exposure time from equation (25.14).

$$t_{\mathrm{D}} = \frac{4f^2}{R} t_{\mathrm{SUB}} \tag{25.15}$$

$$t_{\mathrm{D}} = \left(\frac{4 \times 5.6^2}{0.18}\right) 2\ \mathrm{s} \tag{25.16}$$

$$t_{\mathrm{D}} = 1394\ \mathrm{s} \tag{25.17}$$

$$t_{\mathrm{D}} = 23\ m,\ 14\ \mathrm{s} \tag{25.18}$$

The image on the left shows the setup, while that on the right shows both the two second direct exposure and the 1394 s in-camera exposure. The densities of both should be the same, and they are—almost. It is evident, however, that the in-camera exposure shows slightly less density than the direct exposure. The reason for this should be clear: my simplified version of the Jones–Condit equation ignores absorption of light by the glass in the camera lens.

References

Beaver J Fall 2017 Using new-antiquarian photographic processes to integrate art and science *American Geophysical Union Meeting Abstracts*

James C 2016 *The Book of Alternative Photographic Processes* 3rd edn (Boston, MA: Cengage Learning)

Piper J E 2021 Making cyanotypes in-camera http://www.alternativephotography.com/making-cyanotypes-in-camera/

SCHOTT North America, Inc. 2017 Optical glass data sheets https://refractiveindex.info/download/data/2017/schott_2017-01-20.pdf

Ware M 2016 Cyanomicon II–History, Science and Art of Cyanotype: Photographic Printing in Prussian Blue https://www.mikeware.co.uk/downloads/Cyanomicon.pdf

Part V

The art and science of color

IOP Publishing

Photography (Second Edition)
Physics and art in focus
John Beaver

Chapter 26

The physical basis of color

Most sources of light produce many wavelengths at once; the *monochromatic* (one wavelength only) light of a laser is the exception. Unless one intentionally redirects different wavelengths of light from a source off in different directions, as with a prism or diffraction grating, one usually does not encounter light of only a single wavelength. These single-wavelength 'colors of the spectrum' are our perceptions of individual wavelengths of light. Clearly, we are able to perceive much more than the pure and simple red, orange, yellow, green, blue and violet present in such a spectrum. It is the infinite variety of combinations of these wavelengths that is at the heart of the physical basis of color.

As an example, the pure spectral colors are quite different even from the colors of the rainbow. It is true that a rainbow shows colors that are similar to and in the same order as the colors of the spectrum. But the similarity ends there, as can be seen if one compares them side-by-side (see figure 26.1). Each position in a rainbow is, in fact, a mixing of many different wavelengths at once.

Even a single wavelength of light is 'perceived' differently by different detectors. Figure 26.1 also shows the pure spectrum of the same light source, as photographed by two digital cameras with different light detectors. These digital detectors are not designed for accurate reproduction of single wavelengths of light, and so it should be unsurprising that—when used for this purpose—the results are very different from the perception of the human eye and from each other. Rather they are designed to mimic human color perception as best they can under typical circumstances encountered in photography. Even this simple example demonstrates that we have much work to do in order to untangle the many meanings of 'color' in photography. Color is about much more than simply the wavelength of the light.

26.1 Spectra and sources of light

To understand the complexities of 'color' we need some tools for analyzing light in such a way that different wavelengths can be considered both separately and in

 © IOP Publishing Ltd 2022

Figure 26.1. Left: A photograph, with contrast greatly enhanced, of a rainbow taken with a Nikon D40 digital camera. **Right:** A close-up cross section of the rainbow from the picture on the left can be seen at the top right. The center and bottom images are both of the pure spectrum of the same lightbulb, but taken with two different digital cameras (Canon Powershot G15 in the center and Nikon D40 on the bottom). Clearly a rainbow is very unlike a pure spectrum of individual wavelengths. And the same pure spectrum will appear very different depending on the 'eye' of the beholder. The visual appearance to the human eye of the pure spectrum is quite unlike these two different digital captures, which also differ significantly from each other.

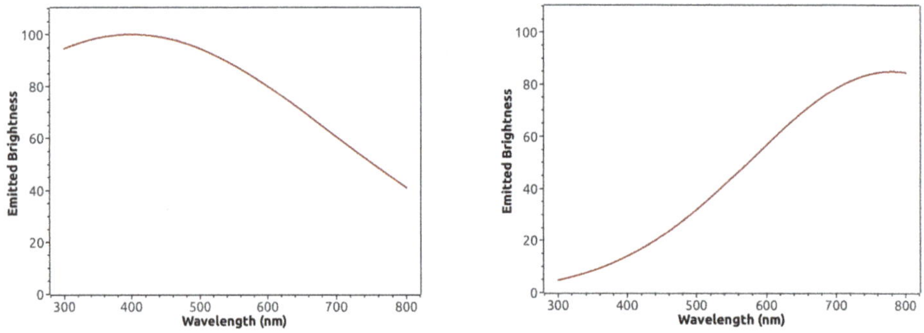

Figure 26.2. Two spectra of two different sources of light. Both of these sources of light emit at all wavelengths in the visible portion of the spectrum. But the light source on the **left** emits more short-wavelength light than long, and so it would appear a pale bluish-white in color. The light source on the **right**, on the other hand, emits more long-wavelength light than short. And so it would appear a pale reddish-white.

combination with each other. We use a *spectrum* as a method for representing the wavelength information of light. The spectrum of the source of light tells us, separately for each wavelength, how much light there is.

The most informative way to represent a spectrum is with graphs, such as those in figure 26.2. In this example the brightness of the light is represented by the vertical axis while wavelength is on the horizontal axis with short wavelength on the left side and long wavelength on the right. The red line thus represents the brightness of the light, separately for whatever wavelength in which we are interested.

The spectrum on the left in figure 26.2 is of a source of light that emits much more short-wavelength light than long-wavelength light; the example on the right is the opposite. These would then appear quite different colors to the human eye. The source of light on the left, judging by its spectrum, would appear much more bluish in color than the one on the right, which would appear reddish. Remember that the human eye is sensitive only from about 400 nm (violet) to 700 nm (red). It is a complex question, however, as to precisely what color sensation would arise when this spectrum of light enters the human eye beyond the vague terms 'reddish' or 'bluish.' Untangling some of these complexities is the concern of much of this and following chapters.

26.1.1 Combinations of multiple light sources

When more than one light shines on a subject, the combined light has a spectrum equal to the sum of the two individual spectra. This means one would add the brightness from each at each wavelength; the result gives the brightness at that particular wavelength for the combined spectrum. Figure 26.3 shows the combined spectrum for the two light sources in figure 26.2.

When interpreting a spectrum, it is important to look carefully at the numbers. Notice that the two spectra in figure 26.2 show, respectively, a brightness of 90 and 18 (in arbitrary flux units) at a wavelength of 400 nm. Thus the combined spectrum shows, at this same wavelength, a brightness of $90 + 18 = 108$. The combined spectrum is brighter at all wavelengths, but more importantly the shape of the combined spectrum clearly differs from either of the two spectra that were combined to make it. And so it would appear a different color, neither obviously bluish nor reddish. Also, notice that the percentage range of brightnesses in figure 26.3 is much smaller than for either of the original spectra. This means that the combined spectrum is more egalitarian; different wavelengths are represented more nearly equal. In fact, a source of light with the spectrum shown in figure 26.3 would appear nearly white.

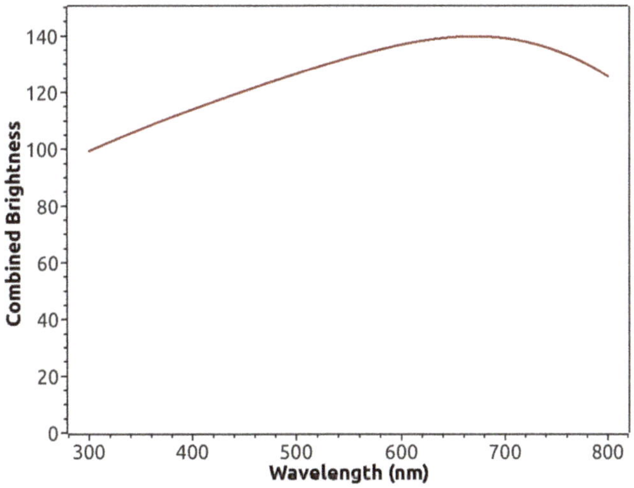

Figure 26.3. The two spectra from figure 26.2 added together. Adding the brightness wavelength by wavelength gives the spectrum for both sources of light shining on the same subject. Even though this source of light emits more light at around 600 nm (the yellow part of the spectrum) than any other wavelength, the color would appear nearly white given the large amount of light emitted at other wavelengths as well.

26.2 Color, light sources and light detectors

The color of a photograph arises from the interaction between the spectrum of the light falling on the light detector and how that detector responds to different wavelengths. Furthermore, if the camera is pointing at an object that is not self-luminous, then the light entering the camera depends on a combination of the spectrum of the light shining on the object and how that object reflects different wavelengths. To understand these interactions, we need to go beyond the general term 'color' and think of light in terms of its spectrum. And so we have three factors that are important for the color of a particular part of a photographic image[1].

1. The spectrum of the source of light shining on the subject.
2. The percentage of light that is reflected by the subject for each individual wavelength. This too can be represented as a graph that is very similar to a spectrum. Instead of representing brightness of light versus wavelength, it is the percentage of light reflected versus wavelength. We call this the *reflection curve* of the subject.
3. The sensitivity of the light detector for each individual wavelength. If one shines light onto the detector, what percentage of the light is actually detected? For every real detector the answer to this question depends on the wavelength, and it can be represented by a graph of percentage of light detected versus wavelength, which is called a *response curve* for the detector.

It is important to note that the second item in the list above is not relevant in all cases. For example, if one takes a picture of a sunset the sky is both the object being photographed and its own source of light; it is not an object reflecting light from some other source. In this case it is just the interaction between the spectrum of the source of light (the sky) and the response curve of the detector that is important.

There is another important factor that I will mention here but then put aside until later. A given detector may be very sensitive to a particular wavelength—530 nm, for example. This means that a large percentage of light of that wavelength would be detected by the detector. But there is also the question as to how the detector interprets this detection. Can it, for example, distinguish between this wavelength and—say—an equal amount of light at 560 nm? Or would it instead detect both wavelengths but lump them together without being able to tell how much of the total was from 530 nm and how much was from 560 nm? These last questions go to the heart of the distinction between black-and-white detectors on the one hand and color detectors on the other. We take up that issue in section 26.6.1 and in even more detail in part VI.

[1] We set aside, for now, the important issue of the relationship between the color of—say—a digital image and the color of a print that is made from it.

26.3 The reflection curve and the reflected-light spectrum

Figure 26.4 shows two hypothetical reflection curves: one for red paint and the other for blue paint. These curves tell the percentage of light that is reflected off the paint for each wavelength. Notice that the blue paint reflects a lot of blue light (short wavelengths) and not much red (long wavelengths); the red paint does the opposite. And so a reflection curve for a surface describes what happens to light that reflects off it. How much reflects? The answer is that it depends on wavelength; the reflection curve is a graph that shows the percentage of light reflected versus wavelength.

Notice that a reflection curve, as in figure 26.4, looks superficially like a spectrum. The reflection curve, however, is hypothetical; by itself it does not say anything about the light reflecting off the object. It simply says, 'whatever light falls on the object of such and such a wavelength—this is the percentage of that light that would reflect.'

To describe the spectrum of the reflected light that actually makes it to the camera, we must combine the spectrum of the light source with the reflection curve of the object off which the light is bouncing. Since a reflection curve is a percentage, we must multiply the two curves together. How does one take 25% of 24? 'Twenty five percent of 24' means literally, '25 divided by 100 times 24'. In this mathematical context the word 'per' means 'divide by,' the word 'cent' means '100' and the word 'of' means 'multiply by'[2].

But neither the spectrum nor the reflection curve is a simple number; they are graphs. Since each is telling what happens for each wavelength, we must multiply the two curves wavelength by wavelength thus producing a third curve. This third curve then represents the spectrum of the light reflected by the object when illuminated with that particular spectrum of light, and I will call this the *reflected spectrum* of the light reaching the camera.

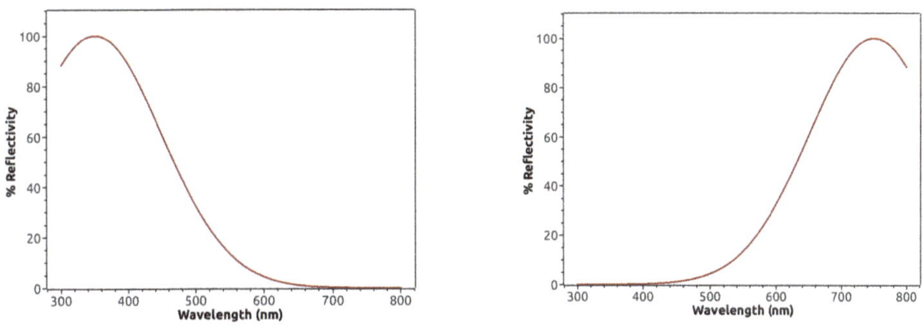

Figure 26.4. Two reflection curves. The one on the **left** is for blue paint while that on the **right** is for red paint. The blue paint reflects a large percentage of 400 nm light but only a very small percentage of 700 nm light. The opposite is true for the red paint.

[2] Instead of percentages between 0% and 100%, a reflection curve can also be portrayed in terms of decimal fractions between zero and one—0.57 instead of 57%, for example. Then one simply multiplies the reflection curve by the spectrum without dividing by 100.

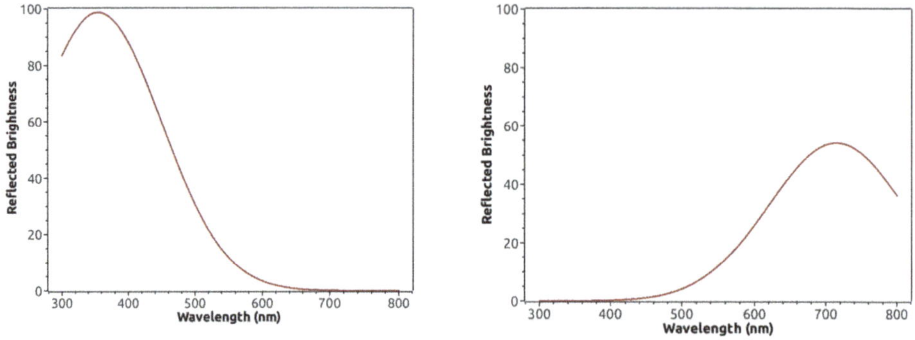

Figure 26.5. The reflected spectra for the two reflection curves of figure 26.4 when illuminated by the bluish light with the spectrum on the left in figure 26.2. The figure on the **left**, then, shows the spectrum of bluish light reflecting off of blue paint, while the figure on the **right** shows the spectrum of bluish light reflected off of red paint. Since the red paint reflects very little blue light, its reflected spectrum is much dimmer.

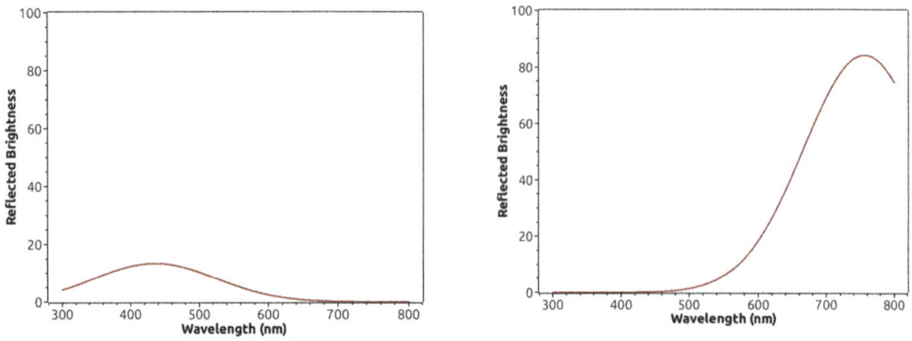

Figure 26.6. The reflected-light spectra for the two reflection curves of figure 26.4 when illuminated by the reddish light with the spectrum on the right in figure 26.2. The figure on the **left**, then, shows the spectrum of reddish light reflecting off of blue paint, while the figure on the **right** shows the spectrum of reddish light reflected off of red paint. Since the blue paint reflects very little red light, its reflected-light spectrum is much dimmer.

Let us take the bluish light with the spectrum shown on the left in figure 26.2 and shine it on both the blue paint and the red paint with the reflection curves shown in figure 26.4. Figure 26.5 shows the resulting reflected spectra. The example on the left is of the bluish light reflecting off of blue paint; unsurprisingly, the reflected-light spectrum also shows mostly short wavelengths. The bluish light reflecting off of the red paint (the right side of figure 26.5) shows mostly longer wavelengths. This is because the light source still had a significant amount of red light; it just had much more blue. It is clear from comparing the two graphs that, when illuminated by bluish light, the blue paint reflects much more light overall than the red paint.

Figure 26.6 shows the opposite case. The longer-wavelength reddish light, with the spectrum shown on the right side of figure 26.2, illuminates the two colors of

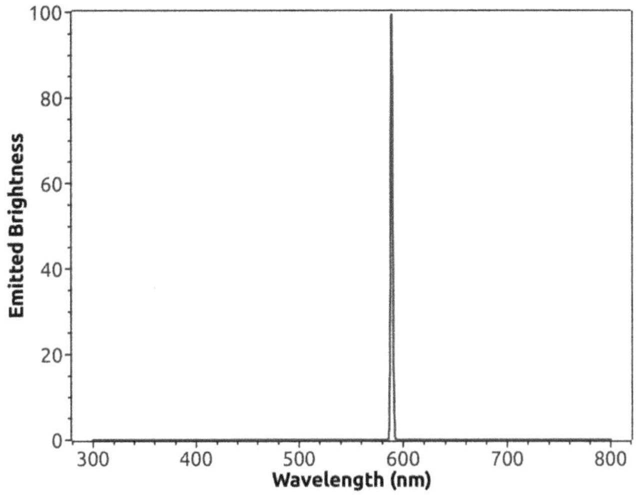

Figure 26.7. The spectrum of a low-pressure sodium lamp. In the visible range only one wavelength (at 589 nm) is emitted. Thus the graph of the spectrum appears as a single spike at that wavelength.

paint. And the opposite happens—the reflected light from the red paint is brighter than that from the blue light.

Notice that, for either case, the reflected-light spectrum is zero at any wavelength for which either the source spectrum or the reflection curve was zero. It matters not whether, at a particular wavelength, the source had zero light or the object reflected zero percent. One hundred percent of zero is still zero and so is zero percent of anything.

Let us consider another example. Consider the spectrum shown in figure 26.7 of a low-pressure sodium vapor lamp. This type of lamp, sometimes used as a home security light, appears an odd yellow-orange color. A glance at the spectrum shows that this source of light is monochromatic; it emits essentially all of its light at only the single wavelength of 589 nm[3]. So what happens if we shine this light on both our red and blue paint? What reflected-light spectrum results?

To figure this out, we do what we just did—we multiply the source spectrum, wavelength by wavelength, by the reflection curve. This is easy to do because the source spectrum is so simple. For the single wavelength of 589 nm, the brightness is 100 (in some arbitrary set of physical units). For all other wavelengths, the brightness is zero. Since zero times anything is zero, the reflected-light spectrum would also have a brightness of zero at all wavelengths except for 589 nm no matter what the reflection curve.

[3] The sodium spectrum is not truly monochromatic; it consists of two very closely-spaced wavelengths: one at 589.0 nm and the other, weaker line at 589.6 nm. The spectrum shown here does not have enough detail to distinguish between these two closely-spaced spikes. Furthermore, sodium does emit very faint light at other wavelengths in the visible spectrum as well as bright light at many wavelengths outside of the visible part of the spectrum.

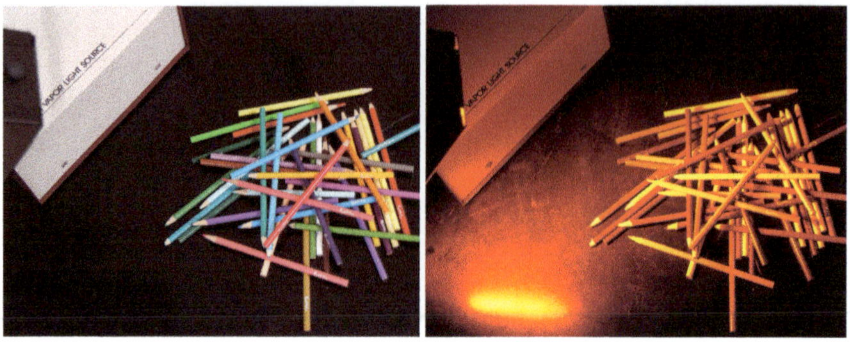

Figure 26.8. A set of colored pencils illuminated by white light (**left**) and the monochromatic light of a low-pressure sodium lamp (**right**). The individual pencils reflect either more-or-less of the single wavelength of the sodium light, but they cannot alter its wavelength. When illuminated by the many wavelengths at once of white light, each pencil reflects different amounts of different wavelengths of light while selectively absorbing other wavelengths.

In this case the reflection curves would look very similar for both red and blue paint. They would both show zero light with a single spike of brightness at the single wavelength of 589 nm. If the spectrum of the light source is monochromatic so too is the reflected-light spectrum no matter what the reflection curve[4]. Thus, both the red and blue paint would appear the same color (yellow-orange) when illuminated with a monochromatic sodium lamp. The difference would be how much light is reflected from the two kinds of paint. A glance at the two reflection curves of figure 26.4 shows that the red paint would reflect a greater percentage of the 589 nm incident light and thus would appear brighter. Figure 26.8 shows two views of a set of colored pencils illuminated by white light (with a spectrum somewhat like that in figure 26.3) on the left and by a sodium vapor lamp on the right. Under monochromatic light the colors disappear: each reflects the only wavelength that is illuminating it. Some appear lighter and some darker, depending on what percentage of 589 nm light they reflect.

Given the image on the right in figure 26.8, we cannot 'go backwards' and figure out the colors of the pencils. The brightness or darkness only tells us how much light is reflected at the single wavelength of 589 nm, and in order to know the color of the pencil one would need information about the other wavelengths as well. And so a pencil may appear dark under the sodium lamp because it reflects little light at any wavelength (a black pencil, for example). Or it may appear just as dark under the same sodium lamp because it is painted with a bright violet paint that happens to reflect very little light at 589 nm (but a lot of light at other wavelengths). Without more information there is no way to tell.

[4] I assume here that *fluorescence* (section 3.8) is not at issue.

26.4 Physical causes of the reflection curve

When light encounters an object some wavelengths reflect, while others do not. The reflection curve details the end result of this process, and this is why objects appear to have color when illuminated by white light. But what happens to the wavelengths that do not reflect, and what are the physical processes that cause some wavelengths to reflect strongly while others reflect poorly?

The answer is that there are several different answers, depending upon the type of interaction between light and matter. Sometimes there is a complex interplay between these different processes, but much of the time we can break it down to one of these: selective absorption, structural color and fluorescence.

26.4.1 Pigments and dyes: color from selective absorption

Pigments are microscopic particles that selectively absorb light to produce color upon reflection. In practice pigments may be suspended in some type of liquid or semiliquid medium to allow them to be applied to a surface. In some cases (watercolor for example), the pigment particles are simply suspended in water. The pigment may remain suspended in the dried medium as a thin layer on the surface of some object—as in paint (see figure 26.9). Or, as in the case of a dye, it may be carried by its medium into the surface of a porous material such as paper, wood or fabric.

Pigments and dyes produce color by way of *selective absorption*. Some wavelengths are absorbed, while other wavelengths are not. What is not absorbed is, instead, reflected. For example, if a pigment absorbs blue and only blue, then it

Figure 26.9. A selection of artist's paints applied to both black-and-white backgrounds. The colors are often named after the chemical nature of the pigments used (e.g., cobalt blue, cadmium red or phthalocyanine green). The combination of the pigment and its medium (e.g., oil, acrylic or water) also affects the *transparency* of the color. Some of the paints shown here have a high transparency and nearly disappear when placed against a dark background.

reflects both red and green. The selective-absorption properties can be described according to a reflection curve, as described in section 26.3, but one could instead say the same thing with its inverse—an *absorption curve.* If, at a particular wavelength, the pigment reflects 80% of the light that falls on it, then it must absorb 20%. And so wherever the reflection curve is high, the absorption curve is low and vice versa.

This selective absorption usually occurs as part of the interaction between an individual photon of light with an individual molecule or atom. Individual photons are absorbed, raising electrons to higher energy states. Since there are only certain energy states allowed, according to the rules of quantum mechanics, only certain energies of photons can be absorbed. But for light, photon energy is proportional to the frequency of the light and so also its wavelength; shorter-wavelength light is made of photons that, individually, have more energy. The precise energies of these electron energy states—and so the particular wavelengths of light that can be absorbed—are complex functions of the atomic and molecular structure of the pigment. And so a slight chemical change can lead to a dramatic difference in the absorption–reflection curve.

Most often, this absorbed photon energy is transferred ultimately to thermal energy, and the electrons return to their low energy states, ready to absorb more photons. But not always; chlorophyll—what makes leaves green—is a good example. Some of the photon energy absorbed by chlorophyll is used to power chemical reactions that ultimately store this photon energy as chemical energy in the form of sugars. This is the ever-so-important process of photosynthesis. See figure 26.10 for a graph of the absorption curves of chlorophyll (there are two types).

Figure 26.10 shows only the visible-light part of the absorption curve, but notice that the curve is lowest in the blue-green part of the spectrum. Since it absorbs less at these wavelengths, it reflects more, and that is the origin of the green in the 'Green, Green Grass of Home.'

In order for light to do something—to cause a physical change in a material—it must be absorbed. Figure 26.11 shows a mixture of potassium ferricyanide and ferric ammonium citrate applied to a piece of watercolor paper. The properties of this are such that it reflects yellow light and so absorbs short-wavelength blue light. But

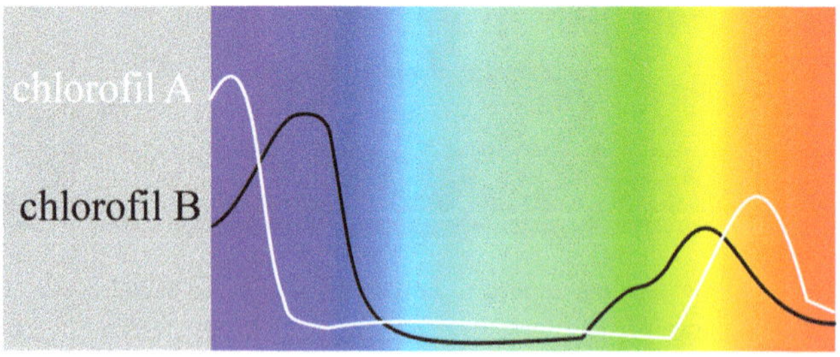

Figure 26.10. The absorption curves for two types of chlorophyll (graphic by Mix321—Own work, Lewińki, Holak, Biologia dla liceum, wyd. Operon, GFDL).

Figure 26.11. Cyanotype sensitizer, made from potassium ferricyanide and ferric ammonium citrate, was brushed onto paper. It appears yellow under white light because it reflects most wavelengths apart from the violet-blue part of the spectrum, which it absorbs. But upon exposure to short-wavelength light (in this case, from a 405 nm laser) chemical changes occur, and ferric ferrocyanide is formed—the pigment Prussian blue. Prussian blue, however, does nearly the opposite: it absorbs most visible wavelengths except for blue-green.

some of the energy from these short-wavelength photons are not converted to thermal energy. Rather they enable a chemical reaction that produces a new chemical: ferric ferrocyanide—the pigment Prussian blue. This is the famous *cyanotype* process invented by John Herschel in 1842 and that we used as a detector in a camera for the cyanonegative photography described in chapter 25. The new pigment that is produced by this exposure to light absorbs the very wavelengths that were reflected by the original sensitizer.

26.4.2 Structural colors: interference and scattering

Structural colors may arise from materials that are completely transparent, with no significant absorption of light. Instead, the phenomena of interference, diffraction and scattering[5]—hallmarks of the wave-like nature of light—cause different wavelengths to be deflected in different directions. Structural colors do not depend on individual molecules absorbing some wavelengths while reflecting others. Rather they result from regular patterns in the microscopic structure of the material—in many cases structures that are too small to be seen with the naked eye but still far larger than individual molecules.

Interference colors are one example. In *thin-film interference*, light reflects off the top and bottom surfaces of a thin transparent film. These reflections may or may not cancel out from wave interference, depending on the details of the thin film and the wavelength of the light. Figure 26.12 shows an interference filter. It is just clear glass with multiple layers of microscopically-thin layers of other transparent materials

[5] Some authors do not include scattering as an example of a structural color. I include it here because it shares many of the properties of structural colors—in particular that color is produced without absorption.

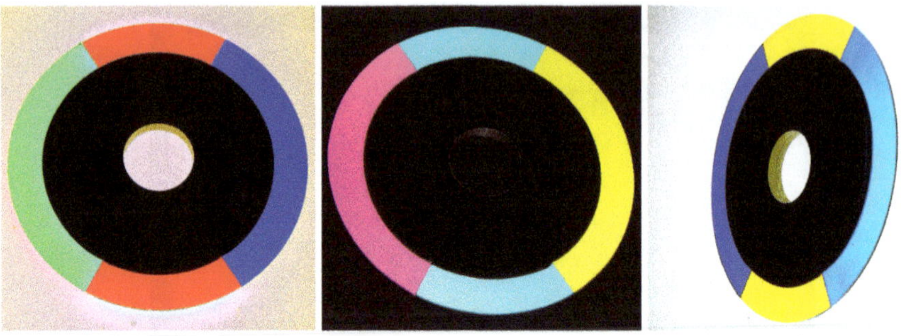

Figure 26.12. Interference colors arise not from selective absorption but rather through wave interference. This wheel of interference filters is seen with light shining through it (**left**), against a black background with light reflecting off of it (**center**), and with light shining through it but as seen from a steep angle (**right**). For light arriving at a given angle to the surface, whatever wavelengths pass through are exactly those wavelengths that are not reflected and vice versa. But this interference effect depends upon the angle, so different wavelengths reflect depending upon the angle of view.

applied to the surface, each with its own carefully-chosen thickness and index of refraction. The result is that some wavelengths interfere destructively—they cancel out. Others do the opposite, reinforcing each other.

For thin-film interference, if a particular reflected wavelength interferes destructively, then that same wavelength—when passing through—will interfere constructively. And the opposite is true: any wavelength that cancels out when trying to pass through will reinforce, interfering constructively, upon reflection. And so whatever wavelengths pass through do not reflect, and wavelengths that reflect do not pass through. The different parts of the interference filter, in figure 26.12, are designed such that red, blue or green interfere constructively when passing through, while not-red, not-blue and not-green interfere constructively upon reflection. We will see, in chapter 28 that there is an important relationship between these transmitted and reflected colors—they are, respectively, the additive primary and secondary colors.

One hallmark of interference is that the details depend upon the geometry of the incident light. Wavelengths that interfere constructively at one particular angle may interfere destructively at other angles. The right-hand image in figure 26.12 is the same filter and backlighting as the first image. But as seen from a different angle, the wavelengths that interfere constructively are different. And so interference colors are often shimmery—they shift when seen from different angles.

Certain feathers from some birds show interference colors as well. Figure 26.13 shows the feather of a mallard as seen from two different angles. Microstructures in the feather split light waves apart, reflecting them from slightly different positions. When they arrive at the viewer, some wavelengths interfere constructively while most cancel out. In this case, short-wavelength blue light interferes constructively. The effect, however, depends critically on the relationship between the angle of view and the geometry of the microstructures in the feather. So with a slight twist, the blue color disappears. This effect is known as *iridescence*, and it is a common way that

Figure 26.13. This duck feather shows iridescence—colors that disappear or change when viewed at different angles. In this case, microscopic patterned structures in the feather cause reflected blue wavelengths to interfere constructively while other wavelengths interfere destructively. But the constructive interference requires a specific range of angles between the viewer and the light source.

blues and greens (and sometimes reds) are produced in the animal and insect world (e.g., see Nassau 1983, pp 265–268 and Greenewalt *et al* 1960).

Wavelength-dependent scattering (section 3.5.1) is another way to make color with materials that do not absorb light. Unlike interference, which depends on a carefully-organized layering of materials at the microscopic level, scattering is a random process. As such it is often considered its own category rather than an example of a structural color.

Rayleigh scattering is responsible for the blue color of the daytime sky. Individual photons of light are deflected by air molecules, and this scattering effect is greater for shorter wavelengths than for longer wavelengths. The blue light one sees in the daytime sky is made of photons from the Sun that would have been missed, but they were instead deflected toward oneself by scattering. Since the shortest wavelengths are more likely to do this, the sky looks blue.

Since Rayleigh scattering arises from extremely tiny particles the scattering is relatively weak, and a large quantity of material is needed to produce a significant color effect. When the particle size is much larger, and so the scattering colors can appear vividly on a small scale, it is often called the *Tyndall effect*. It produces blues when transparent materials contain suspended particles that are uniform in size and somewhat smaller than the wavelength of the light, but still much larger than the individual molecules that cause Rayleigh scattering. Opalescent glass (see figure 26.14) and the blue smoke from the exhaust of a faulty car engine are familiar examples.

Nominally black-and-white silver-gelatin enlarging paper under the right conditions can also produce scattering colors. The transparent gelatin emulsion contains suspended crystals of silver halide with specks of reduced silver, and these are often of the right size to cause scattering colors. For strong scattering colors to appear, the particles must be nearly uniform in size and also the correct size. Sometimes the particle size may lead to warm pinks, oranges and yellows instead of the more familiar Tyndall blues. This is likely best explained by the *Mie scattering theory*

Figure 26.14. **Left:** Tyndall Effect scattering in this transparent piece of opalescent glass causes mostly blue light to scatter backwards when illuminated from above but (**center**) leaves only longer wavelengths to pass through when illuminated from behind. **Right:** The subtle colors in this ephemeral-process photogram arise from scattering by the regularly-sized microscopic crystals of silver halide suspended in the gelatin coating on the surface of the paper.

(see section 3.5). The precise colors that appear vary widely from one brand of paper to the next (see figure 26.14). We explore some of the photographic possibilities of this phenomenon in chapter 31. The so-called chromoskedasic process we describe in chapter 35 strives to create the conditions that achieve the most intense scattering colors in these ordinarily black-and-white papers.

26.4.3 Fluorescent colors

Fluorescent materials absorb individual photons of light at the atomic or molecular level. The energy from the absorbed photon raises an electron to a higher energy level, and in this way the process is similar to selective absorption (section 26.4.1). But the similarity ends there because the excited energy of the electron is used to create a new photon of a different wavelength. Usually a higher energy photon of short wavelength is absorbed, and a lower-energy photon of longer wavelength is emitted. Thus a fluorescent material can, for example, absorb UV light while using some of that absorbed energy to emit its own visible-wavelength light. Figure 26.15 shows fluorescent materials illuminated by a light-emitting diode (LED) bulb of roughly 400 nm wavelength—at the violet edge of the visible spectrum. The fluorescent materials in the picture mostly absorb the short-wavelength light of the LED and then emit their own light in the much-longer-wavelength yellow-green part of the visible spectrum. Some of the materials in the picture are not fluorescent, and so they simply reflect the deep violet light with which they are illuminated. The visible-light picture of the same objects on the right shows, instead, their ordinary pigment properties of selective absorption.

The phenomenon of fluorescence is distinct in a very important way from the other processes described in this section: the phenomenon of fluorescence cannot be

Figure 26.15. Fluorescent colors arise from molecules that absorb high-energy photons (short wavelength) and use this energy to emit their own light at longer wavelengths. The image on the **left** is illuminated with only short-wavelength light at the very violet end of the visible spectrum. The image on the **right** shows the same objects illuminated with white light that contained very little of the short wavelengths needed for fluorescence.

described by a reflection curve. The reflection curve tells, for each individual wavelength, the percentage of the light incident on the subject that reflects. This concept is useful whatever the physical mechanism for that reflection, whether it is due to selective absorption, interference or scattering. The whole point of fluorescence is that the subject can emit light at wavelengths for which no light at all is incident. As such, we will not consider fluorescence further in *Photography: Physics and Art in Focus*.

26.5 The detector response curve

The reflection curve combines with the spectrum of the light falling on the subject to produce an altered, reflected spectrum, and this is the light that actually makes it to the camera. But there is another factor, as no real-world light detector measures all wavelengths equally. Figure 26.16 shows the response curve for two hypothetical detectors. The response curve tells us, for each wavelength, what percentage of the incoming light is actually detected by the detector. It does not take a detective to discover that the detector on the left detects blue light better than red, while the opposite is true for the detector on the right.

We call the example on the left a *blue-sensitive detector*, while that on the right is a *red-sensitive detector*. When light, with whatever spectrum it has, enters the camera it will interact with the detector according to the response curve. We can thus also describe a *detected spectrum*—the spectrum of the light as seen by the detector.

To determine the detected spectrum we perform the same basic operation we used to determine the spectrum of light reflected off an object. The response curve tells, for each wavelength, what percentage of light striking the detector is actually detected. So we must multiply the spectrum of incoming light by the response curve, and we must perform this multiplication separately for each wavelength. The result is another graph—the detected spectrum. We will do this in detail in the next section,

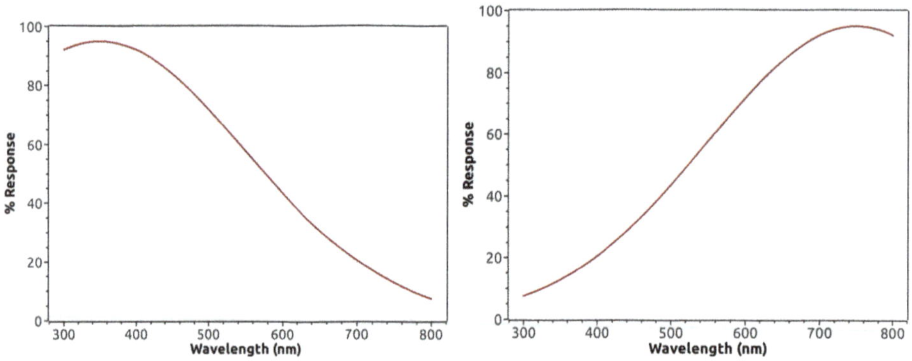

Figure 26.16. Response curves for two different detectors. The detector on the **left** is most sensitive to short-wavelength light, while the detector on the **right** is more sensitive to long-wavelength light. The graphs only show the range of wavelengths between 300–800 nm, but that does not mean the detectors are unresponsive to wavelengths outside this range.

but let us note here that the detected spectrum depends on both the spectrum of the light entering the camera and the response curve. Thus either can greatly affect how the detector 'sees' the world.

26.6 Color and integration

A perfect system for recording a color picture would preserve a full detected spectrum of the light at each point in the picture. But this is not at all practical. Instead one typically compresses the information present in a spectrum into a much simpler form that (hopefully) still captures enough information so as to allow one to reproduce color—at least as perceived by humans.

We discuss more fully in chapters 27 and 28 some of the practical methods for encoding color information. But the basic idea is that the full spectrum of wavelengths is replaced by only three numbers, each of which is a single measure of the overall brightness of the spectrum over a particular range of wavelengths only.

Thus, instead of measuring the brightness of the light at each wavelength between 400–700 nm, we have only three measurements. For example, we could have one number that represents only the total light between 400–500 nm, while another number represents the total light with wavelengths between 500–600 nm and the third between 600–700 nm.

This trio of numbers contains much less information than the original spectrum. This means many different spectra could end up with the same set of three measurements. If done strategically, one can record enough information about the spectrum to at least represent our perception of color with reasonable accuracy. We shall see in chapter 28 that this procedure has much in common with how the human eye records color information.

Just as importantly, any *black-and-white detector* of light does essentially the same thing. The black-and-white detector, at each position in the image, turns the full

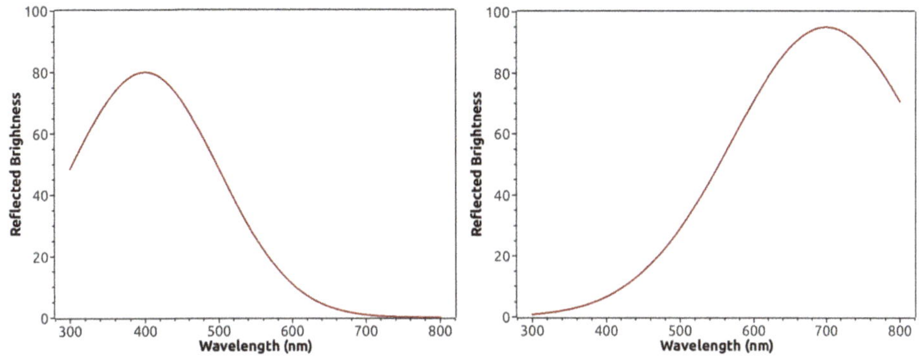

Figure 26.17. Two hypothetical reflected-light spectra entering a camera. When the light of all wavelengths is added up, the light source on the **right** is about 43% brighter than the source on the **left**. Even more importantly, the energy of the two sources is distributed differently over the range of visible wavelengths.

spectrum of light into only one number instead of three. Thus every detector has its own response curve, and the spectrum of light entering the camera interacts with this response curve to produce a detected spectrum. Although the interaction occurs at every wavelength in the spectrum, the detected spectrum itself is not recorded. Most detectors cannot tell, in the end, one detected wavelength from another; they are lumped together into one number if it is a black-and-white detector or three numbers if it is a color detector.

The mathematical process by which a detected spectrum is compressed into one number is called *integration*. It works like this: we multiply, at every wavelength, the reflected spectrum by the response curve to determine the detected spectrum. We then take the detected spectrum and add up all of the area under the curve of the graph. The greater this area, the more total light has been measured by the detector.

Let us do an example, first with the simplest case—a black-and-white detector. Let us take two very different reflected spectra, as seen in figure 26.17. The spectrum on the left clearly would appear more bluish, while that on the right would appear more reddish in color. Notice that, although the two spectra are obviously different in shape, we can also see that the curve on the right 'looks a little bigger' than the curve on the left. What we really mean by this is that when all wavelengths are added together, the spectrum on the right emits a bit more light than the spectrum on the left. This quantity, the total brightness for all wavelengths added together, is represented graphically by the area under the curve of the spectrum. In this particular case, the spectrum on the right has 43% more area under its curve than does the spectrum on the left.

And so how will our detector respond to these two spectra? To determine this, we must consider the response curve of the detector. Let us say that we are using the detector with the blue-sensitive response curve shown on the left in figure 26.16. To determine the spectrum of the detected light, we multiply the response curve by the spectrum of the incoming light. The resulting detected spectra can be seen, for both examples, in figure 26.18.

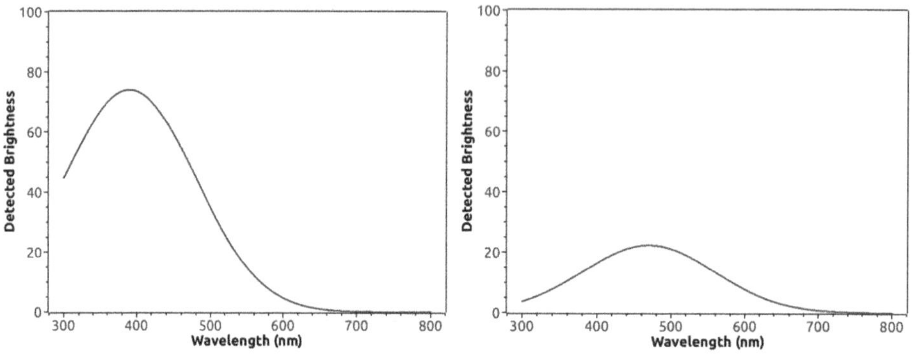

Figure 26.18. Left: The detected spectrum for blue light with a blue-sensitive detector. **Right:** The detected spectrum for red light with the same blue-sensitive detector. When the light of all wavelengths is added up the blue-sensitive detector recorded the blue light as 71% brighter than the red light, even though the red light source was actually the brighter of the two.

But the detector does not actually record this detected spectrum. Instead it simply adds up all of the light, equivalent to the area under the curve in each side of figure 26.18. Clearly, there is more area under the curve on the left, and so this detector would record a larger value for the blue light than for the red light. This detector would 'see' the blue light as brighter than the red light, even though 43% more light from the red source entered the camera.

The process of measuring the total light in the detected-light spectrum by adding up the area under the curve is called integration. In practice the detector does this automatically; in fact, it has no choice in the matter. Although it responds differently to different wavelengths, the typical photographic detector has no way to sort the wavelengths into a spectrum. It simply adds them.

If we know the detected spectrum, as in the left side of figure 26.18 for example, we can use a computer to numerically add up the area under the curve. We also can do a crude approximation graphically by superimposing a grid on the graph, as in figure 26.19, and counting up the squares. Clearly, we would get a different answer if we counted the squares of the same size arranged instead under the graph on the right side of figure 26.18.

Let us now consider the same two spectra of incoming light but detected instead by the red-sensitive detector, the response curve of which is on the right side of figure 26.16. The resulting detected spectra are shown in figure 26.20. Clearly this different detector would 'see' the red light as brighter than the blue light. And so we see that even if the detector ultimately measures only the overall brightness of the light with no color information, color is still crucial.

26.6.1 Color detectors

We can easily extend these ideas to color detectors. A color detector is simply a detector that simultaneously records the picture with three different response curves, thus recording three separate measurements.

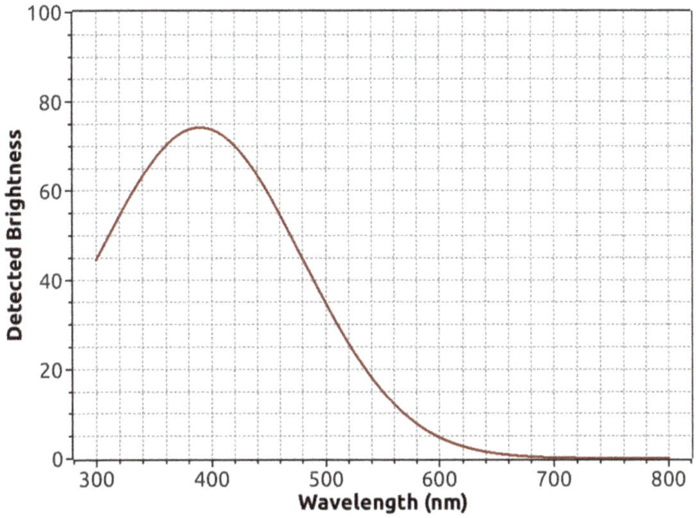

Figure 26.19. Integration adds up the area under the curve of the detected-light spectrum; it turns a full detected-light spectrum into a single number. If one superimposes a grid on the graph, one can then simply count up the squares (using fractions for partial squares) under the curve. It is clear that fewer squares of the same size would fit under the curve on the right side of figure 26.18. For most real light detectors, this integration process is performed automatically as part of the physical process of detecting light.

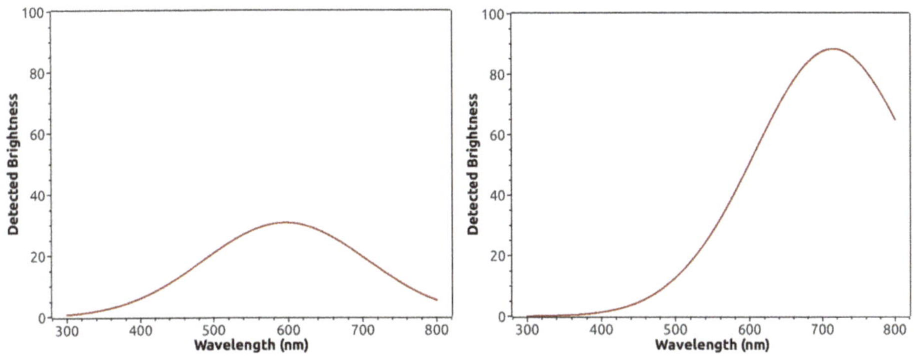

Figure 26.20. Left: The detected-light spectrum for blue light with a red-sensitive detector. **Right:** The detected-light spectrum for red light with the same red-sensitive detector. The detector only records the area under the curve of these spectra. Thus this detector records a larger measurement (nearly four times greater) for the red light than for the blue light.

Our examples in the previous section showed that a red-sensitive detector records red light as brighter than an equal-intensity blue light. We also saw that a blue-sensitive detector does the opposite. And so it is clear that if we measure the light with both detectors, the result tells us something about the overall color of the light, even though we only have two numbers. If the red measurement is much greater than the blue measurement, then we are probably looking at red light. If the opposite is true, well then maybe we are looking at blue light.

If both give the same measurement, maybe it is green light (with a peak wavelength in between red and blue). Or maybe it is white light with all wavelengths equal. In this case, we would not be able to tell whether it is one or the other. Our two detectors alone do allow us to make some educated guesses about the spectrum of the light entering the camera. In particular, we can use it to measure the average *slope* of the spectrum between the two different wavelengths.

We can better sample the true shape of the spectrum by adding a third detector with a response curve in between the other two. This gives us information not only about the slope of the spectrum between the two extremes but also its overall curvature. Most color detectors do in fact use some clever scheme for combining three separate detectors, each with its own response curve, onto the same detector surface. Usually the response curves are somewhat like those shown in figure 26.21 with peaks in the red, green and blue parts of the spectrum. The light interacts with all three simultaneously and produces three separate measurements.

If three is better than two, then why not four? Four simultaneous response curves would, in fact, give more information about the true shape of the spectrum. But for color photography meant to be seen by humans, our eye–brain system of color perception has the final word. We will see in chapter 28 that our eye–brain combination does something very similar. If we choose our system similar to the way the eye–brain does it, then this seemingly-crude scheme can do a pretty good job at reproducing our impressions of color.

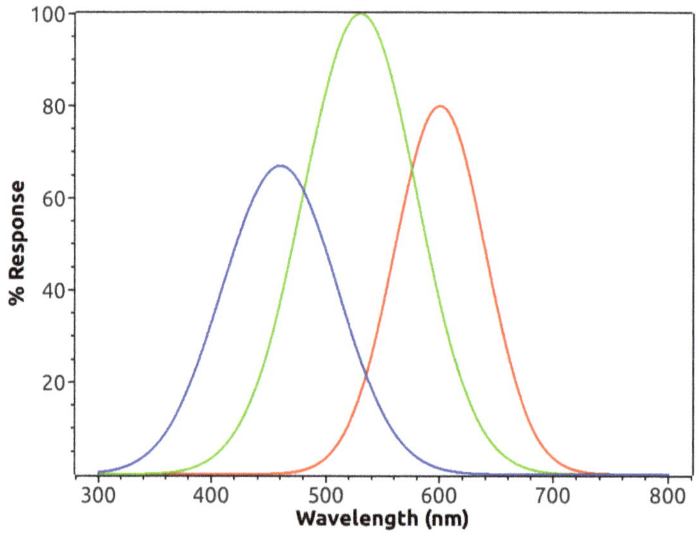

Figure 26.21. Idealized response curves for the separate red-, green- and blue-sensitive detectors in a typical digital camera. In a real digital camera the response curves have somewhat more complex shapes than what is shown here.

26.7 The relation of color to black-and-white photography

A black-and-white detector is simply a single detector that has only one response curve. But it should be clear that all of the issues discussed in the previous sections still apply. In the final picture, what is dark and what is light depends on a combination of the spectrum of the light, the reflection curves of the objects the light shines upon and the response curve of the camera's detector.

This becomes particularly obvious if we take pictures of the same subjects with detectors that have very different response curves. In figure 26.22 I have taken photographs of the same scene with the following detectors.

1. An ordinary color digital camera, designed to show the scene much as it would appear to the eye.
2. An ephemeral-process (EP) photograph that is sensitive from the near-UV through the blue part of the visible spectrum. The response curve extends from roughly 324–450 nm.
3. An ordinary color digital camera but set to a black-and-white mode and using a yellow filter that blocks very short wavelengths. This basically adds together the responses of the separate red, green and blue detectors. The result is a response curve that spans the visible spectrum but excludes deep blue and violet, giving a wavelength response from roughly 450–700 nm that peaks roughly in its middle.
4. The same type of digital camera, but modified so that it also responds to near-infrared wavelengths from 700–1000 nm. A filter was placed over the camera lens to absorb light in the visible part of the spectrum. The overall effect, then, is of a camera that only responds from 800–1000 nm.

The most obvious effect is that the green foliage, made so by the pigment chlorophyll, appears very dark at short wavelengths and very bright at long wavelengths. The daytime blue sky, on the other hand, does the opposite.

Figure 26.22. The scene on the left photographed with three different detectors, sensitive over different ranges of wavelength. **Second from left:** An EP photograph (325–450 nm). **Third from left:** An ordinary digital camera with yellow filter (450–700 nm). **Right:** An infrared-converted digital camera (800–1000 nm). Notice that the green foliage reflects more light at very long wavelengths, but the blue sky is brighter at short wavelengths.

Chlorophyll absorbs light in both the UV-blue part of the spectrum and also the red part of the spectrum. As a consequence, it reflects well at the other wavelengths—the blue-green and infrared parts of the spectrum. Thus we see the foliage appears brightest with the infrared detector and darkest with the UV-blue detector. The blue sky does the opposite because the color comes from Rayleigh scattering, so it is brightest at the shortest wavelengths and darkest at the longest wavelengths.

References

Greenewalt C H, Brandt W and Daniel 1960 Iridescent colors of hummingbird feather *J. Opt. Soc. Am.* **50** 1005–13

Nassau K 1983 *The Physics and Chemistry of Color: The Fifteen Causes of Color* (New York: Wiley)

IOP Publishing

Photography (Second Edition)
Physics and art in focus
John Beaver

Chapter 27

The physiological basis of color

27.1 The wavelength response of the retina

The light-sensitive part of the human eye is called the retina, and it is lined with many thousands of individual microscopic light-sensitive cells. It is known that there are two basic types of these cells, named after their shapes when seen through a microscope. The *rods* are only sensitive to the intensity of light in general and do not seem to take a direct roll in the perception of color. They are, however, necessary for perceiving very dim light, and this is why it is difficult to sense color in extremely dim light.

The *cones*, on the other hand, are connected to color perception. It is not that each cone cell senses color directly. Rather, some of the cones are most sensitive to a range of mostly longer-wavelength light—some to a range of mostly short-wavelength light and some with a wavelength response in between these two. A given part of the retina (and thus a given part of an image) makes use of many of these cells, which are especially close-packed in the central part of the visual field. These cone cells are denoted small (S), medium (M) and large (L) for their wavelength responses. A graph of their approximate response curves can be seen in figure 27.1.

And so, when the eye's lens focuses an image onto the retina, any given spot excites many cone cells, some of which are S, some of which are M and some of which are L. Thinking of the three types of cells as individual detectors with their own response curves, it is not difficult to see how these responses carry information about the color of the light focused on the retina. As an example, the S cone is not affected at all by light of 600 nm wavelength, but the M and especially the L cones are. On the other hand, nearly the opposite is true for light of 450 nm.

One could take any spectrum of light, and use the procedure outlined in chapter 26 to determine the individual responses of the S, M and L cones. Try a different spectrum, and one is likely to get a different answer. And so the idea is that a different combination of S, M and L results in a different color perception.

doi:10.1088/978-0-7503-3703-8ch27

© IOP Publishing Ltd 2022

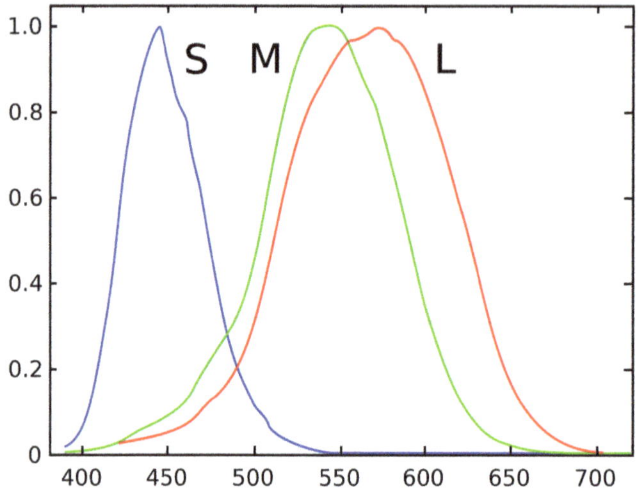

Figure 27.1. The response curves for the three types of cone cells, labeled with S, M and L. Note the similarity to figure 26.1. Copyright © Vanessaezekowitz, Attribution 3.0 Unported (CC BY 3.0), https://commons.wikimedia.org/wiki/File:Cones_SMJ2_E.svg, Cones SMJ2 E.svg.

Thus, the retina has a physiology that seems similar in many ways to the example of the color detector of a digital camera described in section 26.6.1. There too, three different types of detectors, each with its own broad response curve centered at different parts of the visible spectrum, are all exposed to the light simultaneously. It is some combination of those three different images that produces color.

It is tempting to assume that human color perception is essentially the same as the way a digital camera or color film records a color picture. The two certainly do have some things in common, and this is the basis of the three-color model we describe in section 27.2. But the similarity ends there, and the differences are profound; much of color perception is in the brain—not the eye.

This brings up an important point: what is the distinction between the 'physiological' aspect of color and the 'psychological' aspect of color? The answer is that it depends a lot upon whom one asks, since the neuro-physiology of the brain is clearly related to psychology—so much so that there is also the study of 'neuropsychology.' I will not belabor this point, and it is unlikely that I would argue strongly with any reader who takes issue with how I have arranged this material to distinguish between things 'physiological' and 'psychological.'

27.2 The three-color model of color perception

Figure 27.2 shows a photograph of a portion of the liquid-crystal display (LCD) screen of my ordinary computer monitor and a magnified close up of one tiny piece. Looking closely, it is clear that the screen emits only three colors: red, blue and green. The screen has separate light emitters for these three colors, all of which are arranged in a grid-like pattern. Notice, for example, that the yellow star has the red and green spots lit up, while the blue ones are dark, and the white areas have all three

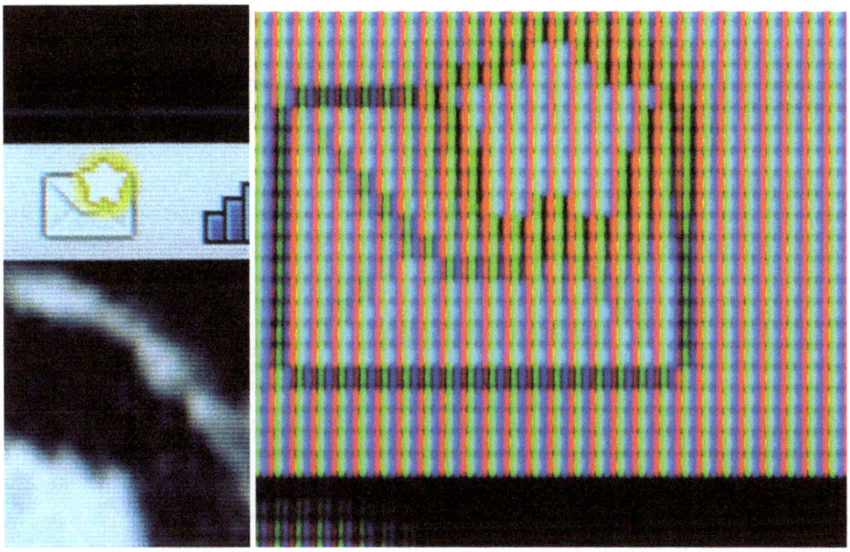

Figure 27.2. A picture of part of an LCD computer screen (**left**) and a close up of the same part of the screen (**right**). All of the colors are synthesized by different combinations of red, green and blue alone. As an example, notice that the yellow star is made of red and green with no blue.

colors lit evenly. At every point on the screen these little red, green and blue dots combine by lighting up in different combinations of brightness. And yet, it produces what I perceive to be a wide (almost full) range of colors, and it is able to simulate subtle color differences.

Combining only red (R), green (G) and blue (B) light in various combinations to make color is called *color synthesis*, and the fact that it works is part of the basis of the *three-color model of color perception*. Another basis is the seemingly-obvious connection to the three-response L, M or S cone cell physiology of the eye just described. In fact it is tempting to move immediately to a simple hypothesis.

> 'The separate R, G and B lights emitted by the computer monitor separately stimulate the L, M and S cones in the retina. The R parts of the monitor stimulate the L cones, while the G parts of the monitor stimulate the M cones and the B parts of the monitor stimulate the S cones. And so RGB light directly simulates LMS cones, and the brain turns those three separate stimuli into color. Thus the brain has some kind of mental "chart" for any given combination of LMS responses; the brain simply looks on the chart and gives us one of the many hundreds of thousands of possible color perceptions.'

Regarding the computer monitor itself, there is such a chart that relates the separate brightnesses of R, G and B to different colors. It is called a *color map*, and we will consider some of its properties in what follows. But these simple assumptions about the brain are very much off-the-mark, as has been demonstrated time and again by neuroscientists, topics we touch upon in chapter 28. RGB ≠ LMS.

Furthermore, there is another problem with the above hypothesis: it is physically impossible to make three different colors of light, each of which separately stimulates only one of the three types of cone cells. This is obvious as soon as one looks carefully at the actual cone response curves in figure 27.1. All three of the responses overlap each other significantly. This is especially true for M and L, which are more similar than they are different. Thus any source of light will trigger responses of more than one type of cone (except for either a single extremely-short or extremely-long wavelength).

The fact remains, however, that RGB color synthesis works pretty well at consistently reproducing a wide range of color perceptions, and so it is the foundation for the technology of color photography, color display and color printing. Exactly how it works in the brain is a different question. We take up that question further in chapter 28, but the short answer is that the answer is unknown—at least in detail. That said, let us proceed with some details of RGB color synthesis, as it has great practical value.

The spectra of the lights typically used for successful RGB color synthesis bear an only superficial resemblance to the LMS cone response curves. The RGB emitted colors usually have a somewhat narrower range of wavelengths emitted than the separate LMS responses. And they are usually spaced more evenly across the visible spectrum with much less overlap between G and R than between M and L.

We can denote the separate brightnesses of R, G and B for a particular color with a trio of numbers. Most commonly, each is assigned a number between zero and 255 with zero meaning no light at all and 255 the maximum amount of light. And so RGB = 0–255–0, for example, would be pure green, while RGB = 50–125–255 would be a little bit of R light mixed with much more green and even more blue. A color map of these possible combinations is a graphical representation of all of the possibilities and what RGB numbers produce them. R, G and B give us three dimensions, and so only some of it can be depicted at once in any two-dimensional graph.

If we encode our RGB colors with 256 possible values separately for R, G and B, this gives us $256^3 = 1.678\,8 \times 10^7$ possibilities. This is called *24-bit color* because a range of 256 possibilities can be represented with eight binary computer bits. There are three of these eight-bit numbers, and so there are 24 binary bits needed to code this many possible colors.

27.3 Additive and subtractive colors

We would like to have a scheme that not only accurately describes color perception in an intuitive yet reproducible way but that also allows us to predict what happens when different colors are combined. But we must make an important distinction between two basic ways of combining colors. If we are talking about combing different sources of light to make an overall color, that is one thing; we are adding light of different wavelengths in order to make a particular spectrum, which gives a

Figure 27.3. The additive primaries—red, green and blue—are in the top row. The additive secondaries—yellow, cyan and magenta—are in the bottom row. Adding any two colors of light in the top row produces the intermediate additive secondary color in the bottom row.

particular color sensation. But if we are talking about light reflected off of surfaces, it is really about what wavelengths of light have been subtracted by absorption. Thus mixing light is *additive*, while mixing pigments (such as paint) is *subtractive*. Not surprisingly, a particular scheme for describing one will give incorrect results for the other.

The computer screen in figure 27.2 is an example of addition of light. Mixing paint, on the other hand, is an example of subtracting light. When one mixes R and G light, one ends up with more light than either the R or G light alone, and yellow results. But if one instead mixes red and green paint, the red paint pigment removes everything except *for* red, while the green removes everything except for green. Thus in this case there is really not much left, and a very dark and muddy color of paint results (the exact color depends on the precise absorption curves of the two pigments).

Since we can synthesize colors by mixing only R, G and B light, we refer to red, blue and green as the *additive primaries*. When two of the additive primaries are combined in equal amounts, we produce cyan, magenta and yellow; these are known as the *additive secondaries* (see figures 27.3 and 27.4).

The rules for subtractive mixing of colors are quite different. Clearly we cannot use red, green and blue as a set of primary colors for mixing paint. As we have seen, mixing red and green paint takes away pretty much all color. But if we think about it a bit, there are three colors that will work as *subtractive primaries*. That is, there is a set of three color pigments that can be mixed in different combinations in order to produce pretty much whatever hue we desire. It is not too difficult to see that the additive secondaries can serve as our set of subtractive primaries. Although red, blue and green will not work as primaries for mixing paint, cyan, magenta and yellow will (see figure 27.5).

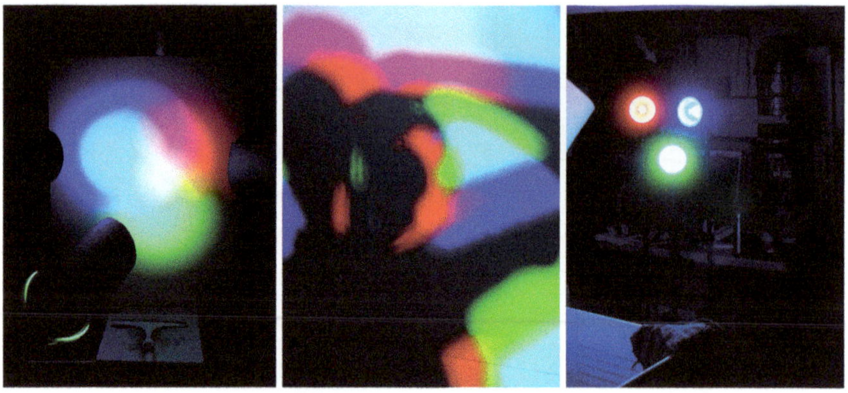

Figure 27.4. Left: Red, green and blue light adds to make white, but in pairs they make the additive secondaries cyan, magenta and yellow. Center: I photographed my shadow on a wall lit by red, green and blue flood lights (seen on the right). Where my shadow blocked all three lights it is dark; where I blocked, none of it is white. The additive primaries can be seen in places where I blocked two of the lights but not the third. The additive secondaries can be seen where I blocked one light, leaving the other two to combine.

Figure 27.5. The subtractive primaries—cyan, magenta and yellow—are in the top row. The subtractive secondaries—blue, red and green—are in the bottom row. Combining any two pigment colors in the top row produces the intermediate subtractive secondary color in the bottom row.

The best way to illustrate this is by example. Let us mix yellow paint and cyan paint. If yellow paint reflects yellow, it means it reflects both R and G light, while absorbing blue. After all, yellow light is simply a combination of R and G light. For light, yellow means 'add red and green,' while for paint, yellow means 'subtract blue.' Cyan light is a combination of B and G light. And so cyan paint must reflect blue and green, while absorbing red. And so cyan light means 'add blue and green,' while cyan pigment means 'subtract red.'

Thus if we mix yellow and cyan paint the yellow will absorb blue, while the cyan absorbs red. Now shine white light on our mix of cyan and yellow paint—white light that combines R, B and G light. The yellow pigment will absorb the blue, while the cyan pigment will absorb the red, leaving only green to be reflected. Thus yellow and cyan pigments mix to make green pigment. It is not difficult to see that magenta and yellow pigment combined will reflect red, since the yellow absorbs blue while the magenta absorbs green, and thus only red remains. Similarly, cyan and magenta combined will absorb red and green and thus reflect blue.

And so we have an interesting reciprocity. The additive primaries are R, B and G light. When combined in pairs, they produce the additive secondaries—cyan, magenta and yellow light. A combination of all three produces white light. The subtractive primaries are cyan, magenta and yellow. When combined in pairs, they produce the subtractive secondaries red, blue and green. Combining all three of the subtractive primaries takes away everything, and produces essentially black. Of course, we often use pigments (in oil paints, for example) that are much more complex than the fundamental subtractive primaries. And there is the added complication of to what degree a given pigment allows light to pass through it and get to other pigments at deeper layers. The mixing of paints is far more complex than what I have laid out here.

27.4 RGB color arithmetic

We can easily explore the relations between additive and subtractive primaries and secondaries with a simple arithmetic. We use a plus sign (+) to represent the addition of light and a minus sign (−) to represent the subtraction of light by a pigment. We then attach these mathematical signs to the letters R, G, B, C, M, Y, W, K[1], to represent red, green, blue, cyan, magenta, yellow, white and black. And so, for example, we represent the addition of R light as $+R$ and the subtraction of G light as $-G$.

We then recognize the following relationships for the addition of light.

$$R + G = Y \tag{27.1}$$

$$G + B = C \tag{27.2}$$

$$R + B = M \tag{27.3}$$

$$R + B + G = W \tag{27.4}$$

This is our statement that the additive primaries—red, green and blue—add together in pairs to make the additive secondaries—cyan, magenta and yellow.

Given these relations, we can see with simple algebra that the following must be true for the subtraction of light.

$$W - R = (R + G + B) - R \tag{27.5}$$

$$W - R = G + B \tag{27.6}$$

[1] It is traditional to use 'K' to represent black in this context presumably because 'B' is already taken by blue.

$$W - R = C \tag{27.7}$$

And so we see that subtractive cyan pigment is given by $C = -R$. When we say a pigment is 'cyan,' we really meant that W light reflecting off of it will appear cyan. This means it is a pigment that subtracts R light and only R light. Non-W light reflecting off cyan pigment may not appear cyan in color. The pigment, however, is still the same; it is a pigment that subtracts R light while reflecting blue and green. We can thus recognize the following relations for the subtractive primary pigments.

$$C = -R \tag{27.8}$$

$$M = -G \tag{27.9}$$

$$Y = -B \tag{27.10}$$

But then what do we mean by a red pigment? It must be the following equations.

$$R = W - G - B \tag{27.11}$$

$$R = W + M + Y \tag{27.12}$$

Thus a red pigment can be made by combining magenta and yellow pigments. The magenta absorbs green, while the yellow absorbs blue, leaving only red to reflect when W light shines on it. It then follows that we must have the following equations.

$$W + M + Y = R + G + B - G - B = R \tag{27.13}$$

$$W + Y + C = R + G + B - B - R = G \tag{27.14}$$

$$W + C + M = R + G + B - R - G = B \tag{27.15}$$

$$W + C + M + Y = R + G + B - R - G - B = 0 = K \tag{27.16}$$

This is our statement that the subtractive primaries—cyan, magenta and yellow—combine in pairs to make our additive primaries—red, green and blue—if we assume the pigments are illuminated by W light. Combining all three subtracts everything, leaving an arithmetic zero or black.

We can also then see the following results for the secondary pigments.

$$R = -G - B \tag{27.17}$$

$$G = -R - B \tag{27.18}$$

$$B = -R - G \tag{27.19}$$

To use these relations, we have to take care whether we are talking about light or pigments. R light is $+R$, while red pigment means $-G - B$. C light is $+ G + B$, while cyan pigment is $-R$. But keeping those distinctions in mind, we can use these relations to answer specific questions about light (even colored light) shining on subtractive pigments. For example, what happens when W light ($R + G + B$) shines on a mixture of the pigments cyan ($-R$) and red ($-G - B$)?

$$(R + G + B) - R + (-G - B) = R + G + B - R - G - B \quad (27.20)$$

$$= 0 = K \quad (27.21)$$

Thus, the answer is no light would be reflected, and it would appear black.

Here is a more complicated example. What happens when C light ($+ G + B$) and B light ($+B$) shine on a mixture of green pigment ($-R - B$) and yellow pigment ($-B$)? The answer is easy. Simply add up all the letters, keeping track of the signs, and see what is left.

$$(+G + B) + B + (-R - B) - B = G + B + B - R - B - B \quad (27.22)$$

$$= G - R \quad (27.23)$$

Clearly G light would be reflected, but what would be the meaning of the remaining '$-R$?' This simply means that although the pigment combination absorbs R light, no R light was shining on it in the first place. And so that fact would make no difference, and the answer is that if one shines a mixture of C and B light on a mixture of green and yellow pigment it would appear green.

We must note another feature of the previous example. Notice that the $+B$'s were canceled by an equal number of $-B$'s. What if instead we had had more $+B$'s than $-B$'s? This is a place where our color arithmetic needs a different rule than ordinary numerical arithmetic. When we say '$-B$' we are implying that the pigment absorbs all blue. Thus it should not matter how much B light is shining on such a pigment; no B light should reflect. So we must modify our color arithmetic so that a single subtractive value of a given color cancels all additive values of that same color. For example, in our color arithmetic, $B + B - B = 0$, not B.

These examples illustrate the limitations of a simple arithmetic for color. Real sources of light and real pigments are not so simple. Table 27.1 gives a summary of the different color arithmetic translations for both light and pigments. To determine what color of light reflects when a source (or sources) of colored light shines on colored pigments, simply follow these steps.

1. Use the table to translate the light source colors and the pigments. For example, translate M light to '$+R + G$' and translate red pigment to '$-G - B$.'
2. Add up the translations but with the rule that a single subtractive value of a given color cancels all additive values of that color. Add the translations as if, for example, $B + B - B = 0$ rather than B. And so if we were to shine B and M light on red pigment, we would add the following values: $B + (R + B) + (-G - B) = B + R + B - G - B = R - G$. Notice that the single $-B$ canceled out both $+B$'s. This is because if the pigment absorbs all blue, then it matters not how much B light shines on it; no blue will reflect.

Table 27.1. The relations between the primary and secondaries for both adding light (left) and for subtracting light with pigments (right). To find the color of the light that is reflected in a given case, simply add the combination of light from the left side to the combination of pigments from the right side and see what is left over.

Name	Light	Pigment
Red	$R \Longleftrightarrow +R$	$R \Longleftrightarrow -G-B$
Green	$G \Longleftrightarrow +G$	$G \Longleftrightarrow -R-B$
Blue	$B \Longleftrightarrow +B$	$B \Longleftrightarrow -R-G$
Cyan	$C \Longleftrightarrow +G+B$	$C \Longleftrightarrow -R$
Magenta	$M \Longleftrightarrow +R+B$	$M \Longleftrightarrow -G$
Yellow	$Y \Longleftrightarrow +R+G$	$Y \Longleftrightarrow -B$
White	$W \Longleftrightarrow +R+G+B$	$W \Longleftrightarrow -0$
Black	$K \Longleftrightarrow +0$	$K \Longleftrightarrow -R-G-B$

3. Interpret the result by using the 'light' side of the table to translate backwards; it is, after all, the reflected light that we are trying to determine. Ignore negative numbers because it simply means that that color of light would have been absorbed, but no light of that color was shining on the pigment. Also interpret, for example, '$2R + B$' as $R + (R + B) = R + M$, or 'magenta with extra red added.' And so, for example, shining a combination of G and C light onto cyan pigment would give the result $G + C - R$. This would translate as reflected light with the color 'cyan with extra green.'

I repeat here the same caveats expressed in the previous section. This three-color arithmetic of light and pigments assumes idealized light sources and idealized pigments. Real light sources and real pigments are far more complex, and accurate results must be analyzed with the more complex procedure laid out in chapter 26. One must compare the precise spectrum of the light source to the reflection curve of the pigment in order to determine the actual spectrum of the reflected light. The three-color model can then be applied to that spectrum of reflected light in order to determine an approximate color perception. So think of this color arithmetic as only a useful starting point for a far more complex reality.

IOP Publishing

Photography (Second Edition)
Physics and art in focus
John Beaver

Chapter 28

The psychological basis of color

28.1 The opponent-process model of color perception

We can easily talk of colors such as 'blue-green' or 'yellow-orange' but what to do with a color name such as 'orange-blue' or 'reddish-turquoise?' In recognition of this basic fact, it has long been proposed that our color perception is based in part upon fundamental opposites—or *opponents*. The most likely candidates are a contrast between something like red and cyan on the one hand and something like yellow and blue on the other. Furthermore, although we have only three different cone types—S (small), M (medium) and L (large)—we seem to have four psychologically fundamental hues: red, green, blue and yellow.

And so it seems that a particular combination of signals from the cone cells is not translated directly into some simple map of color perception stored in the brain. Instead the signals from the three types of cone cells are *preprocessed*. That is, they are combined with each other in various ways to form the foundation of our color perception.

A simplified version is as follows (Conway 2003 and Conway *et al* 2010). The direct responses of the three kinds of cone cells—S, M and L—are preprocessed in the retina by cells called *cone-opponent retinal ganglion cells*. There are four types, and they provide the basic computational comparisons of the direct responses of the cone cells.

1. L-on/M-off
2. L-off/M-on
3. S-on/(L+M)-off
4. (L+M)-on/S-off

We can think of two pairs of ganglion cells. One pair responds to L versus M, while the other pair responds to S versus (L+M). This would seem to have some direct connection to the red-green and blue-yellow opponents that seem to be part of our

doi:10.1088/978-0-7503-3703-8ch28 © IOP Publishing Ltd 2022

color psychology. The first two represent a comparison between the signals of the L and M cones, which have responses most similar to what we call red and green. The last two represent a comparison between the S cone and the combination of L and M together. This would seem to be related to the opponents of blue and yellow, since S has a response close to blue and the sum of the responses of the M and L responses are similar to the sum of red and green, which makes yellow. There is also a similar neural opponent process for black versus white.

Clearly, there is something at the fundamental neurological level that acts in a way analogous to our color psychology that sees blue and yellow on the one hand and red and green on the other, as opposites rather than ends of a continuum, and that also sees yellow as a fundamental hue on a par with red, green and blue. But such a simple and direct relationship has not been born out by the research, and much mystery still surrounds the neural origin of our color psychology (Conway *et al* 2010, Schmidt 2014).

RGB (red, green and blue) synthesis of color is a convenient way to design a computer monitor or a digital color light detector. It makes sense to encode the color information this way in a digital image, but it has little relation to our intuition regarding color. In section 28.4, I describe a common numerical way to represent colors that fits better with our intuitive understanding.

28.2 Yellow without yellow

When synthesizing colors with R, G and B alone, there is mostly a one-to-one correspondence between combinations of RGB and the color perception that results. This is not literally true—if a variation of RGB values is too minute then we cannot distinguish it as a different color, especially for colors of low saturation. But it is not as though a combination such as RGB = 10–125–37 on the one hand and RGB = 230–87–145 on the other look the same as each other.

This general point, however, is not true for the spectrum of the light. For there are many possible spectra that, even when they are qualitatively very different from each other, can nonetheless appear as the same color. The perception of yellow is a good example. We see yellow when the spectrum of light occupies a narrow range of wavelengths between the orange and green parts of the spectrum—what we call the yellow part of the spectrum. But we may also see light as essentially the same color even if it has no light at all of those wavelengths. It only needs to have the right combination of red and green.

Violet is another case. Monochromatic light with a wavelength of 400 nm appears violet. But a color perception very similar can be synthesized with the right combination of red and blue even if neither has any light at wavelengths as short as 400 nm. This is doubly strange when one thinks about it: the R light in this color synthesis is on the opposite end of the spectrum from the short-wavelength B light.

That yellow light can be made from red and green is explained to some degree by simply looking at the SML response curves. Each responds to a broad range of wavelengths, and so there are many different combinations of wavelengths that,

Figure 28.1. Our perceptions of single wavelengths—the colors of the spectrum—form a circle in our color psychology and not a line. Our psychological perception of a color between red and blue—magenta—does not appear in the visible spectrum of single wavelengths.

when added up in relation to the SML response curves, will provide the same combination of SML triggering. That violet can be synthesized with a combination of red and blue, however, is not at all evident from the SML response curves. And it speaks to one of the key features of our color psychology: we perceive color hues not as a spectrum with violet on one end and red on the other but rather as a circle where blue merges to violet, which merges back to red. Figure 28.1 shows an example of such a *color wheel* taken from the color dialog used in the popular image-processing software: GIMP.

The color wheel of hues in figure 28.1 is not the whole story, as evidenced by the triangle inside of it that represents a two-dimensional range of colors—all of which occupy only one place on the color wheel (red). At first glance, the color wheel of hues looks like the visible spectrum of monochromatic colors but with the ends connected to make a circle. There is indeed a resemblance but with an important exception. There is no single wavelength of light that appears magenta, halfway between blue and red on the circle of our color perception.

28.3 Seeing and context

Since we have only three types of cone cells in our retinas, it makes sense that—even if the signals are preprocessed in some as-yet-unknown complex way—any particular combination of stimulation of these three cells always results in the same particular color sensation. That is, we should be able to make some kind of chart such that, for a particular combination of stimulus of the S, M and L cone cells (represented by numbers, for example), the chart would show us the particular color perception that results. Correct?

It turns out that this seemingly-obvious idea is incorrect (Conway *et al* 2010). There is overwhelming evidence, both psychological and neurophysiological, that the color our human eye–brain combination sees very much depends also on the context. There is far more to color perception than simply the stimulation of our cone cells by different wavelengths of light. The exact same spectrum of light can cause very different color perceptions depending upon the context. This basic insight —perception depends not only on the stimulus but also the context—applies to more than just color.

A rather obvious example, clear from simple introspection, is that we unconsciously 'correct' our vision for the amount of light falling on the subject. We see a surface as painted black, rather than white, based on much more than simply how much light reflects off that surface and makes it to the retina of the eye. We also take into account, mostly without conscious thought, how much light is falling on that surface. A person viewing a scene in the world may have many more cues regarding that light than what is available in the direct image of the particular scene being photographed. And so the act of looking at a picture can be very different from the act of looking at that same scene in the world.

A photographer must take this into account because the viewer of the picture has only the picture. That is why we often use incident-light metering in order to adjust for the lighting, whatever it might be in the world, so it appears 'typical' in the photograph. If we want the picture to represent a scene illuminated with either very dim or very bright light, the picture itself must provide the context necessary for our eye–brain combination to make the right assumptions.

As an example, see the two images in figure 28.2; one may recognize them as close-up views of rolls of duct tape. They both look gray, but the right-hand example

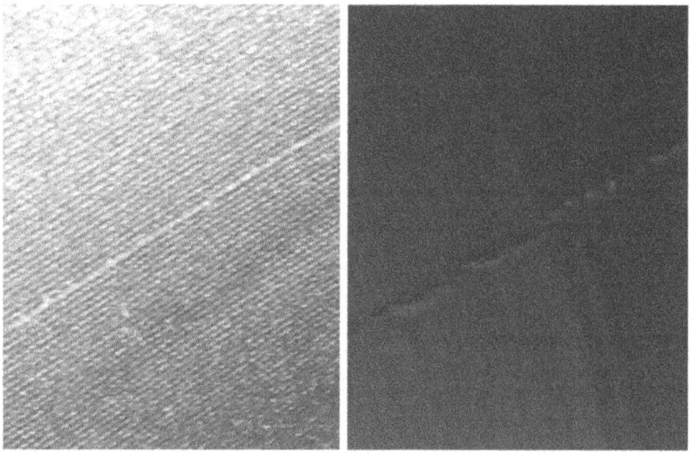

Figure 28.2. Photographs of two rolls of duct tape. Seen out of context they could be pictures of gray duct tape on the **left** and black duct tape on the **right**. The two images are cropped with no adjustments of levels or contrast from the image in figure 28.3.

Figure 28.3. The two pictures of duct tape from figure 28.2 seen in the context of the original scene. Our perceptions of light and dark are dependent upon the context of the light on the subject.

is clearly darker. Maybe it is gray duct tape on the left and black duct tape on the right? Now see, in figure 28.3, the original image from which these details were directly cropped with no additional adjustments of levels. From the context we see a roll of black duct tape on the left, in direct sunlight, with a roll of white duct tape in the shade on the right. We unconsciously use the pattern of light and shadow to make assumptions about the light on the subject and thus to interpret the meaning of the different parts of the image. Seen out of context, we attach very different meanings to those details.

These rather familiar observations about how we perceive levels of brightness also apply to color. The color we perceive depends not only on the spectrum of the light that arrives at the retina, but it also depends critically on unconscious assumptions we make regarding the light that is illuminating the subject. This basic observation about the contextual nature of our color perception reached a worldwide audience in early 2015. A photograph of a dress was posted on social media, and it quickly went viral. Some people saw it as a black-and blue-dress, while others saw the same picture as a white-and-gold dress.

An illustration simulating the same effect can be seen in figure 28.4. On the left is a black-and-blue dress while on the right the dress is yellow and white. When the dress on the left is illuminated by yellowish light, the colors are spectrally identical to those of the dress on the right when it is illuminated with blue light. When the context of the lighting is provided we do not see it that way, and the boxed-off parts of the dresses look for all the world like very different colors (cover all but the joining strips if one does not believe the colors are identical).

We interpret colors in terms of the context of the lighting. If the color of the lighting changes, we still see the colors as the same, even when their literal spectral properties have changed substantially. This basic fact about our color psychology is called *color constancy* (Conway 2003), and it is a very important point for

Figure 28.4. The boxed-off parts of the two dresses have identical colors. In the context of the picture, where it looks as though the dress on the left is illuminated by yellow light while the dress on the right is illuminated by B light, we see those colors as very different. Copyright © Kasuga~jawiki, Editor at Large, Jahobr, Attribution-ShareAlike 3.0 Unported (CC BY-SA 3.0), https://commons.wikimedia.org/w/index.php?curid=59279133, Wikipe-tan wearing The Dress reduced.svg.

photographers. In particular it greatly affects what we perceive as white or gray with no particular hue at all. We consider these issues further in chapter 30; it goes to the heart of the important task of controlling the white balance of a picture.

The famous viral picture of the dress was unusual; the context provided by the picture was ambiguous in just the right way. Some individuals could see it only as the black-and-blue dress on the left of figure 28.4, while others could see it only as the gold-and-white dress on the right.

28.4 'Hue, saturation and value' and 'hue, saturation and lightness'

To understand how we perceive color, it is necessary to understand not only the physical basis of the spectrum of a given source of light, but also how our brains translate that physical sensation into a color perception. The three-color model is probably the simplest way to give a practical description of our color perception, a description that works pretty well most of the time. But it is not a very intuitive way to describe color. It is, for example, nonobvious to most that an RGB value of 255–255–0 (full red, full green, no blue) produces yellow.

There are thus other more-intuitive ways to convey the same basic information as RGB. One of the most popular is called HSV (sometimes called HSB), which stands for 'hue,' 'saturation' and 'value' (or 'brightness' when called HSB). Like RGB,

HSV separates the vast array of possible color perceptions into three categories. The similarity, however, ends there.

- **Hue:** This is the first thing one would typically ask about a 'color.' It answers that most basic question, 'Is it red, green, blue, orange, violet or what?'
- **Saturation:** How 'colorful' is it? Is it a pure color, or is it nearly gray? One can lower the saturation of a paint by mixing gray paint with it.
- **Value (or Brightness):** How dark or light is it?

For a system such as this to be useful, there must be a way to separate the three issues from each other and consider them individually. That is, for a given hue and saturation we should be able to show precisely what we mean by different values and ditto for the other combinations. Furthermore, since practical color systems often use RGB as their foundation, there should be a straightforward way to translate between the two schemes.

It turns out, rather unfortunately, that there is more than one way to set up such a scheme when we look closely at the details. And so there are several systems that differ subtly one from another, the most important alternative to HSV being a system called HSL: 'hue,' 'saturation' and 'lightness.'

Although both HSV and HSL have essentially the same definition of hue, they differ in their meanings of saturation and value (or lightness). And this results in some inevitable trade-offs in 'intuitiveness' between HSV and HSL. In my opinion HSL has a more-intuitive definition of saturation than HSV but a less-intuitive definition of lightness (called value in HSV). It would seem at first glance that the solution would then be to combine both—use HSL's definition of saturation in combination with HSV's definition of value. But that turns out to be logically impossible when one looks carefully into the details; the two competing systems represent a real trade-off and not just a style choice.

In what follows I will use the particular definition of HSV assumed by the popular (and free) image-processing program GIMP, and HSL will not be considered further in *Photography: Physics and Art in Focus*. Once HSV is learned in detail, however, it is straightforward to transfer that knowledge to HSL (or another of the competing color systems) if practical considerations dictate so.

28.5 HSV and RGB

To begin our discussion of HSV consider figure 28.5, made by using the color dialog in GIMP. It shows three rectangles. Each rectangle holds one of these three variables constant, while the other two change from top to bottom on the one hand and left to right on the other.

- **Left:** This rectangle is all of the same *hue*: H = 360°. This is also the same as H = 0° because hue makes a circle that begins and ends at the same point. It is apparent that the starting and ending point for hue is chosen (by convention) to be red.

Figure 28.5. **Left**: Hue = 360° with saturation increasing from bottom to top and value increasing from left to right. **Center:** Saturation = 100% with hue increasing from bottom to top and value increasing from left to right. **Right:** Value = 100% with hue increasing from bottom to top and saturation increasing from left to right.

The directions up and down on this rectangle represent different levels of *saturation*. The top has a saturation of S = 100%, while the bottom has a saturation of 0%. Notice that the bottom is made of simple shades of gray with no color: that is the meaning of 0% saturation.

Left and right on this rectangle represents different *values*. The left side is a value of 0%, while the right side is a value of 100%. Notice that a value of 0% is black, regardless of the saturation.

- **Center:** This rectangle is all of the same saturation: S = 100%.

  The directions up and down on this rectangle represent different hues with 0° at the bottom and 360° at the top. We can see that the hues go around a circle (here stretched out into a line) that begins and ends at the same hue and that goes approximately (but not quite) through the spectral colors of red, orange, yellow, green, blue, indigo and violet in order and then through magenta and back to red.

  The directions left and right on this rectangle represent different values. And so the left side is a value of 0%, while the right side is a value of 100%. Notice that a value of 0% is black regardless of the hue.
- **Right:** This rectangle is all of the same value: V = 100%.

  The directions up and down on this rectangle represent different hues with 0° at the bottom and 360° at the top.

  The directions left and right on this rectangle represent different levels of saturation. The left side is a saturation of 0%, while the right side is a saturation of 100%. Notice that a saturation of 0% is white regardless of the hue but only if (as in this example) V = 100%. A range of different value with S = 0% appears as different shades of gray (see the bottom of the left-most rectangle).

If we look closely at these examples we can see two interesting fact about HSV that may not be obvious at first.

1. If S = 0% then V matters, but H does not matter. To say this a different way, if S = 0% then all hues look the same, but different values produce different shades of gray.
2. If V = 0%, then neither S nor H matters. V = 0% is black whatever the values of S and H.

Now let us relate these values of HSV to RGB. In figure 28.6 I show a GIMP dialog window that allows one to set values for RGB and HSV along with an image of the resulting color (labeled 'Current'). The example on the left shows a selection of pure red represented by RGB = 255–0–0, representing the maximum amount of red but with no green or blue. Note that the values for hue, saturation and value are HSV = 0–100–100. This means there is a hue of 0°, and saturation and value are both set to 100%.

Hue is represented in degrees because it refers to its position on a circular color wheel. The numbers go around the circle with increasing angle from pure red (0°), through pure green (120°), through pure blue (240°) and back to pure red (360°). Exactly at the halfway point between each, at 60°, 180° and 300°, represents equal mixes of these primaries. These are the additive secondary colors yellow (red and green with no blue), cyan (blue and green with no red) and magenta (red and blue with no green). The example on the right in figure 28.6 shows hue set for magenta, at H = 300°, with both saturation and value set to 100%. This corresponds to RGB = 255–0–255 or full red with full blue but with no green.

With both saturation and value at 100%, different hues represent different combinations of at most two of R, G and B. As we start at 0° and work around the color wheel R goes down while G goes up (with B = zero) until there is no more R, and G is at its maximum. As we continue B increases while G decreases (and R = zero), until there is only B and no G. Finally as we work around back to the beginning, B decreases while R increases (with G = zero) until we are back to only R.

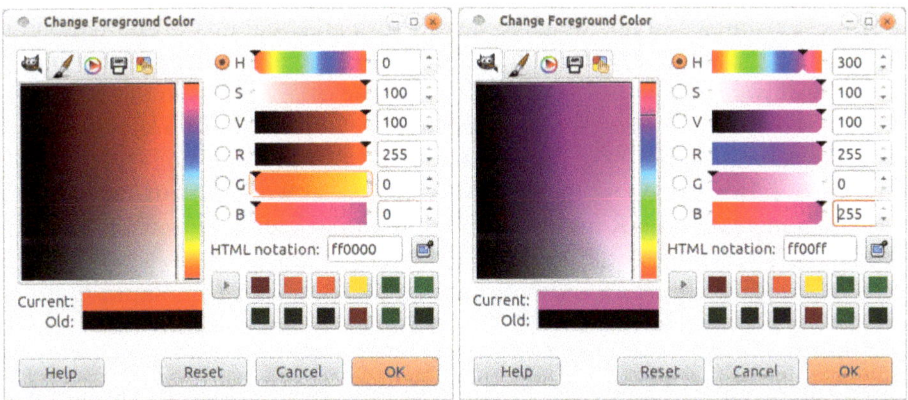

Figure 28.6. Dialog boxes from GIMP showing the effects on RGB of changing hue. The example on the left is pure red (hue = 0°), while that on the right is magenta (hue = 300°). Both have value and saturation of 100%. The chosen color is in the box labeled 'Current.'

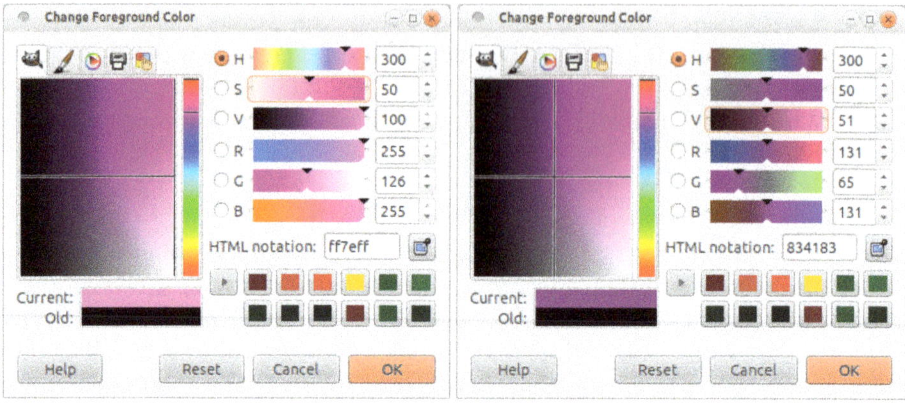

Figure 28.7. Dialog boxes from GIMP that show the effects on pure magenta (RGB = 255–0–255) when saturation is lowered (**left**) and then value is lowered (**right**) from 100%. The chosen color is in the box labeled 'Current.'

Let us see what happens when we adjust saturation and value. The left side of figure 28.7 shows the effect of lowering the saturation of pure magenta from 100% to 50%. Notice that the red and blue stayed the same, both at a full 255, but the green increased from zero. And so the effect of adding green, which was not originally present, leaves the hue unchanged but less intense. It is clear that if we keep lowering the saturation, the green will increase until it is equal to red and blue. Thus eventually we make, for 0% saturation, RGB = 255–255–255, which is white.

On the right side of figure 28.7 we lower the value from the example on the left. So the hue is still 300° (magenta), and the saturation is still 50%, but we have lowered the value from 100% to 51%. In this case all of the RGB values decrease in proportion to each other. So a lower value keeps the same percentages of red, blue and green, but all of the numbers are smaller. Clearly a value of 0%, whatever the hue and saturation, will give black.

For these examples so far, we have started with an equal mix of full red and full blue (255–0–255). What if we had an uneven mix of red and blue? The left image in figure 28.8 shows RGB = 125–0–199. This is a hue of 278°; it is to the blue side of magenta, as it has more blue than red. Because one of the colors (G in this example) is zero, the saturation is 100%. But neither R nor B is at a full brightness (255), so this example has a value less than 100%. It is not shown here, but increasing the value would increase both R and B so that they would keep their same proportions; at a value of 100% blue would reach its maximum of 255.

The right side of figure 28.8 shows the effect of decreasing the saturation, from 100% for the left-hand example to 50% on the right. Before lowering the saturation, we had RGB = 125–0–199 with B as brightest. With the saturation lowered to 50%, RGB changes from 125–0–199 to 162–99–199. Notice that by lowering the saturation both R and G increased in brightness—in such a way that B is still brighter than R, but R is kept greater than the G that has been added. The effect is as if we mixed gray with the color on the left side of figure 28.8. The resulting color

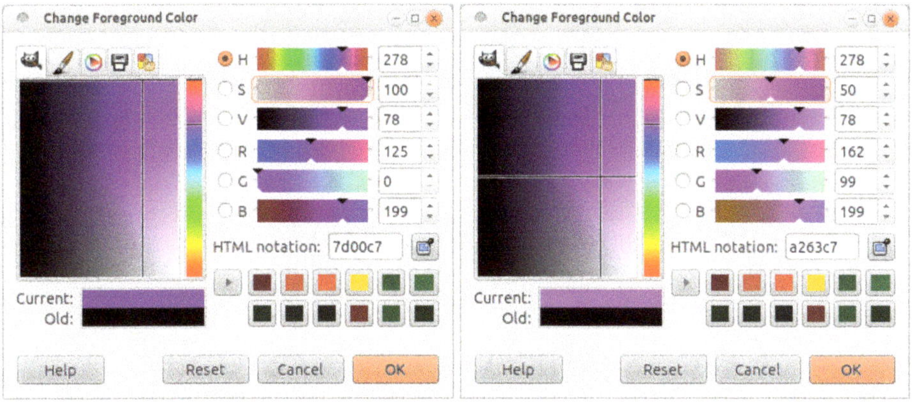

Figure 28.8. Dialog boxes from GIMP showing (**left**) the effect of lowering the value on an uneven mix of red and blue and (**right**) the effect of then lowering the saturation. The chosen color is in the box labeled 'Current.'

looks less intense (and also brighter overall) because the RGB values are more nearly equal to each other. If we decrease the saturation even more, red and green would get brighter still in such a proportion that at 100% saturation we would have RGB = 199–199–199. Thus all trace of the original hue would disappear, and we would be left with a shade of gray.

The best way to understand the relation between HSV and RGB is to try these exercises oneself. The color dialog in GIMP is an accessible option, but there are many other similar interactive tools—both online and as part of image-processing programs.

References

Conway B R, Chatterjee S, Field G D, Horwitz G D, Johnson E N, Koida K and Mancuso K 2010 Advances in color science: from retina to behavior *J. Neurosci.* **30** 14955–63

Conway B R 2003 Colour vision: a clue to hue in v2 *Curr. Biol.* **13** R308–10

Schmidt B P, Neitz M and Neitz J 2014 Neurobiological hypothesis of color appearance and hue perception *J. Opt. Soc. Am.* **31** A195–207

IOP Publishing

Photography (Second Edition)
Physics and art in focus
John Beaver

Chapter 29

Color synthesis in photography

29.1 Color detectors and color pictures

A painter carefully choses colored pigments, mixing them to reproduce the colors they need, and then meticulously applies them to the canvas by hand. It is both a limitation and an advantage that photography is different. Part of the point of a photograph is that the image is captured automatically and mechanistically without selective hand work on every detail. A photograph is not a painting. Nevertheless, there is a rich history of photographers applying paints directly by hand to what would otherwise be a black-and-white photographic print. To use photography to record a color image directly from life is another kettle of fish.

There are really two questions. On the one hand, how does the light detector record information about the different colors of the subject? On the other hand, how does one make a physical photographic image that depicts these same colors when viewed by humans? The two questions are related, especially regarding historical processes for which the recording of the image and the production of the picture are intertwined by a chain of photochemical processes.

A *digital* color image is, in many ways, in an entirely different category altogether. Whatever physical processes were used to record it, the digital image exists as a coded set of numbers that represent values and colors in some chosen *color space*. In the RGB system this is three numbers, at each point in the image, to describe the values of red, green, and blue. However those numbers were acquired the color image then 'exists' independent of whatever means by which a physical, viewable-by-humans picture is made from it.

For a purely photochemical color process, on the other hand, the color information is embodied in some physical photographic medium used as a color light detector. There must then be a carefully-engineered set of physical processes that link every step from the initial capture of the image to the final print. It is a challenge to keep the balance of hues in the final print similar to that of the original subject while also preserving a proper range of values and contrast.

 © IOP Publishing Ltd 2022

The first successful full-color photographs were made surprisingly long ago, and attempts go back to the earliest days of photography. John Herschel's 1840s experiments with *anthotype* were in part motivated by the desire for a color photographic process (Ware 2019, p 19). Although her motivations were apparently more about physics than photography, Mary Somerville's related experiments a few years later were very much about the connections between color and the detection of light (Koo 2013).

We have seen in chapter 27 that it is possible to *synthesize* a wide range of color perception from combinations of as few as three primaries. Three-color synthesis was first demonstrated photographically in an 1861 lecture by one of its early theorists, the physicist James Clerk Maxwell. In a collaboration with the photographer Thomas Sutton, three black-and-white positive photographic transparencies were recorded separately through red, green and blue filters (Longair 2008). These transparencies were then focused onto a screen simultaneously with three projectors, each fitted with the corresponding red, green or blue filter. This is clearly a direct example of *additive color synthesis*.

Maxwell's technique accomplished three-color synthesis in a rather brute-force way; the three colors, from separate exposures, were literally added together by projection. The inconveniences of this method are obvious, so most direct color photographic processes accomplish three-color synthesis by *multiplexing*: the three primary-color images are somehow recorded simultaneously onto the same physical light-sensitive surface.

For a color picture with only a limited palette of primaries to be embodied within a single physical two-dimensional (2D) photographic medium, some of the spatial information must be sacrificed in order to record the color information. Ideally, we want the color information to be recorded within a small enough region of space such that, without using a magnifying glass or microscope, we cannot see the individual primary colors. Instead they all blur together into a more-complex color, synthesized from the primaries. And so the photographic medium must be able to contain finer detail than the actual image. This same point holds, at the level of the physical light detector, for recording of a digital color image as well; the light detector itself must be capable of recording even finer detail than the digital image that is made by it.

In the sections that follow we consider some of the different strategies for synthesizing color on a single 2D surface. Maxwell's and other nonmultiplexing methods are, in a sense, more than 2D; each part of the 2D image has at least three separate exposures from which to synthesize the color at that location. We also consider examples that employ in a clever ways the *thickness* of the photographic emulsion as a third dimension to record color information. Most of these methods use multiple layers to simultaneously record color information. Even so, most such strategies still require the sacrifice of image detail for color information.

There is one important exception: the Lippman process remains perhaps the only direct and nonsynthetic full-color photographic process ever invented. We describe it in section 29.8.1. We also consider, in sections 29.7 and 29.8.2, a couple of ways in which at least a limited palette of colors may arise in a photograph almost by accident.

29.2 Panchromatic light detectors: the secret of color photography

One of the key stumbling blocks to the development of color photography was that early light-sensitive materials only responded to short-wavelength light. Color photography is more than just the existence of colors in the photographic image; they must also correspond to the colors in the world. A red rose appears red because it reflects mostly light of long wavelength, and we perceive those wavelengths as red. If the light detector does not respond at all to long wavelengths, then a photographic image made from it contains no direct information about them. How then is red synthesized in an image that contains no red? It would seem that in order to record and then synthesize the full spectrum of colors in the world, the light detector must be sensitive to at least this full range of colors.

If we consider the perhaps-odd way in which human color perception works, we can see that it is not so simple. RGB color synthesis, for example, can be accomplished with only three separate and limited wavelength ranges even if there are wavelength gaps between the primary colors. And the perceptions of both the shortest and longest extremes of the visible spectrum can be synthesized with some realism even if those wavelengths are not present in the red, green, and blue primaries. As another example, there is a range of wavelengths that we perceive as yellow. But we can synthesize very much the same perception by adding red and green primaries even if they contain no light in that wavelength range.

These caveats aside, a realistic synthesis of colors does require a light detector that spans the bulk of the range of human color perception. Such detectors are called *panchromatic*, and they did not come into their own until the end of the nineteenth century. It was a considerable technical challenge to make detectors with a wavelength response longer than the violet, blue, and blue-green portions of the spectrum. The history of practical color photography is thus intimately connected to finding clever ways to extend the sensitivity of light detectors to longer wavelengths.

29.3 Color photography with sequential exposures

A color image can be produced from three separate black-and-white images, each sensitive over red, green and blue wavelengths respectively. One method is to simply take each of the three exposures in succession. With colored filters, the same detector can be used for all three, as long as it has a broad enough response to wavelength so as to be sensitive across at least most of the visible spectrum.

This technique was first suggested by James Clerk Maxwell, the nineteenth-century physicist who showed that light is an electromagnetic wave. It was first demonstrated at one of his lectures. The three separate black-and-white transparencies were projected back through their respective color filters with all three projections superimposed on top of each other. Thus the light was literally added together by the three projectors.

Figure 29.1. Three photographs were taken with black-and-white instant film, each through a different-colored filter: red (**upper right**), green (**lower left**) and blue (**lower right**), and then they were combined digitally to make a color image (**upper left**).

The results were poor because the photographic emulsions of the day were sensitive mostly only to ultraviolet (UV), violet, blue, and blue-green light. It has been speculated that the only reason it worked even as well as it did is because of a coincidence: many red pigments also reflect (or transmit) UV as well as red.

The filters—typically red, green and blue—selectively absorb light so as to allow only the desired range of wavelengths through. They can be mounted on a wheel that is rotated between exposures to bring the different filters into play. See figure 29.1 for an example of the basic idea using black-and-white instant film. I took three separate pictures, one after the other, each with a different-colored filter placed over the lens of the camera. The three black-and-white negatives were scanned, and the resulting three digital images combined to form a single color image.

Alternatively, different detectors could be used instead if they have different sensitivities to wavelength. This process has long been used in astronomical research. A good example is the Palomar Observatory Sky Survey. Large-format glass photographic plates were exposed in a specially designed wide-field telescope. Each image was made twice, once with a photographic emulsion, sensitive in the red part of the spectrum, and then again with a different emulsion that was sensitive in the blue part of the spectrum.

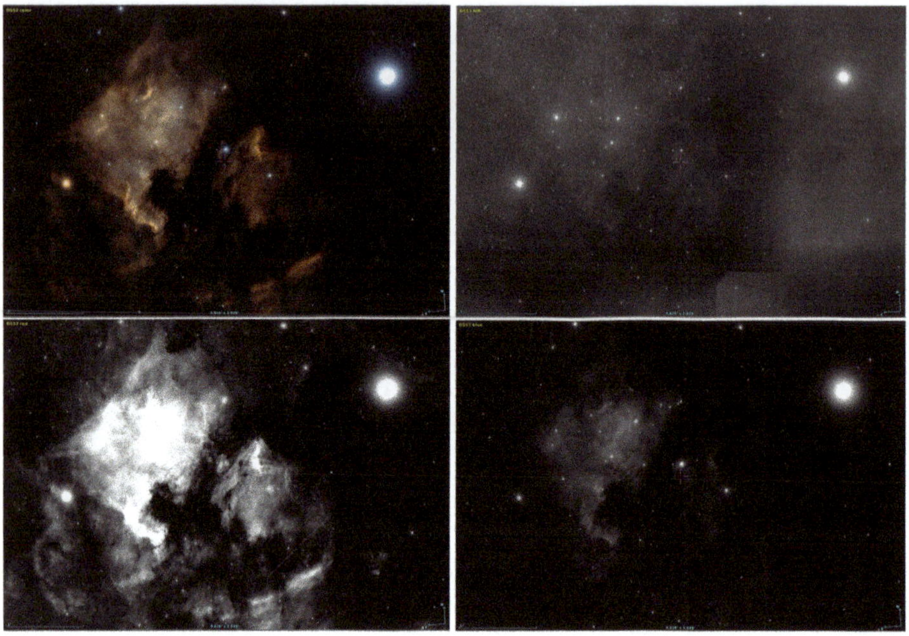

Figure 29.2. Images of the North America Nebula made with detectors sensitive to four different ranges of wavelength, and a false-color image (**upper left**) made from them. (Images made with the Aladin Sky Atlas, DSS-II: [Bonnarel *et al* 2000, p 40 and Lasker *et al* 1996].)

A sort of duo-color picture could be made from these two black-and-white images, but that was not really the point. Rather, the point was the additional science that can be done with the color information. These photographic plates have been scanned and digitized, and newer ones have been made with better emulsions (and coverage of the southern skies) with a third added in the near-infrared part of the spectrum. This Digitized Sky Survey II (DSS-II) is accessible to anyone for research. Figure 29.2 shows a part of the sky in the Sword of Orion as imaged by DSS-II. The upper left 'color' image was produced digitally from the three individual exposures: near-infrared at upper right, red at lower left and blue at lower right.

The image at upper left is really *false color*; it does not represent the colors the human eye would perceive when viewing the subject[1]. It is a somewhat arbitrary mapping of near-infrared (not even visible to the eye!), red and blue exposures into an RGB additive image. Figure 29.3 shows the wavelength responses of the exposures used as red, green and blue in the false-color image.

The 'blue' exposure covered from the deepest violet at 400 nm all the way to about 570 nm in the yellow-green part of the spectrum. The 'green' exposure covered

[1] This is apart from the fact that the dim glow of the nebulosity in the image is too faint to adequately trigger human color vision when viewed directly.

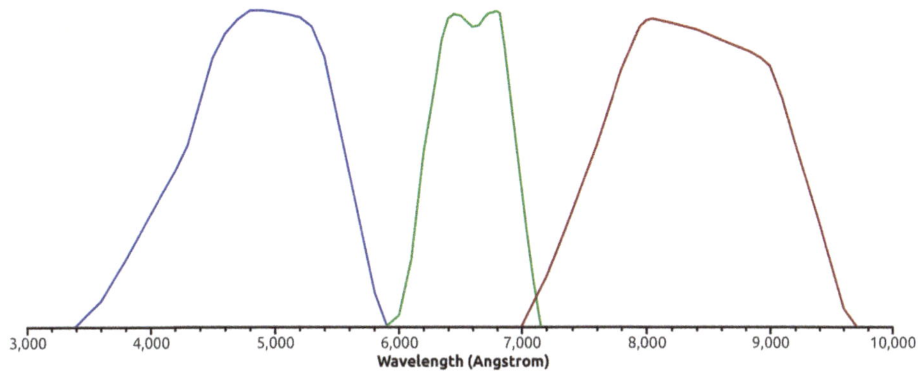

Figure 29.3. Wavelength responses of the three exposures made for figure 29.2 (data from Ažusienis and Straižys 1969).

from 600–700 nm—wavelengths that appear orange and red to the eye. The 'red' exposure was completely in the invisible-to-the-eye, infrared part of the spectrum.

29.4 Continuous tone versus raster color

The disadvantage of sequential exposures to make a color image is obvious: the subject may move between exposures. Figure 29.1 demonstrates this; even though the subject is a landscape and the camera was mounted on a tripod the pattern of sunlight and shadows shifted between exposures, producing an odd color pattern as the three primaries combined in complex and unpredictable ways. This can be an interesting and useful effect in the proper context, but it does limit the usefulness of the technique.

Most modern color photography is multiplexed—the separate primary-color images are recorded simultaneously onto the same physical light detector. But multiplexing is also used for most methods of displaying a color image to be viewed directly. This means that the three primary colors must be displayed in a tiny pattern that is (hopefully) too small to see with the naked eye at a comfortable viewing distance. The combination of primary colors may thus blur together to synthesize more-complex colors.

There is an important distinction regarding how color is synthesized from a limited number of primaries in a physical photographic medium.

- **Continuous tone:** Each spot of a particular primary color may take on a full range of values from light to dark.
- **Raster:** Each colored spot is from only a small, limited set of colors, strategically spaced in a pattern on a white (or transparent) background.

In a continuous-tone color image, complex hues are synthesized by combinations of primaries, each of which at a given point in the image may take on different values. In theory, a microscopic grouping of only three tiny colored spots—one for each

Figure 29.4. Left: A simple four-by-four space raster pattern using only three colors, spaced on a white background. **Right:** When this pattern is repeated at a microscopic scale, a complex color is synthesized.

primary in a three-color system—is needed to produce the full range of hue, value and saturation (HVS) at any point in the image.

The most obvious example of a continuous-tone color image is that produced by an ordinary computer monitor. Millions of different combinations of HSV can be synthesized by a simple repeating pattern of tiny triplets of red, green, and blue spots—each of which can vary from black to full brightness. See figure 27.2 in section 27.2 for an example.

An important example of a raster color image is an inkjet print; tiny droplets of a limited set of colored inks are placed on a background of white paper. In such an image a particular combination of HVS is controlled by the density and spacing of these tiny colored spots—too small to see individually—on the otherwise white (or transparent) medium. To synthesize colors over a large color space, each point in the image must be subdivided to contain enough space for a large number of color spots.

And so each point in the image contains—hopefully at a level too small to see with the unaided eye—a pattern of many (possibly overlapping) colored dots with a controlled amount of white space scattered between them. Figure 29.4 shows a simple example. A set of only three possible colors are scattered on a grid-like *raster pattern* of only four-by-four spaces. When the raster pattern on the left in figure 29.4 is repeated many times on a microscopic scale, the color on the right is synthesized.

Clearly the ability to synthesize a large number of colors in this way depends upon the number of different colors in the limited set. But it also depends upon the possible number of spots in a given tiny region of this raster pattern. Another important consideration is whether or not the colored spots are transparent and may overlap each other, allowing for *subtractive synthesis*. We consider this possibility more fully in section 29.5.

Without the extra dimension of continuous tones for each primary color, a raster color needs more physical space to synthesize the same color space. For this reason raster colors usually employ more than just three primaries. A high-end modern inkjet printer, for example, may use as many as a dozen different-colored inks to synthesize its full color space.

29.5 Additive versus subtractive synthesis

In chapter 27 we described how colors can be synthesized either by selectively adding red, green and blue light to darkness, or by selectively subtracting light from white and employing the subtractive primaries cyan, magenta and yellow. It is thus tempting to conclude that additive synthesis is appropriate whenever we are combining separate sources of colored light, while subtractive synthesis applies to pigments—which selectively absorb (and thus subtract) light passing through or reflecting off them. There is some truth to this, but it mostly misses the point regarding the distinction between additive and subtractive synthesis.

Consider the following comparison. There are two basic strategies for making a flat-screen computer monitor or television. One method is to employ an array of tiny red, green and blue light-emitting diodes (LEDs). These lights are too small to see individually at any given point in the image; instead they blur together to the human eye, and we see the combination of those emitted colors. Clearly, this is an example of additive color synthesis.

But there is a different approach to accomplish the same end. With a liquid-crystal display (LCD) monitor, a flat LED surface provides white-light illumination from behind a layer of tiny, separate LCD colored filters, which selectively absorb wavelengths from the white light passing through to the viewer. At the microscopic level, light is subtracted from white to produce color. Despite their origin in light-absorbing pigments, the viewer of the screen sees the same thing: an array of tiny colored lights that blur together to synthesize colors. And so both LED and LCD monitors use the additive primaries of red, green and blue.

Let us also reconsider the example of the raster color depicted in figure 29.4. The example uses tiny squares of cyan, magenta and yellow, in combination with white, to synthesize complex color. Is this additive or subtractive synthesis? It turns out that we can describe it as either one or the other.

We can, for example, see this as a tiny surface that would reflect a certain amount of white light, but the pigmented squares selectively subtract light. We can represent a white square as an addition of one unit each of red, green and blue. And so, without the pigments, we would have 16 (R+G+B) because there are 16 squares total. The pigmented squares subtract light from this total. Cyan subtracts red, yellow subtracts blue and magenta subtracts green. And so, given the number of squares of each pigment in the raster, we have the following equations.

$$16(R + G + B) - 2R - 4G - 6B \tag{29.1}$$

$$=14R + 12G + 10B \tag{29.2}$$

We can also think of it a different way; we see the sum of whatever colors of light are reflected off each square. White light reflecting off each white square reflects a single unit each of red, green, and blue, while light reflecting off a cyan square reflects a single unit each of blue and green light. Similarly, each yellow square reflects single units of red and green, while each magenta square reflects a unit each of red and blue. And so adding all of these together, we have the following equations.

$$4(R + G + B) + 2(B + G) + 4(R + B) + 6(R + G) \tag{29.3}$$

$$=14R + 12G + 10B \tag{29.4}$$

We see that a simple raster pattern, even if made from pigments applied to white paper, can be analyzed as either subtractive or additive synthesis. If this is so, then which set of primaries should we use if we are to synthesize colors in the manner of figure 29.4? The answer is that it depends upon what aspect of the color one wants to compromise. For it is impossible to synthesize the full range of colors with this type of a raster.

If we choose the subtractive primaries cyan, magenta and yellow, then it is impossible to make red, blue or green at full saturation. Red is the subtractive secondary, made from combining yellow and magenta in equal portions. To demonstrate, let us consider a ten-by-ten 100-square raster, to put the final answer in terms of percentages. If we fill half of the squares with magenta (-G) and half with yellow (-B), we then have, when we shine white light upon it, the following.

$$100(R + G + B) - 50G - 50B = 100R + 50G + 50B \tag{29.5}$$

$$= 50R + 50(R + G + B) \tag{29.6}$$

$$= 50R + 50W \tag{29.7}$$

This is red but of only 50% saturation—what we usually call pink.

We could, instead, try using red, green and blue as the primaries—but that is out of the frying pan and into the fire. If, for example, we want to make yellow, the best we can do is to fill half the squares with red (-G-B) and half with green (-R-B).

$$100(R + G + B) - 50G - 50B - 50R - 50B = 50(R + G) \tag{29.8}$$

$$= 50Y \tag{29.9}$$

This is yellow with 100% saturation but only 50% value, so it is a dark and dingy yellow. We see then that the color synthesis of figure 29.4 is neither clearly additive nor subtractive in principle, and this is not an issue of whether the colors arise from white light shining on subtractive pigments or from a set of individual colored lights adding together. Figure 29.5 shows the famous painting *A Sunday on La Grand Jatte* by Georges Seurat. The detail on the right is from a relatively tiny part of what is a huge painting. It is, in effect, a raster pattern of a limited palette of paint spots; when viewed from a distance, more-complex colors are synthesized.

A pointillist painting with its raster of opaque pigments effectively synthesizes only a limited color space, and it is a type of synthesis that is neither unambiguously additive nor subtractive. So how then, for example, does an inkjet printer produce a

Figure 29.5. Left: *A Sunday on La Grande Jatte*, Georges Seurat 1884–86. **Right:** a tiny detail from the painting. Georges Seurat, Art Institute of Chicago, CCO Public Domain Designation.

full color space with only dots of colored ink on white paper? In this important case (and some others we will review), the color synthesis should most definitely be considered subtractive, and it is indeed the subtractive primaries that are most likely to be used.

In contrast to figures 29.4 and 29.5, an inkjet printer uses pigmented spots placed on the white paper that are transparent, and they are allowed to overlap each other in layers. It is the overlapping of transparent pigments on the same substrate that allows for true subtractive synthesis.

Thus, when a magenta spot is placed on top of a yellow spot, the magenta absorbs green through which only red and blue passes. The yellow then absorbs blue, so only the red makes it through the pigments. If the spots of color are not allowed to overlap, or if they are opaque (so only the upper-most layer is visible), then we can produce only the limited subtractive (or additive) synthesis of figure 29.4. Figure 29.6 shows a photogram of feathers made with three overlapping layers of cyan, magenta and yellow pigments. The pigments are at least somewhat transparent, so the image shows a degree of subtractive synthesis of blue, red and green. Had the pigments been more transparent, a fuller color synthesis would have resulted.

Modern color photographic emulsions, whether used to make a print or a color transparency, use multiple transparent layers to employ subtractive synthesis. Unlike the raster color of an inkjet print, a color film emulsion is continuous tone. The left-hand image in figure 29.7 shows a magnified view of a color film transparency. The center image shows part of the same transparency at much higher magnification, and only the film grain is apparent—tiny overlapping clouds of cyan, magenta and yellow dye along with some black spots of dark silver.

The right-hand image in figure 29.7 is a view through a microscope (at much lower magnification than the film examples) of a digital inkjet print. It too consists of spots of cyan, magenta, yellow and black. But it is a computer-controlled regular pattern rather than a random one. The dots appear in regular lines because of the physical mechanism of the printer itself, but the position of the dots within a line appears more random. And crucially they are allowed to overlap each other to achieve true subtractive synthesis.

Figure 29.6. Three separate exposures in transparent layers of cyan, magenta and yellow synthesize the subtractive secondaries of red, green and blue.

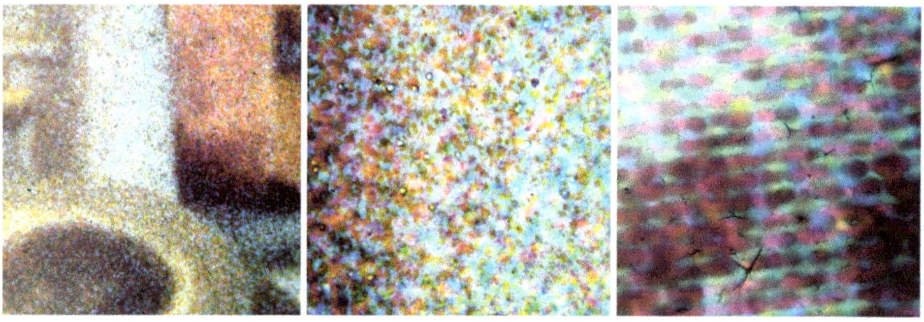

Figure 29.7. Left: Detail of a color film transparency, as seen through a microscope. The film grain is just barely visible. **Center:** At much higher magnification only the film grain is visible—small clouds of cyan, magenta and yellow dyes. **Right:** A magnified view of an inkjet print, made of a regular pattern of cyan, magenta, yellow and black dots.

29.6 Multiplexed color detectors

We here consider three methods—two contemporary and one historical—for recording three-color images simultaneously onto one flat light detector. In this section we work backwards in time; this is a case where complex technology has made things conceptually simpler.

29.6.1 Digital detectors and the Bayer mask

A digital camera uses a *charged-coupled device (CCD)* or a *complementary metal-oxide semiconductor (*CMOS) *array* to detect light. We describe such detectors in detail in chapter 38, but such a detector is essentially a rectangular grid of rows and columns of separate light detectors. An electronic signal is measured at each light-sensitive site, and a virtual image is reconstructed.

These digital detectors cannot, by themselves, distinguish between different colors of light. Thus they are, at their most basic, black-and-white detectors. Although the individual detectors cannot tell the difference between different wavelengths of light, they are sensitive to a broad range of wavelengths.

A CCD or CMOS array multiplexes three separate color images by placing colored filters on the detector itself over the individual pixels. The trick is that some pixels are covered by microscopic red filters, while others by blue and others by green. If we arrange these *microfilters* in a regular pattern, groups of pixels can be combined to form a color image.

In practice, this is accomplished with what is called a *Bayer filter* or *Bayer mask.* The typical arrangement can be seen in figure 29.8. What is called a single 'pixel' in a color image from a CCD includes information from the measurements of a square of four neighboring pixels on a black-and-white CCD covered with a Bayer mask. Two of those pixels are located behind green parts of the Bayer mask, while the other two pixels are behind, respectively, red and blue sections of the mask.

One could simply combine these four measurements to produce one color RGB pixel in the final image. But this would mean, for example, that a 3000 × 2000 pixel CCD array would produce a picture with only 1500 × 1000 pixels. Instead, sophisticated computer *demosaicing* routines are employed to better take advantage of the information available in the image. This is usually performed, by the microcomputer circuitry built into the camera, immediately after the picture is

Figure 29.8. An illustration of the layout of a Bayer mask array of colored microfilters covering the light-sensitive pixels (represented in gray) of a CCD array detector. Copyright © Cburnett – Own work, Attribution-ShareAlike 3.0 Unported (CC BY-SA 3.0), https://commons.wikimedia.org/w/index.php?curid=1496858, Bayer pattern on sensor.svg.

taken. But many cameras also have the ability to store the image data in an unprocessed, so-called '*RAW*' format, allowing the user to employ their own chosen demosaicing routine once the image has been downloaded to a computer.

The one-red, one-blue, and two-green system of the Bayer mask directly translates to an RGB encoded image, but it is not the only possibility. Some digital camera manufactures use more-complex systems, using four different-colored filters instead of only three, to better take advantage of the wavelength response of the detector. These four colors are then processed internally to make a three-color RGB image.

29.6.2 Multilayer color emulsions

Modern photochemical color film and print materials use silver-gelatin emulsions to detect light. There are three different emulsions, each with its distinct range of wavelengths to which it is sensitive. These three emulsions are arranged in layers: a blue-sensitive layer on top and green- and red-sensitive layers underneath. Since the green- and red-sensitive layers are also inherently sensitive to blue light as well, a yellow filter layer is typically placed under the top-most blue-sensitive layer to block blue light from reaching the underlying layers (Stroebel and Compton 2000, pp 319ff).

The emulsions are semitransparent; wavelengths that are not absorbed are not detected. To some extent, the opposite is true as well. Thus wavelengths of light that are not detected by upper layers can travel to lower layers that are sensitive to those wavelengths.

Upon development of the layers of emulsion the dark silver that is produced is replaced by colored dyes, and the dark silver is bleached away. The emulsions and corresponding dyes are cleverly engineered such that each dye couples to only one of the layers. To produce colors through subtractive synthesis, cyan (-R) dye is coupled to the red-sensitive layer, yellow (-B) to the blue-sensitive layer and magenta (-G) to the green-sensitive layer.

This seems as though it would produce colors that are complementary to how they appear in the world. And it would, but for an important fact: this is a negative process. The development of the color medium directly produces not only hues that are the complements of those in the subject but also reversed values. Thus there must be a second step to make a positive image; a negative of a negative is a positive. This second step reverses the values to a positive, and it reverses the hues as well[2].

A bright red subject, for example, produces a dark cyan $(-R)$ place on the negative, while a dark green subject produces a bright magenta $(-G)$ area. A second exposure of the dark cyan $(B + G)$ area of the negative exposes both the green and

[2] This second step may be a direct photograph of the negative, or it may be accomplished chemically through *reversal processing* (see section 34.2).

blue layers and thus makes a bright area that has both magenta ($-G$) and yellow ($-B$) dyes. When illuminated with white light, these colors combine by subtractive synthesis to make bright red—the color of the subject: $R + G + B - G - B = R$. Similarly, the exposure of the bright magenta area of the negative yields dark cyan ($-R$) dye in the red-sensitive layer of the positive and dark yellow ($-B$) dye in the blue-sensitive layer, and these combine with white light to make dark green—just as in the subject: ($R + G + B - R - B = G$).

29.6.3 Autochrome Lumière process

It is interesting to note that the modern digital camera produces a color image in a way that is strikingly similar to one of the first commercially successful methods of color photography. The *autochrome process*, invented over 100 years ago by Louis and Auguste Lumière, employed a single black-and-white emulsion on a glass plate, undercoated with a thin layer of tiny transparent globules of potato starch. The glass photographic plate was then exposed in a camera through the backside of the glass. Thus the image focused by the lens first passed through the layer of starch before reaching the light-sensitive emulsion.

The trick was that, before being coated onto the glass plate, the starch was dyed in three separate batches of red, green and blue and then mixed together. And so the starch globules acted like tiny colored microfilters during the exposure, very much like the Bayer mask of a digital camera. Only instead of a regular array, it was an essentially random arrangement of red, green and blue starch filters.

After exposure, the photographic emulsion was then reversal processed so as to form a positive transparency. When lit from behind, the black-and-white image is viewed through the same microfilter starch globules that were used to record the image, so a color image appears. Surviving autochrome images are today highly prized because they are one-of-a kind direct positives, they have a remarkably subtle range of tones and hues and they are some of the only directly-recorded color images from the early 1900s. Figure 29.9 shows a photomicrograph of a tiny detail in an autochrome image. An autochrome is a color transparency, so we are seeing light shining through from below. Notice it is made entirely of red, green and blue spots that are not allowed to overlap each other. Clearly this is an example of additive color synthesis, and this is why the additive primaries of red, green and blue are used.

One of the challenges faced by the Lumiére brothers was to find an emulsion that was sensitive not only to blue and green light, but also to red light. It does no good to filter red light if the detector is not sensitive to it. In recent years a few individual photographers have taken up the considerable challenge of duplicating the complex, commercial autochrome process at home. But they do have the advantage that panchromatic emulsions are now commonplace and easily available.

29.7 Accidental duo-color in black-and-white emulsions

The early experiments in anthotype by John Herschel were motivated in part by the search for what might be called the Holy Grail of the early days of photography.

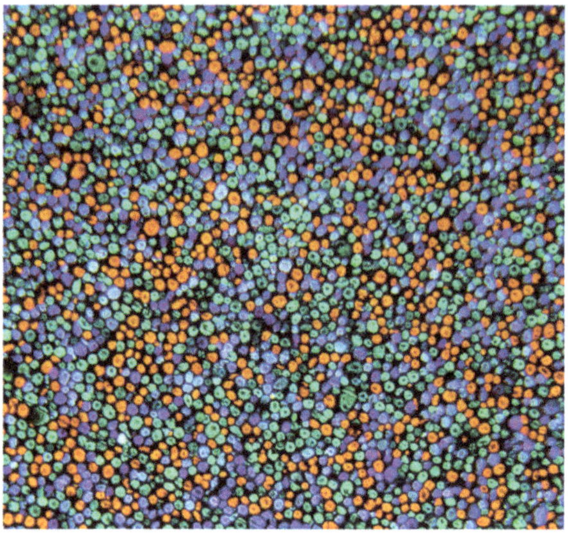

Figure 29.9. A photomicrograph of the tiny different-colored starch grains in an autochrome plate. Copyright © Janke, CC0 1.0 Universal (CC0 1.0) Public Domain Dedication, https://commons.wikimedia.org/wiki/File:Microphoto_of_Autochrome_plate.jpg, Microphoto of Autochrome plate.jpg.

Can we find some material that, when exposed to light, changes so as to directly mimic both the brightness and the color of the light that fell upon it? Frustration set in early, as it turned out to be far easier to find materials that respond as a negative, turning darker with greater exposure to light. But also, the end product of most photochemical materials is a monochromatic picture.

I here describe a method that stumbles over something of a Grail that is perhaps at least one-quarter full and allows one to directly make a color picture from colored light—sort of. Ordinary silver-gelatin emulsion on paper, designed to make developed positive black-and-white prints in the darkroom from black-and-white transparent negatives, will print out if exposed to enough light. But the printed-out silver is very different from the developed-out silver and often shows colors rather than the familiar black or dark brown.

These printing-out colors arise not from selective absorption of pigments, but rather from wavelength-dependent scattering; we describe the physical mechanism of *Mie scattering* in section 3.5.2. The colors that arise are often subtle, of relatively low saturation and they are connected to the physical size and uniformity of the light-sensitive crystals that make up the photographic emulsion.

There have been hundreds of different enlarging papers produced over the years, and most employ at least two different light-sensitive salts. These salts have somewhat different ranges of wavelength response; for at least a few papers, they seem to produce nearly complementary scattering colors.

Such a paper can thus directly turn different colors depending upon the color of light that falls upon it. There is a catch; enlarging papers are typically of extremely-low sensitivity when printing out. This is by design, as we shall see in chapter 35.

Figure 29.10. Color filter tests of EP exposures with 12 different enlarging papers. For each test, the paper was exposed through a yellow filter (Hoya K2) on the **left** and a UV filter (Schott UG11) on the **right**. Most of the papers respond similarly to the light through the two filters, but the first and last papers in each group show marked differences in hue when exposed through the two filters.

By using the technique of ephemeral-process (EP) photography described in this book, however, the sensitivity can be greatly increased by use of an accelerator; it can be made fast enough even for exposures in a camera.

For figure 29.10,[3] physics student Gabriella Cutié Rodriguez exposed several different types of accelerated enlarging paper to the same source of light but through two different-colored filters.

The circles on the left side of each exposure were made through a yellow ‘K2’ filter, of the sort used by photographers to darken the blue sky in a black-and-white picture. It blocks only wavelengths shorter than about 475 nm, in the blue-green part of the spectrum, while passing longer wavelengths. The circles on the right side of each exposure were through a Schott UG11 UV filter. It blocks all light that has a wavelength longer than about 375 nm, in the near-UV part of the spectrum. The wavelength responses of the two filters can be seen in figure 29.11, superimposed upon the approximate overall wavelength response of the enlarging paper itself.

It is clear from figure 29.10 that the scattering colors produced are either blue or orange, depending on the variety of paper, and this has been born out with similar experiments with many other paper varieties as well. This should be unsurprising; uniform crystal size can result in colors through Mie scattering, as discussed in section 3.5.1. We further consider this process regarding so-called *chromoskedasic* pictures in section 35.7.

[3] First presented at the symposium *Pixels, Pallettes & Perception*, held 2–3 March 2018 in Madison, Wisconsin.

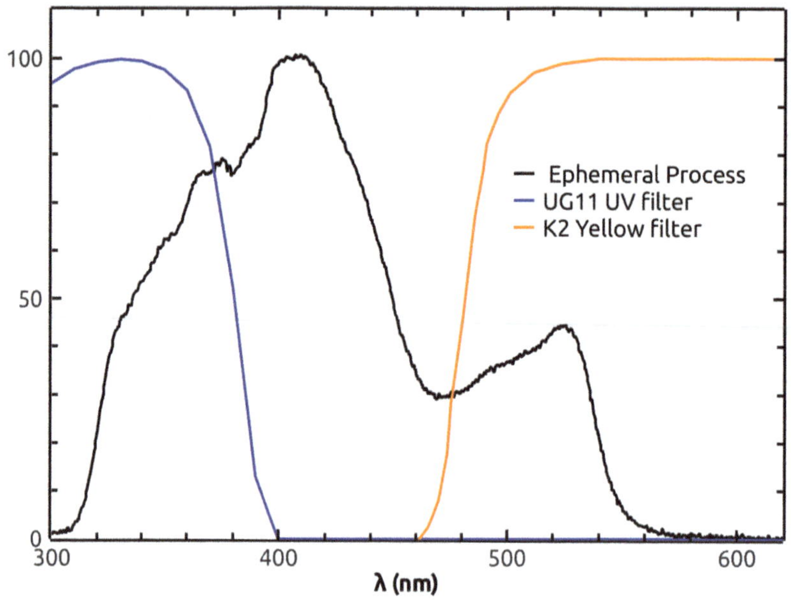

Figure 29.11. The detected spectrum of a type of EP paper as illuminated by sunlight compared to the transmission curves of the two filters used for figure 29.10 (filter transmission data from SCHOTT and HOYA filters).

For most of the varieties of paper shown in the test, the same scattering color resulted whatever the color of the light. But there are two tests (the first and the last) in figure 29.10 that are different: the exposure through the yellow filter came out a pinkish-brown color, while that through the UV filter camera out a bluish color. This is especially evident if the paper is scanned and the contrast and color saturation is digitally increased (figure 29.12).

Presumably, the two different wavelength responses in these special papers arise from the somewhat different response curves of silver chloride on the one hand and silver bromide on the other. Both are sensitive to the shortest wavelengths, but the response of silver bromide extends to longer wavelengths than silver chloride (Stroebel and Compton 2000, pp 194, 195). A reasonable guess is that, for certain papers, the size of the silver halide crystals is uniform, producing strong Mie scattering colors. But the silver chloride crystals with their shorter-wavelength response are smaller and thus produce mostly blues, while the silver bromide crystals are larger and so produce warmer tones.

This behavior means that it is possible to use such a light-sensitive material to make a color picture, even though it was intended by the manufacturer to be used only for black-and-white prints. Figure 29.13 shows an example alongside an image of the same scene made with an ordinary digital camera. The color of the EP photograph looks surprisingly realistic, but a close inspection shows that it not full color. Rather it is a *duotone* image with only blue, orange and various combinations

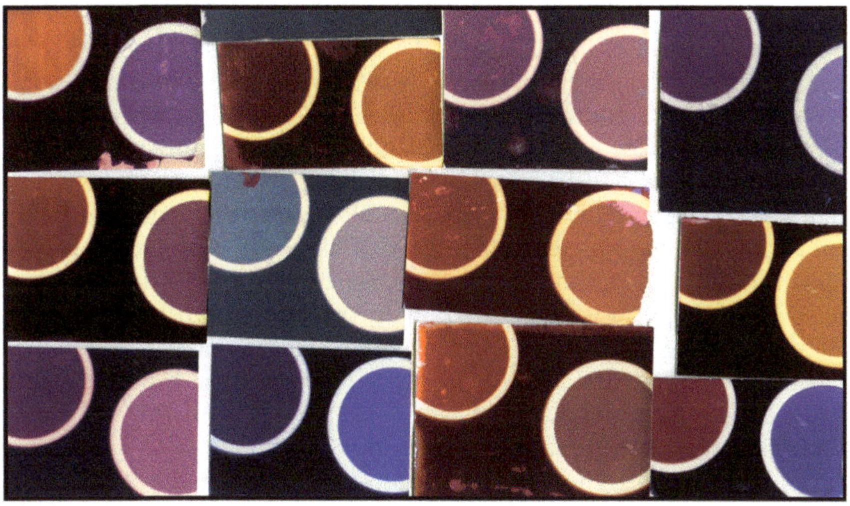

Figure 29.12. The same as figure 29.10 but with the contrast stretched and the saturation increased.

Figure 29.13. The image on the **left** is a full-color picture taken with an ordinary digital camera that separately records the image in three different colors (red, green and blue). The picture on the **right** is the same scene imaged with EP photography. It is only a duotone image, consisting of various mixes of two tones only (blue and orange). But for this particular scene, the image looks (to humans anyway) almost as natural as if it were full color.

in between. The duotone colors just happen to match up well to most of the actual colors in the scene—in particular the blue sky and the blond-colored wood fence in the background.

Figure 29.11 shows that the EP paper is sensitive only to wavelengths shorter than about 540 nm, which is in the green part of the spectrum. It is not sensitive at all to the yellow, orange and red parts of the spectrum. So how is it that reds and yellows are visible in these pictures, and they correspond to reds and yellows of objects in the world?

The answer seems to be partly luck. Recall from chapter 28 that just because something looks yellow, it does not mean that it reflects light in the yellow part of the spectrum. The yellow filter used for the exposures in figure 29.10 is a case in point. The transmission curve in figure 29.11 shows the filter transmits very little light at wavelengths shorter than a cutoff at about 480 nm—in the blue-green part of the visible spectrum. But it transmits light at essentially all visible wavelengths longer than that—green, yellow, orange and red light. The EP paper, however, knows nothing of those longer wavelengths; it only responds to that little bit between about 480 nm–540 nm. Apparently this particular EP paper turns reddish-orange when exposed to only that long-wavelength tail end of the paper's response curve. Shorter wavelengths, on the other hand, make the paper turn more blue in color.

And so it would appear to be luck that many things in the world have spectra that look something like these two filter responses—either blocking long or short wavelengths and so either making the paper turn blue, reddish-orange or some mix of the two. More examples of images made from scans of EP negatives made directly in cameras can be found in chapter 31.

Note that it is the longer wavelengths—associated with the tail end of the spectra of yellow, orange and red objects in the world—that make the paper turn reddish-orange. And it is the short wavelengths—associated with blue objects in the world—that make the paper turn bluish in color. And so the duotone hues correspond directly to those in the world. But regarding the values, the paper responds as a negative; brighter light makes the paper darker. To make a natural-looking image from the scanned negative, it must be digitally inverted to a positive. Ordinarily, this process would also reverse the hues to their complementary colors: blue would turn to yellow and vice versa. But since the hues are already in the correct direction, inverting the image from a negative to a positive makes the colors backward. For these pictures I have inverted the values only, leaving the hues as they were.

And so we have found our Holy Grail—a way to make color pictures directly from a single light-sensitive material and its direct response to light without the use of filters, coupled dyes or added pigments. It is still, however, a negative process and not—as is the daguerreotype for example—straight from camera to wall. And it is only a two-color rather than a full, three-color process. Thus our Grail is only one-quarter full.

I would be remiss if I failed to point out that I am far from the first to notice the diverse printing-out colors of silver-gelatin enlarging papers. It has long been familiar to the practitioners of several alternative processes (described in chapter 35): lumen, chemigram, chemilumen and chromoskedasic. And although I have not found it described in print, there is clear evidence, in the work of at least some practitioners of solargraphy, that they too are aware of the fact that something of a real color picture can be made in this way with the choice of the right paper.

29.8 Nonsynthetic color in photography

All of the color photographic processes described so far are *synthetic colors*; a complex color arises out of the combination of a two or three basic primaries. But there are at least two examples of color in photography that arise more directly.

29.8.1 Lippmann process color photography

Gabriel Lippmann won the 1908 Nobel Prize in Physics for his invention of one of the earliest color photographic processes. In some ways, Lippmann photography is the only true color direct photography ever invented. Other methods—such as autochrome, the chromogenic process and the Bayer mask of a contemporary digital camera—rely on a three-color process. Red, green and blue colored filters sample three parts of the spectrum of the light. Colors are synthesized, taking advantage of the three-color nature of human color perception. Lippmann photography was never made commercially viable, and it was overshadowed by the autochrome process patented at about the same time. But it has been revived by at least a few contemporary photographers, notably Hans Bjelkhagen; his overview of the process provides much of the information below (Bjelkhagen 2017).

Lippmann photography uses no colored dyes or filters. The photograph is made with a black-and-white emulsion and consists of an extremely fine-scale *wave interference pattern* of only black and white. In this way it is like a hologram. The hologram, however, uses wave interference to record distance information, while the Lippmann photography uses interference to record wavelength information. When a Lippmann plate is viewed correctly the same full spectrum of wavelengths is visible as the original scene.

To accomplish this, Lippmann used a silver emulsion of extremely fine grain so the microscopic interference pattern could be recorded in detail. It was also necessary that the emulsion be panchromatic—sensitive to a broad range of wavelengths. One could not record the color red, for example, if the emulsion is not even sensitive to red wavelengths.

The wave interference is produced in a clever manner. The emulsion is coated onto a glass plate to act as a stable and transparent base, a feature common for the black-and-white photography of the day. But the Lippmann plate is positioned in the camera such that the emulsions faces away from the camera lens. The emulsion is then backed by a mirror, so light passes through the microscopically-thin emulsion twice. The reflective backing of the emulsion must be in perfect contact, and Lippman performed this feat by putting the glass plate in a special holder that allowed for a layer of shiny liquid mercury to contact the emulsion. The mirror-like backing of mercury allows the incoming and outgoing waves to interfere with each other and produce either a dark or light band depending upon whether the two waves interfere constructively or destructively.

The key is that the condition for constructive or destructive interference depends not only on distance within the emulsion but also on the wavelength of the light. Thus, color information is recorded as well. When the developed plate is viewed with the proper lighting the same interference conditions are produced, and one sees the same wavelengths that caused the interference during the exposure.

Although both holography and Lippmann photography produce images in a black-and-white emulsion with wave interference, there is a fundamental difference. The interference pattern of alternating dark and light bands is on the 2D surface of the hologram emulsion. In a Lippmann plate, however, the interference pattern is

Figure 29.14. The intense blue of the morpho butterfly is a structural color. Copyright © Didier Descouens – Own work, Attribution-ShareAlike 4.0 International (CC BY-SA 4.0), https://commons.wikimedia.org/w/index.php?curid=15656403, Morpho didius Male Dos MHNT.jpg

within the thin layer of the emulsion itself. At each point in the Lippmann plate image, dark and light bands of constructive and destructive interference alternate with each other in the tiny space from the front of the emulsion to its back. Where red light exposed the plate, these bands are farther apart, and they are closer together where blue light exposed the plate. When light reflects off the developed plate, the interference bands produce these same wavelengths in the reflected light.

And so the color is produced not by selective absorption and reflection by dyes or filters. Rather the color in the Lippmann plate arises from the wavelength-dependent constructive interference of a microscope regular pattern of physical obstructions. These are called *structural colors*, and it is how many blues and greens are formed in the animal kingdom—the intense, iridescent blue of the morpho butterfly being one of the most famous examples (see figure 29.14).

The interference that occurs in the making of a Lippmann plate is similar to what is known as *Bragg interference*. Crystalline materials consist of regularly-spaced arrangements of atoms and when illuminated by light of a wavelength that is roughly similar to this spacing, an interference pattern appears. One can analyze this observed interference pattern in order to determine the atomic structure of the crystal. The relation between the spacing of the alternating layers that reflect light (d) and the wavelength of the light (λ) is given for Bragg interference by the following equation (ignoring angle factors).

$$d = \frac{\lambda}{2} \tag{29.10}$$

For interference in a Lippmann plate, the same relationship holds but for one difference: the wavelength of the light must be divided by the index of refraction (n) of the emulsion layer. As discussed in chapter 3, light slows down when it enters a transparent material, and n is the factor by which it does so. A consequence of this is that the wavelength of the light is decreased in the material by that same factor (as compared to air). And so for Lippmann photography, we have the following relationship.

$$d = \frac{\lambda}{2n} \tag{29.11}$$

Thus we see that longer wavelengths produce a more widely-spaced interference pattern. If we choose the middle of the visible spectrum $\lambda = 550$ nm and an index of refraction of 1.4 for gelatin (choosing something between that of water and glass), equation 29.11 gives $d \approxeq 200$ nm. Thus the photographic emulsion must have a grain fine enough to be able to record details down to significantly less than this size. That is very small indeed, and producing such an emulsion is one of the technical challenges of Lippmann photography.

The spacing between atoms in a crystal is much smaller than this, closer to only 0.1 nm. And so we can solve equation 29.10 to determine what wavelength would be needed to observe Bragg interference for a crystal. The answer is clearly about 0.2 nm, and this is in the x-ray part of the electromagnetic spectrum. And so *x-ray diffraction* is one of the principal tools by which scientists have probed the structure of crystals.

29.8.2 Color in daguerreotype

We consider the daguerreotype process in more detail in chapter 35. One of the oldest photographic processes, daguerreotype produces an image on a polished silver plate by selective scattering of light. It is notoriously difficult to faithfully reproduce an image of a daguerreotype because its appearance depends so much on the angle and quality of light illuminating its surface.

The scattering of light off the surface of a daguerreotype may produce subtle structural colors, perhaps through a form of Mie scattering. On occasion these colors seem to correspond somewhat to those in the subject. Some examples by contemporary daguerreotypists can be found in Rexer (2002), especially the example by Jerry Spagnoli on page 46 and by Robert Shlaer on page 48. The blue sky appears bluish in contrast to the nearly complementary hues of the foreground.

It seems, however, that this effect is a coincidence; gold toning has selectively turned overexposed highlights to a bluish tone. And it is the sky that is the most overexposed part of the images because it is a highly blue-sensitive process. The subtle colors of daguerreotypes was appreciated early on and reportedly exploited in 1850 by Levi L. Hill. A relatively-unknown photographer at the time, Hill published a pamphlet claiming to have perfected a full-color daguerreotype process from life with no colors applied by hand. The description of his highly-complex (and rather dangerous) process was vague, and he never produced convincing examples. Recent attempts to reproduce Hill's process have shown no convincing evidence of colors from life. This is unsurprising given that the daguerreotype process is sensitive to only blue and violet light (Mcardle 2016).

References

Ažusienis A and Straižys V 1969 Improved determination of the response curves and parameters of the UBV system. Summary of results. *Sov. Astron.* **13** 316

Bjelkhagen H I 2017 Lippmann colour photography http://www.alternativephotography.com/lippmann-colour-photography/

Bonnarel F, Fernique P, Bienaymé O, Egret D, Genova F, Louys M, Ochsenbein F, Wenger M and Bartlett J G 2000 The ALADIN interactive sky atlas. A reference tool for identification of astronomical sources *Astron. Astrophys. Suppl. Serv.* **143** 33–40

Koo K 2013 Mary Somerville and Margaret Huggins: a collaborative voice at the embryonic stages of nineteenth-century astronomical specializations *PhD Thesis* York University Toronto, Ontario

Lasker B M, Doggett J, McLean B, Sturch C, Djorgovski S, de Carvalho R R and Reid I N 1996 The Palomar–ST ScI Digitized Sky Survey (POSS–II): preliminary data availability *Astron. Soc. Pac. Conf. Ser.* 101*Astronomical Data Analysis Software and Systems* J Barnes 88

Longair M S 2008 Maxwell and the science of colour *Phil. Trans. R. Soc.* **366** 1685–96

Mcardle J 2016 Colourreotype? https://onthisdateinphotography.com/2016/10/22/october-22-colourreotype/

Rexer L 2002 Maxwell and the science of colour *Photography's Antiquarian Avant-Garde: The New Wave in Old Processes* (New York: Harry N. Abrams)

Stroebel L and Compton J 2000 *Basic Photographic Materials and Processes* 2nd edn (Waltham, MA: Focal Press)

Ware M 2019 Argyronomicon: Silver Photographs on Paper: Chemical History of Their Invention, Deterioration, and Conservation https://www.mikeware.co.uk/downloads/Argyronomicon.pdf

IOP Publishing

Photography (Second Edition)
Physics and art in focus
John Beaver

Chapter 30

Filters

A filter is a piece of glass that fits over the front of the camera lens and selectively absorbs some wavelengths of light while allowing others to pass. Thus we can use a filter to alter the spectrum of light before it reaches the detector. There is a different but related meaning of the word 'filter' in photography—the use of digital processing to alter, according to a prearranged scheme, a digital photograph. In some cases such a *digital filter* might do very much the same thing a glass filter would—only it does it after the picture has already been taken. But most digital filters are quite unlike anything one can do with a piece of glass placed over the camera lens.

The most common types of glass filters work by selective absorption. Since the filter absorbs some wavelengths of light better than others, it also transmits some wavelengths better than others. Whatever is not absorbed is transmitted. If the filter absorbs 90% of the light at a wavelength of 550 nm, then only 10% of the light at that wavelength enters the camera. Thus we can describe the properties of a filter with a *transmission curve*, a graph of the percentage of light transmitted versus wavelength. There are other types of filters for which the light not transmitted is reflected (rather than absorbed) either through thin metal coatings or thin-film interference. But the use of such filters is less common in photography.

As is the case for a reflection curve, the filter cannot create light that is not already there; it can only subtract light and not add to it. Furthermore, it can only do this separately for each wavelength. One cannot, for example, use a filter to take away some of the blue light and make it red. One can, at best, only take away some of the blue light while leaving the red light alone.

Thus a filter always decreases the total amount of light entering the camera, often at least somewhat for all wavelengths. And so the use of a filter usually requires one to alter the exposure to compensate. If one is using a through-the-lens meter built

 © IOP Publishing Ltd 2022

into a camera, then this is not much of a problem; the meter is looking through the filter too. If a separate light meter is used, then one must apply an exposure correction. This is called the *filter factor*, and it is often printed on the side of the filter. Unfortunately, filter factors are usually written as the factor rather than the number of exposure steps by which the exposure must be increased. Thus a filter factor of two means that only half the light overall gets through the filter, and this corresponds to a change of one exposure step—not two.

There is a special kind of filter called a *neutral-density filter* that decreases all wavelengths equally. Such a filter, then, would decrease the total light but not alter its spectrum. A neutral-density filter is useful, for example, in a circumstance in which one wants to use a large aperture with a slow shutter speed in bright light. But most other filters do alter the spectrum of the light entering the camera, and that is the point. We can use a filter to accomplish an effect similar to changing the color of light falling on the subject. Or, to look at it a different way, the use of a filter can have an effect similar to using a detector with a different response curve. In either case, it alters not only the color of the light, but also what subjects appear darker or lighter than others.

30.1 Filters and black-and-white photography

Filters are widely used in black-and-white photography. Although no color information is recorded by a black-and-white detector, it still has its own response curve that interacts with the spectrum of the light entering the camera. By using a filter to alter the spectrum of the light entering the camera, one can exercise some control over what parts of the picture are rendered dark or light. See figure 30.1 for some examples of filters designed for black-and-white photography.

Colored filters are used in black-and-white photography more often than in color photography. Strongly colored filters can greatly restrict some colors (rendering those objects darker) while passing other colors (rendering those objects relatively

Figure 30.1. A selection of filters intended for black-and-white photography.

Figure 30.2. The same black-and-white picture taken through three different color filters. The scene in color is portrayed on the **left**. The next three pictures are the same scene taken with black-and-white film through blue, green and red filters, respectively. Notice especially the brightness of the sky compared to the trees and grass.

brighter). The seemingly-black filter on the upper right in figure 30.1 is for technical applications; it passes only nonvisible ultraviolet (UV) light. The deep blue filter on the lower left is unusual; most filters for black-and-white photography are reds, oranges, yellows or greens. Figure 30.2 shows an example of the same scene photographed with color film on the left, and then with black-and-white film through (from left to right) blue, green and red color filters. Unsurprisingly, the blue sky appears bright through the blue filter but dark through the red filter. In this particular case, the green filter probably provides the most 'natural' rendering of the scene.

Even the more 'natural-looking' image through the green filter is still not the same as just a black-and-white version of the color picture. Notice in particular the larger plants in the foreground (mullein, I believe) compared to the more yellowish grasses. In the color picture they are of nearly the same value, and so they blend in with each other. The black-and-white picture through the green filter, on the other hand, better separates them from each other, giving the picture four layers (mullein, grasses, trees and sky) instead of just three (mullein or grasses, trees and sky).

This illustrates why filters are probably more commonly used for black-and-white photography than for color photography. A deep red filter makes the blue sky much darker, causing the white clouds to stand out in stark contrast. But if one were to use this same red filter with a color detector, the entire picture would be red! With black-and-white photography, on the other hand, one can use strong filters to drastically alter the spectrum of the incoming light, thus increasing or decreasing the contrast between objects of different colors.

The most commonly used filter for landscape black-and-white photography is a light yellow-green. This darkens the blue sky somewhat, making the clouds stand out more, and it also lightens up green foliage. A typical yellow-green filter of this type has a filter factor of less than two, and so only a small exposure compensation is needed.

Another filter commonly used for outdoor black-and-white photography is deep red, often with a filter factor as high as four (and so two full steps of exposure compensation are required). Apart from making the blue sky appear very dark in comparison to clouds, a deep red filter greatly diminishes the effect of atmospheric haze, which can greatly lower the contrast in a distant landscape.

30.2 Filters and color photography

In color photography, filters are typically used simply to change the overall color balance of the picture. Remember that the spectrum of light entering the camera depends not only on the way different objects in the picture reflect light (because of their different reflection curves), but just as critically it depends on the spectrum of light illuminating those objects.

In table 30.1 I have listed some common examples of typical lighting sources. Each has its own characteristic spectrum, and in section 30.2.1 I describe some useful ways to categorize these light sources. But the reflected-light spectrum from a given object will be different depending on the source of light that illuminates it. Thus, depending on the source of light illuminating the subjects in our picture, all of the colors will be altered from how they would appear with a different source of light.

Common sources of 'white' light vary significantly in their spectra—far more than it seems from our everyday use of those light sources. The basic reason for this should be clear from the discussion in section 28.3. The *color constancy* property of our eye–brain automatically adjusts for different light spectra illuminating things in the world. But the light detector in a camera is not a brain, and so it needs some help from us in order to make a photograph look natural.

What if, for example, the colors produced by our camera's particular detector look natural whenever our subject is illuminated by daylight? The colors of these

Table 30.1. Examples of different lighting sources. Each of these has its own characteristic spectrum and associated color temperature. Data adapted from Stroebel and Compton (2000, p 10) and https://en.wikipedia.org/wiki/Color_temperature http://www.en.wikipedia.org/wiki/Color_temperature.

Lighting Source	Typical Color Temperature (K)
Candle	1800
Household tungsten bulb	2800
Tungsten photo flood	3200
Moonlight	4100
Compact fluorescent	3000–5000
Tube fluorescent	5000
Horizontal daylight	5000
Vertical daylight	5500–6000
Electronic camera flash	6000
Daylight (overcast)	6500
Clear blue sky	20 000
Neon sign	NA

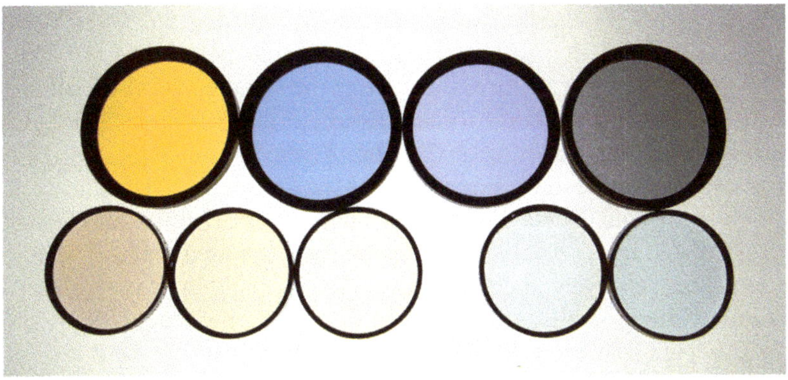

Figure 30.3. Filters for color photography are mostly used to change the overall white balance of the picture. The upper-left filter is to turn much-more-bluish daylight into the color of light emitted by a tungsten lightbulb. The two blue filters on the top are for doing the opposite. The bottom row of filters are for warming the color (the three on the lower left) or cooling the color (the two on the lower right). The filter on the upper right is neutral density (gray), but it polarizes the light as well as dims it.

same subjects will then look very unnatural, as imaged by our detector, if they are illuminated instead by the incandescent tungsten light of an ordinary household lightbulb. The resulting difference between the two photographs will be far greater than our own subjective experience of seeing the subject in the world when illuminated by these two different types of light.

Since a filter selectively absorbs some wavelengths more than others, we can use a filter to 'correct' for a particular source of light. In the above example, we could use a filter that alters the spectrum of the light from a tungsten bulb so that it looks like daylight. Then our detector—which in this example gives the right result for daylight —will record the image under tungsten light as if the subject had instead been illuminated by daylight. This example describes a very common use of filters in color photography; we will consider other examples as we go along. See figure 30.3 for some examples of filters for color photography. Notice that they are, for the most part, less 'colorful' than the filters for black-and-white photography shown in figure 30.1. A filter for color photography is typically used to make one variety of 'white' light look like another, and so the colors are of very low saturation.

30.2.1 Color temperature and white balance

Heat a solid to a high enough temperature and it will emit its own light. This *thermal radiation* is emitted by the objects all around us. But the thermal spectrum of a room-temperature object is dominated by wavelengths at about 10 000 nm—roughly 20 times the wavelength of visible light. To emit visible light, the temperature must be proportionally higher, as measured on the absolute temperature Kelvin (K) scale. Room temperature on that scale is about 300 K. Twenty times that is 6000 K—roughly the temperature of the surface of the Sun, which is, needless to say, a very good emitter of visible light.

An emitter of purely thermal radiation (often called a *blackbody*) produces a particular broad, hump-shaped spectrum of light that peaks at a particular wavelength. The higher the temperature, the shorter the wavelength at which most of the light is emitted. For a blackbody to peak in the middle of the visible spectrum, it must be roughly the temperature of the Sun. The hot filament of an incandescent lightbulb emits mostly thermal radiation, but the temperature is significantly lower than that—only about 3000 K. Such a light source emits most strongly at wavelengths longer than visible light, in the near-infrared part of the spectrum; what we see is the relatively smaller percentage of visible light it also emits.

The color of direct sunlight on a clear day is a complicated case; it has been altered by Rayleigh scattering in the air, for example. And so the direct sunlight is lacking in the blue that has been scattered away—the same wavelengths that makes the sky blue. On a cloudy day, this is all mixed back together, and it usually seems to have no color at all—as white as anything can be. Although we must always be mindful that we see color in context, the 5770 K blackbody of the Sun is about as white as it gets.

A blackbody hotter than the Sun (the star Vega is a good approximation) emits the peak of its spectrum at wavelengths shorter than that of the Sun. And so, compared to the solar spectrum, it emits more short-wavelength light and less long-wavelength light; it appears more bluish in comparison to sunlight. The cooler filament of an incandescent lightbulb on the other hand (or the cooler star Betelgeuse), emits more long-wavelength and less short-wavelength light compared to the Sun. And so the incandescent light would appear more reddish in color. Seen in isolation, all three of these examples appear pretty-much white to our eye–brain, which seems to automatically see an almost-white source of light as white unless there is a different color context with which to compare it. The light of a quartz-halogen incandescent flashlight may look brilliant white at night. But in the daytime, when compared to direct sunlight, it seems distinctly yellow.

Many sources of light do not emit like a blackbody—fluorescent lights, light-emitting diodes, the blue sky and neon signs are all good examples. For these *nonthermal* sources of light, there is no simple connection between temperature and their spectra; indeed the spectra of these nonthermal sources of light are often far more complex than the simple hump shape of a blackbody spectrum. But for any source of 'white-ish' light, we can still make an overall comparison of this complex nonthermal spectrum to the overall effect of an ideal blackbody spectrum. We can ask the question, 'If we were to compare the apparent color of this (for example) fluorescent light to various blackbody thermal sources of different temperatures, what blackbody temperature would appear the same color?' This is the basis of the concept of *color temperature.*

> The color temperature of a source of light is the temperature of the perfect thermal blackbody light source that would appear the same color.

Roughly speaking, a more bluish source of light has a higher color temperature than a more reddish source of light. If the source of light is too different from that of

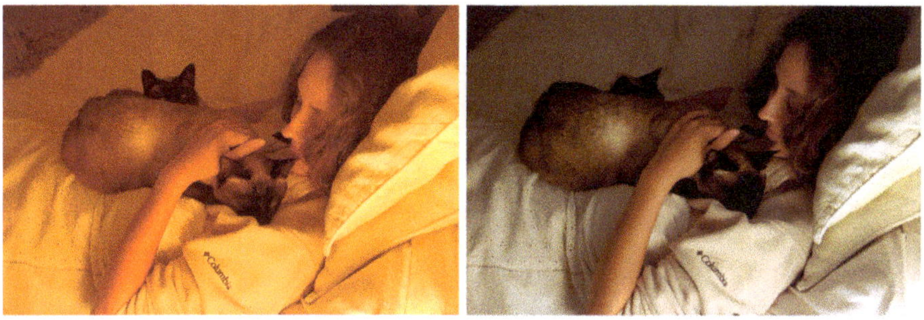

Figure 30.4. Left: An image taken in tungsten light 2700 K with a detector optimized for daylight 2700 K. **Right:** The picture retaken in the same light but with the detector optimized for tungsten light.

a blackbody, then the concept of color temperature is ill-defined and not at all useful. Although a neon sign appears an odd sort of reddish-orange, the concept of color temperature is inappropriate to describe its color; color temperature is most useful for differentiating between sources of light that are nearly white. If two almost-white sources of light are placed next to each other the one of lower color temperature will look redder, while the one of higher color temperature will look bluer. See table 30.1 for examples of sources of light and their approximate color temperatures.

Regarding the spectrum of white light—we know it when we see it. But it is always context dependent. In natural scenes, with a lot of information about the source of light, we humans seem to make similar assumptions when we look at a given picture. Our eye–brain automatically makes some kind of correction for the light shining on the subject, so we see objects as having color constancy even when the reflected spectrum is quite different. The detector in a camera, however, does not know what kind of light is shining on the subject unless we tell it[1].

If we use a detector that will give the correct color balance when the subject is illuminated with light of a particular color temperature, then the color balance will look unnatural if the subject is instead illuminated with a different source of light. Figure 30.4 shows a picture taken in 2700 K tungsten light from household incandescent bulbs. In the example on the left, a detector optimized for daylight was used. Since tungsten light is of a much-lower color temperature, the picture appears too yellow. The picture on the right was taken in the same light but with the detector response curve altered so as to properly render tungsten light.

Figure 30.5 shows the opposite effect. Both pictures were taken in daylight, but for the photo on the left the detector response curve was mistakenly optimized for tungsten light. Expecting a more yellow light source, the detector rendered the picture too blue.

[1] Many contemporary digital cameras use complex algorithms that try to identify objects in the scene and use this information to make reasonable assumptions about the lighting.

Figure 30.5. Left: An image taken in 5500 K daylight with a detector optimized for 2700 K tungsten light. **Right:** The picture retaken in the same light but with the detector optimized for daylight.

For photochemical detectors such as traditional film, the response curve is a fixed property of the detector. There are versions of film that have a response curve designed for the color temperature of an incandescent light (called tungsten film[2]), and films designed instead for daylight. Use one with the light intended for the other, and the examples in figures 30.4 and 30.5 result.

With the photoelectronic detectors in a digital camera, on the other hand, it is possible to alter the response curve internally (within limits). And so the same detector can properly record an image according to the illuminating spectrum of either figure 30.5 or 30.4. Choosing a detector response curve appropriate to the spectrum of the illumination, to properly render colors of objects, is called setting the *white balance*. Most digital cameras have preset response curves for common sources of light, and offer these as choices. Typical examples are daylight, cloudy daylight, tungsten, fluorescent and electronic flash. Most also have an option for auto white balance. This feature makes reasonable assumptions about the typical colors in a scene, so the algorithm's success depends upon how typical the scene is. Cameras with more elaborate features allow for the white balance to be defined manually by the photographer from picture to picture. The photographer takes a picture of a white card illuminated by the same light that illuminates the subject to be photographed. The camera then uses that photograph to internally define the meaning of 'white' for the subsequent picture of the subject.

30.2.2 Filters and color temperature

The transmission curve of a filter alters the spectrum of the light that passes through it. One can thus use a carefully-designed filter to change the spectrum of the light to match the expectations of the light detector—an important task for film-based photography. The most common example is the distinction between the high-color temperature of daylight and the much-lower color temperature of the tungsten light from an incandescent, indoor photo floodlight. There are special filters designed

[2] Most tungsten films were designed for the color temperature of a particular type of quartz-halogen flood lamp.

specifically to alter the spectrum of the light entering the camera from one of these to the other. The #80 filter, for example, is blue in color, and it is made to alter the spectrum of tungsten light so that it looks like daylight. Subtle variants—80A, 80B and 80C—are designed for specific types of tungsten flood lights. The #85 filter (also with slightly-different 'B' and 'C' options) does the opposite: it makes daylight look like tungsten light. As such, it looks orange in color. The first filter in figure 30.3 is an 85B filter, while the next two are different #80 filters.

Sometimes one wants to alter the picture as if the subject were illuminated by a light source of only a slightly-different color temperature. There are many special filters for this. If a given filter is designed to make the subject appear as if illuminated with light of a slightly-higher color temperature than 'white,' it is called a *cooling filter*. If the filter makes the subject appear as if illuminated by a slightly-lower color temperature source, it is called a *warming filter*.

Yes, it is the warming filter that makes the subject look as though it were illuminated with light of lower color temperature. That sounds absurd, but the terms 'warming' and 'cooling' come from our emotional response to color, which turns out to be opposite the physical relationship between color and temperature. The three filters on the bottom left of figure 30.3 are warming filters, while the two on the lower right are cooling filters.

The color temperature of direct sunlight varies greatly with the angle of the Sun. Low-angled sunlight must pass through more air on its way to the eye, and so more short-wavelength light is scattered out of the line of sight by Rayleigh scattering. This loss of short wavelengths lowers the color temperature of the sunlight—dramatically so if the Sun is just rising or setting, or if there is the right kind of haze in the atmosphere. There is a particular angle of sunlight, most typically from one to two hours before sunset or after sunrise, for which we humans seem to find the color temperature of sunlight to be particularly appealing. And so a warming filter can be used to simulate this *magic hour* or *golden hour* at other times of the day.

As a further example of the context dependency of color—and especially of what we see as white—notice from table 30.1 that the color temperature of moonlight is lower than that of sunlight. This means that moonlight should appear 'warmer' than sunlight, but we mostly experience the opposite effect. Moonlight is 'cold' and 'silvery,' while every child uses a yellow crayon to portray the Sun. The color temperature of moonlight (even when the Moon is high overhead) really is slightly lower than sunlight because the reflection curve of the lunar surface is, on average, non-gray. There are even some lunar soils that are noticeably orange in color when seen up close. But the color temperature of moonlight is still much higher than even the hottest quartz incandescent bulb, which seems brilliantly white when it is the sole source of light. One obvious reason that direct sunlight seems so yellow is that we see it in the context of a blue sky.

30.3 Polarizing filters

The filter on the top right in figure 30.3 seems to have no particular color at all, and it is in fact designed that way. It is a neutral-density filter—a filter with a transmission curve that is constant over the visible spectrum. Such a filter is sometimes useful for reasons having to do with exposure as discussed in chapter 22. This particular example, although it does act as a neutral-density filter, has a different purpose: it is a *polarizing filter*.

Brightness and wavelength are not the only ways in which two sources of light can differ from each other. Light is an electromagnetic wave, a changing pattern of electric and magnetic fields that travels through space. These fields point perpendicular to each other in space, and the light wave travels in a direction perpendicular to both. But this leaves open an infinite number of possibilities. For example, if an electromagnetic wave is traveling directly upward, the electric field must point perpendicular to that—i.e., horizontally. But does the electric field point horizontally North, South, East, West or one of the infinite possibilities in between? Any one particular choice of orientation of the electric and magnetic fields is called a *polarization* of the wave, as described in section 3.9.

A polarizing filter passes light of one polarization while blocking light with the polarization 90° to it. The polarization direction that is selectively blocked depends upon the orientation of the filter. Thus a polarizing filter attaches to the camera lens with a mount that twists, so the photographer can vary the angle of polarization that is blocked by the filter.

Most sources of light are unpolarized, and for such light the use of a polarizing filter makes little difference, except as a neutral-density filter. But there are two important sources of polarized light in everyday scenes.

1. The daytime blue sky on a clear, sunny day is strongly polarized at angles 90° from the Sun.
2. Light reflecting at shallow angles off nonmetallic reflective surfaces—water, for example, or the varnish on a table top—may be highly polarized in directions crosswise to the reflecting surface.

We do not see the effects of polarization directly because to the human eye or to the light detector in a camera one polarization looks just the same as any other. A polarizing filter cleverly employs long-chain molecules to effectively block light polarized in one direction while allowing polarizations at right angles to pass. For unpolarized light, which has all polarizations at once, this means the polarizing filter blocks half the light.

Since the light from shallow reflections and the daytime sky are polarized, a polarizing filter can be used to selectively block them, making them darker in contrast to unpolarized light. A common use is to darken the blue sky relative to the clouds in a color picture (see figure 30.6). For black-and-white photography one can simply use a deep red filter to darken the sky. But this option is unavailable for color photography, unless one wants everything to be deep red. Since the blue sky is strongly polarized, one can simply rotate the polarizing filter until the sky is at its darkest. Objects in the picture that are not strongly polarized (clouds for example) will remain unaffected, so contrast is increased.

Figure 30.6. A polarizing filter was used to greatly darken the blue sky and reduce the glare off of the rock surfaces. The polarizing filter was oriented so as to reduce the glare off the (mostly) vertical surfaces, but this increased the glare on the horizontal surfaces.

A polarizing filter may also be used to reduce glare in order to make colors more vibrant. Glare essentially mixes skylight (since it is a mirror-like reflection) with the diffusely-reflected light from the subject, and this added white light reduces the saturation of the colors. A polarizing filter can be used to reduce this glare because light glancing off a surface is often strongly polarized (e.g., see figure 30.6).

30.4 'Color' in astronomy

In section 28.1 we considered the opponent-process model of color perception. The brain does not directly process the individual red, green and blue sensitive stimuli. Rather these signals are preprocessed, and combinations are sent instead. There may be a physical 'logic' to this; it is sometimes a good way to get the most out of the information available. Astronomers, with no conscious awareness of advances in the neurophysiology of the human retina, have long taken a similar approach.

It is very important for an astronomer to observe the spectrum of an astronomical body; the spectrum gives us the biggest clues as to the source's physical nature. An astronomer will often use a telescope to gather the light from a star or galaxy and then disperse that light with a spectrometer. The problem, however, is that taking a spectrum is very difficult and time consuming. And what if one wants to analyze the light of, for example, the thousands of stars that make up even one single star cluster? Often it is simply impractical to gather a full spectrum of so many astronomical objects, especially if they are very faint.

Astronomers thus often settle for the next best thing—what is called *filter photometry*. Instead of sending the light through a spectrometer, it is sent through a color filter through which only a particular range of wavelengths may pass. This process is then repeated through different colored filters. Thus the astronomer takes

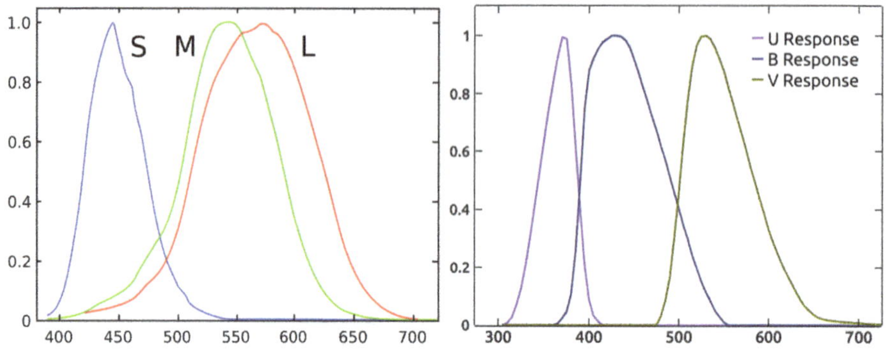

Figure 30.7. **Left:** The wavelength responses of the S, M and L wavelength cone cells in the human retina. Copyright ©Vanessaezekowitz, Attribution 3.0 Unported (CC BY 3.0), https://commons.wikimedia.org/wiki/File:Cones_SMJ2_E.svg, Cones SMJ2 E.svg. **Right:** The wavelength responses for the U, B and V filter system used by astronomers (data from Ažusienis and Straižys (1969)).

multiple pictures of the same object with each picture representing the overall amount of light for a different part of the spectrum.

If an astronomer were to do this through red, green and blue filters, the astronomer would be doing something very much like what is done by the cone cells in the human retina. Instead the filters are carefully chosen based on the particular kinds of astrophysical questions the astronomer is trying to answer, but the basic idea is the same. In fact the most common system traditionally used by astronomers involves three filters that, coincidentally, have much in common with the small, medium and large (S, M and L) cone cells of the human eye. But the responses are at shorter wavelengths overall; they are called U (for UV), B (for blue) and V (for 'visual'—i.e., the middle of the human visual part of the electromagnetic spectrum). If more filters are included—there is no reason there has to be only three—then even more information about the spectrum of the object can be inferred. And so this system is often extended to include R (for red) and I (for infrared). See figure 30.7 for a graph that shows the wavelength responses of the U, B and V filters next to the responses of the cone cells of the human eye.

As is the case for the human eye, astronomers do not analyze these colors in terms of the individual measurements from the filters. Instead, they use combinations of these measurements and such a combination is called, somewhat metaphorically, a *color index* or just simply a 'color.' And so, instead of publishing the measurements obtained through the filters U, B and V, the astronomer publishes V, U−B and B−V.

Compare this to the preprocessing of the signals from the three types of cone cells, S, M and L. The brain seems to interpret not S, M and L separately, transforming various combinations into colors and brightnesses. Rather, the brain seems to act upon something more like M−L, S−(M+L) and S+M+L (Bevil 2003, Conway *et al* 2010, Brian *et al* 2014). The latter combination gives the overall brightness, with no color information in and of itself, and so is analogous to the astronomer's V magnitude. The other two combinations contain color information, and they are analogous to the astronomer's color indexes U−B and B−V. The color indexes represent the excess of one filter measurement over another, and so represent how that part of the light spectrum is *tilted.*

We can see how the color index relates to what we normally think of as color. For example, a reddish star would give a stronger signal through the yellowish V filter than through the bluish B filter, while a bluish star would do the opposite. Of course, when talking about a U or I filter, these are outside the range of human perception altogether. And so a 'color' such as U-B would have no direct relation to any human-perceived color. But it is not difficult to see that the idea is the same, and that it tells something about the nature of the source's spectrum (its tilt).

Astronomers sometimes go one further and calculate even more complex combinations of the basic measurements, producing a 'color–color.' There is another filter system used by some astronomers—called the Strömgren system—that uses direct measurements through four filters, called u, v, b and y. But again, the individual measurements are never even published. Instead, astronomers work with the following: y, b−y, (u−v)−(v−b) and (v−b)−(b−y). The first of these (y alone) gives the overall brightness of the spectrum, the second (b−y) gives the tilt of the spectrum at a particular wavelength, while the remaining two give the *curvature* of the spectrum over two different wavelength ranges. For this particular example, information about the brightness, distance, size, temperature and age of the star can be determined as well as the amount of intervening dust that happens to lie in our line of sight to the star.

See figure 30.8 (data from observations made for Beaver *et al* 2013) for an example with the star cluster NGC 6705 in the constellation Scutum. It shows a plot of y versus b–y for all of the many hundreds of stars in the star cluster. This basic 'color-magnitude diagram' is one of the most important tools of any stellar astronomer. It shows the relation between the overall brightness of the star (top is

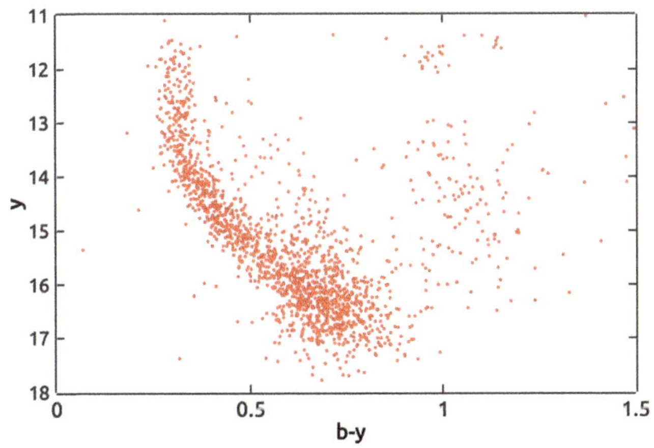

Figure 30.8. A 'color-magnitude' diagram for the star cluster NGC 6705. Each point represents an individual star in this cluster of many hundreds. The vertical axis is the overall brightness (through the y filter), while the horizontal axis is the color index (b−y).

bright; bottom is dim) and its overall color, and thus its *temperature* (bluer, and so hotter on the left; redder, and so cooler on the right). Indirectly it tells us the size of the star. The little group of stars on the upper right of the diagram are called *giants*. They are both cool and very luminous, and the only way to be both of those at once is to be very large.

And so we start out with four separate measurements, and we end up with four different numbers that are combinations of what was directly measured. But these odd mathematical combinations give more useful information about the spectrum of the observed star than do the individual measurements of u, v, b and y alone.

There is another important reason that astronomers traditionally do not publish the direct measurements through their filters but instead only publish color combinations. For technical reasons, a color such as B−V is sometimes easier to measure accurately than the individual B and V brightnesses. Accurate measurement of the individual filter brightnesses requires a complex process of calibration. But in many situations when measuring the difference between B and V, much of that calibration partially 'cancels out.' This leaves us with a measurement of B−V that is, surprisingly, more accurate than the measurements of either B or V alone. And the eye–brain combination apparently includes specific physiological structures—cone-opponent retinal ganglion cells—that directly measure differences in the cone responses rather than measuring the individual responses directly (see section 28.1).

References

Ažusienis A and Straižys V 1969 Improved determination of the response curves and parameters of the UBV system. summary of results *Sov. Astron.* **13** 316

Beaver J, Kaltcheva N, Briley M and Piehl D 2013 Strömgren-Hβ photometry of the rich open cluster NGC 6705 (M 11) *Publ. Astron. Soc. Pac.* **125** 1412

Conway B R, Chatterjee S, Field G D, Horwitz G D, Johnson E N, Koida K and Mancuso K 2010 Advances in color science: from retina to behavior *J. Neurosci.* **30** 14955–63

Conway B R 2003 Colour vision: A clue to hue in v2 *Current Biol.* **13** R308–10

Schmidt B P, Neitz M and Neitz J 2014 Neurobiological hypothesis of color appearance and hue perception *J. Opt. Soc. Am.* **31** A195–207

Stroebel L and Compton J 2000 *Basic Photographic Materials and Processes* 2nd edn (Waltham, MA: Focal Press)

Chapter 31

Color experiments with black-and-white photography

In section 29.7, we described *ephemeral-process photography* (EP) and found that certain varieties of black-and-white silver-gelatin enlarging papers can show scattering colors that relate to the wavelengths of the light used to expose it. Thus we have the possibility of a sort of two-tone color photography from a nominally black-and-white emulsion. Enlarging paper prints out with painful slowness when it is dry. When using the technique of EP photography, however, the speed can be vastly increased by the application of an accelerator. More details about EP photography can be found in chapters 25 and 48. Here we consider several ways to use EP photography for making color pictures.

31.1 In-camera color EP photographs

Figure 31.1 shows two examples of color EP photographs made with negatives exposed in a camera. The colors in the images, although adjusted digitally, all derive directly from the original paper negatives. We notice that the technique is particularly effective for rendering low-saturation blues, yellows, oranges and reds.

The images are blurry around the edges because they were exposed using only a simple double-convex lens in a homemade camera. I used such a crude lens because it needed to be fast—about $f/3$—in order to record the image in the reasonable amount of time of about one hour in this case. An ordinary camera lens of this speed, and one that will also cover the 13 cm × 13 cm format used for the negatives, is expensive and difficult to come by.

Ordinarily even a slower lens would record the image in about 15 minutes or so, but in this case I added a filter that blocks many of the short wavelengths for which this paper is sensitive. We saw in section 29.7 that the right choice of paper will record both blues and oranges, at different ranges of wavelengths. But most of the

doi:10.1088/978-0-7503-3703-8ch31

© IOP Publishing Ltd 2022

Figure 31.1. Both of these pictures were made directly from EP negatives exposed in cameras.

Figure 31.2. EP photography naturally produces colors that are backwards upon inversion from a negative to a positive (**left**). For the image at **right**, I have rotated the hues back 180° around the color wheel.

range of wavelengths prints out as blue; only the longest wavelengths in the sensitivity range print out with the contrasting orange color. Thus I have used a filter that blocks much of the short wavelengths to bring the two colors more into balance.

Figure 31.2 shows another example, this time photographed at $f/4.7$ for about two hours. The left-hand example is a straightforward digital positive from the scanned paper negative. The colors are distinct, but they are backwards; the corn was yellow and not blue. This process produces a negative regarding light and dark, but the colors correspond to the colors in the world and not their complements. For the right-hand example I have reversed the values from a negative while leaving the colors as they were. Notice the absence of green, even as the reds, oranges, blues and yellows render well (the scene contained green grass in the background).

Figure 31.3. Color pictures made from in-camera EP negatives.

Figure 31.3 shows two more examples, and here I have chosen subjects with colors that render well with this process. Sand, some types of skin tone, or red or yellow flowers all work well when combined with blue sky or blue-painted objects. A landscape of green trees and grass on a cloudy day is a poor choice and will likely come out virtually monochrome.

31.2 Color EP photograms

Photograms are an especially good technique for exploring the color possibilities of EP. Objects can be chosen in part for their color filtering properties in order to more strongly emphasize the often-subtle EP colors. The source of light can also be better controlled than when taking EP pictures directly with a camera. For the image *Blue Monarch* on the upper right of figure 31.4, it seemed fitting to leave the colors reversed rather than switching the hues back to something more natural.

The example at upper left was illuminated by a single-wavelength 405 nm laser. But the mineral crystals fluoresced, producing their own light at longer wavelengths. The example at lower left was exposed twice with two different types of light, each producing the complementary color of the other. The example at lower right was exposed with a light-emitting diode flood lamp.

31.3 EP contact negatives from color prints

Figure 31.5 shows another way to get very strong colors. The EP paper was exposed in contact with a positive print, made from an ordinary camera, thus making an EP negative. The negative is then scanned, digitally inverted and adjusted in the same ways as for in-camera negatives. But the reds, yellows and blues in these *chromogenic prints* correspond very well to the spectral responses of these colors in some EP papers, and very intense colors are possible. For more-detailed instructions on how to perform one's own EP photography experiments, see chapter 48.

31.4 Color ephemeral prints

Figure 31.6 shows an EP contact print made by shining light directly through a 35 mm color negative. I deliberately chose an enlarging paper that produces strong

Figure 31.4. EP photograms may make especially intense colors if the right combination of paper, objects and light source are chosen.

Figure 31.5. The image was made from an EP negative exposed not in a camera but in contact with a color positive print from an ordinary camera.

Figure 31.6. Left: An EP print made in contact with a color film negative. **Right:** An EP print made in contact with a digital negative.

duotone colors in order to make a color print from the color negative. It is kept in a little match box because it is, of course, still sensitive to light; to look at it is to damage it. Even the match box has too many light leaks to keep the image safe for long-term storage. So it and several others are kept in a larger light-tight box; I take them out every now and then to peak at them in dim light.

Notice that the hues are backwards, as expected. Furthermore, I used a bit of a trick. An ordinary color film negative has a strong orange mask, essential to achieve a proper color balance for its intended use of producing a positive print in the darkroom with color print paper. The orange mask, however, blocks most of the light that the black-and-white enlarging paper is sensitive to for EP photography. Here I have used *cross-processed* film. An emulsion meant for reversal processing, to be projected directly as a positive transparency, was instead processed as a negative. The result is a high-contrast negative without the orange mask.

For more control, one can use a negative transparency that is printed digitally with an inkjet printer onto transparency film. This allows one to reverse the hues on the negative, so they come out correctly on the print. See the right side of figure 31.6 for an example.

High-quality transparency films are available, and inkjet prints on these films are often used to make contact prints for alternative processes such as cyanotype or platinum printing. Furthermore, one can print the negative large to make a large print, and it can be made from any digital image regardless of its origin. See figure 31.7 for an example that compares the original digital image and an EP print made from it.

Unlike a film negative, the color and contrast can be controlled digitally at the negative stage in order to produce good results with a particular combination of light source and enlarging paper. The contrast—in both hue and value—can be enhanced by carefully stacking two identical negatives for the contact print. For the examples I have tried, the most-intense colors on the print come from the blues and

Figure 31.7. A digital image (**left**) can be inverted to a negative and the hues reversed before printing on transparency film with an inkjet printer. An EP contact print (**right**) can be made from the negative.

Figure 31.8. Left: A color transparency printed with an inkjet printer from a digital negative image. The greens were removed from the image before printing. **Right:** Direct EP contact print made from two layers of the transparency at left.

yellows in the negative; greens produce the least-intense colors. One can take advantage of this by reducing the green in the color negative before printing; more intense duotone colors result in the print. See the right side of figure 31.8, and the negative transparency used to make the print at the left.

Part VI

Light detectors

IOP Publishing

Photography (Second Edition)
Physics and art in focus
John Beaver

Chapter 32

Types of detectors

Part of the essence of photography is that light falling on a surface is recorded. This is the *graphy* in photography. A permanent record is made of the intensity (and maybe color) of light that fell on each point of a given surface. To do this one must have some material that undergoes a physical change when light interacts with it. This material must have a surface on which the image can focus, and there must be some way to record different physical responses at different positions on the surface.

Throughout *Photography: Physics and Art in Focus* I use the word *detector* to represent, in a general sense, any light-sensitive surface used to record an image. It could be the retina of the human eye, traditional photographic film, a digital detector in a digital camera or a light-sensitive material that is used to make a print. Ultimately, there are three crucial features of every light detector.

1. Light must cause some physical change to occur in the detector.
2. Once the exposure is completed, there must be a way to stop further physical changes from occurring. Or, barring that, there must be some way to 'read' the results of the detector and transfer those results to some permanent, non-light-sensitive storage.
3. There must be some way to tell, after the exposure is complete, which part of the detector underwent what physical change. That is, we need to be able to record separately the physical changes that occur at different locations on the detector. Otherwise, we would have not an image but rather only a single measure of brightness.

Light interacts with some physical material. That is, some of the light disappears, and the material changes in some way. The idea that some materials change visibly upon exposure to light is far from modern, since there are many common everyday examples. For example, many dyes and pigments fade with light over time; when a

doi:10.1088/978-0-7503-3703-8ch32

© IOP Publishing Ltd 2022

piece of furniture shields part of a painted wall from light, its outline is visible when it is moved years later.

My usage of the word detector is somewhat unconventional in that I use it to refer to any two-dimensional (2D) surface used in photography to make a record of an exposure to light—whether or not its use is for the original capture of an image in a camera or instead to make a permanent photographic print. We can divide most photographic light detectors into two broad categories: photoelectronic and photochemical. The retina of the human eye (see section 32.5) is more complex than either category and includes aspects of both.

32.1 The physics of photons

The physics of light is described in some detail in chapters 2–5, but here I recap and expand upon some concepts that are especially relevant to light detectors.

Light has a *wave-like* nature; it does all of the basic things that waves do. But it also has a *particle-like* nature; in many circumstances it transfers energy in discrete clumps at particular points in time and space. Such a particle of light is called a *photon*. These aspects of light seem at first glance to contradict each other, but they do not; the two concepts are unified by *quantum physics*. Photon interactions are of particular importance for understanding photographic light detectors, but the wave-like properties of light play a role too. Below I outline some of the most basic principles of the physics of light as it pertains to photographic light detectors.

- Light is an electromagnetic wave: a changing pattern of the electric and magnetic properties of space that travels through space at the speed $c = 3.0 \times 10^8$ m s^1.
- The perceived color of light is related to its *wavelength*: the distance between successive maxima in the electric (or magnetic) fields at a given point in time. It can also be described by its *frequency*: the number of electric field maxima that pass a given point in space per second. Frequency is measured in *Hertz* (Hz) or repetitions per second. Wavelength is a length, and so it can be measured in meters (m); however, the wavelengths of visible light are submicroscopic, and so the nanometer (nm), or 10^{-9} m, is more commonly used. The human eye is sensitive over the range approximately 400–700 nm. We perceive 400 nm light as violet and 700 nm light as red.
- There is an inverse relationship between wavelength (λ) and frequency (f) for light. And the speed of light (c) relates the two.

$$c = \lambda f \tag{32.1}$$

$$f = \frac{c}{\lambda} \tag{32.2}$$

$$\lambda = \frac{c}{f} \tag{32.3}$$

- Light carries both energy and momentum. When light interacts with matter, one or both must be transferred. It is the transfer of *energy* that is important for understanding light detectors; in order for light to be detected, some energy must be transferred from the light to the detector.
- The smallest unit of energy transferred by light is the photon. A photon interacts with an individual atom or molecule at a particular point in time.
- When light interacts in a wave-like manner, resulting in wave phenomena such as interference and diffraction, we can determine the frequency of the light. If that same light is allowed to interact in a particle-like manner, we can measure the energy of those photon interactions. There is a direct relationship between the energy (E) of the individual photons and the frequency (f)of the light of which they are a part.

$$E = hf \tag{32.4}$$

$$= \frac{hc}{\lambda} \tag{32.5}$$

  Here h is a tiny constant called the Planck constant.
- Photons of higher energy correspond to light of shorter wavelength (and higher frequency). Photons of lower energy correspond to light of longer wavelength (and lower frequency).
- The *brightness* of a source of light is related to the number of photons, while the frequency (and wavelength) of the light is related to the energy of each individual photon.

32.2 Photoelectronic detectors

A *photoelectronic detector*, upon exposure to light, directly produces some kind of measurable electronic response. Most digital photographic detectors are of this type, and they are usually an array of rows and columns of separate light-sensitive sites, called *pixels* ('picture element').

There are different ways in which light can interact with conductors or semiconductors and cause a measurable electrical effect. For example, individual photons can remove electrons from their atoms. If this happens in an evacuated glass tube, the electrons can be accelerated through an applied voltage to make a measurable electric current. This procedure is not ordinarily used by photographers to record images, but devices called *photomultipliers* employ this principle for the precise measurement of very faint light (mostly by astronomers and particle physicists). Photomultipliers can easily detect individual photons.

Other devices can be designed for which light alters its electrical resistance (or conductance). A *photoresistor* becomes less resistant to electrical current when light shines on it. This can be used in a circuit with a current meter and a battery to make a light meter—an instrument for measuring the brightness of light. Indeed, many older cameras used photoresistors for their internal light meters. Photoresistors, however, are not particularly useful for recording images.

A photovoltaic cell turns light directly into electrical energy. A voltage is produced, and power can be delivered when it is illuminated by light. Very-low power photovoltaics (the metal selenium is a good example) were used for some of the first electronic light meters. But like photoresistors, photovoltaics have not been particularly useful for forming images.

Finally, semiconductor *photojunction devices* use the interaction between light and a boundary between two dissimilar semiconducting materials. These devices can be manufactured in arrays of microscopic, separately light-sensitive pixels. We will examine two types of photojunction photographic imaging devices in chapter 38.

32.3 Photochemical detectors

A *photochemical detector* can be thought of as any material for which the exposure to light produces some kind of chemical change. If these changes are visible to the eye (or can be made visible by further chemical processing), then the material can be used as a detector. There are two main categories of photochemical detectors.

1. A *latent image* is formed upon exposure to light. A change occurs at the molecular level, but no visible change is apparent until the detector undergoes a chemical treatment called development. In this case we say the detector *develops out*. Ordinary silver-based color or black-and-white film photography is of this type.
2. The detector *prints out*; visible changes occur in the detector as it is exposed to light with no chemical development needed. Many antiquarian and alternative processes are of this type.

There is an important complication with most photochemical detectors, and this was an important stumbling block in the history of their invention. There must be some way to stop the detector's sensitivity to light after the image is formed. Otherwise, the act of looking at the picture would continue to expose it to light. This process is traditionally called *fixing* the image, and it usually requires some kind of additional chemical treatment of the detector.

Another important aspect of photochemical detectors is the nature of the particular *base* or *substrate* that is used. To what type of flat surface is the light-sensitive material attached? From what is that substrate made, and by what means does the light-sensitive chemical attach to it? Is the base transparent or opaque? If it is opaque, is it black or white? If the substrate is transparent, one can shine light through it perhaps to use lenses to enlarge and project the image onto another surface.

Two different approaches are most common for attaching a photochemical detector to its substrate. The simplest is to allow the light-sensitive chemical to penetrate into the fibers of a semiporous surface, such as paper, in essentially the same way that the dyes soak into the paper in a watercolor painting. The second technique is to allow the light-sensitive chemical to form microscopic crystals, and to then suspend these in some sort of gelatinous material, which is then applied to the top of the substrate as a thin coating. This is called an *emulsion*, and most modern photochemical detectors use this method.

32.3.1 Negative and positive

Since photochemical detectors may result in a visible change in the detector upon exposure to light, there is the possibility of a direct physical relationship between the picture and the original exposure. That is, a greater exposure to light leads to a greater visible effect on the detector. But this leads to an important question: does a greater exposure to light cause the detector to become visibly brighter, or does it instead make the detector visibly darker? A *negative process* results when the detector is rendered darker by greater exposure to light. A *positive process* is, of course, the opposite.

If our immediate task is to produce a picture for the wall, then we most likely want bright areas in the picture to represent more light in the world. It would seem, then, that the detector of choice would be one that uses a positive process. It would allow for the picture on the wall to be the actual detector that was exposed in the camera. We call this a *direct positive*, and we will describe a few photochemical detectors that work this way.

But the majority of photochemical processes suitable for photographic detectors work as a negative process. For those processes we must take some extra steps in our journey from the camera to the gallery wall. We shall see, however, that negative processes have some distinct practical advantages over positive processes. There are two traditional approaches to get from a negative image to a positive image, and we consider them in turn.

32.3.1.1 Negative on negative

The most straightforward way to get a positive image from a negative process is to recognize that a negative of a negative is a positive. If one simply repeats the negative process, but uses the result of the original exposure as the source for a second exposure, then a positive image results.

For example, one can expose a sheet of 8 × 10 inch light-sensitive *enlarging paper* in a pinhole camera. Upon processing, a negative image results; the brightest features in the scene become the darkest areas on the paper. This sheet of paper with its negative image can then be placed (in a darkroom) image-side-down directly onto another sheet of the same kind of paper with its light-sensitive side facing upward. One can then simply expose the second sheet by shining light through the back of the negative image. Upon processing the second sheet, a positive image appears because it is exposed the least (and so appears the brightest) wherever the original negative was darkest.

This is called a *contact print*, and it is an easy way to make a positive from a negative. Clearly, the print is the same size as the negative. So if the negative is made in a camera, it must be a very large camera if one wants a large print. If the original negative is made on a transparent film, then one can expose the light-sensitive print paper by projecting light through the negative with a lens—sort of a camera in reverse. Such a device is called an *enlarger*, and it is the most common traditional method for making large prints from small negatives.

The negative-on-negative process has the obvious practical advantage that it separates the process of recording an image in a camera from the process of making a print to hang on the wall. It, therefore, opens the possibility of making a large number of prints from a single image capture. It also allows for adjustment of the image in the step of making the print, and so maybe one can get a good print out of a not-so-good image capture. For these reasons, the negative-on-negative approach has dominated much of the history of photochemical photography.

32.3.1.2 Reversal processing

There is another traditional way to get a positive image out of a negative photochemical process, and it is called *reversal processing*. It is sometimes possible to process the light-sensitive material in a clever two-step way by first producing a negative and then reversing the negative to a positive directly in the same piece of light-sensitive material. So we have, in the end, something that is much like a direct positive—the picture on the wall is the same object that was in the camera.

The history of reversal processing is almost exclusively associated with silver-gelatin emulsions, and so we take up some of the technical aspects more fully in that context (chapter 34).

32.3.1.3 Digitally-printed negatives

Finally, there are not-so-traditional ways to get a positive image with a negative process. Any digital image can now be easily made into a transparent negative with image processing software and digital printing. The digital image, inverted to a negative by the software, is printed onto acetate with an inkjet printer at the same size as the intended print. This negative can then be exposed in contact with a negative-process light-sensitive paper to make a positive print. Some photochemical printing materials are sensitive only to ultraviolet light, and this makes it very difficult to expose them with an enlarged projected image. And so they must be contact printed, and for that a negative is needed that is the same size as the final print. It is increasingly common for photographers to use digital inkjet negatives for this purpose.

The opposite is also possible. One can record a digital image of a photochemical negative through scanning for example. The image can then be converted to a positive through digital processing, and a digital print is made.

32.4 Basic photochemistry

There is a sense in which most photochemical detectors are also photoelectronic—a photon is absorbed, and some transfer of electric charge results between atoms and molecules. Chemistry is about the electrical attraction between the positively-charged nucleus of an atom and its negatively-charged electrons. The nuclei do not change (if they do, it is called a *nuclear reaction*). But it is possible for the electrons, especially those that are most distant from the nucleus, to transfer from one atom to another thus taking negative charge away from one and adding it to another.

An atom or molecule that has a net electric charge is called an *ion*[1]; it is an *anion* if it has a net negative charge and a *cation* if it has a net positive charge. The process of an atom or molecule decreasing its number of electrons is called *oxidation*, while the process of increasing its number of electrons is called *reduction*.

The oxidation state is related to the net charge of an ion, and it is denoted by a superscripted number with a following plus or minus sign affixed to the chemical symbol[2]. For example, 'Ag^{+}' is a silver (Ag) atom that has lost one electron; taking away -1 is the same as adding $+1$, and the number 'one' is assumed if no other digit follows. Neutral silver, with just the right number of electrons to balance its internal charge, is simply denoted 'Ag' with no superscript.

Like charges repel and unlike charges attract, so an isolated atom wants to be neutral with just the right number of electrons to balance the positive charge of its nucleus (it is the positive charge of the nucleus that gives an atom its identity). Because of the intricate rules of quantum physics, however, there is more to it than that, and when atoms are put together with other atoms other factors come into play.

Certain molecular combinations of atoms are only stable if they are ions—if they are either missing or have extra electrons. A good example is nitrate, which is a nitrogen atom bound to three oxygen atoms. But this only happens if it can find an extra electron, and so form NO_3^-, a nitrate cation. Silver on the other hand, is happy as just ordinary Ag. But in the right context, it is more quantum-mechanically happy to give up one of its electrons and instead be Ag^+. Put Ag^+ together with NO_3^-, and they stick together (unlikes attract) to form the neutral compound silver nitrate ($AgNO_3$).

A compound such as silver nitrate does not mean that one particular Ag^+ is stuck specifically to one particular NO_3^-. Instead, it forms a *crystalline solid*—a regular arrangement of alternating silver and nitrate. It can also be dissolved in water to form an *aqueous solution* in which the ions are separated from each other by the water molecules, constantly associating and dissociating with no ion belonging to any other ion in particular. Compounds that will do this easily are called *soluble*, while those that will not are called *insoluble*.

Some atoms or compounds have electron structures that make it easy for them to lose electrons, donating them to some other chemical species in a chemical reaction. The act of losing electrons is called oxidation, but the atom or compound that does so is called a *reducing agent*. In the opposite sense, an atom or compound that easily accepts electrons—and so is likely to 'steal' them from some other chemical species —is called an *oxidizing agent*. When an oxidizing agent accepts electrons, it is reduced in the process.

---

[1] A *molecule* is often defined to be a bonded-together group of atoms that is electrically neutral. And so, for example, a more proper name for a similar structure with a net electric charge is *polyatomic ion* or *molecular ion*.

[2] It is, strictly speaking, only the net charge if all bonds to the atom are ionic.

And so what does all of this have to do with light? In certain circumstances the absorption of a photon—a particle of light—can provide the energy needed for the transfer of electrons from one chemical species to another and cause chemical changes to occur. When this happens, it does so on a one-on-one basis. A photon (or several photons in a very short period of time) transfers its energy all at once and at one particular place. If these chemical changes enabled by the absorption of light produce some visible change, then they can be used as the basis for a photochemical detector.

One of the great feats of early twentieth-century modern science was the discovery of fundamental physical laws that provide explanations for the mostly already-known laws of basic chemistry. The principles of quantum physics allow one to calculate the electron structures of atoms and also the rules by which those structure may change when a given atom is combined with other atoms. These rearrangements of atomic electron structure involve transfers of energy, and this is the ultimate source of the energy transfers associated with chemical reactions.

Why, for example, is nitrate one of the ways in which nitrogen and oxygen like to arrange themselves? The question can be answered on the one hand in terms of pre-twentieth-century concepts of chemistry such as valence, the periodic table of elements, acids, bases, etc and the complex chemical rules that relate these concepts to each other. The fundamental laws of physics, however, were too incomplete in the 1900s to provide an underlying physical explanation for these chemical rules. It was not until the development of quantum mechanics in the first third of the twentieth century—a revolution in the fundamental understanding of matter and energy—that physics was up to the task of providing an explanation for many of the facts that chemists already knew (Chomsky 2000, pp 110–111).

32.5 The eye as a detector

The eye looks like a camera. There is a lens, a dark box (albeit nearly-spherical in shape), and a light detector it its back, the retina, upon which an image is focused. The *retina* is covered by a matrix of millions of light-sensitive cells, seemingly similar to the light-sensitive pixels on the detector of a digital detector. These cells are 'wired' to the brain by the optic nerve. It is tempting to think of the eye as simply a biological version of a modern digital camera, but the differences are profound.

The English word 'photography' was apparently coined by John Herschel, astronomer and one of the pioneers of photography[3]. The combination of 'photo' (light) and 'graphy' (writing) implies that we are 'writing with light,' and indeed that was literally the first photographic method. Centuries before photographic detectors were invented, the *camera obscura*—a dark box with a lens to focus an image—was used to trace images by hand.

An even more convenient arrangement was patented in 1807 by William Hyde Wollaston, three decades before the first successful photographic process. The *camera lucida* used a clever combination of mirrors and prisms that allowed one

[3] The French word 'photographie' was coined independently, at about the same time, by Hèrcules Florence.

Figure 32.1. A camera lucida uses an arrangement of mirrors and prisms to superimpose an image onto a piece of drawing paper. The image is then traced by hand (graphic credit: By Unknown—illustration from the *Scientific American Supplement*, January 11, 1879, Public Domain).

to see a virtual image of the subject superimposed onto a piece of drawing paper, as in figure 32.1. This allowed the artist to mark key points in the image—corners, ends of lines or locations of important features—in order to draft an exactly photo-realistic drawing.

John Herschel was an avid user of the camera lucida (see Evans *et al* [1969] for many examples). Its use influenced both Herschel and, the more-famous pioneer of photography, William Henry Fox Talbot. Both were intrigued by the idea of making a permanent photographic record, but they were somewhat frustrated by the imperfection and tedium of the hand-drawing process. And so the idea of making a direct, permanent recording of the image formed by a lens was already 'in the air' by the 1830s when the first chemical photographic detectors were invented.

The physical process by which the retina detects light is complex and multi-stepped. There are both chemical and electrical changes as well as changes in the physical structure of individual molecules. The retina is elaborately structured, and it is inside out; the light-sensitive part is on the back surface, not the surface facing the eye lens. Thus the light focused by the lens must pass through several layers of only semitransparent cells before being detected. Figure 32.2 illustrates this structure. The two basic types of light-sensitive cells—called *rods* and *cones* because of their shapes—are on the right of the diagram; the light focused by the eye lens enters from the left.

As is the case for most light detectors, we must appeal to the particle-like nature of light to understand even the simplest aspects of the process. Within the rod or

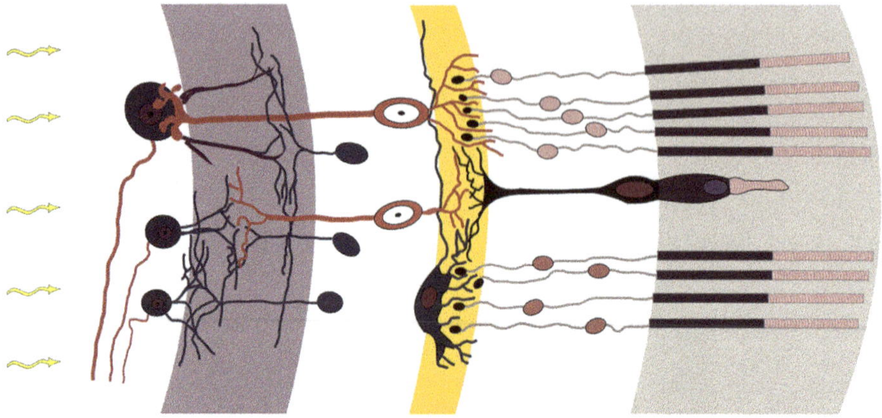

Figure 32.2. A diagram of the structure of the human retina. Photons focused from the eye lens enter from the left. The light-sensitive rod and cone cells are on the far right side of the diagram. Copyright © derivative work Fig retine bended.png: Anka Friedrich, Attribution-ShareAlike 3.0 Unported (CC BY-SA 3.0), https://commons.wikimedia.org/w/index.php?curid=7550631, Retina-diagram.svg.

cone cells, individual photons are absorbed, and this triggers a complex set of chemical pathways that ultimately leads to neurological stimulus. The stimulus is transmitted to other parts of the brain by neurons (seen at the far left edge of figure 32.2) that ultimately bundle together into a nerve cord—the optic nerve. I say 'other parts' of the brain because the retina and optic nerve are considered to be a part of the brain and not separate organs; the retinal tissue develops directly from part of the embryonic forebrain (Helga 2003).

The rods and cones are elongated in shape, and they are structured like a stack of thousands of thin disks. Each disk is separately light-sensitive. Since the photons arrive grazing along these stacks of disks, there is a high probability that an individual photon will interact with one. And so the rods and cones are highly-sensitive to light, apparently able to detect even single photons. The retina overall, however, is much less efficient; the majority of photons are absorbed by the overlying layers of the retina and so never make it to the layer of the rods and cones (Helga 2003 and Holmes 2016).

The key to the light sensitivity of the rods and cones is the molecule *retinal*. It couples to different proteins called *opsins* to form *photopigments*—*rhodopsin* (also called *visual purple*) in the rods and three different types of *photopsins* in the cones. The opsins are permanently located in the rods and cones, but the source of the retinal necessary for the light sensitivity is a layer of tissue called the *retinal pigment epithelium*, located just outside the retina (the far right-hand edge in figure 32.2).

The visual process is initiated when an individual photon interacts with the retinal part of a photopigment molecule. The photon disappears, but its energy does not; it is used to raise an electron in the retinal to a higher energy level, and this causes the molecule to change shape in a process called *photoisomerization*.

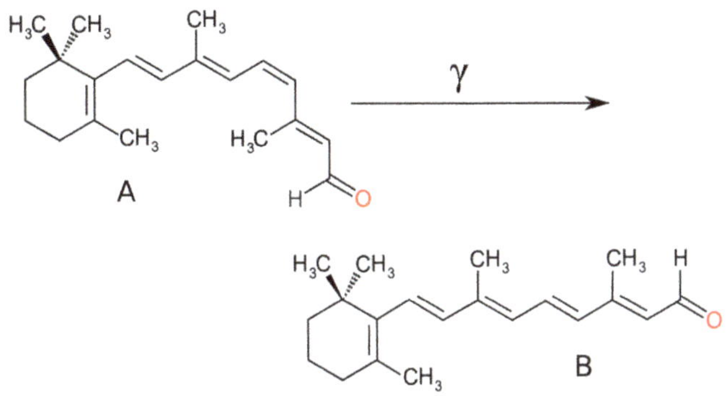

Figure 32.3. Photoisomerization of retinal. Copyright © RicHard-59 – own work, Attribution-ShareAlike 3.0 Unported (CC BY-SA 3.0), https://commons.wikimedia.org/w/index.php?curid=18461428, RetinalCisandTrans.svg.

An *isomer* is a particular way in which a given set of atoms can be arranged to form a molecule. As an example, the molecules allene and propyne are both made of three carbon atoms and four hydrogen atoms. But these same atoms can link together in two different ways. And although these two molecules have the same combination of atoms, they have different geometrical shapes, and this gives them somewhat different chemical properties. And so *isomerization* means simply that the atoms of a molecule rearrange themselves in some way; the molecule changes shape but still contains exactly the same atoms as before. Thus photoisomerization is a change in shape of a molecule brought about by exposure to light. Figure 32.3 shows the two isomers of retinal, the second being the result of photoisomerization.

Once the retinal changes to a different isomer, a completely new set of chemical pathways lie before it, and this triggers a complex chain of electrochemical events that eventually leads to a nerve impulse traveling along the optic nerve to the brain.

At a superficial level, the retina of the eye is similar to the film or digital detectors one would use with a camera. In both cases the detector is covered by many individual microscopic detectors, each of which is separately sensitive to light. For the retina of the eye, these are special light-sensitive cells. But for most of the detectors that we will consider, the process of going from absorption of photons to the formation of an image is relatively straightforward. Not so for the human eye! This book is not the place to work through the complex details of human vision, but I list some of the more striking differences. Further details can be found in, for example, Helga (2003) and in Kolb *et al* (2018) and regarding color vision in Conway *et al* (2010) and Schmidt *et al* (2014).

1. There are two basic kinds of light-sensitive photoreceptor cells in the eye, and they are not distributed uniformly. The cone cells are responsible for color vision, and they are very densely concentrated at the center of the retina (in a region called the *fovea*). For humans the cone cells come in three types,

sensitive to three overall parts of the spectrum—short wavelengths, middle wavelengths and long wavelengths. The rod cells are more densely arranged at the periphery of the visual field and they are far more sensitive to light than the cones. They play a key role in night vision.

2. The electronic signal from the detector in a digital camera is typically processed by a computer; some of this is performed after the exposure by microprocessors within the camera itself. But the light detector in the eye—the retina—elaborately *preprocesses* the direct signal from the rods and cones before signals are transmitted along the optic nerve, in ways that are only partially understood. This is perhaps not too surprising when we recall that the retina *is* part of the brain. This preprocessing is quite sophisticated, and some of it occurs via the very neural wiring in two layers of specialized cells and neurons that lie over the layer of rod and cone photoreceptors (Helga 2003).
3. Color photographic detectors make use of colored dyes or filters to render sets of light-sensitive elements responsive to different ranges of wavelengths in order to ultimately synthesize color. In this general sense, the retina is similar—it too has three types of cone cells, sensitive to different ranges of wavelength. However, the similarity ends there. The neural wiring of the retina combines the signals from the different cone cells to make opponents —direct comparisons of photoreceptor responses rather than the responses themselves.
4. Unlike photographic detectors, the retina has two separate visual systems—*photopic vision* for relatively bright light and color vision and *scotopic vision* for dim light. These two systems involve different preprocessing structures as well as distinct photoreceptors. The high-sensitivity photopigment rhodopsin, used by the rod cells, is bleached by light. Upon exposure to bright light, this high-sensitivity pigment is destroyed, leaving the lower-sensitivity cone photopigments to take over. Rhodopsin is then regenerated in dim-light conditions. Thus the retina can modify its own sensitivity according to the intensity of light. Any given photochemical detector has a far more limited *dynamic range*. Some photoelectronic detectors, however, do have a sensitivity that can be adjusted according to the lighting conditions.
5. The retina engages in *parallel processing*. The photostimuli from neighboring rod and cone cells are combined with each other in complex ways, in real time, to rearrange and compress the data (in ways still only partially understood) so as to maximize the efficiency of many visual tasks, such as detection of edges and the direct perception of contrast.
6. The structure of the retina enables it to process direct and peripheral vision in different ways. A spot of light focused on the retina is processed differently by specialized *ganglion cells*, regarding whether it is focused to the center or periphery of the visual field (e.g., see Helga 2003, figure 10).
7. The chemical and neurological processes in the retina constitute a cycle. The photoisomerizaton of retinal by photons leads to a chain of chemical and neurological responses that circle back upon themselves to their starting

point, allowing for the detection and interpretation of new photons. This cycle takes longer for dim light than for bright light (Kalloniatis and Luu 2018), allowing for better detection of dim light.

I list some of these weird ways in which the retina–brain work (insomuch as they are understood, and I understand them) because I find them fascinating. Also, these facts have implications for photography as an art. Seeing is mostly in the brain. This means that, as practitioners of 2D visual art, perhaps our task is not really to make a literal visual mapping of some part of 'the world' onto a flat surface. Perhaps, instead, we are simply trying to somehow cause interesting things to happen in the brains of people looking at our pictures. And so, how are we to best do that?

32.6 The wavelength response of detectors

The photographic detection of light is an *interaction*—light encounters matter, and both are transformed. For nearly all detectors, it is the light that undergoes the bigger change—it disappears. But the energy it carried does not, and it is this transfer of energy that allows for some physical change to occur in the detector.

The detection of light is about much more, however, than the overall absorption of energy. For most photographic detectors, it comes down to an interaction between individual atoms or molecules and individual photons of light. And there is more than one way this can all add up to a particular amount of absorbed energy.

1. Very many photons per second per square meter are absorbed, but each photon individually has relatively low energy. This corresponds to light that is bright but of long wavelength
2. Fewer photons per second per square meter are absorbed, but each photon has relatively higher energy. This corresponds to light that is dim but of short wavelength.
3. Some of the light arriving at the detector is not absorbed at all. It is instead reflected, transmitted or scattered.
4. Some of the light that is absorbed does not contribute to the photographic effect; it is absorbed but not detected.

An individual photon is either absorbed by matter, and all of its energy transferred in some way, or it is not. And so there must be at least one photon absorbed per light-absorbing unit of the light detector (an individual molecule, for example). This is one way to state the so-called Stark-Einstein law (Ware 2019, p 45).

When considering the details of any photon–matter interaction, the energy of the photon is always important. Photons of different energies are not equally likely to interact in that particular way; therefore, the particular interaction happens with greater frequency for some wavelengths of light than others. This leads to the principle that a material can only respond to the wavelengths it absorbs. If particular wavelengths are completely reflected then no photons corresponding to that energy are absorbed and thus no effect can be produced (Ware 2016, p 377).

References

Chomsky N 2000 *New Horizons in the Study of Language and Mind* (Cambridge: Cambridge University Press)

Conway B R, Chatterjee S, Field G D, Horwitz G D, Johnson E N, Koida K and Mancuso K 2010 Advances in color science: from retina to behavior *J. Neurosci.* **30** 14955–63

Evans D S, Deeming T J, Evans B H and Goldfarb S (ed) 1969 *Herschel at the Cape: Diaries and Correspondence of Sir John Herschel* (Austin, TX: University of Texas Press)

Helga K 2003 How the retina works *Am. Sci.* **91** 28–35

Holmes R 2016 Seeing single photons *Phys. World* **29** 28–31

Kalloniatis M and Luu C 2018 Temporal Resolution *Webvision: The Organization of the Retina and Visual System* ed H Kolb, R Nelson, E Fernandez and B Jones (Online: Self-Published) https://webvision.med.utah.edu/book/part-viii-psychophysics-of-vision/temporal-resolution/

Kolb H, Nelson R, Fernandez E and Jones B (ed) 2018 *Webvision: The Organization of the Retina and Visual System* (Online: Self-Published) https://webvision.med.utah.edu/

Schmidt B P, Neitz M and Neitz J 2014 Neurobiological hypothesis of color appearance and hue perception *J. Opt. Soc. Am.* **31** A195–207

Ware M 2019 Argyronomicon: Silver Photographs on Paper: Chemical History of Their Invention, Deterioration, and Conservation https://www.mikeware.co.uk/downloads/Argyronomicon.pdf

Ware M 2016 Cyanomicon II–History, Science and Art of Cyanotype: Photographic Printing in Prussian Blue https://www.mikeware.co.uk/downloads/Cyanomicon.pdf

IOP Publishing

Photography (Second Edition)
Physics and art in focus
John Beaver

Chapter 33

The characteristic curve

I use the term *density* to mean the measurable physical effect on the detector, caused by the *exposure* to light. Exposure can be related to the total light energy transferred to the detector per surface area. Photographers have their own way of describing this (chapter 22), but it can be expressed in physical units of joules per square meter (J m^{-2}) or number of photons of light per square meter that strike the detector. Since the illuminance of the light falling upon the detector is a flow or *flux* of this energy, it is expressed not in J m^{-2} but rather in J m^{-2} per second. And so we have a familiar relationship.

$$\text{exposure} = \text{illumance} \times \text{time} \tag{33.1}$$

For a photoelectronic detector, exposure results in an electrical signal of some kind, so greater density means a larger signal. For a photochemical detector, density represents the visible change in the detector. It becomes darker (for a negative process) or lighter (for a positive process), and the degree to which this happens is the density.

My usage of the word density in this general sense is somewhat unsatisfactory because the word is usually only applied to photochemical detectors; the word *signal* is more commonly used in regard to photoelectric detectors. So density, in the case of photochemical detectors, usually has a more technical meaning and is defined according to a particular mathematical relationship. Furthermore, most photochemical detectors are developed out, and so most of the density is a direct result not of the action of light during the exposure but rather the chemical reactions during development. For a typical photochemical detector, the relationship between density and exposure is complex and indirect.

A graph that shows, for a given detector, the quantitative relationship between density and exposure is called the *characteristic curve* for the detector. It tells, for any given exposure, what density results. But before we can plot density and exposure together on a graph, we must agree upon their precise mathematical definitions.

 © IOP Publishing Ltd 2022

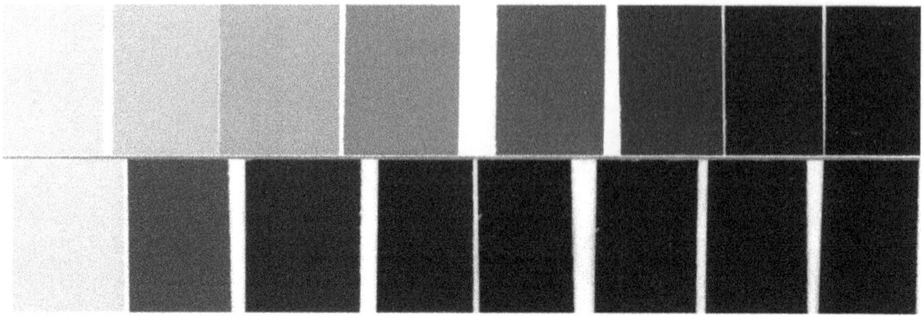

Figure 33.1. The same overall range of exposures applied to silver-gelatin enlarging paper but with two different choices for the exposure steps in between. Although the top set of exposures appears to be in roughly equal steps, it is actually the bottom set for which equally-stepped exposures were applied, with each strip receiving 9.07 seconds more exposure than the previous. For the top strip, each exposure was a factor of two greater than the previous.

Figure 33.1 shows a scan of a particular photochemical detector—black-and-white enlarging paper used in the method of ephemeral-process photography. This is a negative process, so the paper turns darker with greater exposure. I did the experiment twice, each time giving a range of different exposures to see the range of densities that would result. The illuminance was the same for both, and so I varied the exposure by varying the time according to equation (33.1). For both of the strips, the lightest portion was exposed for 1/2 second, while the darkest portion was exposed for 64 seconds.

Even though both examples in figure 33.1 show the same total range in exposure—and the same total range in density—I have broken up this range in two different ways. The example on top appears to be in roughly equal steps, as if I had added the same exposure each time from one to the next. The bottom strip, on the other hand, appears as though I made the biggest step in exposure between the first and second, adding progressively less exposure each time.

What were the actual exposures I used? The answer may be surprising; it is the opposite of how it looks. I added exposure in equal steps for the bottom example and not the top. It was made with this list of exposures (in seconds): 0.5, 9.6, 18.6, 27.7, 36.8, 45.6, 54.9, 64. That is, each exposure was 9.07 seconds longer than the previous. For the top example, on the other hand, in order to produce a set of densities that appear in equal steps, I used the following list of exposures: 0.5, 1, 2, 4, 8, 16, 32, 64. And so a set of densities that appear to be roughly in equal steps was produced by a set of exposures each of which is twice as great as the previous.

The basic result shown in figure 33.1 is typical of photochemical detectors. In order to achieve what appears as the next added step in density, it is necessary to multiply the previous exposure by some factor. For this reason, photographers work in terms of what they call *exposure steps* that are actually successive factors of two. The name is somewhat misleading; imagine walking up a staircase for which the distance doubled with each step.

A sequence of numbers such as 1, 2, 4, 8, 16, 32, etc can be written instead as follows: 2^0, 2^1, 2^2, 2^3, 2^4, 2^8, etc. This sequence follows the function $y = 2^x$. This is

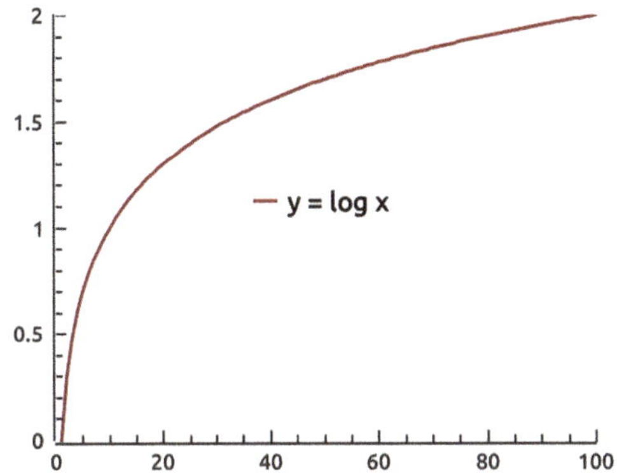

Figure 33.2. A graph of the common logarithm ($\log x$) is defined in terms of powers of ten, so it depicts ten as one and 100 as two.

known as an *exponential function*, and its inverse is called a *logarithmic function*. Photographers base their exposure steps on an exponential function with a base of two, but other bases are also commonly used. So, for example, we can set up an exponential function as follows.

$$y = 10^x \tag{33.2}$$

And so values for x of 1, 2, 3, 4, etc simply represent y values of 10, 100, 1000, 10 000, etc or the power of ten. The inverse of this particular exponential function is called a *common logarithm*, and it is defined as follows.

$$\log(10^x) = x \tag{33.3}$$

And so $\log 10 = 1$, $\log 100 = 2$, $\log 1000 = 3$, etc. A graph of the common log function can be seen in figure 33.2.

The common logarithm is also called a *base ten logarithm* because it is the inverse of the base ten exponential function. Logarithms have interesting and useful mathematical properties, the most important of which are the following equations.

$$\log(ab) = \log a + \log b \tag{33.4}$$

$$\log a^b = b \log a \tag{33.5}$$

Because of equation (33.5), logarithms 'undo' exponents, turning them into simple factors. Equation (33.4) shows that logarithms turn multiplication into addition. For this reason, published tables of logarithms were important in precomputer times because addition (and subtraction) is easier to perform by hand than multiplication (or division). This led to the old joke that certain types of snakes only reproduce when placed on a wooden table[1].

[1] Even adders can multiply on a log table.

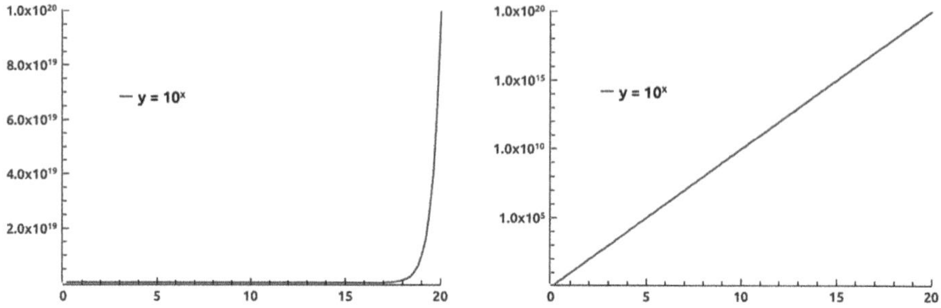

Figure 33.3. The exponential function $y = 10^x$ plotted on **(left)** an ordinary linear scale and **(right)** a logarithmic scale. The logarithmic scale straightens the exponential function and makes a large range of values more manageable.

The left side of figure 33.3 shows the exponential function $y = 10^x$ plotted versus x, while the right-hand graph shows that same function plotted with a *logarithmic scale* for the y-axis. On a logarithmic scale, each tick mark of the graph represents not an amount but rather a factor. For the example shown here, each tick is ten times greater than the tick below it. Since $\log(10^x) = x$, this is simply a straight line. And so use of a logarithmic scale on the vertical axis of a graph has the effect of turning an exponential function into a straight line. Notice how the logarithmic scale compresses an enormous range in values to a much smaller scale. Most of the direct graph of the exponential function, as shown on the left, is almost useless; at the scale of the graph, 90% of it is either indistinguishable from zero or nearly vertical.

We can also define logarithms with bases other than ten—for example a *base two logarithm*, written $\log_2 x$, is the inverse of the base two exponential function that photographers have chosen for their exposure steps. It would then represent not powers of ten but rather powers of two. It is, however, common to use the base ten 'common logarithm' even so. The base ten logarithm has the convenience that it is directly related to our base ten numbering system, and there is a very simple relationship between logarithms of different bases. Equations (33.4) and (33.5) are written for common logarithms, but they hold true no matter what the base.

$$y = 2^x \tag{33.6}$$

$$\log_2 y = x \tag{33.7}$$

$$\log_{10} y = \log_{10} 2^x \tag{33.8}$$

$$= x \log_{10} 2 \tag{33.9}$$

$$\log_{10} y = \log_2 y(\log_{10} 2) \tag{33.10}$$

$$\log_{10} y \approx 0.301 \log_2 y \tag{33.11}$$

Here I have made the base ten of the common logarithm explicit for clarity. Thus logarithms of different bases are simply proportional to each other, and to convert

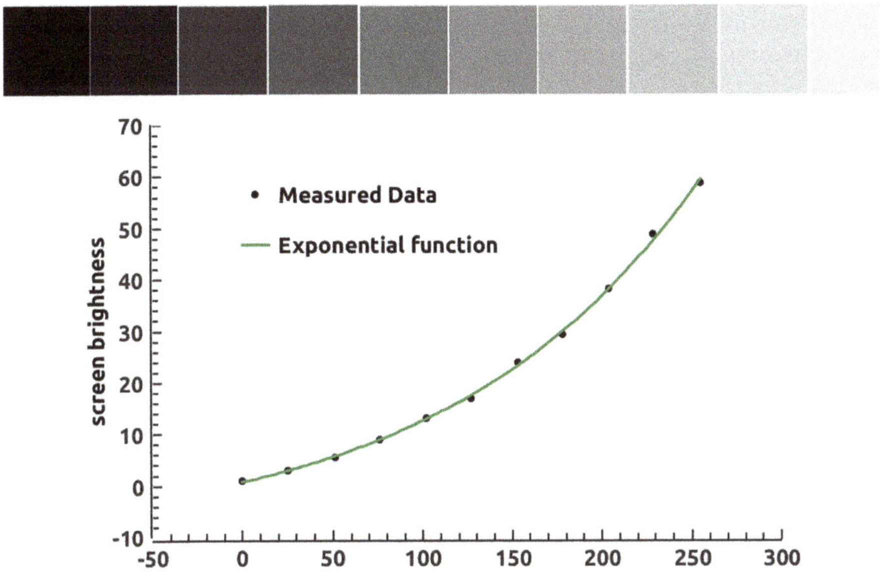

Figure 33.4. **Top:** A digital grayscale image produced so as to appear in roughly equal steps of brightness on a computer monitor. **Bottom:** A graph of the measured brightnesses of the grayscale sections, plotted on an additive scale. An exponential, rather than linear, function fits the data (green curve). This illustrates that we see levels of brightness in multiplicative, rather than additive, steps.

from base two to base ten, a simple factor of roughly 0.3 is all that is required. And so we have an important result.

> The photographer's factor-of-two steps appear as even increments of approximately 0.3 on a common logarithmic scale.

Our goal is to make an appropriate graph that relates density to exposure—the characteristic curve of our detector. And so we see from figure 33.1 that—for the photochemical detector shown in the example at least—the appropriate horizontal scale for our graph is not exposure (E) but rather $\log E$. But what is the appropriate measure of density? To answer this question we must consider not only the physical nature of the detector in question but also the way in which the human eye–brain combination perceives steps in lightness and darkness.

Figure 33.4 shows another set of gray levels that appear in roughly equal steps in brightness. This is an image created digitally rather than from a negative-process photochemical detector, and so I show the levels arranged from darkest to lightest. When I display this image on my computer monitor, it appears to me as roughly equal steps in brightness. In a darkened room I used a light meter to measure the brightness of the monitor for each of the gray levels shown in figure 33.4, and the result can be seen graphed at the bottom.

The graph of brightness levels shows an increasing slope, like some type of exponential function—and indeed, the data is well fit by one (the green curve). This crude experiment illustrates the well-known fact that our vision is better described by a logarithmic scale; what appears to us as equal steps in brightness is more accurately described as equal factors (Adler *et al* 2014). And so photographic density might also be better defined according to a logarithm of measured levels of brightness. This was first recognized in the early days of silver-gelatin photography. In the late 1800s Ferdinand Hunter and Vero Charles Driffield made the first quantitative measurements of the density of exposed photographic emulsions, and we still use their definition today (Ferguson 1920).

One way to quantitatively measure the 'darkness' of a piece of exposed and developed photographic film is to shine light through it and measure by what factor the light has dimmed while passing through the film. A device that does this is called a *densitometer*; it directly measures the *transmittance* (T) of the film. The transmittance is simply the fraction of light that is transmitted, the rest having been absorbed by the film. For a negative process, greater exposure means less transmittance, and so it makes sense to define density as $1/T$ rather than T. However, this simple measure would have little relationship to how we perceive steps in brightness. So, instead, the photographers' definition of density (D) also incorporates a logarithm as shown in equation (33.12).

$$D = \log(1/T) \tag{33.12}$$

It appears that it is appropriate for both density and exposure to be described in terms of logarithms of the physically measured quantities. For this reason, the characteristic curve of a photochemical detector has, since Hunter and Driffield, been portrayed as a graph of D versus $\log E$. Since density is also defined in terms of a logarithm, the characteristic curve is a *double-log* or *log–log* plot, with a logarithmic scale on both axes.

But there is another important consideration—the physical process by which exposure causes the light detector to undergo its physical change. For a negative-process photographic film, the exposure to light ultimately causes the transmittance of the film to decrease. And so what is the mathematical relationship between the exposure on the one hand and the inverse of the film transmittance on the other? How will that relationship appear on the log–log plot of our characteristic curve?

We have already seen that an exponential function appears as a straight line on a graph that has a logarithmic y-axis. But an exponential function is a strongly-curved line on a double-log plot. And so is there a mathematical function that appears straight on a log–log plot? And does the mathematical relationship between $1/T$ and E follow such a relationship for a real photochemical detector? The answer is 'yes' to the first question and 'sort-of yes' to the second.

A double-log plot straightens out not an exponential function but rather a *power law*. A power law is defined by the following equation.

$$y = ax^{\gamma} \tag{33.13}$$

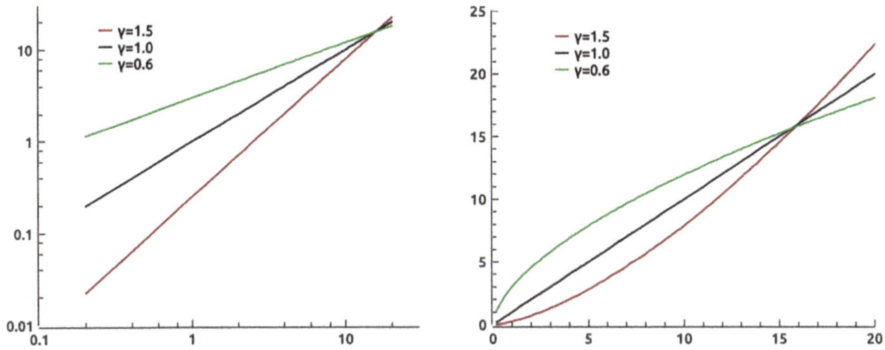

Figure 33.5. **Left:** Three power-law curves plotted on a double-log graph. They all appear as straight lines; the greater slope corresponds to the higher power, gamma (γ). **Right:** The same mathematical functions plotted on an ordinary linear scale for x and y. For $\gamma > 1$ the curve is concave upward, while the opposite is true for $\gamma < 1$. A straight-line, linear curve corresponds to $\gamma = 1$.

Here a and γ (the Greek letter gamma) are constants. Notice, as with an exponential function, there is an exponent involved—but for a power law it is the *constant* that is the exponent and not the variable x. We can easily see that this function will appear as a straight line on a log–log plot by taking the logarithm of both sides.

$$\log y = \log(ax^{\gamma}) \tag{33.14}$$

$$\log y = \log a + \log x^{\gamma} \tag{33.15}$$

$$\log y = (\log a) + \gamma \log x \tag{33.16}$$

$$y\text{–axis} = \text{intercept} + \text{slope} \times x\text{–axis} \tag{33.17}$$

If we use a log–log plot to graph a power law, with $\log y$ plotted versus $\log x$, then we get a straight line with a slope equal to the exponent, γ, and a y-intercept equal to the logarithm of the scale factor a. The left side of figure 33.5 shows the log–log plots of three different power laws—each with a different γ. They all appear as straight lines on this double-log graph, the steeper line belonging to the larger value of γ. The right side of figure 33.5 shows these same three functions plotted without the logarithmic axes, and here we see a qualitative difference. A power law with a $\gamma > 1$ curves upward, while $\gamma < 1$ produces a downward-curving graph. A value of $\gamma = 1$ leads to a simple straight line.

It turns out that the relationship between E and $1/T$ for developed-out photographic film is often related approximately by a power law but with two important exceptions. First, equation (33.13) has the property that $y = 0$ when $x = 0$. But this is not the case for the relationship between film transmittance and exposure; there is a *minimum density* that occurs even with zero exposure. This is called the *fog density* or D_{min}. And so a more appropriate mathematical relationship would be represented in equation (33.18).

$$y = ax^{\gamma} + b \tag{33.18}$$

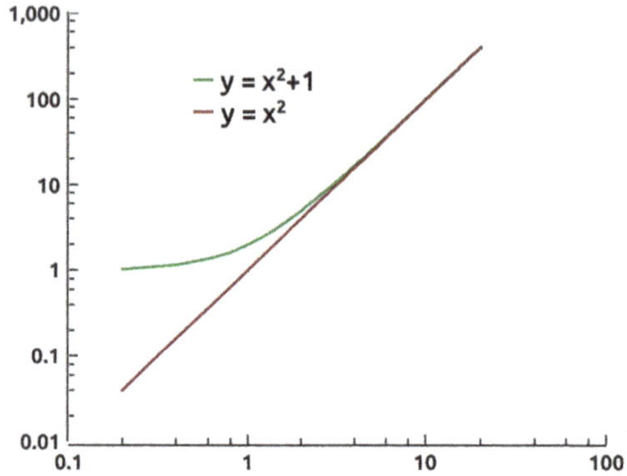

Figure 33.6. A simple power law (red curve) appears as a straight line on a log–log plot, unless it also includes an additive constant (green curve). The characteristic curve of photographic film has a 'toe' like the green curve because the film absorbs some light even with zero exposure.

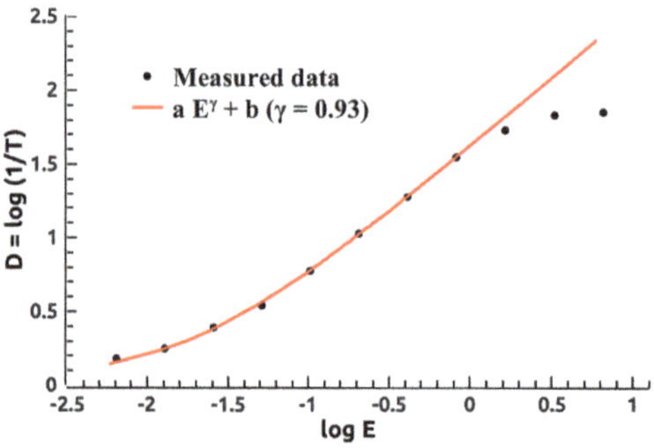

Figure 33.7. The characteristic curve for real densitometer data from Stroebel and Compton (2000, p 96). The red curve is a power law, with $\gamma = 0.93$, fit to the toe and straight-line part of the data. There is also a shoulder because real photographic film has a maximum density that is not exceeded with increased exposure.

Here b is the value of y for $x = 0$. Taking the logarithm of both sides of equation (33.18) cannot be simplified to a straight line like equation 33.13. But it is clear that for large enough values of x, the addition of the constant b would make an insignificant difference. And so we expect equation (33.18) to appear very much like a straight line on a log–log plot for large values of x but to deviate significantly from that at small values. Figure 33.6 shows log–log plots of power laws with $\gamma = 2$ both with and without an additive constant. We would, therefore, expect our characteristic curve to appear more like the green curve: with a straight-line part but also a flat *toe* corresponding to the expected D_{min} of our photographic film.

Figure 33.7 shows a D versus $\log E$ curve for some real densitometer data from Stroebel and Compton (2000, p 96). Notice that the data points are separated on the horizontal axis by $\log E = 0.3$, corresponding to a difference of 1.0 if we had used a base two logarithm instead. Thus the different exposures were made in the photographer's traditional steps, corresponding to factors of two.

Plotted on top of the measured data in figure (33.7) is the mathematical function of equation (33.18) with values of a, γ and b chosen so as to best fit the data. It clearly has a toe and a *straight-line part* as we would expect from a power law with a D_{min}. But there is also a *shoulder*: the density increases more slowly than the power law at large exposures, appearing to level off to a maximum density.

It should be unsurprising that the characteristic curve of a real photographic film has a maximum-density shoulder or D_{max}. There is, after all, only so much of the photochemical reactant to be darkened by light and the chemical development process. Once the film is as dark as it can get, one would expect greater exposure to have no measurable effect.

The true mathematical shape of the characteristic curve for photographic film cannot be summarized by any simple formula. It can only be modeled accurately by a complex statistical analysis that includes all aspects of the process, from exposure to light through chemical development and fixing, to the final measurement by the densitometer. But the toe and straight-line portions are often well fit by a power law with a D_{min}. Some special kinds of technical film have a D_{max} that is too high to measure with a typical densitometer, so the measurable part of the characteristic curve shows only the toe and straight-line part with no shoulder (e.g., see Stroebel and Compton 2000, p 109, figures 4–28 and 4–29). In those cases the measured part of the characteristic curve appears very similar to a simple power law plus constant, as in the green curve of figure 33.6.

33.1 The characteristic curve and photoelectronic detectors

Whether negative process or positive process, developed out or printed out, a photochemical detector produces a visible image in itself. Expose a roll of black-and-white silver-gelatin negative film in a camera, develop and fix the film, and then hold it up to the light. The image is right there, visible in the very physical object that interacted with the light. Not so for a photoelectronic detector. Only an electronic signal is produced by exposure to light. The visible image must be produced from that signal by other means: either a lighted screen or some kind of printing process. This means that we can, somewhat independently of how the detector responds to light, transform the electronic signal into something that corresponds to the logarithmic manner in which we perceive levels of brightness. For a photoelectric detector, the characteristic curve—a graph that relates exposure (the cause) to density (the effect)—can be connected more directly to the physical relationship between exposure and signal. Thus for a photoelectric detector there is little motivation for the nearly 150 year-old $\log(1/T)$ versus the $\log E$ graph of Hunter and Driffield.

For these reasons photoelectric detectors are more commonly described by the direct relationship between exposure and the resulting electronic signal. For the most-important photoelectric detectors we consider in this book that response is approximately *linear*, corresponding to the following graph.

$$y = ax + b \tag{33.19}$$

Note that equation (33.19) is the same as equation (33.13) for the special case of $\gamma = 1$.

Photographers often use the term 'linear' in a different sense—to signify the straight-line part of the D versus $\log E$ characteristic curve of photochemical film emulsions. This usage is both wrong and right. First, why is it wrong? We have already seen why: the response is, at its simplest, not linear but rather a power law. Furthermore, the toe of the curve is part of the same power law as the straight-line part, yet photographers call only the straight-line part 'linear.' If γ is not equal to one, then no part of the characteristic curve is linear. Ironically, if $\gamma = 1$ then not only the straight-line part but also the toe of the characteristic curve is linear. From a perspective of the physical relationship between exposure and the actual darkening of the film, the photographers' common use of the term 'linear' is incorrect.

But, as is the case for many commonly-used 'incorrect' terms, there is a method to the madness. If we are using a photochemical detector not for scientific measurements but rather for making pictures, then we have already seen that it is density—as defined by photographers with a logarithm—that is the point.

References

Adler M, Mayo A and Alon U 2014 Logarithmic and power law input–output relations in sensory systems with fold-change detection *PLOS Comput. Biol.* **10** e1003781

Ferguson W B (ed) 1920 *The Photographic Researches of Ferdinand Hurter and Vero C. Driffield* (London: The Royal Photographic Society of Great Britain)

Stroebel L and Compton J 2000 *Basic Photographic Materials and Processes* 2nd edn (Waltham, MA: Focal Press)

Chapter 34

Silver-based photochemical detectors 1

34.1 Black-and-white silver-gelatin emulsions

The most common type of black-and-white photochemical detector is the *silver-gelatin emulsion* (also called gelatin-silver emulsion). This is the basic photochemical process used both for ordinary black-and-white film and for making black-and-white prints in the darkroom. It is also the foundation for the *chromogenic* color film and printing process we describe in section 34.3.

Microscopic[1] crystals of a *silver-halide salt* are suspended in a thin, transparent layer of gelatin, which is coated onto the surface of the plastic film or print paper. The word 'halide' refers to a column (group 17, second from the right) in the periodic table of elements called the *halogens*. As with other columns in the periodic table, the halogens are grouped together because they would require the same number of electrons to complete their outermost electron shell. For the halogens the outermost shell is only one electron shy of being filled, so they easily accept electrons from other chemical species. Since they all have this in common the halogens all have similar chemical properties, and this is why they are grouped together. Examples of halogens include fluorine (F), chlorine (Cl), bromine (Br) and iodine (I).

Halogens form halide salts with elements that easily donate electrons, such as those that have only one electron in their outermost shell. Silver (Ag) is an example of such an atom. Since atoms like to have their outermost shell filled, everyone can be happy. Thus, a silver atom gives up one electron to a halogen and now both have their outermost shell filled (Ag had one extra while the halogen had one too few). In addition, the silver now has a positive electric charge (since it lost an electron), while the halogen now has a negative charge (since it received an extra electron). The two stick together due to the electrical attraction of unlike charges.

[1] The size of the crystals varies in silver-gelatin emulsions, but the smallest crystals are somewhat too small to be seen with an ordinary light microscope.

doi:10.1088/978-0-7503-3703-8ch34 © IOP Publishing Ltd 2022

Examples of such silver-halide salts are silver chloride (AgCl), silver bromide (AgBr), silver iodide (AgI) and silver fluoride (AgF). Three of these in particular are of interest for photography, as they share similar physical and light-sensitive properties: silver chloride, silver bromide and silver iodide. Like many ionic compounds, they may form an ordered and ongoing three-dimensional alternating pattern of silver and halogen ions—a crystal.

It is easy to make a silver halide by mixing the compound silver nitrate ($AgNO_3$) in an aqueous solution with a halide salt such as potassium chloride or sodium chloride (KCl or NaCl), potassium bromide or sodium bromide (KBr or NaBr) or potassium iodide or sodium iodide (KI or NaI). Both silver nitrate and the halide salt are soluble. In solution, a halide salt such as potassium chloride dissociates into the ions K^+ and Cl^-. Similarly, silver nitrate dissociates in solution into Ag^+ and NO_3^-. But when the two compounds are dissolved together into the same solution, the Ag^+ has a greater affinity for the Cl^- than the NO_3^- from which it came, and the K^+ has a greater affinity for the NO_3^-.

This would not matter so much if all of the ions stayed in solution. But the combination of the silver and halide ions—silver chloride in this example—is markedly insoluble in water. And so solid crystals of silver chloride form and sink to the bottom of the container, leaving a solution of—in this example—potassium nitrate. The overall reaction works like the following equation (using, as an example, KCl as the halide salt).

$$AgNO_3 + KCl \longrightarrow KNO_3 + AgCl \tag{34.1}$$

Silver chloride is sensitive to light by mechanisms we discuss later. But producing a fine sludge of silver chloride in this simple way is not very useful for photography. There is a trick—let the reaction take place in a thick solution of gelatin, so the silver-halide crystals remain suspended and separated from each other as they form. The gelatin is mostly transparent, and it allows the crystals to grow separately to a consistent size. This thick liquid gelatin, with silver-halide crystals suspended in it, can then be applied as a thin coating onto film or paper and allowed to dry.

And so a silver-gelatin emulsion is a transparent gelatin matrix of tiny crystals of some combination of silver chloride, silver bromide and silver iodide (other, more-complex silver salts are sometimes used as well). The silver-halide crystals are sensitive to light by way of the following process (Stroebel *et al* 2000, pp 198–200 and Ware 2019, chapter 23).

1. Photons are absorbed at an imperfection in the silver-halide crystal. The energy of the photons is transferred to negatively-charged electrons in the crystal. This higher energy level makes the crystal electrically conductive.
2. Electrons migrate to the imperfection site, and neutralize (reduce) Ag^+ to make simple elemental silver (Ag). It is this ordinary silver that makes the dark part of the image (the silver-halide crystals are mostly white).

3. When used for developing out, this image of silver is far too faint to be visible; there may be as few as only four reduced silver atoms in a given crystal. The image is then amplified by a chemical developer, which further reduces Ag^+ to silver by chemical means—but only in the presence of existing silver. And so a *developed-out* image forms that is made of millions of times more silver than the original latent image originally reduced by the photons. The step of chemical development turns the invisible latent image into a visible 'blatant' image.
4. Ordinarily, the remaining silver halide would then be chemically removed (using a fixer) in order to render the image no-longer sensitive to light.
5. If the image is *printed out*, rather than developed out, then millions of times more photons must be absorbed in order to make a visible image because there is, by definition, no chemical development to amplify the image. This means that printing-out processes inevitably require far more light to make an image.

In a photographic emulsion these crystals of silver halide are microscopically small. But each is still comprised of an enormous number of individual atoms. When photons strike these atoms, there is a small possibility for any individual photon that it will be absorbed, and in the process a few atoms of metallic silver are freed from their halide companions. These silver atoms would appear dark, but in an ordinary exposure there are far too few of them to see.

The more photons that arrive at a given silver-halide crystal, the more likely it is that at least several silver ions have been reduced to metallic silver, producing a tiny metallic silver speck in that crystal. In regions of the emulsion that received more light, a greater percentage of silver-halide crystals contain metallic silver atoms, and those crystals that received more light are likely to contain a greater number of such atoms. And so we have a *latent image*, which is invisible at this point. This latent image can persist in the emulsion for many years, especially if it is kept at a sufficiently low temperature. A few years ago I successfully developed images—albeit degraded—from a roll of exposed film I found in a camera made in the early 1940s; the automobiles in the pictures were consistent with that time period.

The process of making the latent image visible in an exposed silver-gelatin emulsion is called *development*, and a (usually liquid) chemical that does this is called a *developer*. The developer accomplishes this feat by chemically reducing the silver halide to metallic silver, eventually turning the entire silver-halide crystal dark. The developer will do so only wherever some free silver atoms already exist. And so only those silver-halide crystals that contain the latent silver atoms from the exposure to light are reduced to dark silver metal. Thus only areas that were exposed to light turn dark, and a visible *negative* image is formed.

The development process is more-or-less rather than either-or. Put the emulsion in the developer solution and, as time passes, silver-halide crystals with more latent silver atoms will develop first, while crystals with fewer latent silver atoms will take longer to develop. Thus more and more of an exposed part of the emulsion will be reduced to silver, making that part darker and darker. But those parts that received more light during the exposure become darker faster during development. Thus an *amplification effect* occurs.

The trick is to develop the exposed emulsion just enough that the dark areas are as dark as needed, while leaving the unexposed areas still white (or clear if it is a transparent piece of film). But since the emulsion must be developed only for a set length of time, the development must then be stopped by placing it in a *stop bath* (usually a weak acid solution) that chemically neutralizes the developer.

The final step is to remove the remaining still-sensitive-to-light silver halide left in the emulsion, or otherwise the image would not be permanent; it would gradually turn dark all over if exposed to light. This process is called *fixing* the image, and a chemical that accomplishes this is called a *fixer*. The most common approach is to use a fixer that dissolves the remaining silver-halide crystals, allowing them to be washed away. It is sometimes underappreciated just how important to the history of photography was the discovery of a practical method for fixing silver-based photographic emulsions. The astronomer and photography pioneer John Herschel is usually given credit for first solving this problem with his discovery that sodium thiosulfate (sometimes called 'hypo') can be used as a fixer for silver-halide photography (Ware 2019, section 7.5.4).

Silver-gelatin emulsions are made for different purposes; the two most common uses are image capture in a camera and producing a photographic print to hang on a wall. Emulsions for image capture are usually designed with higher sensitivity and a greater range of wavelength response. Black-and-white photographic film, used for in-camera image capture, usually contains a combination of silver chloride, silver bromide and silver iodide, along with organic dyes that extend the sensitivity to longer wavelengths. On the other hand, black-and-white photographic *enlarging paper*, used to make prints in the darkroom from film negatives, is much less sensitive and typically has a wavelength response that extends only to shorter than the blue-green part of the spectrum[2]. These emulsions use only combinations of silver chloride and silver bromide, with no silver iodide.

The silver-gelatin process is most easily used as a negative process because the exposed areas turn dark as compared to the relatively-transparent gelatin matrix. Ordinarily, the same process is used again in a second step to make a positive print from the original negative. The original image is captured on a transparent film medium so that light can be transmitted through it. A piece of white paper coated with the emulsion is then exposed by projecting light through the transparent negative, and a positive paper print appears. A negative of a negative is a positive. This extra step has some advantages; the second exposure and development can be carefully chosen to adjust the range of tones that appear on the final print. It is possible to get a good print from a mediocre negative.

[2] This has the advantage that one can handle them under an orange 'safelight,' so the photographer can see to work without exposing the light-sensitive paper.

34.2 Direct positives and reversal processing

Although silver-gelatin emulsions are now most commonly used as a two-step negative-upon-negative process, there is a way to make a positive image in only one step. The fully-developed silver transmits little light but is somewhat *reflective*, and this reflective property can be enhanced by carefully-chosen development and toning. Thus we can instead view the emulsion against a black background; with proper lighting the more-exposed areas of developed silver appear brighter than the black background showing through the lesser-exposed areas of clear gelatin. And so a positive image appears.

Such a *direct positive* is the foundation for the *dry-plate tintype* process we describe further in chapter 35. Several other historical direct-positive processes work in a similar way. One disadvantage to this approach is that it is difficult to achieve bright highlights, so the tonal range is far more limited than a print on white paper made from a negative.

There is another way, however, to make a positive print from the very silver-gelatin emulsion that was exposed in the camera. *Reversal processing*[3] produces a positive print with a full tonal range on white, exposed silver-gelatin paper. The paper emulsion is exposed in the camera and then developed, which brings out a negative image in the ordinary way. Instead of fixing the image—removing the remaining undeveloped silver halide—the developed silver is bleached away. This makes the negative image disappear, and what remains is undeveloped silver halide.

But this remaining silver halide is not uniform. Where the original exposure was large most of the silver halide was developed to silver and then bleached away, leaving relatively little silver halide remaining. The opposite is true for areas that received very little exposure. For these regions most of the original silver halide remains because very little was exposed, developed and bleached.

One can thus expose the emulsion to a bright uniform light. Simply turn on the lights in the darkroom, for example. This fully exposes all of the remaining silver halide. The print is then redeveloped, and a positive image appears. Figure 34.1 shows an example of a print made with ordinary silver-gelatin printing paper exposed in a large format camera and then reversal processed. As is the case with a direct positive, the picture on the wall is the same object that recorded the light in the camera.

The image in figure 34.1 is reversed left to right (see the numbers on the clock) because the viewed surface of the print was facing the camera lens during the original exposure. When making a print from a negative, on the other hand, the lens-facing emulsion side of the negative is placed face-down while exposing the print. A normal orientation results.

[3] The terms 'direct positive' and 'reversal processing' are sometimes used inconsistently by photographers. In this book I use 'reversal processing' only as described in this section. I use 'direct positive' to refer to all other photographic processes for which the final positive print is the same object that was originally exposed to light.

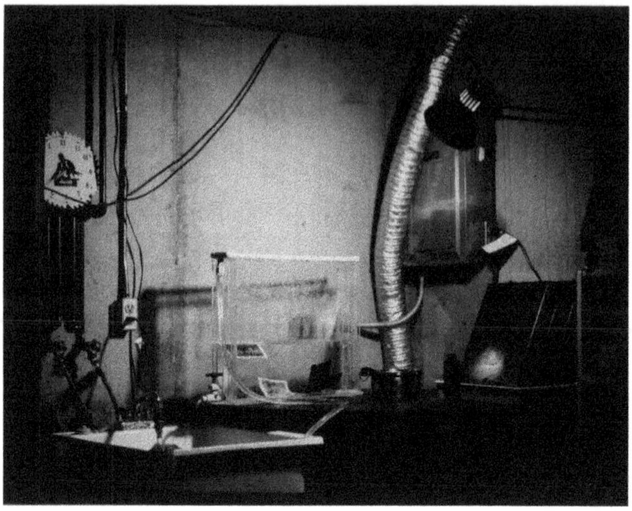

Figure 34.1. A direct and straightforward scan of a 9 cm × 12 cm positive image on reversal-processed enlarging paper.

Reversal processing can be used to make 35 mm color transparencies for direct projection with a slide projector. The technique can also be used in the darkroom to make an enlarged negative transparency from a film negative.

34.3 Chromogenic color emulsions

The typical color film process is called chromogenic or *C-type*. As with black-and-white photography, the chromogenic process uses silver halides as the light-sensitive material. But the chromogenic emulsion uses subtractive color synthesis (chapter 29) to record a color negative image (or a positive color image with reversal processing).

There are typically three layers of different silver-halide emulsion, each with a different wavelength response curve. The layers are sensitive, respectively, to only red, blue or green light—the additive-primary colors.

The exposed silver in each layer is developed (which makes it black), but it is ultimately removed in a two-step process. During development, a colored dye is chemically attached to the silver halide in each layer. This is called *dye coupling*. Each additive-primary layer is coupled to its respective subtractive complement (the subtractive primary colors). And so a cyan-colored dye is coupled to the red-sensitive layer, a magenta dye is coupled to the green-sensitive layer and a yellow dye is coupled to the blue-sensitive layer.

After development the black silver is bleached away, and the remaining undeveloped silver halide is removed with a fixer (these two steps are often done at the same time, combining the bleach and the fixer into a bleach-fix or *blix*). What remains are the colored dyes, and they are the negative of the colors that were exposed by the three silver-halide layers (London *et al* 2005, p 171).

This same chromogenic process is used to make color print paper, which can then be exposed by light projected through a color negative.

34.4 Instant film

The Polaroid corporation was the first to market systems of silver-gelatin processes for which the light-sensitive material and chemical processing are self contained in the same package, so no wet processing in a darkroom is required. These films require special cameras; many of Polaroid's designs were marvels of ingenuity. I will not go into the complex and amazing history of these films and cameras in detail, but they fall into two basic categories: *peel-apart* and *integral films*. In later years Fuji manufactured both kinds of instant films as well. As of this writing only integral films are manufactured, but there are still existing old stocks of peel-apart film (in my refrigerator, for example).

34.4.1 Peel-apart instant films

Peel-apart instant film consists of a silver-gelatin negative that is exposed in the camera. It is then, by a clever mechanism inside the camera, matched up face-to-face with a transfer print material. This paper is not light sensitive, but it has a chemical developer incorporated into it. The exposed negative and transfer paper are then pulled together through rollers, which first crush a paper package of gelled processing chemicals. Upon passing through the rollers on the way out of the camera, the processing gel is spread out between the negative and the print transfer paper, both of which have opaque backsides.

Outside of the camera, this negative-gel-print sandwich is allowed to process for a minute or two; a positive print is then peeled off and the negative (with the processing gel stuck to it) is usually discarded. The processing gel contains a silver-halide solvent and an activator that triggers development of the latent image in the negative to reduce it to silver. The silver-halide solvent dissolves the remaining silver halides in the negative, which then diffuse through the gel to the print transfer material. The silver halides are then made dark by the developer incorporated into the transfer print material. Since the light areas of the negative are the regions with the least reduced silver, they are the regions with the most silver halide. In those regions more silver halide diffuses to the print, and that part of the print is darker. The net result is a positive print. A variation of this *diffusion transfer process* incorporates a multilayered emulsion and colored dyes to make a color print (Stroebel *et al* 2000, p 210).

34.4.2 Integral films

Integral films are complex multilayer emulsions with all components of the silver-based photochemical process included into one thin package with a transparent plastic cover on the top through which the resulting image is visible. After exposure, the chemistry is activated by passing the film package through a set of rollers (built into the specialized cameras) that crush microscopic capsules of chemicals.

The positive image appears gradually after a few minutes. Polaroid originally made only color versions of this film in small sizes of a few inches on each side. Both color and black-and-white versions are now available as of this writing.

Both integral and peel-apart instant films produce a final positive print. Because the peel-apart negative is placed face-to-face with the print transfer paper to make the final print, the original mirror-reversed negative image is reversed again back to a normal orientation. Not so for integral films; they produces a left-right reversed image like other direct-positive processes. Polaroid's solution was to design cameras that incorporate a mirror to reverse the image optically before it arrives at the film.

References

London B, Upton J, Stone J, Kobré K and Brill B 2005 *Photography* 8th edn (Englewood Cliffs, NJ: Prentice-Hall)

Stroebel L, Compton J, Current I and Zakia R 2000 *Basic Photographic Materials and Processes* 2nd edn (Waltham, MA: Focal Press)

Ware M 2019 Argyronomicon: Silver Photographs on Paper: Chemical History of Their Invention, Deterioration, and Conservation https://www.mikeware.co.uk/downloads/Argyronomicon.pdf

IOP Publishing

Photography (Second Edition)
Physics and art in focus
John Beaver

Chapter 35

Silver-based photochemical detectors 2

35.1 Daguerreotype

Daguerreotype was the first commercially-successful photographic process, introduced by Louis Jacques Mande Daguerre in 1839. It is in some ways one of the most physically-complex methods to make a picture; the precise mechanism behind the production of a daguerreotype image is still not fully understood.

A highly-polished silver-plated sheet of metal is first exposed, in a special dark box, to a halogen gas (iodine, chlorine, bromine, or some combination of the three). This results in the formation of silver halide directly onto the polished silver plate. The plate is then exposed in a large-format camera and a latent image of metallic silver forms in a manner somewhat like that of the silver-gelatin process. This latent image is then made visible, usually by development with mercury vapor.

But the daguerreotype process differs in a fundamental way from the silver-gelatin process and indeed most other photographic processes. In the silver-gelatin process, the reduced metallic silver absorbs more—and so reflects less—of the light that falls on it. And so a negative image results when the emulsion is placed on a white or clear background; the reduced silver reflects less light than the white paper base in a silver-gelatin print, for example. The daguerreotype process relies instead on a physical change in the direction in which light reflects off the developed plate.

The unexposed parts of the daguerreotype consist of highly-polished silver— it is in essence a mirror, so when illuminated a mirror-like *specular reflection* occurs. But areas of the plate that have been exposed to light and reduced back to silver metal undergo a physical change in structure, and they scatter light that reflects off of those areas. A somewhat more *diffuse reflection* results from the exposed and developed areas (Stroebel *et al* 2000, pp 203–204). The daguerreotype is illuminated and viewed such that the reflection off the mirror-like surface of the unexposed regions misses

 © IOP Publishing Ltd 2022

the viewer altogether. But the diffusely-reflecting exposed regions reflect light in all directions, some of which are toward the viewer. Thus the exposed areas appear bright, while the unexposed areas appear dark.

There is still controversy regarding the exact physical process by which this happens. It is often stated that the Mercury mixes with the silver to form an *amalgum*, which is dull instead of shiny (e.g., see Stroebel *et al* 2000, p 204). But this has been called into question most particularly by the fact that there is a variation of the daguerreotype (called the Becquerel process) that uses no mercury at all (James 2016, p 18).

The traditional daguerreotype process is very difficult and quite dangerous because toxic gases are used. By 1860 other processes were discovered that were more practical, and so its use became increasingly rare. Nonetheless, there is nothing quite like a daguerreotype. The image has a shimmery quality that is captivating. Because of this appeal, in recent years there has been something of a Renaissance in the use of daguerreotype, and techniques have been worked out to make it less dangerous (Rexer 2002, p 28ff). Two important contemporary photographers using the daguerreotype process are Jerry Spagnoli and the late Chuck Close.

35.2 Callotype, salted paper, and albumen print

In 1834 William Henry Fox Talbot publicly described a photographic process he dubbed *photogenic drawing*. Sized[1] paper is treated with a weak solution of sodium chloride (table salt) and then silver nitrate. When exposed to light it turns dark, and thus objects may be laid upon it and their negative shadows recorded when the *salted paper* is placed in sunlight. An additional bath of sodium chloride makes it much less sensitive to light, and so the image can be viewed—at least briefly—without significant further darkening.

This is the barest outline of the salted paper process; the details are far more complex. See James (2016, chapter 5) and Ware (2019, section II) for more on the historical evolution of the process. One key ingredient is the sizing of the paper—Talbot's paper was sized with gelatin—to prevent the chemistry from permeating too deeply into the paper fibers. The paper fibers themselves are also key; they help take up excess chlorine, as the action of light reduces the silver chloride to dark, elemental silver.

The salt print is a printing-out process, and this greatly limits its sensitivity. For this reason it was used by Talbot to make photograms and contact prints in direct sunlight. It is still used today by art photographers, often as a method for making a contact print from a negative that was produced by some other process.

By 1840 Fox Talbot had greatly improved upon his salted paper with the *calotype* process—the first *developed-out* photographic process on paper. A latent image made by the action of light is chemically amplified afterward in the darkroom. The daguerreotype, invented about the same time, also uses chemical amplification.

[1] Paper is often coated with a semipermeable *sizing* (often starch or gelatin) for added strength and to make it less absorbent.

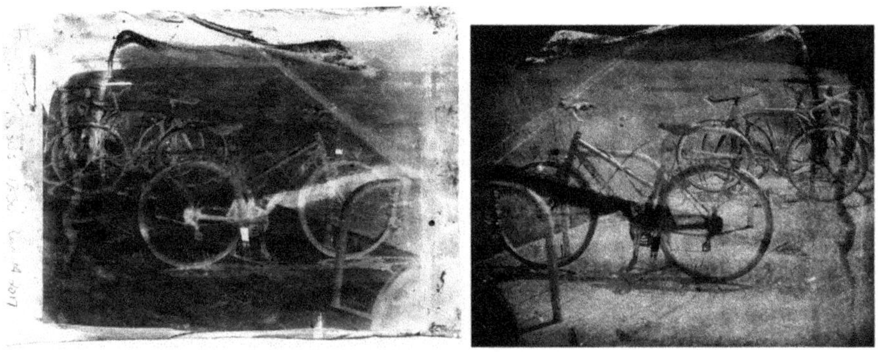

Figure 35.1. *Bicycles*, ©Brittonie Fletcher, calotype negative and salt-print positive 2017. Used by permission.

Although Daguerre's process was more commercially successful it used the action of gases, while Talbot's was a wet-chemistry process. Both used potentially hazardous chemicals, but liquids are easier to control than gases; nearly all photography since daguerreotype has used the chemistry of aqueous solutions.

The calotype process is complex, but it uses silver iodide as its light-sensitive salt rather than silver chloride. It was sensitive enough to record an in-camera negative on a sunny day in a matter of minutes. Once chemically processed, the paper negative was oiled or waxed to make it semitransparent; it could then be used to make a positive calotype contact print with sunlight. It is more likely that the simpler salt-print process was used to make a print, since exposure time was not of a concern.

Talbot's calotype process was initially overshadowed by daguerreotype, much to the inventor's frustration. But it paved the way for much of the future of photography for a simple reason: an unlimited number of prints can be made from a single in-camera exposure of the negative.

Prints from a calotype negative are inherently soft due to the inevitable texture of the paper fibers. Although considered a disadvantage by many, some photographers reveled in the painterly look of a salt print made from a calotype negative. The partnership of Hill & Adamson studio made thousands of calotype negatives in Scotland in the 1840s (Cooke and Kinneberg 2014, p 219). Figure 35.1 shows a calotype negative and a salt print made from it by the Edinburgh-based photographer Brittonie Fletcher.

By the late 1850s most photographic prints from negatives were *albumen silver prints*. Albumen is a protein made from egg whites with properties similar to gelatin. In the albumen print, the salt—sodium or ammonium chloride—is mixed with the albumen first and then coated onto the paper and allowed to dry, producing a smooth and glossy finish. When dipped in silver nitrate, silver chloride forms on the surface of the smooth albumen rather than inside the paper fibers. An albumen print can thus produce a more finely-detailed image with a wider tonal range. Figure 35.2 shows an albumen print by Brittonie Fletcher. Notice the fine detail present, clearly

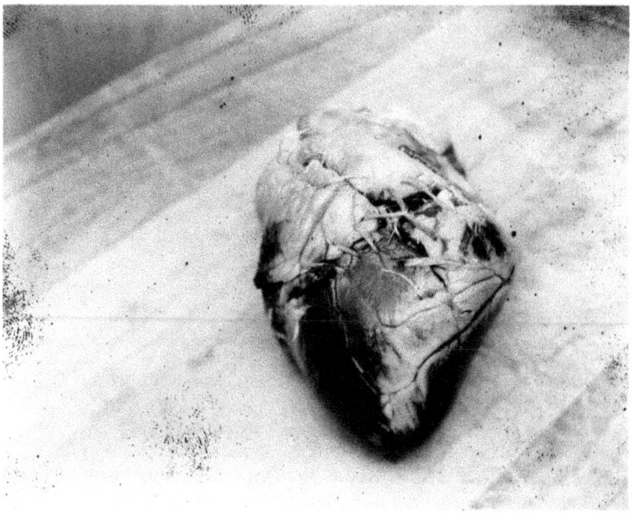

Figure 35.2. *Goat Heart*, ©Brittonie Fletcher, 20 × 24 inch albumen print 2008. Used by permission.

showing the fingerprints Fletcher left (intentionally) on the ultra large-format negative.

Like Talbot's salt print, this is a printing-out process. As with most silver processes, the excess silver halide must ultimately be removed, or the print will continue to darken with exposure to light. By the 1850s John Herschel's effective sodium thiosulfate 'fixer' was in widespread use, essentially the same method of fixing silver-gelatin prints up to the present.

35.3 Wet collodion, ambrotype and tintype

In the two decades before the invention of the silver-gelatin process, silver-halide emulsions were made instead with *collodion*—a sticky liquid made of gun cotton dissolved in ether. The collodion is mixed with halide salts and then poured by hand onto a glass plate that will fit in the camera. If tipped carefully, the collodion flows over the surface of the plate and then quickly thickens.

Once the collodion is thick enough to no longer flow, the coated plate is dipped—in the dark—into a tank of silver nitrate. The reaction of the silver nitrate and the halide salts then forms, on the surface of the collodion, crystals of light-sensitive silver halide. The now-sensitive plate can then be loaded into a light-tight plate holder and exposed in the camera.

This is called the *wet collodion* or *wet-plate process*. The plate must be exposed in the camera and then developed before the collodion completely dries. To do wet-plate photography in the field one needs a portable darkroom to prepare the plates immediately before exposure and develop them immediately after.

Wet collodion dominated photography for about 20 years, and then it was largely replaced by dry silver-gelatin emulsions that could be prepared ahead of time. Despite its inconvenience, it is capable of finely-detailed negatives that can be used to make many prints. The sensitivity is moderate—an exposure in daylight with a typical lens takes a few seconds. But earlier processes required exposures of several minutes.

The collodion plate can be developed in a different way that leaves the reduced silver metal as a dull-whitish color instead of black. If the back of such a collodion plate is painted black, we then have whitish developed silver on a black background. Thus the collodion process can be also be used to make a direct positive—called an *ambrotype* in the United States and a *collodion positive* in the United Kingdom.

A collodion positive can also be made by pouring the collodion onto a preblackened metal plate instead of glass, and this is usually called a *tintype*. The name is somewhat misleading because the plates were almost always made of black-enameled iron rather than tin. For that reason, some photographers instead use the term *ferrotype*. But that term has its own problems (the same word, as a verb, names a particular method for putting a gloss finish on a silver-gelatin print).

A tintype can also be made by hand coating a silver-gelatin emulsion onto the same enameled black plate. It too can be developed so as to make the developed silver whitish instead of black in order to make a direct-positive image. Using silver gelatin instead of collodion has the advantage that plates could be coated in the darkroom well ahead of time and developed in the darkroom long after the exposure was taken. And so the photographer did not require a portable darkroom as in the wet-plate process. Tintypes were relatively easy and inexpensive to make. Although the wet-plate version was less convenient from one point of view, there was a commercial advantage to the fact that they could be exposed, developed and sold on the spot. See figure 35.3 for examples of collodion images by Gregg McNeill, all of the same subject—a collodion negative, and both collodion positive (ambrotype) and tintype, along with an albumen print made from the collodion negative. Notice that the print is reversed left-to-right as compared to the negative, which must be placed emulsion-side-down onto the light-sensitive paper when making a print. The print, however, shows the same orientation as the subject; it is the negative, when seen from the emulsion side, that is reversed left-to-right.

Within two decades of its introduction, most uses of wet-plate photography were eclipsed by the dry, silver-gelatin developed-out process. But it is now quite popular among art photographers who are attracted by the 'poured' look of the emulsion. The process and equipment required are much simpler than daguerreotype, and it can be used as both a negative and direct-positive process.

Julia Margaret Cameron often made albumen prints from wet collodion negatives. She was criticized at the time for her acceptance of 'flaws' in her emulsions, and this is in part why her photographs are so prized today[2]. Probably the most famous contemporary practitioner of wet-plate photography is Sally

[2] See, for example, Julia Margaret Cameron's working methods.

Figure 35.3. The same subject photographed with collodion negative (**top left**), an albumen print made from it (**top right**), ambrotype (**bottom left**) and tintype (**bottom right**). Photographs by Gregg McNeill, DarkBox Images 2022, used by permission of a model made by Ian Chrichton.

Mann; her retrospective book *Sally Mann: A Thousand Crossings* (Mann *et al* 2018) contains many examples.

35.4 Lumen process

The so-called *lumen process* is a contemporary take on printing out with silver. Instead of using papers specifically designed for printing out, ordinary black-and-white *enlarging paper* is used. This paper is designed to be developed out, not printed out, and as such many varieties are quite slow indeed when used for this unintended purpose. Direct sunlight is used for the exposure, and photograms are more common than positive prints from negatives. Exposure times are many hours or even days long, depending on the particular variety of paper used. In most cases the lumen print is chemically fixed in the ordinary way to make a permanent print.

When enlarging paper is printed out, subtle tones and hues often appear that are not present when the paper is developed out. Furthermore, the printing-out speed depends critically on the level of humidity and moisture, so botanical photograms are especially popular; the added sensitivity from the moisture of the leaves can produce dark outlines and a halo effect. The final step of fixing often greatly reduces the darkness (requiring an initial overexposure), lowers the contrast and mutes the colors, giving lumen prints a very particular and appealing look. Enlarging paper

Figure 35.4. Left: A fixed lumen print made by placing feathers and leaves directly onto a piece of enlarging paper and exposing it to sunlight and a 405 nm laser. **Right:** Scan (slightly enhanced in contrast) of an unfixed lumen photogram exposed with sunlight.

that is well past its expiration date and no longer suitable for developing out may still work perfectly when printed out. Of the many hundreds of varieties of enlarging paper produced over many decades, even those that are nearly identical when developed out may produce strikingly different results when printed out. As such lumen photographers often use found papers that are decades old. Figure 35.4 shows (at left) one of my own lumen prints made with an enlarging paper that expired in 1975. The flowers and feathers were exposed in sunlight, but the meandering dark squiggles were added afterwards using techniques described in section 35.5. I left the lumen print on the right unfixed, and this gives more pronounced colors; the price is that the original lumen print is still sensitive to light. Here we have its digital capture. For the particular paper chosen, the colors correspond somewhat to the real colors of the light that exposed it, as described in more detail in section 29.7.

35.5 Ephemeral process

What I call ephemeral-process (EP) photography uses printed-out silver-gelatin enlarging paper to make a picture (Beaver 2017 and Fudge and Beaver 2017). In this respect it is much like lumen process (section 35.4), but there are many differences. In particular lumen prints are usually fixed, while EP usually is not, and this is the source of the 'ephemeral' in EP. But also EP uses what I call an *accelerator*, a liquid that is brushed on just prior to exposure that allows the printing-out process to occur up to many hundreds of times faster[3]. And so only seconds of exposure to direct sunlight may be required for an EP photogram. Because of this much higher

[3] The same term is also used by photographers in a different way (see Stroebel *et al* 2000, p 21).

sensitivity, EP photograms can be made easily with artificial light, and it is practical for daylight photography as the light detector in a camera (although exposures of many minutes are required).

After exposure the paper is washed and dried, and it returns to something like its initial very-low sensitivity. It can then be scanned to capture the image. The scanning process exposes the paper further and so damages the image. But if done carefully, the damage can be insignificant. Since the paper is exposed mostly only where the accelerator is applied, one can control what parts of the paper are exposed and what parts are not. The ideal enlarging paper for EP prints out very slowly when dry but is very fast when the accelerator is applied. This provides the greatest contrast between the accelerated and unaccelerated areas.

It is important to note that the accelerator is not a developer. A developer has its own reducing agents that change the silver halide to silver metal, and when developing a silver-gelatin emulsion the developer acts in the dark after the emulsion has been exposed to light. The EP accelerator does not work in this way; it has no effect at all if it is applied in the dark after the paper has been exposed to light.

35.5.1 How does it work?

The description in section 34.1 of silver-gelatin emulsions leaves out one important detail in the chain of events that leads to the formation of an image. When light reduces the Ag^+ to ordinary silver, electric charge must be conserved. And so the inevitable consequence is that some halide will also lose its excess negative charge and become a neutral halogen—chlorine, bromine or iodine, depending on which silver salt is involved. If nothing is done with the halogen, then it will have a strong tendency to simply find those silver atoms and reform new silver halide. This reverse reaction happens at a lower rate, and so the overall effect is still to reduce Ag^+ to silver, but it has the effect of greatly reducing the sensitivity to light.

In practice, this is unimportant for the extremely faint latent image that is meant to be developed out. There is so little halogen released (because so little silver is reduced) that the gelatin matrix surrounding the crystals can accept the halogen easily. During development millions of times more halogen is released, but the developer solution contains its own ingredients to deal with that. For a printed-out image, however, millions of times more silver is reduced, and there is little means to take up of this halogen—this is the prime factor that slows down the whole process. Because of this, silver-gelatin emulsions for printing out are ordinarily made differently than those intended for developing out.

Recall that silver nitrate is combined with a halide salt to make crystals of silver halide. If the two compounds are not mixed in a balanced proportion, then an excess of one or the other will remain. Printing-out emulsions are made with an excess of silver nitrate, while developing-out emulsions are made with an excess of halide salt (Eaton 1957, p 21). When printing out, this excess silver nitrate can then accept the liberated halogen (this also makes new silver halide). Developing-out emulsions are made with an excess of halide salt to prevent this from happening. Otherwise, the paper would eventually turn dark on its own even if kept in the dark. And so printing-out papers often have a rather short shelf life and are usually prepared immediately prior to use (Lawless 2005).

For printing out, the take up of halogen by the excess silver nitrate is greatly accelerated by the presence of water. Thus it is well known that printing-out emulsions are 'faster' in conditions of high humidity (Lawless 2005). If one simply wets the printing-out paper it becomes more sensitive, but photographers rarely do this. In most cases of printing out a photographer is making a contact print with a large negative, and the wet paper would damage the negative.

So how does EP photography fit into all of this? The difference is that we are printing out with papers designed for developing out. These papers are mostly designed to be poor at accepting liberated halogens, and so they print out very, very slowly. The simple addition of water greatly accelerates their printing-out speed because the halogen can then go into solution. The reaction (with chlorine as an example) is as follows.

$$4\ Cl + 2\ H_2O \longrightarrow 4\ HCl + O_2 \tag{35.1}$$

But in order for this to happen with maximum efficiency, one must get rid of the oxygen (O_2), or this reaction will simply undo itself (albeit at a slower rate). The oxygen can simply bubble out of the solution as a gas, but it can be disposed of far more efficiently by a chemical compound that easily reacts with oxygen to make a different, stable compound.

Sodium sulfite (Na_2SO_3) is such a compound. It reacts with oxygen to make sodium sulfate (Na_2SO_4) in the following way.

$$2\ Na_2SO_3 + O_2 \longrightarrow 2\ Na_2SO_4 \tag{35.2}$$

Indeed, sodium sulfite is one of the ingredients of Dektol, the most commonly used developer for enlarging paper partly because it is a preservative but perhaps also for this other reason. Ascorbic acid, better known as vitamin C, will also work for this purpose. So a simple water solution of either sodium sulfite or ascorbic acid makes for an effective accelerator for printing out enlarging paper. The two accelerators are not exactly equivalent in their action with all enlarging papers because sodium sulfite has other effects as well (Stroebel *et al* 2000, p 215). More details on the preparation of EP accelerators can be found in chapter 48.

Since we want a paper that, when dry, prints out very slowly, it would seem that developing-out papers with a large excess of halide salt might have the right properties. It is difficult to know for sure because the chemical formulas of commercial developing papers are proprietary. But to test this theory we have made six different silver-gelatin papers—three each of silver chloride and silver bromide (AgI is ordinarily used only for film and not for enlarging papers). Emulsions of each of the three silver halides were prepared in three different ways—with a balance of salt and silver, with a 100% excess of salt and with a 100% excess of silver. Figure 35.5 shows an exposure test of all six papers, both dry and with a sodium sulfite solution (Fudge and Beaver 2017). Nearly the entire paper was covered with emulsion, but accelerator was applied only to a small area. The accelerated areas achieved a similar

Figure 35.5. Exposure test with simple silver-halide papers. The **left** column is silver bromide, while the **right** column is silver chloride. The **top** row was prepared with a 100% excess of silver nitrate, while the **bottom** row was prepared with a 100% excess of halide salt. The **middle** row was made with a balance of silver and salt. All were exposed simultaneously with a broad-spectrum, high-intensity light-emitting diode lamp.

density whatever the mix of silver and salt. As expected, the papers with an excess of salt did not darken at all in the amount of time required to darken the accelerated area. The examples with a near balance of salt and silver showed slight but noticeable darkening in the unaccelerated areas.

Not shown in this test is the additional fact that unexposed examples of both papers with an excess of silver nitrate eventually fogged in the dark over the course of a couple of days. But the papers with an excess of salt resisted fogging.

The formulas of commercial developing-out print papers are far more complex than this test, so it is unclear how to generalize these simple results. But tests of different commercial papers often show markedly different tones—both wet and dry—when printing out. Some individual papers give very different tones depending on what is applied to them. In figure 35.6 we have compared several different types of papers, exposing them all to sunlight. The papers were all exposed to achieve a range of densities by varying the exposure time. For each paper there were four tests: dry, wet with deionized water, wet with a sodium-sulfite solution and wet with a sodium-chloride solution.

For all of the papers, the greatest acceleration occurred with the sodium-sulfite solution (which aids in the removal of halogen) and the least with the sodium-chloride solution, which adds chloride (but not bromide) ions.

35.6 Chemigram

Although there are historical antecedents, the *chemigram process* was explored most intentionally by Pierre Cordier beginning in the 1950s. It is a cameraless process that uses ordinary silver-gelatin enlarging paper. Over the years a wide range of

Figure 35.6. Six different kinds of enlarging paper were exposed to direct sunlight in two-second intervals. The tests on the **left** column were brushed with deionized water, the **middle** column with a solution of sodium sulfite, and the **right** column with a solution of sodium chloride (table salt).

variations have been developed around the basic concepts. Thus we can think of chemigram as a set of tools.

Selective applications, often by hand with brushes, of developer and fixer are combined with exposure to light. Masks are also often selectively added by hand, and these may selectively block exposure to light. If the mask is waterproof (an oil, for example), then it may resist the action of developer or fixer in the region where it was applied.

All of this occurs in real time—the developer which blackens preexposed regions, the fixer which prevents further sensitivity to light, the exposure to light and the masks that block light or resist chemistry, and it may be dissolved over time by the liquid in the developing tray or added by the artist with a brush. See figure 35.7 for an excellent example by Brittonie Fletcher.

Typically, the paper is moved sequentially from developer to fixer and back again. The room lights are left off, and then turned on and then off again. Areas that receive no developer may then print out, and the photolytic silver may produce colors as in the lumen process, while other areas may develop out to a neutral tone. The possibilities are nearly endless, and producing a chemigram is an exercise in choreography and accident (see for example Anderson 2012 and 2013).

35.7 Chromoskedasic

The term *chromskedasic* was coined by Dominic Man-Kit Lam and Bryant W Rossiter (Lam and Rossiter 1991 and Lam and Baran 1991). They describe a cameraless photographic process that has some things in common with chemigram and lumen process but acts upon an otherwise-ordinary developed (but unfixed) silver-gelatin print on enlarging paper. The term is meant to suggest color that arises not from pigments but from the scattering of light.

Figure 35.7. *Rocks* from Erratics, ©Brittonie Fletcher, chemigram 2019. Used by permission.

The process uses silver-gelatin enlarging paper, and it may begin with an exposure in the darkroom in the ordinary way with light projected through a negative by an enlarger. It can also be used with no exposed image content to produce abstract photographic 'paintings.' The paper is treated with an *activator* solution (typically the strong base sodium hydroxide) and then (or sometimes simultaneously) with a *stabilizer*, which acts as a weak fixer. The activator and stabilizer steps may take place in the presence of room lighting or with safe lighting.

Much of the intent is to maximize the different scattering color tones that arise naturally in the printed-out silver. In sections 5.1 and 29.7 we discussed wavelength-dependent Mie and Tyndall scattering, which can produce colors when the scattering particles are uniform in size. Smaller particles can produce a bluish color, while somewhat larger particles can produce warm tones instead. The chromoskedasic technique apparently produces relatively uniform silver grains, the size of which can be controlled somewhat by the artist and even varied in different parts of the image.

Apart from the resulting scattering colors, the technique often results in significant *silvering out* (also called *silver mirroring*). Reduced silver comes to the surface of the gelatin emulsion, producing a reflective, metallic sheen (Henn and Wiest 1963 and Di Pietro 2004). This can happen naturally to silver-gelatin prints over long periods of time; it is usually considered a form of degradation. In the chromoskedasic process it is often sought after as another artistic tool.

Since some of the chromoskedasic process is carried out in room lighting, the activator solution may then result in a limited sabattier effect. When this particular result is sought after, the process is often called *chromoskedasic sabattier*. Further refinements of the process are described by Megan Crawford (2018). See figure 45.1

in chapter 45 for some excellent examples of chromoskedasic-sabattier prints made by the UK-based artist Ky Lewis.

References

Anderson C Z 2013 The chemigram https://www.alternativephotography.com/the-chemigram/

Anderson C Z 2012 *Experimental Photography Workbook: A Manual of Analog Black & White Darkroom Practice* (London: Routledge Taylor and Francis Group)

Beaver John 2017 Using new-antiquarian photographic processes to integrate art and science Am. Geophys.Union Fall Meeting Abstr. New Orleans 13 Dec 2017

Cooke T and Kinneberg C (ed) 2014 *The Photography Book* 2nd edn (London: Phaidon Press Limited)

Crawford M 2018 Chromoskedasic sabattier: a step-by-step guide https://www.alternativephotography.com/chromoskedasic-sabattier-a-step-by-step-guide/

Di Pietro G 2004 A local microscopic model for the formation of silver mirroring on black and white photographs *Proc. Int. Conf. on Metals Conservation (4-8 December 2004)* 126–36

Eaton G T 1957 *Photographic Chemistry in Black-and-White and Color Photography* 2nd edn (Orlando, FL: Morgan & Morgan)

Fudge H and Beaver J 2017 Ephemeral process photography: fresh ideas for artistic expression using old techniques *Research in the Rotunda (Madison, WI)*

Henn R W and Wiest D G 1963 Microscopic spots in processed microfilm: their nature and prevention *Photogr. Sci. Eng.* **7** 253–61

James C 2016 *The Book of Alternative Photographic Processes* 3rd edn (Boston, MA: Cengage Learning)

Lam D M-K and Baran A J 1991 Painting in color without pigments *Sci. Am.* **265** 136–8

Lam D M-K and Rossiter B W 1991 Chromoskedasic Painting *Sci. Am.* **265** 80–5

Lawless L 2005 Making P.O.P. https://unblinkingeye.com/Articles/POP/pop.html

Mann S, Greenough S, Kennel S, Als H, Daniel M, Faust D G, National Gallery of Art and Peabody Essex Museum 2018 *Sally Mann: A Thousand Crossings* (New York: Abrams, Inc.)

Rexer L 2002 *Photography's Antiquarian Avant-Garde: The New Wave in Old Processes* (New York: Harry N. Abrams)

Stroebel Leslie, Compton John, Current I and Zakia R 2000 *Basic Photographic Materials and Processes* 2nd edn (Waltham, MA: Focal Press)

Ware M 2019 Argyronomicon: Silver Photographs on Paper: Chemical History of Their Invention, Deterioration, and Conservation https://www.mikeware.co.uk/downloads/Argyronomicon.pdf

IOP Publishing

Photography (Second Edition)
Physics and art in focus
John Beaver

Chapter 36

Nonsilver photochemical detectors

I describe here several photographic processes that are well out of the mainstream of commercial photography. Many of these are antiquarian—they predate not only digital but also the standardized silver-gelatin film and printing processes that immediately predated the advent of digital photography. These processes are finding increasing use in contemporary art photography. This overview of nonsilver alternative photographic processes is nowhere near comprehensive; for more details I highly recommend James (2016).

36.1 Cyanotype and Van Dyke processes

Cyanotype is one of the oldest of photographic printing processes invented in 1842 by the English astronomer and polymath John Herschel. It uses relatively inexpensive iron salts to make a print that is Prussian blue in tone.

In its traditional form a liquid sensitizer, made by mixing solutions of potassium ferricyanide and ferric ammonium citrate, is brushed onto watercolor paper and allowed to dry. Upon exposure to ultraviolet (UV) light, the ferricyanide is reduced by photons to ferrocyanide and ultimately combines with the ferric ion from the ferric ammonium citrate (the detailed reactions are complex). The result is ferric ferricyanide—Prussian blue. The single letter difference—ferro instead of ferric—means that the former involves Fe^{3+}, while the latter is Fe^{2+}. Thus an electron is gained (reduction), and it is the action of photons that enables this.

The final result, Prussian blue, involves both states of iron. Four Fe^{3+} (ferric) ions are bound to three combinations of the Fe^{2+} (ferrous) ion with six CN^{-} (the *cyanogen* ion, comprised of carbon and nitrogen). The formula can be written $(Fe^{3+})_4[Fe^{2+}(CN^{-})_6]_3$. This mixed state of ionization is unusual, and it is responsible for the deep blue color. It absorbs red light easily, the low energy from the long-wavelength photons causing electrons to skip from the ferrous to the ferric ion (Ware 2016, p 67). The red light is absorbed, but the rest of the spectrum—blue and green—is reflected.

doi:10.1088/978-0-7503-3703-8ch36

© IOP Publishing Ltd 2022

There is a lucky feature that makes all of this practical as a light detector. The potassium ferricyanide and ferric ammonium citrate used for the sensitizer are soluble in water. The Prussian blue that results from the action of light, however, is not. After exposure one needs only to wash away the remaining sensitizer, and the Prussian blue stays in the paper, yielding a negative blue-tone image.

Cyanotype is not an emulsion; the sensitizer goes into the fibers of the paper rather than forming a layer that sits on top. It is sensitive to violet and near-UV light only, with its peak sensitivity near the short-wavelength edge of human vision. Thus it must be exposed either in a special UV light box or to direct sunlight. It is a printing-out process and so has a very-low sensitivity to light compared to developed-out silver-gelatin emulsions.

36.1.1 New cyanotype

There are many accounts in print of the cyanotype process, but by far the most comprehensive—covering every imaginable aspect of the process and its history in engaging and erudite detail—is the book *Cyanomicon II* by photographer and chemist Mike Ware (2016). Ware is, in fact, the originator of what is likely the most significant innovation in the cyanotype process since its discovery—what is called *new cyanotype* (James 2016, pp 212–6). He greatly improved both the light sensitivity and the tonal range of the traditional process by substituting some of the ions used (ammonium and oxalate instead of potassium and citrate). This also has the practical advantage that all of the ingredients of the sensitizer mixed together are stable. The traditional formula requires the potassium ferricyanide and ferric ammonium citrate to be stored separately and mixed only just before use. The left-hand image in figure 36.1 is an 8 × 10 inch new cyanotype print on watercolor paper;

Figure 36.1. An 8 × 10 inch format negative was used to make a print with Mike Ware's new cyanotype formula.

Figure 36.2. The same negative used for figure 36.1 was used to make a Van Dyke print.

it is a contact print from a negative of that same size exposed in a large format camera.

36.1.2 Van Dyke brown process

Soon after the discovery of cyanotype, John Herschel realized it could be adapted by using the light sensitivity of the iron salts to make prints from other metals. Many others added to this knowledge over the years, using this foundation of iron-salt sensitivity to make prints with silver, gold, mercury, platinum and palladium. The Van Dyke process, invented in 1889, couples the sensitivity of iron salts to silver to produce a brown-tone print (Ware 2016, p 50). Although it is silver that forms the dark areas of the print, it is the iron salts—rather than salts of silver—that provide the light sensitivity. Thus Van Dyke printing is far more closely related to cyanotype than to the silver processes described in chapters 34 and 35. Figure 36.2 shows a Van Dyke contact print made with the same negative as was used for figure 36.1. Sunlight—the most convenient source of the necessary UV light—was used for the exposure.

36.2 Platinum and palladium printing

Platinum printing is the most expensive, and according to many the very best, of all black-and-white printing processes. Although the final image is made of platinum (one of the most expensive of metals), regarding its light sensitivity, it is like cyanotype—an iron-based process. The details are different, but—as is the case for cyanotype—ferric salts of iron are reduced by absorption of photons to ferrous salts. This happens in the presence of salts of platinum, initially producing a very faint negative image. The faint image is then chemically developed to produce a strong image made of platinum (James 2016, chapter 13).

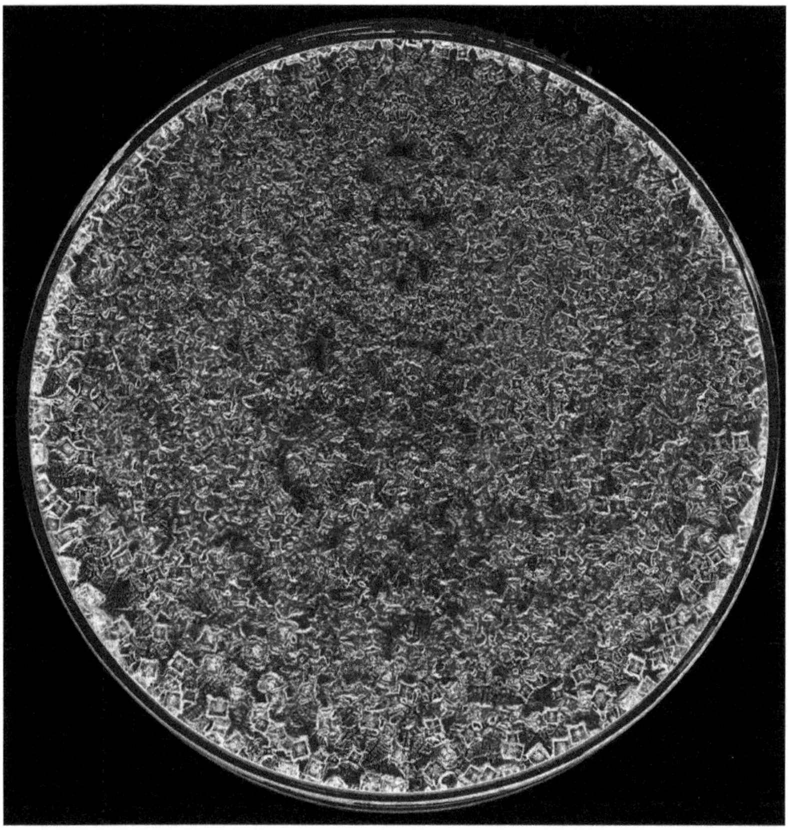

Figure 36.3. *Closed*, ©Brittonie Fletcher, platinum/palladium print from cliché verre negative 2010. Used by permission.

In skilled hands, the platinum printing process can produce the most vibrant range of values with a lush and complex tone. The main difficulty is twofold: it is difficult to master, and it is very expensive. The element palladium is chemically similar to platinum, but sometimes it is less expensive and can be substituted for platinum to make a *palladium print*. Many practitioners exert further control over the tonality by using a strategic mix of both metals. Apart from its unparalleled tonal range the other prized advantage of platinum or palladium printing is that, when done correctly, it is considered to be one of the most permanent or *archival* of all black-and-white printing processes. See figure 36.3 for an example by Brittonie Fletcher.

Fletcher combined this printing process with the technique of *cliché verre*; the 'negative' was made by hand on a transparent medium.

36.3 Gum bichromate

Gum bichromate (sometimes called gum dichromate) is, in some ways, the conceptually simplest of all photographic processes. It is also one of the most

Figure 36.4. Left: An awful example of a four-color gum-bichromate print by the author, using four separate exposures with CMYK pigments. **Right:** *Edible Mapleton*, ©Brittonie Fletcher, 12-layer CMYK gum bichromate 2022. Used by permission.

difficult to master if one wants excellent results. Watercolor pigment is mixed with a gum arabic base and made light-sensitive with potassium or ammonium dichromate. This sensitizer is then brushed onto watercolor paper and contact printed in the same manner as platinum or cyanotype. Exposure to UV light makes the gum arabic insoluble, thus trapping the suspended pigment. The print is then gently rinsed with warm water, and the unexposed portions wash away. The final print is, in essence, a photographically-printed watercolor.

In order to achieve good results, multiple exposures in layers are needed in order to build up proper contrast. The precise mixture of the gum arabic, dichromate, and pigment must be carefully controlled in conjunction with the exposure and final washing. It has the advantage, however, that virtually any pigment can be used to make the image. In fact one can even make color prints by using separate exposures with four color-separated black-and-white negatives, each using pigments of cyan, magenta, yellow, and black (CMYK, see sections 27.4 and 29.5).

It is very easy to make a gum-bichromate print, but it is very difficult to make an excellent one. Figure 36.4 at left shows a very bad example—my only attempt to make a four-color gum bichromate print (my black-and-white attempts are slightly less awful). At right is an excellent example of a CMYK gum bichromate by Brittonie Fletcher (profiled in section 45.2), using multiple layers of each pigment for a total of 12 separate exposures.

36.4 Anthotypes and chlorophyll prints

Most photochemical light detectors become darker with greater exposure to light, resulting in a negative image. But it is well known that many things respond in the opposite sense, through *bleaching or fading*. Most color dyes fade with light over time. This is, in fact, one of the key difficulties for the archival preservation of color photography. Whether it be a color negative, a color slide or a color print, the dyes that make those colors are susceptible to fading if exposed to light over a long period of time.

Leave a color film negative sit out on a table, exposed for weeks to bright afternoon sunlight, and disaster is sure to ensue. The dyes in the negative will fade. This may not be too disastrous if everything fades by the same amount, for one could then correct this in the process of making a print. But what if, for example, one of the negatives is left half-covered for those weeks by a copy of Noam Chomsky's *The Logical Structure of Linguistic Theory* (which one has been meaning to read)? A harsh line will likely be evident crossing the negative; the uncovered part having faded while the covered part, protected by the opacity of Chomsky's tome, remains intact. It may be possible to scan one's damaged negative and fix it with clever use of Photoshop or GIMP, but the task will likely set one to swearing.

A clear lesson to be learned from this thought experiment is that one should always quickly clean up one's mess. Also this means that color dyes are sensitive to light, so they could possibly be used as a photographic detector.

36.4.1 Anthotype

An *anthotype* is a print made by the bleaching action of light on (usually) vegetable dyes on paper. The history of the anthotype goes back to the earliest days of photography, first described in detail by John Herschel—the inventor of cyanotype (Ware 2016, p 42). The paper is coated with the dye, typically made from flower petals or fruits, and allowed to dry. Objects or a film transparency are placed upon it, and it is then exposed to direct sunlight. After a prolonged exposure of at least several hours and even as much as several weeks, the uncovered areas fade while the dye remains unfaded in the covered areas (James 2016, p 42ff).

Thus the anthotype is a direct-positive process. But the sensitivity is far less than even a VLS printing-out process such as cyanotype. And so its use as a detector in a camera is impractical. The anthotype, once completed, is still sensitive to light, so viewing the image would eventually destroy it. The extremely-low sensitivity is inconvenient when making the image, but it means the resulting photograph can withstand quite a bit of additional exposure without degrading significantly. Even some of John Herschel's anthotypes from the 1840s still survive today (Ware 2016, p 42).

In order for light to have any effect, it must be absorbed. If a particular dye is red, it means that it reflects red light well. Thus it is unlikely that it will be easily bleached by red light for the simple fact that little red light is absorbed. The opposite is true for a blue dye, and this leads to the basic notion that dyes are bleached primarily by light that is a complementary color to that of the dye. This is likely true for visible light, but it has no bearing on bleaching by UV and infrared light that are not visible to the human eye and so are not related in any direct way to the perceived color of the dye. Figure 36.5 shows my first rather-poor attempt at an anthotype. I stained a

Figure 36.5. My own mediocre example of an anthotype photogram using the juice from a pressed purple iris.

piece of watercolor paper purple by rubbing it with a wilting purple iris. I then layered objects on it, held in place by a sheet of glass, and exposed it to direct sunlight for about 12 hours (over two days). It is still sensitive to light, but the act of scanning it to make the image in this book was insignificant compared to the original exposure.

If I were to frame it and display it in a moderately-dim hallway, with 1000 times less intensity than sunlight (and probably several thousand times less UV light), it would take many months to fade noticeably due to the additional exposure to light (the dyes may be unstable for other reasons). An excellent contemporary example of an anthotype, by the artist Erin Woodbrey, can be seen at left in figure 45.6 in chapter 45.

36.4.2 Chlorophyll prints

Chlorophyll is the UV-sensitive pigment that both allows a plant to produce energy from sunlight and also makes the leaves green. Pluck a green leaf from a plant, and the leaf no longer has access to the plant's system of vascular circulation of water and nutrients. The leaf will dry and become brittle, of course. But also the green chlorophyll will fade upon prolonged exposure to light. Thus one can use a chlorophyll-laden green leaf to make a photographic image in the same manner as an anthotype but with no need to brush on a dye (James 2016, p 56f).

This is called a chlorophyll print, and the artist Binh Danh was the first to produce a successful body of work using this idea. He renders the leaf with its image more permanent by casting it in resin to preserve its delicate physical nature and also perhaps because the acrylic resin blocks a large percentage of UV light. Figure 36.6 shows a chlorophyll print on a leaf by Brittonie Fletcher.

Figure 36.6. *Is This the Path to Sustainability?*, ©Brittonie Fletcher, chlorophyll print from digital transparency 2020. Used by permission.

References

James C 2016 *The Book of Alternative Photographic Processes* 3rd edn (Boston, MA: Cengage Learning)

Ware M 2016 Cyanomicon II–History, Science and Art of Cyanotype: Photographic Printing in Prussian Blue https://www.mikeware.co.uk/downloads/Cyanomicon.pdf

IOP Publishing

Photography (Second Edition)
Physics and art in focus
John Beaver

Chapter 37

Reciprocity failure and solarization

37.1 Reciprocity failure

It is desirable that the density of a photographic emulsion depends only on the exposure and not the details of how that exposure was made. Exposure is the product of illuminance and time. The *law of reciprocity* states that the density achieved depends only on the exposure and not specifically on either illuminance or time individually. This means one could (for example) quadruple the illuminance while reducing the exposure time by the same factor, and it would still have the same effect on the detector. When the law of reciprocity holds, a photographer's job is easier.

Reciprocity failure means that a different density results when using a long exposure time at low illuminance compared to a short exposure time at high illuminance—even if both combinations are the same exposure. The most commonly-encountered example of reciprocity failure relates to silver-gelatin emulsions. It is well known that silver-gelatin emulsions are, in effect, less sensitive at very low levels of illuminance. Because of this, a greater exposure is needed to achieve the same result (the same density) whenever the illuminance on the detector is very low.

There are two likely causes for very low illuminance of the image plane: either a very tiny aperture (for example in pinhole photography) or very dim scene lighting (for example in night photography). In either of these cases (or a combination of both), a greater exposure is needed than what one would expect from tests of the detector at higher levels of illuminance. Thus one must compensate for the lower sensitivity of the detector by choosing a longer exposure time than would be used if reciprocity failure were not a factor. Reciprocity failure greatly complicates the life of the silver-gelatin-film–based astrophotographer or pinhole photographer.

But what is the physical cause? Why is the sensitivity of a silver-gelatin emulsion reduced at very-low levels of illuminance? To answer these questions, we must recall

doi:10.1088/978-0-7503-3703-8ch37 © IOP Publishing Ltd 2022

an important detail regarding how a latent image is formed in a silver-gelatin emulsion. Recall from chapter 34 that the first step is for a stable speck of silver to form, from the action of photons, at a crystal sensitivity sight. It is estimated that at least four Ag^+ ions must be reduced by photons to elemental silver before the speck is stable. Otherwise the affinity the silver has for the halogen, which is also formed in the process, causes them to simply retrade that electron and again form, for example, Ag^+ and Cl^- ions (Stroebel *et al* 2000, p. 200).

A very-low level of illuminance means that very few photons arrive per unit of time at a given sensitivity speck. This means that very few silver ions are reduced to silver per time unit. At extremely-low levels of illuminance, it essentially never happens that photons succeed in reducing four silver ions to silver in the same place in a short enough time for them to not, in the mean time, convert back to ions. And so a stable speck is never formed.

At high levels of illuminance, on the other hand, none of this is a problem. Photons reduce silver ions to silver a such a high rate that they essentially never have time to revert back to ions before a stable speck of four silver atoms is created. This is the ordinary use of silver-gelatin emulsions in which reciprocity failure is not a factor. Reciprocity failure is the case in between. The illuminance is low enough that the natural 'undoing' of the reduction of silver ions to silver is a factor, but it is still possible to record a latent image.

37.1.1 Reciprocity failure and very-low sensitivity photography

One can expose a piece of silver-gelatin enlarging paper in a pinhole camera on a sunny day and develop the paper negative after the exposure is finished. The exposure time required to get a successful photograph can be many minutes long.

In chapter 24 we discussed the case of very-low sensitivity (VLS) photography—taking pictures with VLS detectors such as cyanotype or printed-out silver-gelatin emulsions. Ephemeral-process (EP) photography is a good example, and it provides a useful comparison because the same type of enlarging paper is often used for pinhole photography. In the case of EP photography, however, the paper is not developed; it is left to print out instead with an accelerator applied to increase the speed. The exposure time on a sunny day for EP photography is comparable to that of pinhole photography.

The successful pinhole photographer knows that reciprocity failure is important. If the light meter says an exposure time of five minutes is needed, it will certainly take much more. Pinhole photography is a textbook case of a situation where reciprocity failure is a significant, or even dominant, effect. If the sky becomes hazy, and the light meter now says an exposure time of ten minutes is needed, then reciprocity failure will be an even bigger effect.

The EP photographer uses the same light detector (in a printing out rather than developing out mode) as the pinhole photographer and needs similar exposure times. Yet reciprocity failure is of no consequence. Why?

The answer is obvious if one recalls that reciprocity failure is not about exposure time, but it is about *image-plane illuminance*. Reciprocity failure is a large effect whenever the light falling on the silver-gelatin emulsion in the camera is very, very dim. That is true for pinhole photography because of the tiny aperture. The long exposure time is not the cause of the reciprocity failure; it is the consequence of the very-low image-plane illuminance (which is the cause of reciprocity failure).

In EP photography, on the other hand, the long exposure does not result from low image-plane illuminance—far from it! The EP photographer typically uses a fast lens in broad daylight, and the image-plane illuminance is about as high as it gets in photography. Instead, the long exposure times in EP photography result from the VLS of the printing out process. But the high image-plane illuminance means that reciprocity failure is not a factor.

37.2 Solarization

On occasion the relationship between exposure and density—the characteristic curve—can be far stranger than the typical examples reviewed in chapter 33. Consider, instead, a characteristic curve that is shaped like a hump. When looking at such a graph from the left to right, increasing exposure would cause greater density as expected. After the peak of the hump, however, density would decrease with increasing exposure. Or what if the characteristic curve were the opposite of that—a 'u' shape. In that case, the density would decrease with increasing exposure until a minimum density is reached at some particular midpoint exposure. Then, as the curve turns upward, density would increase with exposure. For either a hump-shaped or a u-shaped characteristic curve, there are particular densities that are caused by both small and large exposures.

If by density we mean the darkness of a silver-gelatin emulsion, the hump-shaped characteristic curve would represent an ordinary negative for the parts of the image that received the least exposures in which the characteristic curve slopes upward. But the downward sloping side of the hump, corresponding to the highest exposures, would be a positive image. And so the image would be part positive and part negative. The same would be true for the u-shaped characteristic curve; it would be part positive and part negative. The difference is whether it is the shadows or the highlights that have been reversed to a positive.

A picture that shows one of these relations between density and exposure is usually said to be *solarized*, and there is a very particular look to it. Figure 37.1 shows an example. The image on the left is much as the scene appeared in life, while the image on the right has been solarized. In this case, the solarized effect has been accomplished with digital image processing on an ordinary digital image. The hump-shaped curve superimposed on the solarized right-hand image is something like a characteristic curve. But rather than a relationship between the exposure of a detector and the density that results, this is an *adjustment curve*. It represents changes made to the brightness levels in the image on the left in order to produce the image on the right.

We will consider adjustment curves in more detail in chapter 42, but they have much in common with a characteristic curve. In particular the hump shape of the adjustment curve means that two different brightness levels in the left-hand image

Figure 37.1. The image on the **left** has been digitally solarized to produce the image on the **right**. The brightest range of values has reversed from a positive to a negative while leaving the darker areas as a positive image.

Figure 37.2. A color image can also be solarized. Not only do the values of the highlights reverse to a negative, but the hues of the highlights also switch to their compliments.

can result in the same brightness level in the solarized image. Notice that some of the darker tones in the solarized image are also darker on the original. But some of the darker areas in the solarized version were actually the brightest areas on the original.

The adjustment curve works in the opposite sense of the silver-gelatin film characteristic curve, which shows an upward slope for a negative image. An upward sloping adjustment curve therefore represents a positive image, while a downward slope results in a negative image. For the image in figure 37.1 the shadows are left as a positive, but the highlights have been reversed to a negative.

A color image can be solarized as well. Although the details are more complex, it is usually the case that not only do the values at the high end reverse to a negative, but also the hues associated with those parts switch to the opposite side of the color wheel. See figure 37.2 and note that the reversed parts have turned from a bright yellowish to a dark bluish.

There are two ways in which silver-gelatin emulsions can be made to respond with an effective characteristic curve that gives a solarized look. The names for these effects, however, are confusing. The preferred technical term for the effect that most commonly produces a solarized look in a silver-gelatin emulsion is not 'solarization.' Rather, the 'correct' term is the *Sabattier effect*. Nonetheless, most people still use the term solarization when they see the Sabattier effect in action. The Sabattier effect results in a u-shaped characteristic curve in the silver-gelatin emulsion.

I also describe below another process by which silver-gelatin emulsions can result in something like a solarized look, but the characteristic curve is hump-shaped instead—the opposite of the Sabattier effect. It is used much more rarely than the Sabattier effect, and it more often than not affects only a small part of a picture. Nonetheless, the preferred technical term for this effect is *true solarization* (Stroebel *et al* 2000, pp 124–6, 201–2). The reversal in true solarization is the opposite of that in the Sabattier effect. The Sabattier effect reverses the areas of the emulsion that received the least exposure to light. True solarization, on the other hand, reverses only the regions that received a very large amount of overexposure.

37.2.1 Sabattier effect

The Sabattier effect produces a solarized look during the development phase of the silver-gelatin image. The emulsion is exposed in much the normal manner (but with a somewhat reduced exposure). While it is in the process of being developed, just as the visible developed image is beginning to appear, it is given a brief exposure to bright, diffuse light. Under the right conditions this second exposure to light will, as the development proceeds, darken the unexposed areas while leaving the previously developed areas relatively unaffected.

The process is difficult to control, and it works best with particular combinations of emulsion, developer, initial exposure to the image and the details of how it is flashed with light during development. As such, its most common use is when making a silver-gelatin positive print from a negative under controlled conditions in the darkroom; see figure 37.3 for an example. Both silver-gelatin prints were made from the same negative. The left-hand example was made in the ordinary way, while for the right-hand example I applied the Sabattier effect. Notice that the highlights on the print have reversed to a negative, while the shadow areas have remained a positive. But remember that this is at the stage of a print from a negative. And so the unreversed shadow areas on the print are those that received the most exposure.

It is also possible to solarize an in-camera negative. Figure 37.4 shows two unconventional examples. The image on the right, by photographer Teresa Patrick, was made by employing the Sabattier effect on silver-gelatin print paper in the darkroom, using the ordinary techniques for solarizing a print. But, rather unconventionally, the paper was not exposed by light projected through a negative in order to make a positive print, as in figure 37.3. Instead, Patrick exposed the 18 × 24 cm sheet of enlarging paper directly in a pinhole camera and then brought it into the darkroom for development.

Figure 37.3. Left: A silver-gelatin print made in the darkroom from a film negative. **Right:** The Sabattier effect was used to make a solarized print from the same negative. I used a different developer and enlarging paper and flashed the print with light while it was only partially developed. Some of the highlights (dark areas on the original negative) reversed to a negative on the solarized print.

Figure 37.4. Left: *Emma no. 1*, John Beaver 2006. **Right:** *Yard and House*, Teresa Patrick 2005, used by permission.

Ordinary development would have produced an ordinary negative from which a positive print could have been made by making a contact print with a second sheet of enlarging paper. Instead Patrick solarized the original exposure such that about half of the image reversed from a negative to a positive. She then toned the image and framed it. The piece of paper went directly from the camera to the wall—a sort-of direct positive but with a solarized look.

The Sabattier effect causes the low-exposure areas of the emulsion to reverse and become high-exposure areas. If this is performed at the step of the final print then highlights reverse to become shadows. When the Sabattier effect is instead performed on the negative, the same thing happens. If a positive print is then made from the solarized negative, however, the roles are switched. In the final print it is instead the shadows that are reversed to highlights.

The left-hand image in figure 37.4 was made by solarizing a negative and then making a positive print from the solarized negative. The dark-haired model was wearing dark clothing and lying on a black cloth, all of which switched to become bright areas of the picture while leaving the skin tones unaffected. With a combination of experience and luck, pleasing effects can sometimes result and highlight important parts of the picture.

37.2.2 True solarization

In true solarization areas exposed to a very-high illuminance reverse. This most often happens during the exposure of the film negative in the camera, and the direct image of the Sun focused onto the film is the usual culprit. Sometimes it is accidental, as in the left-most image in figure 37.5, where the image of the intensely-bright surface of the lightbulb reversed to a negative in this Fuji instant-film photograph.

The image on the right has a similar effect, but it did not involve a silver-gelatin emulsion at all and so is technically not an example of true solarization. Rather this is one of my early cyanonegative photographs (chapter 25); I exposed cyanotype paper for several minutes in a homemade camera using a fast simple magnifier as a lens. The direct image of the Sun literally burned a hole in the paper during the exposure.

I am somewhat embarrassed by this picture, not because of burning a hole in the negative; I expected that would happen, and wanted to see what the effect would be. Rather it is because of the name I picked for the image. I named it *Black Sun* after the novel by Edward Abbey. It was a year or so later that I discovered Abbey had named his novel after the famous photograph, 'The Black Sun, Tungsten Hills, Owens Valley, California,' by Ansel Adams—one of the most famous examples of true solarization in a photograph. I plead an embarrassing amount of ignorance, rather than unspeakable arrogance, for accidentally naming one of my pictures after a photograph by Ansel Adams.

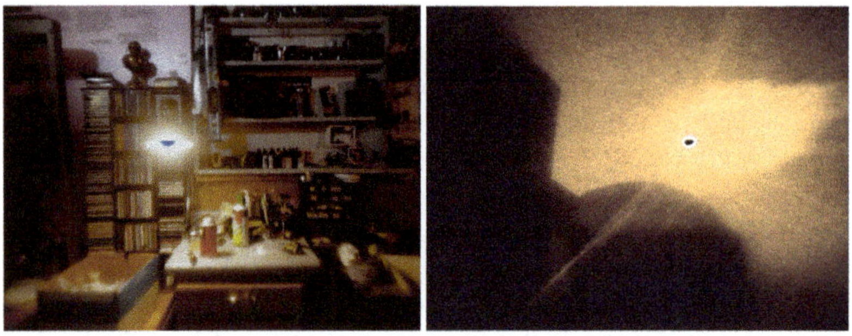

Figure 37.5. Two examples of accidental true solarization. **Left:** While testing a new lens with instant film, the intense illuminance from the lightbulb, combined with the particular circumstances of the instant-film development, reversed both the values and hues. **Right:** This cyanonegative photograph–a direct image of the Sun–resulted in an effect much like true solarization because the focused image of the Sun literally burned a hole in the cyanotype paper detector.

37.2.3 Mackie lines

If one looks closely at some of the detail in the solarized image in figure 37.1, the border areas between reversal and nonreversal to a negative reveal a strange detail. One part of this area is enlarged as the left-hand image in figure 37.6. This border region seems to be outlined by a bright line, as if someone outlined the edges with white paint. These are called Mackie lines, and they are one of the hallmarks of solarization.

In the right-hand image of figure 37.6, a detail from the left-hand image in figure 37.4, the Mackie lines are dark instead of bright. This is because this image was made by applying the Sabattier effect to solarize a negative rather than a positive. The Mackie lines were white on the negative, but now the reversal to a positive has made them black.

Many authors, for example London *et al* (2005, p 300) and Stroebel *et al* (2000, p 202), suggest that the Mackie lines result from the retardation of development by the physical migration of chemical byproducts in those areas during the solarization process. These chemical 'edge effects' may be a factor, but there must be more to it than that. For even pure digital processing—with no silver-gelatin emulsion involved—can easily produce the same effect.

See, for example, the left-most image in figure 37.6, which was solarized digitally from an ordinary digital photograph. The Mackie lines are very clear even though no chemistry was involved. I performed this solarization effect not by clicking on a 'solarize' filter in Photoshop (which could be specifically coded to simulate the look of Mackie lines) but rather by a simple adjustment of the brightness values of the image in a way that is similar to how the Sabattier effect changes the characteristic curve.

Consider a region, such as the shadow under the lizard, that has a very sharp boundary between light and dark. Upon solarization, the dark area will remain

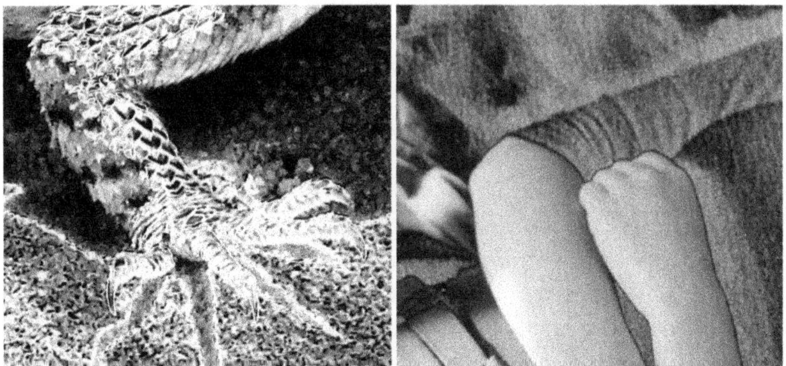

Figure 37.6. The regions on the border between reversal and nonreversal are outlined. These are called Mackie lines, and they give solarized images a graphic arts feel.

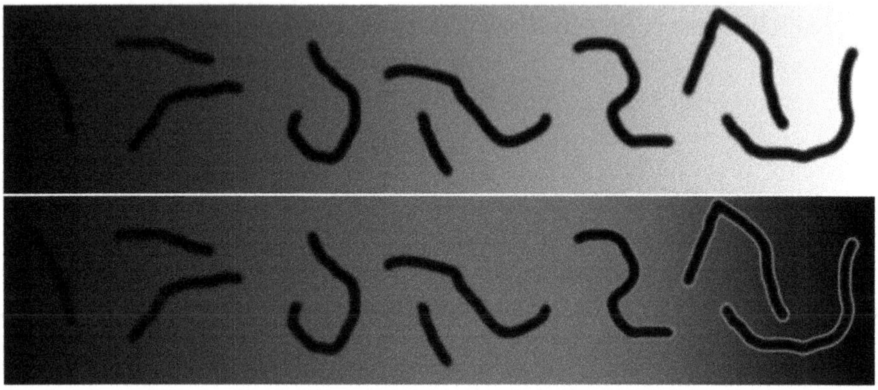

Figure 37.7. The Mackie-line phenomenon appears even in this simple digital image. The **top** image was created with a soft-edge paintbrush tool in GIMP. When the brightness levels are transformed (**bottom** image) in a way similar to the characteristic curve produced by the Sabattier effect, Mackie lines naturally appear at the soft borders of the marks in the reversed areas.

dark, while the light area switches to dark. So it was dark next to light, and now it is dark next to dark.

But there is usually, in any image, a transition between a dark area and a neighboring light area—a narrow region where there are shades of gray in between. What happens to these shades of gray? Some of these shades of gray are dark enough that the solarization process does not reverse them, but they are still light enough to contrast strongly with both the unsolarized dark areas and the solarized light areas that have switched to dark. Thus there is a narrow region that is brighter than the regions on either side of it and so shows up as a line marking the boundary between the solarized and unsolarized area—a Mackie line.

Further evidence for this explanation for Mackie lines is illustrated by the simple experiment shown in figure 37.7. For the top image I used GIMP to make a simple grayscale image, and then I drew over it with black lines made with a soft-edged brush tool. The dark marks have a range of gray levels at the edges. I then transformed the brightnesses of the pixel values to make the highlights negative while leaving the shadows unchanged, in a manner much like the Sabattier effect. Mackie lines appear just as one should expect.

37.2.4 Negative or positive?

The Sabattier effect can be employed at the development stage of either the negative or positive, and both choices have advantages and disadvantages. Solarizing a film negative in the darkroom is much more difficult than solarizing a print, and so it is done less often. But some of the best solarized images—most famously many of those by Lee Miller and Man Ray—were made by solarizing the negative.

The Sabattier effect naturally leads to an image with a very low range of tones. It is easy to make the darkest areas of the print or negative very dark. But the lightest areas are never white; the best one can hope for is a light-tone gray. And so using the

Sabattier effect at the stage of the print has a disadvantage: the final print will inevitably lack tonal range in the highlights. When the Sabattier effect is used to solarize a negative, this is also true for the negative. The added step of making a positive print from the solarized negative, however, allows the possibility of producing a full tonal range in the final print.

This distinction is evident in the two images in figure 37.4. I was able to accomplish the near-full tonal range in the left-hand image only because of adjustments I made in the process of making a positive from the solarized negative. The original negative had a very poor tonal range indeed. Teresa Patrick's direct print solarization on the right, however, has a much smaller tonal range; she was able to extend it somewhat with the added step of chemical toning.

Apart from the question of tonal range, there is another factor in choosing whether to apply the Sabattier effect directly to a final positive or instead to a negative that is then inverted to a positive. Which does the photographer want reversed—the shadows or the highlights? Does the photographer want the Mackie lines to be white or black?

Applying the Sabattier effect with no additional step of inverting the image from negative to positive will produce a final image with normal shadows and reversed highlights and with white Mackie lines.

Applying the Sabattier effect and then inverting the image from negative to positive will produce a final image with reversed shadows and normal highlights and with black Mackie lines.

It is important to note that the issue here is not, strictly speaking, whether one is dealing with a positive or negative. It is whether or not, after the Sabattier effect is applied, an additional reversal step (such as making a positive print from a negative) is performed.

For example consider the image in figure 37.8. It has reversed highlights and white Mackie lines, suggesting that I made this by applying the Sabattier effect to a piece of enlarging paper while making a final positive print in the darkroom. Instead, I produced this image by applying the Sabattier effect to an instant-film negative. In this particular scene the figures and the trees were dark, while the sky and the ground were bright. Those dark areas would develop out light on a normally-processed negative, while the sky and ground would be dark. But when I solarized the negative, the light areas of the negative (corresponding to the darker figures and trees) reversed from a negative to a positive. The brighter ground and sky remained a negative.

I could have then applied the usual additional step of reversing the film negative to a positive print. But the Sabattier effect had already turned the important parts of the picture (the figures) into a positive image. Instead I simply scanned the negative and adjusted the contrast. And so I have a final image that looks like I solarized a print in the darkroom even though it was made by solarizing a film negative. The additional step of scanning the negative to make a digital image allowed me to increase the tonal range of the image through digital image processing before making a (digital) print.

Figure 37.8. *Examining the moon*, John Beaver 2007.

There is another important question, and it goes to the heart of much that we will explore in chapter 44: do we want an image, or do we want an object? Part of the appeal of using the Sabattier effect in the darkroom to solarize a print is that the process is only partially under the artist's control, and so much is left to chance. And every print is different. But if, instead, one makes 100 identical prints from the same solarized negative then the element of chance comes in only at the level of the image and not the final print. This sometimes leads to a tough choice. Do I want a more-compelling image even if it is only virtual, or do I want a more-compelling object (a unique solarized print, for example) even if it is inferior as an image?

References

London B, Upton J, Stone J, Kobré K and Brill B 2005 *Photography* 8th edn (Englewood Cliffs, NJ: Prentice-Hall)

Stroebel L, Compton J, Current I and Zakia R 2000 *Basic Photographic Materials and Processes* 2nd edn (Waltham, MA: Focal Press)

IOP Publishing

Photography (Second Edition)
Physics and art in focus
John Beaver

Chapter 38

Photoelectronic detectors

38.1 Light meters

Here we consider detectors that utilize the direct liberation of electrons by the action of light. Many photochemical detectors—an individual silver halide crystal in a gelatin emulsion for example—also rely on the liberation of electrons by photons. For the silver halide crystal, this leads to chemical changes that result in a darkening of the material. In this chapter we, instead, consider materials for which the liberated electrons are measured electronically. This electronic signal is then used to produce an image in some other, often roundabout, way.

The simplest of such detectors measure the light received at only one location. To form an image, on the other hand, one must record the light separately over a two-dimensional (2D) surface. As such, electronic light detectors were first used in photography not for making images, but as light meters for setting exposure.

38.1.1 Selenium cells

Selenium is a semiconductor with chemical properties similar to sulfur. When deposited as a thin layer onto a sheet of metal, it can be used as a light detector. A second, transparent conducting layer is placed on top of the selenium to make a *selenium cell.*

A selenium cell can be used in a *photoconductive* mode: a voltage is applied by a battery, and a current flows in the circuit depending on the amount of light falling on the cell. But it can also be used as a *photovoltaic* cell: it creates its own voltage when light strikes it. Thus all that is needed to make a light meter is the selenium cell and a sensitive electric current meter.

Selenium-cell electronic light meters began to appear by the late 1920s and were used for photographic metering by the early 1930s. The Weston Model 617 of 1932 is a good early example. The selenium cells directly powered a sensitive electric current meter, and they were calibrated with scales to set photographic exposure.

doi:10.1088/978-0-7503-3703-8ch38

© IOP Publishing Ltd 2022

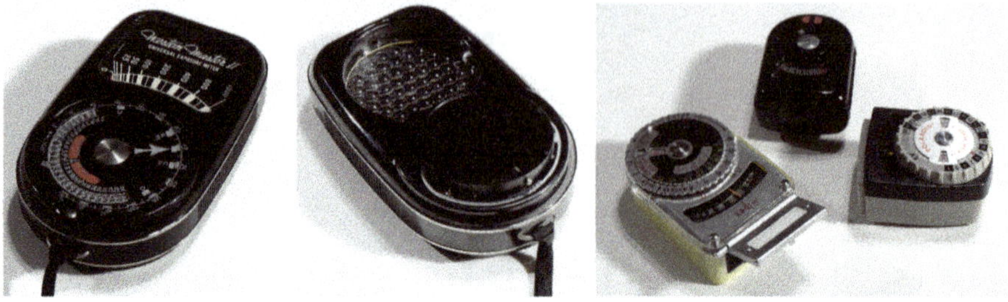

Figure 38.1. **Left** and **Middle:** Front and back views of Weston Master II universal, one of the most important early hand-held photographic selenium-cell light meters, introduced in 1945. **Right:** By the 1950s and 1960s, more powerful selenium cells allowed for more compact meters. The two Polaroid meters on the right were meant to clip onto the top of a camera.

Figure 38.2. Two selenium meter cameras. **Left:** Voigtländer Vitomatic II (1958–62). **Right:** Voigtländer Dynamatic II Deluxe (1961–62). The Dynamatic is an early example of automatic exposure, even though it used no outside power source.

By the late 1930s selenium cells had become powerful enough that small, hand-held light meters could be made, or they could even be incorporated into cameras. The Weston Master II introduced in the mid-1940s, seen at left in figure 38.1, is a good example of a palm-sized hand-held meter. Its selenium cell, the window of which can be seen in the center image, measured about four centimeters across. Several examples of much smaller meters from the 1950s can be seen in the right-hand image in figure 38.1. Two of these are small enough that they were meant to clip onto the body of a camera as an accessory.

The built-in selenium cell in the camera on the left in figure 38.2 can be seen as the rectangular window on the left of the camera's front. The meter is directly above it on the top; adjusting the aperture and shutter controls matches a marker with the needle when the exposure is correctly set. The range of scene brightness that can be accurately metered is quite limited compared to modern light meters, but it is adequate for most outdoor conditions. Although at least 60 years old, the selenium meter on this camera still works and is reasonably accurate.

The camera on the right in figure 38.2 contains an even smaller selenium cell, located in a hidden window inside the ring surrounding the camera lens. But it was still powerful enough to power a current meter that actually controlled the aperture of the lens automatically once the shutter speed was set. A complex system of mechanical linkages connected the camera aperture directly to the meter needle once the shutter was pressed.

38.1.2 Photoresistors

A *resistor* can be used to control the flow of current in a circuit that has a separate source of voltage. Put a battery in a simple loop circuit with a resistor, and a larger resistance means a smaller current will flow in the circuit. And it is this electric current that is the most directly-measurable electrical effect.

A *photoresistor* is a device that has a resistance that decreases with the amount of light incident upon it. Thus we can easily make a light meter with a photoresistor, a voltage source and a current meter. Because the power to physically move the meter comes not from the incident light but rather from the battery, a photoresistor light meter is potentially far more sensitive to dim light than a photovoltaic device such as a selenium-cell meter.

Figure 38.3 shows a photoresistor-based hand-held light meter (center) and a camera (right) with automatic exposure controlled by a photoresistor. Both used the most common example of a photoresistor—the cadmium-sulfide photocell (left).

A cadmium-sulfide cell has a very high resistance in the dark. When light is incident upon it, absorbed photons can excite electrons into a set of energy levels called a *conduction band.* The unexcited electrons are not free to move around from atom to atom, and so the material is an electrical insulator. But the conduction-band electrons are relatively free to move; the greater number of electrons excited to the conduction band, the better the material conducts electricity.

Figure 38.3. Left: A small cadmium-sulfide photoresistor. Copyright © Nevit Dilmen, Attribution-ShareAlike 3.0 Unported (CC BY-SA 3.0), https://commons.wikimedia.org/wiki/File:LDR_1480405_6_7_HDR_Enhancer_1.jpg, LDR 1480405 6 7 HDR Enhancer 1.jpg. **Center:** A hand-held meter that uses this same type of photocell. **Right:** A camera from the late 1970s that featured automatic exposure metered by a cadmium-sulfide photoresistor.

In a typical cadmium-sulfide photocell, like that shown in figure 38.3, the two electrodes are attached to opposite sides of a conducting surface that is broken by a 'river' of the photoresistive material. The zig-zag pattern allows for a greater surface area for light to be absorbed, thus increasing its sensitivity.

A light meter based on a cadmium-sulfide cell is simple to make. The resistance varies with incident light in a nearly linear fashion, and the cell responds to a range of wavelengths that is similar to that of typical photographic emulsions and detectors. In principle all that is needed is a battery, the photocell and an electric-current meter with an appropriate scale. As such, cadmium-sulfide light meters began to replace selenium-cell meters by the 1960s.

In practice, however, a photoresistor-based meter has two significant disadvantages. For one thing, cadmium-sulfide cells respond rather slowly to large changes in the amount of light. When using such a meter, one must sometimes wait a second or two until the meter has fully adjusted to the light.

Probably the biggest disadvantage, however, is that the simple photoresistor light meter circuit requires a battery that maintains a perfectly steady voltage even as it ages. The current in the circuit depends upon the battery voltage just as much as the resistance of the photocell. There is a battery chemistry that is ideal; the mercury cell. It maintains a constant voltage of 1.35 V even as it depletes. When the battery is fully depleted, it simply stops working altogether. The commonly used alkaline-type battery, on the other hand, has a voltage that gradually decreases as the battery depletes. Thus most cadmium-sulfide light meters required a mercury battery for readings that were at all accurate.

I write this in the past tense because it is now very difficult to find a mercury battery. Because of the environmental toxicity of mercury their manufacture and sale has been banned in most countries. This has, unfortunately, greatly complicated the use of many otherwise-wonderful vintage cameras and light meters. There is an alternative battery chemistry that somewhat matches the properties of mercury cells —called a zinc-air battery. But the slight differences in electrical properties and much shorter lifetime (especially in arid environments) means that it is hit or miss whether a zinc-air battery will work accurately in a beloved vintage camera. The readily available silver-oxide battery also maintains a near constant voltage but its voltage is significantly higher (1.55 V), so a light meter meant for the mercury battery will give inaccurate results (unless it is recalibrated) if a silver-oxide battery is used.

Some cadmium-sulfide light meters used a more complex circuit, called a *resistance bridge*. With this circuit the photocell does not directly control the current to a meter in order to measure the light. Instead, a precisely-calibrated variable resistor is adjusted (usually by hand) such that the current no longer flows through a branch of the circuit between it and the photoresistor. The current meter is not used to read the light value; instead, the bridge circuit is balanced by adjusting the variable resistor until the current meter reads zero. If we know what resistance is required to balance the bridge, we can then calculate the resistance of the cadmium-sulfide cell and thus the amount of light falling upon it. A battery is needed to provide a current to show when the bridge is unbalanced, but its precise voltage is not critical. Thus a silver-oxide battery may work well in a resistance-bridge light meter even if it was originally designed for a mercury battery.

38.1.3 Silicon photodiodes

A silicon photodiode is in some ways similar to the light-emitting diode described in chapter 5 but used in the opposite sense. It consists of a p–n junction: two different semiconductor materials placed in contact with a potential energy gap between them. The p-type material wants to be positive, while the n-type material wants to be negative. With no external voltage applied, it reaches an equilibrium with a voltage difference between the two sides. A voltage in one direction causes a current to flow, while a *reverse bias* (voltage in the opposite direction) leads to no current flow. This is the 'diode' in 'photodiode;' it can be used as a one-way valve for electric current.

But in a photodiode the absorption of photons of light—and the energy they carry—changes the situation. With no external voltage applied the absorption of light can cause current to flow, and the photodiode acts something like a more-efficient selenium cell. This is the foundation for the silicon solar cells used in solar panels.

For measuring light, photodiodes are often used in a photoconductive mode. A reverse bias is applied; in the absence of light, no current flows. The absorption of photons allows current to flow in proportion to the amount of incident light. A silicon photodiode can thus be used as a light meter in much the same way as a photoresistor, but with three important advantages.

1. A silicon photodiode is more sensitive than a photoresistor of the same light-intercepting area.
2. The photodiode responds to light more quickly than a cadmium-sulfide photoresistor.
3. The reading of a silicon photodiode light meter has little dependence on the battery voltage used to provide the reverse bias. This is in contrast to cadmium-sulfide photoresistor meters, which require a precise battery voltage.

Because of these advantages, most in-camera light meters from the late 1970s onward have used silicon photodiodes. In recent years, these have been replaced by the more-sophisticated charged-coupled device (CCD) or complementary metal-oxide semiconductor (CMOS) arrays as described in section 38.3.2. See figure 38.4 for a picture of the silicon photodiode of a modern hand-held light meter. It has a relatively large area so as to intercept more power from the incident light for greater sensitivity.

38.1.4 Bolometers

A bolometer detects light by way of its absorption and direct conversion to thermal energy. Many materials have an electrical resistance that varies strongly in relation

Figure 38.4. The large-area silicon photodiode from a contemporary light meter.

to temperature. Such materials are often used to make electronic temperature sensors, but they can also be used to detect light.

One of the earliest and most successful arrangements is the Langley bolometer (Richards 1994). A strip of resistive material is coated with a dark substance (lampblack, for example) so that it readily absorbs light and causes a corresponding rise in temperature. The resistance is then compared electronically to an identical strip that is protected from the incident light.

Most light detectors require the absorption of individual photons by individual atoms or molecules in the material. As such, there is usually a lowest detectable photon energy and thus a longest detectable wavelength of light. Since bolometers rely only upon the direct conversion of electromagnetic energy to thermal energy, they are sensitive to long wavelengths of light that are invisible to most other detectors. Thus bolometers are most commonly used for detecting infrared and longer wavelengths. Since the temperature changes are very tiny, they are usually cooled to very low temperatures with liquid nitrogen or liquid helium.

Bolometers have been widely used in astronomy as a means for measuring the brightness of stars at long wavelengths. Because a bolometer is, in principle, sensitive to any and all wavelengths of light, a *bolometric magnitude* is a measure of the total electromagnetic radiation of all wavelengths arriving from a star[1].

38.1.5 Pulse-counting detectors

A pulse-counting detector separately records the arrival of individual photons of light. Such an approach works well for measuring very low levels of illuminance, and so pulse-counting detectors have been used primarily for scientific and technical applications.

[1] In practice, no real telescope–bolometer combination can measure this directly, least of all because many wavelengths of light are blocked by Earth's atmosphere.

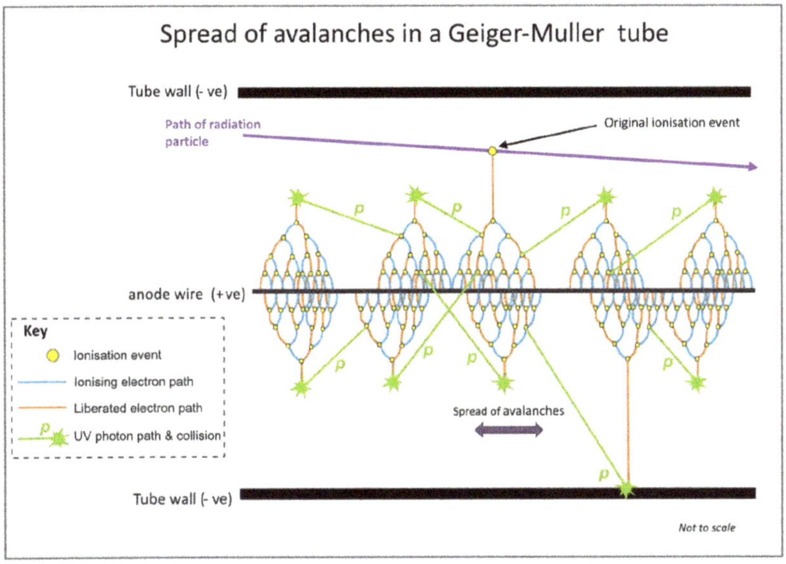

Figure 38.5. A representation of the avalanche process by which a Geiger–Müller tube detects a gamma ray photon. Copyright © Dougsim, Attribution-ShareAlike 3.0 Unported (CC BY-SA 3.0), https://commons.wikimedia.org/w/index.php?curid=22417438, Spread of avalanches in G-M tube.jpg.

The most familiar example is the *Geiger–Müller tube*, commonly called a 'geiger counter,' It is used to detect *ionizing radiation*: very high-energy photons in the gamma ray part of the spectrum and other high energy particles produced by radioactivity. A low-density gas is contained within a metal tube. A wire passing through the long axis of the tube maintains a high voltage difference with the tube. Ordinarily, the voltage is just a bit too low for a spark to jump the distance between the metal tube and the central wire. When a gamma ray or other radioactive particle passes through the tube, however, it ionizes the gas along its path and produces a brief conducting path. A sudden 'avalanche' of ionization occurs; liberated electrons accelerated by the high voltage collide with new atoms and ionize them. In some cases ultraviolet photons are emitted in the process, and these then travel to other parts of the tube and cause secondary ionization avalanches.

In this way, an ionization of a single atom leads within a few microseconds to a strong pulse of electric current between the outer tube and the inner wire. The current pulse contains far more energy than the original photon that started the ionization avalanche, and so it is easily detected and counted with circuitry. A diagram of the process can be seen in figure 38.5.

A *photomultiplier tube* can be used to detect individual photons of light in the visible range. A photon strikes a metal *photocathode* target inside a glass vacuum tube, and electrons are liberated by the photoelectric effect (see chapter 4). A voltage is placed between the photocathode (given a negative charge) and a second metal electrode (called a *dynode* and given a positive charge). The electron is accelerated by the resulting electric field toward the dynode and strikes it with enough energy to

Figure 38.6. An RCA 1P21 photomultiplier tube.

liberate more electrons from the dynode surface. These secondary electrons are then accelerated toward a second dynode, which is given an even greater positive voltage than the first, and even more electrons are liberated. This process is repeated through several chains of dynodes, and an easily-measurable pulse of current results from even a single photon[2].

A good example is the RCA 1P21 photomultiplier tube (see figure 38.6), which is sensitive over the range of about 300–700 nm and uses nine dynodes to multiply the signal. This particular type of photomultiplier was used by astronomers as part of the foundation for the multicolor photometric system described in section 30.4 (Bessell 2005).

The low-light sensitivity of pulse-counting detectors is of no great advantage if the light is already bright. And at high levels of illuminance, they have a big disadvantage: if two photons arrive within a characteristic *dead time*, they are counted only as one. Thus pulse-countering detectors significantly undercount photons at high levels of illuminance where it becomes increasingly likely that multiple photons will arrive within the detector's dead time.

38.2 Scanning photoelectric imagers

There is a way to record a 2D image focused by a lens by using a detector that is essentially just a light meter—something that only measures light at one location. We can use the light meter to scan back and forth across the image plane and so record different parts of the image at different times. The full image can then be reconstructed from the way the scanned signal varies with time. This approach has two obvious disadvantages for ordinary still photography.

[2] Only about 10% of photons are detected at all.

1. Most of the light is wasted because at any given moment light is arriving undetected at most parts of the image.
2. It can be problematic to record a rapidly-moving subject because different parts of the image are measured at different times.

38.2.1 Electronic video tubes

There is one important application in which a continuously scanned image is more to the point—video. An ordinary analog video or television feed, after all, works in the same way. The image continuously scans across and down the 2D image, repeating the scan many times per second, faster than what the eye–brain can perceive. For depiction of continuous motion, there are advantages to recording different parts of the image at once.

One strategy is to focus an image onto a photocathode inside a vacuum tube. The photoelectric effect creates what amounts to an image of electrons in the vacuum next to it. This 2D electron image is then scanned with an electron beam, producing a varying electric current that is used as the signal for the video feed. The electron beam is manipulated back and forth by a changing magnetic field produced by coils of wire in order to scan the image. RCA's *Vidicon* tube is a well-known example.

In the case of the *image dissector*, on the other hand, it is the entire electron image that is manipulated by magnetic fields to scan it across an aperture, so only a tiny part of the image is measured at once.

Image dissectors were revived for use in astronomy in the 1970s and early 1980s and were used in conjunction with an *image intensifier* to make an intensified image dissector scanner (IIDS). An image intensifier electronically amplifies the brightness of an image; focus a faint image onto one side of the image intensifier, and an amplified bright image appears on the other. A photocathode creates an electronic image, which is then accelerated by a high voltage to impact a phosphor screen and produce light.

In the IIDS, the image dissector is used to scan the intensified phosphor image from the image intensifier. The IIDS was used not in a video mode, but rather the signal was allowed to build up over time to increase the ratio of signal to noise. They were used mostly not for 2D images but rather to scan the one-dimensional (1D) output of a spectrometer. By the late 1980s the IIDS was largely replaced for this purpose by CCD detectors (section 38.3.2)[3].

38.2.2 1D array scanners

An array of many individual light detectors can be arranged in a line. If this 1D array is then carried on a track to move slowly across some flat object, the image can be recorded with time and a 2D image reconstructed. The array could be made from a row of tiny photodiodes all manufactured onto the same silicon chip, or it could be a long and narrow CCD or CMOS array as described in section 38.3.2.

[3] I was a graduate student in the astronomy department at Ohio State University at just the time the IIDS spectrometer was being replaced by a CCD spectrometer.

The Epson V700 scanner, just one of many examples, images six lines at once, each with 20 400 rows. In principle, only one such row would be needed to scan a black-and-white image. The rows, however, are covered with red, green and blue microfilters in order to simultaneously record a three-color image. Two rows are used for each color presumably to reduce noise in the final scanned image.

Such a device must, of course, be controlled by a computer, and the output is a digital image. Although similar scanners have been produced to scan the image focused by a large-format camera, the technology is used most often to make finely-detailed digital images of flat, physical media.

38.3 2D photoelectronic imagers

We have seen that there are many ways to produce an electronic response to light. To simultaneously record all parts of a 2D image in this way, however, has its challenges. Light focused onto a photocathode inside a vacuum tube, for example, produces a corresponding 'image' of electrons. But how do we both simultaneously and separately measure the electrons from different parts of the image?

We could make a 2D array of photodiodes, each separately measuring the light at a different part of the image. But even if we made a square of only 100 by 100 photodiodes, we have $100^2 = 10\,000$ wires in order to separately measure the signal from each. It is a poor image indeed that can distinguish only 100 points in each dimension.

38.3.1 Electrophotography

A selenium cell produces an electric charge in the presence of light. So why not simply use a lens to focus an image onto a sheet of selenium and then measure the pattern of electrical charge that develops on its surface? The problem is that there is no easy way to do this. It is a simple task to measure the total amount of charge accumulated on the whole piece of selenium, but there is no easy way to tell how much of the charge came from which location. Thus we have an excellent light meter but not a detector suitable for recording images.

It turns out that there is a rather roundabout way to record an image with a selenium detector. It works like this (Stroebel *et al* 2000, pp 206, 207).

1. Focus an image onto a sheet of selenium so that different parts accumulate varying amounts of negative electric charge.
2. Dust the sheet of selenium with a very fine black powder (called *toner*) that will stick to the metal everywhere there is a buildup of electric charge. If done just right, wherever the charge is greater, more toner will stick.
3. Use a high voltage (about 1000 V) to give a sheet of paper a positive electric charge.
4. Roll the paper onto the sheet of selenium so that the negatively-charge toner sticks to the positively-charged paper (opposites attract).
5. Heat the paper up so that the toner bakes onto it.

The bulk and weight of a mechanism to reproduce these steps renders the process impractical as a way to make a picture with an ordinary camera. It is, however, the basis for the traditional photocopy machine. Instead of a flat sheet of selenium, a rotating drum is typically used. An image focused by a lens is scanned onto the drum as it rotates.

Most contemporary photocopiers works in two steps. The image is first captured digitally by a scanner, and the resulting digital image is used to expose the selenium drum with computer-controlled lasers.

38.3.2 CCD and CMOS array detectors

The most common type of detector in a digital camera is either a *CCD* or *CMOS array*. CCD and CMOS are acronyms for, respectively, charge-coupled device and complementary metal-oxide-semiconductor. Fortunately we can arrive at a basic understanding of what these detectors do without diving headfirst into the pit of technical details regarding the precise meaning of these terms.

The word 'array' means that the detector consists of many tiny individual detectors that are arranged in an evenly-spaced rectangular pattern. Unlike the selenium sheet used in a photocopier or the random pattern of silver-halide grains in a silver-gelatin emulsion, a CCD or CMOS array is actually many individual detectors (typically millions) arranged in a regular grid of rows and columns.

The two types of detector arrays, CCD and CMOS, work in slightly different ways, but the overall effect is much the same. Photons are absorbed at the location of a particular light-sensitive site (called a pixel), and this produces an excess negative electric charge there. This negative charge is then measured, and the result is converted to a digital number that represents the brightness of the light at that particular pixel.

Instead of measuring the electric charge directly and separately from each pixel, a CCD transfers the charge from pixel to pixel in a sort of 'bucket brigade' fashion. The electric charge is passed from one pixel to the next along rows and then down the last column one row at a time. Thus the array is 'read out' one pixel at a time. As the tiny bits of electric charge from each pixel leave a CCD array, they are amplified to make a measurable voltage. It is a simple task for a computer to determine which measurement corresponds to what particular location on the array, and thus a picture can be reconstructed.

And so after the exposure is finished, an enormous number of pixels can be measured separately, one at a time, without the need for an equal number of individual electrical contacts. To illustrate this idea, consider a simplified CCD array of only 3×3 pixels. Figure 38.7 illustrates an example of the different steps in the charge transfer process.

The nine pixel values of the exposed image are represented in the square at the upper left. The square to its right represents the first step in the charge transfer process; successive steps are represented by the other squares in order from left to right and top to bottom. Each measured pixel comes out, one at a time, at the lower

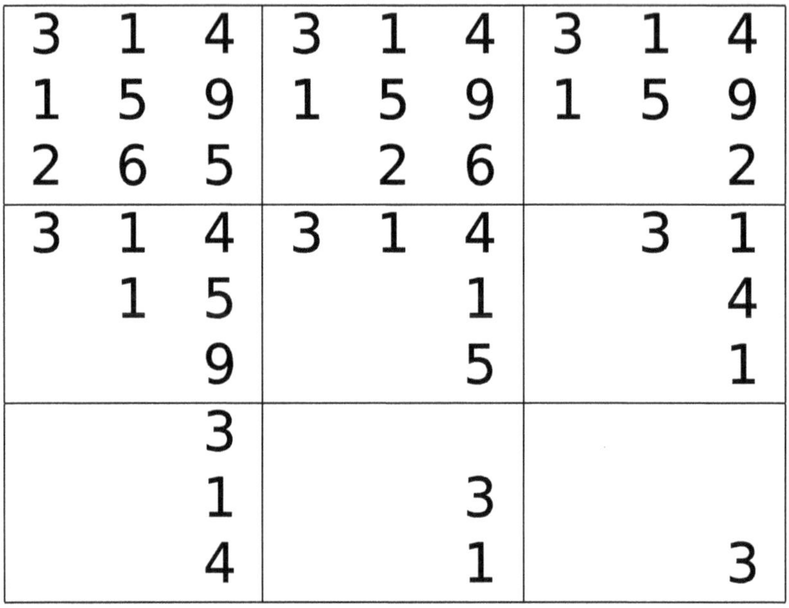

Figure 38.7. A simplified example of charge transfer in a CCD array, showing the nine steps in reading out the values in this imaginary nine-pixel CCD.

right of the CCD. The bottom row shifts right, and then the second from the bottom row shifts right and then down, followed by the top row. The pixels values thus come out one at a time in the order 5, 6, 2, 9, 5, 1, 4, 1, 3. A computer is needed to reconstruct the 2D image from this 1D string of pixel values.

Real CCD arrays may take different approaches to this overall theme. Some transfer the values only along rows to a blank region that is kept in the dark. And thus the light-sensitive part of the CCD array can be used again while the previous image is being read out pixel-by-pixel from the covered area. Other CCD arrays are read out in sections rather than the entire chip at once.

CMOS arrays differ in that they are read out one row at a time; one row is read out while the other rows are being exposed. And so different parts of the array are, in effect, exposed at different times. This can lead to odd effects for still photographs of rapidly moving subjects, but it makes CMOS arrays more practical than CCD arrays for recording video. For this reason, CMOS arrays are increasingly common in digital cameras that double as a video camera.

38.3.3 The physics of CCD arrays

The light-sensitive sites on a CCD array are capacitors: individual microscopic spots on a silicon chip that can store electric charge in the form of electrons. When a bias voltage is applied, they become sensitive to light. The energy from absorbed photons raises electrons to higher energy levels, which allows them to move freely and gather at the capacitor site. The more photons that arrive at that part of the CCD array, the more electrons accumulate. By strategically altering the voltages applied to the different capacitor pixel sites, they can be induced to pass their charge to a neighboring site in order to read out the image.

The CCD is made from layers of silicon-based semiconductor materials that have been strategically tainted (*doped* is the technical term) with impurities. In addition there are tiny pieces of conducting metal interlaced throughout the structure of the CCD chip to form microscopic wires and electrodes.

A semiconductor, as its name implies, has properties midway between those of conductors and insulators. While the positively-charged nuclei of atoms remain intact, the negatively-charged electrons can sometimes transfer from one atom to another. In a good conductor—a copper wire is an example—some of the electrons do not belong to any particular atom (one each per atom, in the case of copper). They can move around freely on the surface of the wire, randomly colliding with each other much like the atoms or molecules in a gas. These *conduction electrons* respond easily to external electric fields, and thus conductors are very good at easily transferring electric charge (and energy) from one place to another. An insulator, on the other hand, is the opposite of a conductor; it is difficult to transport electrons, which remain stuck to their individual atoms.

If a steady current of electric charge is desired, a good conductor is an excellent tool for the job. But a photoelectronic light detector depends on the build up of static charge. There needs to be some way for that charge to move and gather at the light-sensitive site. On the other hand, we want it to stay in place until the exposure is finished. For a good conductor, any static charge rearranges itself almost instantly until all of the electrical forces are balanced. If electrons are photoelectrically removed from their atoms on a conductor to gather at one part of the image, they will not stay there. An insulator has the opposite problem; there is no easy way to make the charge move in order to measure it. A semiconductor provides the right balance: the charge can be better controlled, allowed to build up in one place during the exposure but then encouraged to move in order to measure it after the exposure is completed.

There is more to it than that, as semiconductors have many unique and complex properties. In particular when two different semiconductor materials are placed in contact with each other, the door is opened to many unusual phenomena. Not only can some of the electrons move from atom to atom, but the empty spaces they leave behind begin to have many of the properties of particles. And so we have negatively charged electrons, and positively-charged *holes* that interact with each other in complex ways that can be fully described only by the strange rules of quantum physics.

For most photographic purposes CCD and CMOS arrays are sensitive to too broad of a wavelength range because they are sensitive well into the near-infrared part of the electromagnetic spectrum. Many camera lenses do not focus well at these wavelengths, and even when they do their focus point is different for infrared than it is for visible light. Thus if the full range of wavelengths were used, the picture would be blurry.

Figure 38.8. The CCD array detector from a digital single-lens reflex camera. This particular CCD array has 3000 × 2000 pixels on a silicon chip with dimensions of 23.7 × 15.6 mm. The individual pixels are far too small to see at this scale, and the green color is from a glass filter covering the sensor. The filter not only protects the delicate sensor, it also blocks near-infrared light to which the sensor is particularly sensitive.

This problem is usually solved by placing a special filter directly over the detector—a filter that blocks infrared light while passing visible light (see figure 38.8). Equipped with this filter, a typical CCD array passes light of wavelengths stretching across the visible part of the spectrum.

A photoelectronic detector such as a CCD or CMOS array is, in a sense, a complex machine. This structure of different types of semiconductor and metal conductor is built up to form a patterned flat surface of several layers—like overlaying floor plans of a multistory building. This overall flat, but still three-dimensional, design is engineered to produce the array of light-sensitive capacitor sites (the pixels) that produce the signal from the incident photons. It also includes the semiconductor structures needed to receive input signals and to transfer the charge and amplify the signal as the array is read out.

The manufacture of CCD and CMOS arrays has in many ways overtaken traditional photochemical processes. It is perhaps ironic that photochemical photographic techniques are a crucial part of the manufacturer of such digital detectors. The elaborate microscopic structures in the silicon chip of a CCD or CMOS array are vacuum deposited one layer at a time. The detailed structure of each layer is realized with *photolithography*. A light-sensitive coating is applied to the silicon wafer, a patterned mask is overlaid, and it is exposed to light and then processed. The light alters the coating so that it can then resist chemical etching or the vacuum deposition of additional layers. And so etching can dig out tiny patterns in the surface, or additional patterned layers can be added. One can compare this most high-tech and modern of industrial processes to the antiquarian gum-bichromate process we described in chapter 36.

References

Bessell M S 2005 Standard photometric systems *Annu. Rev. Astron. Astrophys.* **43** 293–336

Richards P L 1994 Bolometers for infrared and millimeter waves *J. Appl. Phys.* **76** 1–24

Stroebel L, Compton J, Current I and Zakia R 2000 *Basic Photographic Materials and Processes* 2nd edn (Waltham, MA: Focal Press)

IOP Publishing

Photography (Second Edition)
Physics and art in focus
John Beaver

Chapter 39

Three-dimensional photography

39.1 Stereo photography

Our brains use many cues in order to perceive depth; atmospheric perspective, converging lines and the overlap of objects are obvious examples. But there is another way the eye–brain combination may perceive depth that has no counterpart in a single, flat image. We have *binocular vision*—two eyes spaced roughly five centimeters apart that see the world from two slightly different vantage points. Our brains can combine this information to produce a sensation of a three-dimensional (3D) world (Kalloniatis and Luu 2018).

This ability is called *steropsis*, and it has been recognized since the earliest days of photography that one can trick the brain into producing the same effect simply by taking two pictures simultaneously of the same scene, but from two slightly different places. If these two photographs are then placed in a special viewing device, such that each eye sees only one of the two images, then a perception of depth can be achieved that is similar to that of viewing the real scene in-person. Figure 39.1 (top) shows a *stereo pair*—two photographs of the same scene but taken from two slightly different vantage points. For this particular example, I simply took a picture, stepped a couple of feet to the side, aimed my camera at the same place and then took a second picture. Looking closely at a detail of the two images (bottom), it is clear they are different. The vertical edge of the foreground rock, for example, overlaps the background cliff differently. The right-hand view was taken by stepping to the right, and you can see more detail to the left in the background cliff, as you partially see around the foreground.

This is called *stereo photography*, and its popularity has waxed and waned in cycles over the past 150 years. Special cameras have been made that use dual lenses and shutters to record both images simultaneously onto different parts of the same roll or sheet of film. Some of these, such as the made-in-Milwaukee Stereo Realist

doi:10.1088/978-0-7503-3703-8ch39

© IOP Publishing Ltd 2022

Figure 39.1. **Top:** A stereo pair of photographs, made by stepping slightly to one side before taking the second picture. **Bottom:** A detail showing the shift, between the two images, in foreground compared to background.

Figure 39.2. A David White Stereo Realist camera. Copyright © Kim Scarborough, Attribution-ShareAlike 2.5 Generic (CC BY-SA 2.5), https://commons.wikimedia.org/wiki/File:Stereo_Realist.jpg, Stereo Realist.jpg.

(figure 39.2) were moderately popular, and working models can still be found on the used market. The Stereo Realist and several other competing models used 35 mm slide film in a half-frame format. The two lens–shutter assemblies are the outer two circles on the front of the camera; the viewfinder to frame the picture is halfway between. Two little square windows at bottom left and bottom right are the rangefinder windows for focusing.

Figure 39.3 (left) shows a large format Speed Graphic camera modified as a stereo camera. The dual lens and shutter assembly was modified from an old passport camera. At the center, there is a stereo pair taken with the camera, using Polaroid

Figure 39.3. **Left:** A Speed Graphic camera for 4 × 5 inch film with a dual lens and shutter mounted for stereo photography. The lens–shutter assembly is modified from an old passport camera. **Center:** An example of a stereo pair photographed with the camera. **Right:** My ugly-but-functional stereoscope for viewing the stereo image.

Figure 39.4. From **left** to **right:** A popular Holmes Stereoscope (Copyright © Davepape, Public Domain, https://commons.wikimedia.org/wiki/File:Holmes_stereoscope.jpg, Holmes stereoscope.jpg) from the nineteenth century; several vintage stereo cards; a Viewmaster stereo viewer and the special circular stereo cards that go with it; a viewer for 35 mm half-frame stereo transparencies, such as produced by a Stereo Realist camera.

instant print film. Both images were recorded simultaneously onto the same print, but they must be cut apart and reassembled in reverse order to produce a proper stereo view.

To view a stereo pair, each eye must separately see the two images in the pair. A device to accomplish this is called a *stereoscope*. My ugly-but-working version, cobbled together with hobby plywood and a pair of cheap reading glasses, can be seen at right. Clearly, this approach lends itself most simply to stereo pairs that are separated by the same distance as the human eye; thus the individual images are quite limited in size.

Figure 39.4 shows (left) a type of stereoscope popular in the nineteenth century. The dual images were printed onto cards, often in large quantities by photo-reproduction techniques. In my own small collection of vintage stereo cards (center), there are brief stories on the back. Many are staged scenes of drunken husbands being scolded by their wives.

Figure 39.5. The two images in figure 39.1 have been combined with GIMP to form a red–blue stereo image that can be printed any size and viewed with special glasses (**right**).

At the right in figure 39.4, there are two popular formats of stereo viewers. The Viewmaster uses circular cards with positive transparencies that rotated in the viewer to bring into view a different stereo pair. When the viewer is pointed at a source of light to back-illuminate the film transparencies, the image is viewed through a pair of lenses. The right-most viewer in figure 39.4 accepts individual cards with half-frame 35 mm stereo pairs on positive transparency film, such as that produced by a Stereo Realist or competing camera.

There is another approach to viewing a stereo pair. Instead of optically redirecting each eye to a separate picture, we may use properties of light to separate the two images, even if they are placed on top of each other. The simplest way is to use color. Color one image red and the other blue, for example. Print (or project) the two images on top of each other and then place similar-colored filters over each eye —red for one eye and blue for the other. This works, although the separation of the two images is less complete, and the dueling colors can be confusing, even if one is attempting to reproduce only a monochrome image. But it has the advantage of convenience that such stereo pairs can be printed at any size and can be viewed with simple and inexpensive dual-color glasses.

Figure 39.5 shows a dual-color stereo image made with GIMP by combining the two images in figure 39.1. At the right, there is a pair of inexpensive red–blue glasses by which to view the image.

If the images are projected onto a screen, a better separation of overlapping images can be accomplished using *polarization*. Project the two images with separate projectors, each passing through a polarizing filter—but with the filters rotated 90° relative to each other. Then view the combined image with glasses that contain polarizing filters also crossed with respect to each other. Each filter blocks the polarization projected through the crossed filter and allows the other image to pass. This is clearly advantageous for color stereo images, but it requires a projection screen that maintains the polarization of light that reflects off it.

Some popular accounts of our binocular vision compare it to the astronomer's use of trigonometric parallax to measure the distances of stars or the double image in a rangefinder camera. In both cases, there is a view of the same object from two different vantage points—just as with our binocular vision. When the nearer object is seen against a more-distant background, the two views are in slightly different directions; measurement of this *parallax angle*, along with some simple trigonometry, can then be used to find the distance to the object.

Our eyes do indeed have to point inward—they must cross slightly—in order to converge on a nearby object, as compared to a distant one. This is a factor in the brain's processing, but the perception of depth that forms in the brain is clearly far more than simple mental trigonometry. Stereo photography is a demonstration of this complexity—the perception of depth in a stereo image can be quite striking, even though the eyes looking into the stereo viewer point nearly in the same direction to view the two images.

39.2 Lenticular photography

A lenticular lens is an arrangement of many narrow, parallel cylindrical lenses, usually molded from transparent plastic. Figure 39.6 depicts an example. The cylindrical lenses act like magnifying glasses for the back of the lenticular sheet.

If an image is placed on the back of the lenticular lens, views from different angles see different portions of the small region behind each cylindrical lens. It is thus possible to interlace two images, with alternating parallel strips behind each cylindrical lens. If we alternate strips of red and green for example, as illustrated in figure 39.7, then one vantage point sees only green while the other sees only red.

But we can also interlace two separate photographs in this same manner, such as the two images in a stereo pair. Such a *lenticular photograph* is yet another way to view a stereo image. If the size and shape of the cylindrical lenses are chosen carefully, each eye sees a separate image when holding the lenticular photograph at an ordinary viewing distance. This has the advantage that no special viewing glasses

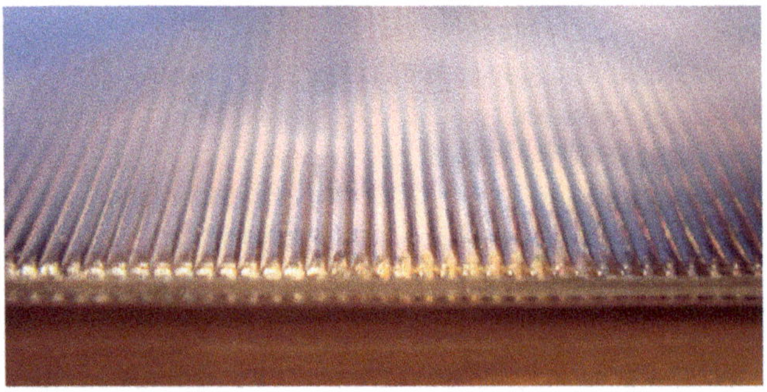

Figure 39.6. A close up photograph of a lenticular lens. Copyright © World Imaging – own work, Attribution-ShareAlike 4.0 International (CC BY-SA 4.0), https://commons.wikimedia.org/wiki/File:Close_up_of_the_surface_of_a_lenticular_print.jpg, Close up of the surface of a lenticular print.jpg.

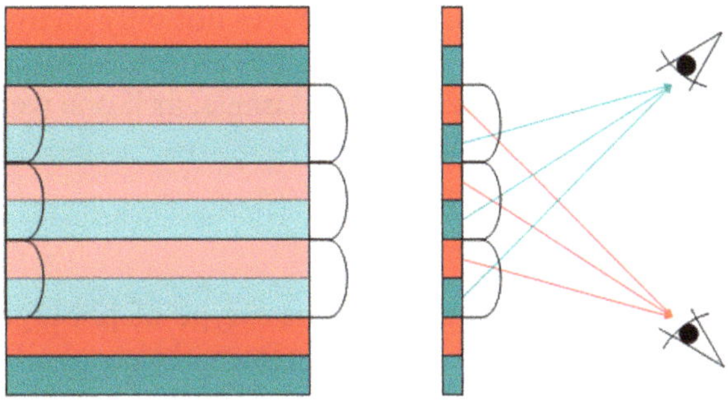

Figure 39.7. An interlacing of red and green strips, when viewed through a lenticular lens, produces a red image as seen from one angle and a green image as seen from another. (Attribution-ShareAlike 3.0 Unported (CC BY-SA 3.0), https://commons.wikimedia.org/wiki/File:How_a_lenticular_lens_works.png, How a lenticular lens works.png.)

Figure 39.8. Left: A Nimslo 3D lenticular camera from the 1980s. Copyright © John Alan Elson, Attribution-ShareAlike 4.0 International (CC BY-SA 4.0), https://commons.wikimedia.org/wiki/File:Nimslonew.jpg, Nimslonew.jpg. **Right:** View of the image plane, showing the four separate image areas produced by the lenses. Copyright © John Alan Elson, Attribution-ShareAlike 4.0 International (CC BY-SA 4.0), https://commons.wikimedia.org/wiki/File:Nimslonew.jpg, Nimslonew.jpg.

or apparatus are needed. There is, however, clearly a trade-off with image detail; only half of the picture space is used for each of the two images.

The basic arrangement of lenticular photography can accommodate the interlacing of more than two images—four for example—each seen with its own viewing angle. Thus we can make a *lenticular camera*, with four lenses side-by-side instead of the two that make a stereo camera. The Nimslo camera of the 1980s is an example; see figure 39.8.

Each lens of the Nimslo produces a separate image on the 35 mm film, from its separate vantage point. A lenticular print can be made with specialized equipment by exposing the print material to light from the four negatives, projected from four angles by separate projectors. If the light from these projectors passes through a lenticular lens in front of the light-sensitive paper, an interlaced image results. Think of figure 39.7, but with the eyes replaced by projectors.

A lenticular photograph made with four lenses instead of two is more than a stereo photograph. It is really *three* separate stereo images, all recorded from slightly different angles—combining respectively the first-second, second-third, and third-fourth lens combinations. Thus one sees not only an image that produces the 3D effect of stereopsis, but one can—albeit to a quite limited degree—shift the angle of the print and see around to a different angle, as if one were moving around a real 3D object.

39.3 Holography

At a superficial level, a hologram seems to be a type of photograph. A flat light-sensitive material is used to record light coming from objects. And when one looks at the processed hologram, an image of that object is visible. But there is an important sense in which a hologram is not a photograph at all. For if one looks closer and closer at a hologram, with greater and greater magnification, it becomes apparent that it contains only complex alternating patterns of light and dark. And this pattern at a given place on the surface of the hologram is unrelated to light and dark areas of the image.

A hologram contains no image on its surface at all; rather, the fine-scale pattern of light and dark is an *interference pattern*. When viewed in the proper way, the interference pattern reconstructs the 3D structure of the light reflecting off the subject. A given part of the image—the left eye of a portrait subject, for example—is not located at any part of the hologram in particular. Instead, any given location on the hologram contains a complex interference pattern that carries information about the *entire* image, as seen from that particular location. The interference pattern at a different location on the hologram reconstructs the entire image as seen from that different vantage point. And so the hologram contains 3D information about the subject.

This is possible because the process of making a hologram takes advantage of more than just the intensity of light coming from the subject. The hologram also uses the light wave's *phase*—the precise location in space, at a particular moment in time, of a given peak in the wave.

In a hologram's simplest form, the coherent light of a laser—spread out by lenses—is used to illuminate the subject. But part of the same beam is also split off with a partially reflective beamsplitter and sent directly to the detector. Thus some of the laser light comes directly from the laser source, and some comes from reflections off the subject. These two waves interfere with each other—adding together if the wave peaks happen to match up but canceling out if the peaks lie opposite each other. This produces a fine-scale interference pattern. When the hologram is illuminated by the same laser light but without the object in place, the interference pattern produces a 3D image of the subject.

Modern holograms are more sophisticated than the simple example described here, but they still involve some kind of geometrical arrangement between light source, beam splitters and the photographic detector; a camera is not usually involved at all. The hologram may be encoded not by a pattern of dark silver grains but rather by changes in the index of refraction or scattering properties of a transparent material, resulting in a brighter hologram that can be viewed in ordinary

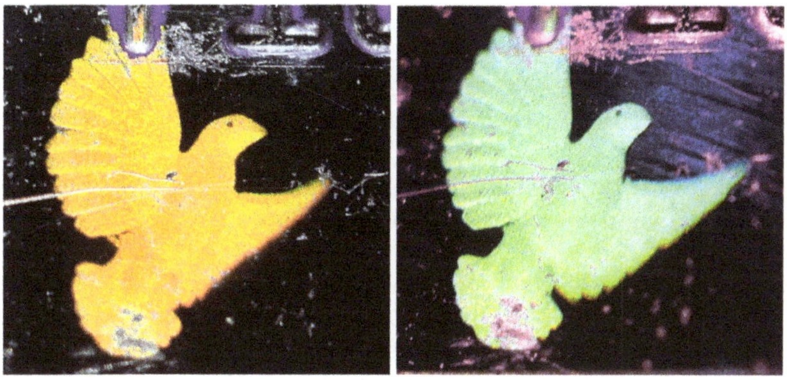

Figure 39.9. The same security hologram photographed from two different angles.

incoherent, polychromatic light. The wavelength dependence of the interference produces colors when seen at different angles.

Since such a hologram is embodied in a microscopic pattern of physical changes in the medium, it does not lend itself to photographic reproduction; a picture of a hologram is not a hologram! Thus holograms are often used as security features in currencies and credit cards. Figure 39.9 shows two views of the same credit-card security hologram. Notice that the head of the dove is clearly depicted as a 3D object seen from different vantage points.

Thus one could photograph a hologram with a stereo camera, and a stereo pair would result. But this would be only one possible stereo view; the hologram allows one to see stereo views from a range of angles.

In recent years, the word 'hologram' has been appropriated to describe unrelated processes that mimic the idea of a hologram. For example, an ordinary image can be projected—often outdoors—onto a large, suspended, thin film that is partially transparent and partially reflective. The image focused onto the film thus seems to be literally suspended in a physical space. This is usually combined with video to enhance the 3D effect. Such 'holograms' are usually performances carried out at night or in dim stage lighting.

39.4 Light-field photography

When light strikes a given place on an ordinary detector such as a silver gelatin emulsion or a digital charged-coupled device (CCD) array, the detector only records the *brightness* of the light that hit that part of the detector. The use of a colored mask allows one to also record some wavelength information. But when a CCD or piece of film detects a photon, the detector cannot tell from which direction the photon came.

But digital detectors now exist that can tell not only the brightness of the light striking some part of the detector but also how much of that light came from each part of the lens. This is done with a special mask, analogous to the Bayer mask that turns a CCD array into a color detector. The mask places tiny microlenses over assigned clusters of the detector's light-sensitive sites. A camera that works in this way is called a *plenoptic camera*, and crude attempts go back as far as Gabriel Lippmann in 1908 (Ng *et al* 2005). Consumer plenoptic cameras first became available in 2012.

The overall effect is that each pixel in the image is actually a tiny very-low–resolution photograph of the back side of the camera lens. Each pixel contains information not only on how much light arrived at that part of the detector, but it can sort out that light by which part of the lens it came from (Ng *et al* 2005, section 1.3).

This information, combined with known properties of the lens in the light-field camera, can be used to reconstruct a 3D model of the objects in the world from where the light rays arrived. It also means that a light-field camera can focus after the picture was taken. A disadvantage is that resolution is sacrificed in order to gain this additional depth information.

A light-field image is not really a picture in the ordinary sense; one must use specialized image processing to produce a conventional digital image that can then be printed or viewed on a screen. In so doing, one can choose not only which distance is in best focus, but also the overall depth of focus. Thus a single exposure from a plenoptic camera can produce an infinite variety of photographic views of the same subject.

An exposure from a plenoptic camera is still, however, taken from a range of vantage points at most only the size of the camera lens. Thus, although it has 3D information about the subject, it does not produce the same 3D effect as a stereo camera. The stereo camera with its two widely separated lenses can partially see around obstacles, and thus duplicate the information needed to produce the effect of stereopsis in the eye and brain.

References

Kalloniatis M and Luu C 2018 Perception of depth *WEBVISION: The Organization of the Retina and Visual System* ed H Kolb, R Nelson, E Fernandez and B Jones https://webvision.med.utah.edu/

Ng R, Levoy M, Brédif M, Duval G, Horowitz M and Hanrahan P 2005 Light field photography with a hand-held plenoptic camera *Stanford Tech Report CTSR 2005-02*

Part VII

Photography as an art and the meaning of digital

IOP Publishing

Photography (Second Edition)

Physics and art in focus

John Beaver

Chapter 40

The digital and the analog

One might say that analog techniques are, as a matter of principle, more 'precise' than digital. Analog is smooth, continuous and infinitesimally divisible, whereas digital is discrete and thus inherently limited in its precision. One could offer, as an example, the set of smoothly curving grooves on a vinyl long-play (LP) recording, as opposed to the digitally encoded sound information in a compact disc or digital sound file recording. The argument is that there is a clear limit to the amount of information in the digital file, while the grooves on the vinyl LP, since they are smooth and continuous, contain an essentially infinite amount of information.

But this view of digital versus analog is incorrect, at least as a general principal. It may turn out to be true in a particular instance, but the opposite could be true as well; the devil is in the details. Although any particular instance of digital encoding has limited precision, the digital process in general does not. If, in a particular case, it is the method of digital encoding that is limiting precision, one need only change to a different digital encoding to increase the precision. Furthermore, any particular instance of analog encoding, despite its appearance of an infinitesimally dividable range of possibilities, is also limited in precision.

It is the physical nature of the act of measurement that ultimately decides how much information is potentially available, not whether it is recorded as digital or analog. And finally, many measurement processes that seem at first glance to be analog, with a smooth and continuous range of possibilities, are actually more like digital if one looks closely enough.

40.1 Pixels and granularity

One aspect of the digital is its *discreteness*—that it can be subdivided only so much until the smallest unit is reached. This fact is obvious for a digital detector such as a charged-coupled device (CCD) array; a given example has only so many pixels. Indeed, the basic concept of discreteness is built in to the very word *pixel*—which is

 © IOP Publishing Ltd 2022

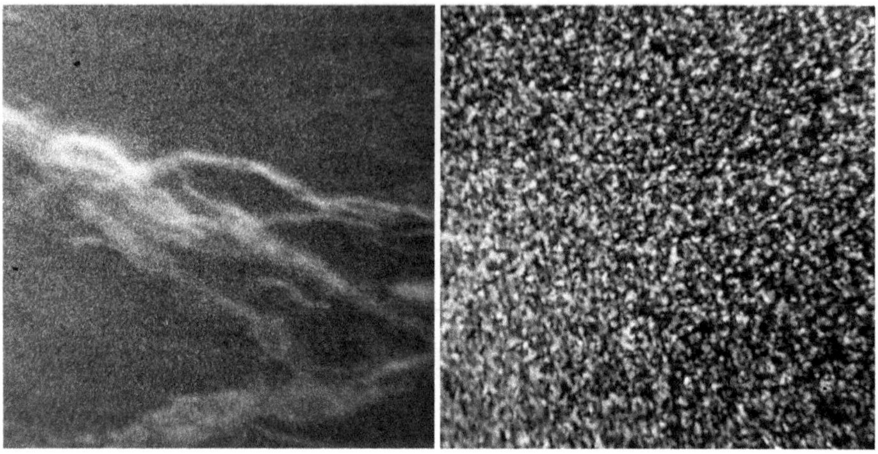

Figure 40.1. **Left:** Detail of a black-and-white film negative, as seen through a microscope. The film grain is just barely visible. **Right:** At much higher magnification, only the film grain is visible, with no hint of image detail.

simply shorthand for 'picture element.' By definition, there is no such thing as half a pixel.

But let us consider ordinary black-and-white film. We have already seen that a black-and-white silver-gelatin emulsion consists of light-sensitive crystals of silver halide suspended in a thin layer of gelatin. Each exposed and developed silver-halide crystal acts as a sort-of 'atom' for the picture. That is, a given crystal is only one shade of gray; it cannot be, for example, half black and half white. This means that the size of the crystals (and the spacing between them) determines the smallest amount of detail that can be imaged on the film. If the crystals are bigger (and farther apart), then less detail can be imaged in a given square centimeter of the film.

We call the microscopic pattern of developed silver-halide crystals *film grain*, or simply *grain*. The quantitative measure of film grain is called *granularity*, and a subjective determination of its overall effect on the picture is called *graininess* (Stroebel and Compton 2000, p 258). See figure 40.1 for an example of a silver-gelatin film emulsion as imaged through a microscope. The left-hand image is at lower magnification, and one can see both a part of the image detail and a hint of the film grain. A detail of the same image is shown on the right, at much greater magnification. No image detail is visible at this scale, only the pattern of the silver specks.

Figure 40.2 shows an image from a digital camera, along with a tiny detail, greatly magnified. The individual pixels are visible, and they are portrayed as simply square blocks. The digital pixels have no particular shape—they are simply image brightness values at different locations on a two-dimensional grid. Most would agree that this sort of magnified detail is both less interesting and more unappealing than the analogous detail in the silver-gelatin emulsions—or a pointillist painting by Seurat. But regarding the amount of information contained within each, the comparison is not so straightforward.

Figure 40.2. The grid-like pattern of pixels at the right is a from a tiny detail of the digital image at left.

40.2 Resolution

An important aspect of any detector is how much spatial detail is present in the image captured by it. If one were to look closer and closer at the image on a piece of film with magnifiers of higher and higher magnification, at what point would there be no more detail to be seen? Or, regarding a digital image, how much information is really there? At what point will zooming in on the image yield no additional detail?

This concept of *resolution* is complex and prone to confusion, and we must make a distinction between the total information contained in a picture and the information per surface area. It is the type of detector material that usually limits the latter. The former is related both to the type and the size of the detector. Confusingly, the word 'resolution' is often used inconsistently, sometimes meaning the total amount of information and other times meaning the information per surface area. The process of printing adds even more complications.

For an image encoded digitally, the concept of resolution is potentially straightforward so long as we agree upon terms. Let us consider, for example, a digital image that has dimensions of 1000×1500 pixels. This means the image could be subdivided by 1000 parts in the horizontal direction and 1500 parts in the vertical direction. What this means in practice depends upon a critical question—what will we do with it?

Suppose, for example, that we mean to print the digital image at a particular physical size. Clearly, the ratio of width to height is 2:3. Let us say we use this digital image to make a print that is 24 cm high; the width must then be 16 cm if it is to be the same shape as the digital image. At this scale, the vertical dimension thus contains 1500 discrete picture elements stretched over 24 cm—62.5 pixels per centimeter. Since we have scaled the horizontal dimension in proportion, it too has 62.5 pixels per centimeter.

How would such a print look, in practice? Since there are 62.5 pixels per centimeter, two adjacent pixels would be separated by $1/62.5 = 1.6 \times 10^{-2}$ cm $= 0.16$ mm. At a typical viewing distance, this might be discernible but barely so. Thus our digital image contains enough information to make a print at this size that is adequate to fully represent a detailed image. If, on the other hand, we were to use the same digital image to make a print that is six times that size in each dimension, then two adjacent pixels would instead be 0.96 mm apart. One could inspect the image up close and see that there is no more to see.

It is clear that apart from the detail in an individual print—which also depends upon how large the dimensions are—there is a measure of how much information is contained within the image itself. This is not about pixels per centimeter—something that only has meaning in the context of a particular chosen print scale—but rather the total number of pixels. If we multiply the number of pixels in each dimension —1000×1500 in our example—we have the appropriate measure. Thus the image in our example has 1.5×10^6 pixels, or 1.5 megapixels.

40.3 Signal and noise

The fundamental quantum nature of light becomes particularly apparent when a detector is subject to very low levels of illumination. Individual photons of light are either detected or not; there is no in-between. A detector can detect, at a given location, one photon or two photons or three photons. But it cannot detect 1.3578 photons. This basic fact means that the mathematics of light detection is a problem of statistics.

Along with any measured signal from a light detector comes noise—statistical variations in the measurement that cannot be predicted ahead of time and in and of itself carries no direct information about the image. Even though noise is, by its very nature, unpredictable, there is still much we can say about it. In section 40.3.1 we work through a nonphotographic example.

40.3.1 Pennies and Poisson

Over the course of the year 1997, I collected, from random sources, 1080 pennies and put them in a box. Apparently I had little else to do back then, so I sorted the pennies by year, counting up the number of 1984 pennies on the one hand and 1985 pennies on the other. A graph of this data—pennies collected versus year—could then be called a *penny spectrum*. Figure 40.3 shows such a penny spectrum for my box of 1080 pennies.

What conclusions can one make from this data? Notice that, for example, there is a definite 'hump' of pennies around the year 1982, and there is a big drop in the number of pennies between 1970 and 1971; I collected 13 in 1971 but only six in 1970. Can I then, for example, conclude that there really were $\frac{13}{6} = 2.17$ times more 1970 pennies in circulation than 1971 pennies? To answer that question, we must first ask ourselves a different one: if I had separately collected another random set of 1080 pennies, would it have made exactly the same penny spectrum?

The short answer to this question is likely not. If the events are random, then only the probabilities remain the same, but the specific result will likely be only approximately the same. The detailed answer to this question is one of the most fundamental problems of statistics. It is beyond the scope of this book to work

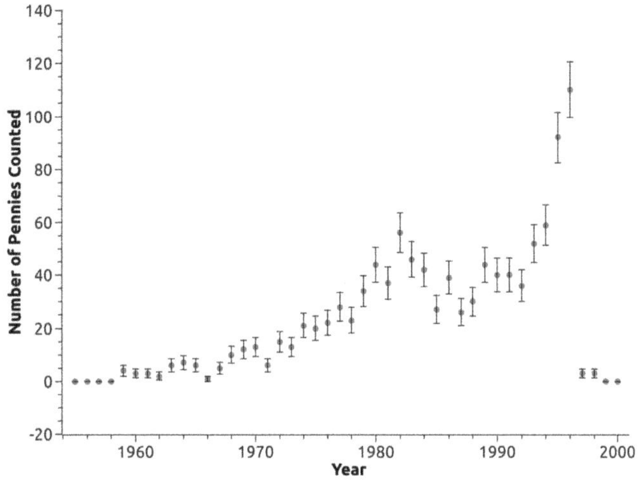

Figure 40.3. A graph of 1080 pennies collected in 1997, with number collected plotted versus year.

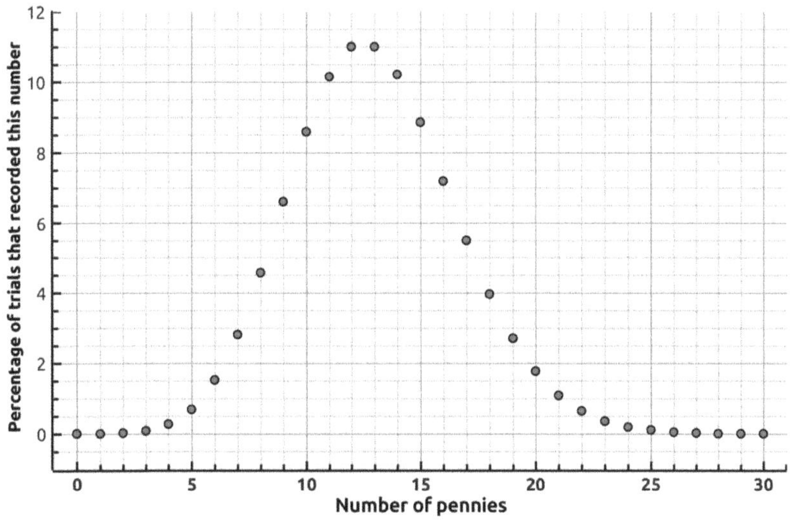

Figure 40.4. A Poisson distribution that peaks between 12 and 13 events.

through that answer in detail (see, for example, Bevington 1969), but it is not difficult to understand the basic result.

Let us imagine I had collected not one or even two boxes of 1080 pennies, but that I had collected 10 000 such boxes. I could then, for example, count how many 1970 pennies—out of the total of 1080—were in each box. With these numbers, I could make a different kind of graph, called a *relative frequency distribution*. Each point on the graph tells the percentage of the 10 000 boxes that contained that particular number of 1970 pennies. Figure 40.4 gives an example, corresponding to our hypothetical tallying of 1970 pennies collected out of a huge number of 1080 penny samples. The distribution looks like a slightly asymmetrical hump, and the peak of the distribution (in my contrived example) lies between 12 and 13 pennies. This peak represents the experimental result that occurred most frequently. But even though 12 or 13 pennies were counted most frequently, this accounts for only 22% of the pennies counted. Clearly, for most of the 10 000 collections, either more or fewer than 12 or 13 pennies from 1970 were counted.

One can see from figure 40.4 that it was very common to count, for example, 10, 11, 12, 13, 14 or 15 pennies. But it was very rare to record as many as 30 pennies or as few as three. A frequency distribution such as this is important because repeating the same measurement many times tells us after the fact how likely it was to have obtained any one particular result. In this case there was, for example, a high probability of recording 12 pennies from 1970 but a very low probability of recording 28 pennies.

And so what would have happened if I had made one additional collection of 1080 pennies, for a total of 10 001 collections? How many 1970 pennies would it have contained? There is no way to know! The distribution only tells us the likelihood, not what happens. But a careful analysis of this common situation of counting randomly-occurring events shows that, in the limit of more and more repetitions of the measurement, a frequency distribution like what is shown in figure 40.4 would result. This Poisson distribution is an idealized mathematical distribution that describes the results of independent counts of a set of random events, in the limit as the number of trials approaches infinity.

An important feature of the Poisson distribution is its width, which clearly tells us something about the uncertainty in the expected results if we were to repeat the measurement again. If the distribution is very wide, it means that measurements either much greater or much less than our most-likely peak value are still very likely. But if the distribution is very narrow, then the opposite is true; it is very *un*likely to measure a value much different than the peak value. In figure 40.4, for example, we may note that the peak of the distribution, at between 12 and 13 pennies, is about 11%. At half of this peak, 5.5%, the distribution stretches from about 8.5 to 17 pennies, so we could say the distribution has a full width, half maximum (FWHM) of 8.5, or ±4.25.

From my single box of 1080 pennies, all I can say is that 13 out of 1080 were from 1970, so the best I can do is to generalize that result—to claim that when I collected these pennies, my best estimate is that 13 out of every 1080 pennies (1.2%) in circulation were from 1970. But the Poisson distribution for our thought experiment of performing the same measurement many, many times tells us something else—it tells us the uncertainty in our best estimate. Because we counted only 13 pennies for 1970, we should expect that count to be representative of the total number of such pennies to within about ±4.25. This uncertainty is about 33% of our count of 13. If we use the 13 counts to form conclusions about the number of 1970 pennies in circulation, we should only expect our conclusions to be accurate to within ±33%.

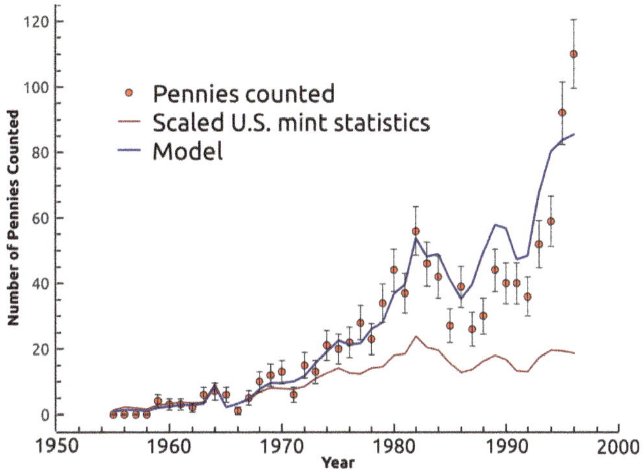

Figure 40.5. The same data as figure 40.3, but with Poisson statistics error bars added. The red line shows the number of pennies put into circulation each year by the US Mint, scaled to fit the left end of the data. The blue line shows the same Mint statistics as modified by a simple model that also includes pennies being taken out of circulation.

A full statistical analysis of the Poisson distribution indicates that a better measure of its width, and thus the statistical uncertainty in our number of counts, is not the FWHM, but rather what is known as the *standard deviation*, σ (the Greek letter sigma). It has a particular mathematical definition, the details of which are not important for this discussion, but it turns out to be quantitatively similar to the FWHM. For the Poisson distribution shown in figure 40.4, the standard deviation is approximately $\sigma = 3.5$.

Happily there is a very simple relation between a particular number of counts on the one hand and the expected Poisson-statistics uncertainty on the other. The best estimate for the uncertainty in the number of counts is simply the square root of the number of counts. For my single box of 1080 pennies, I counted 13 from 1970. Poisson statistics tells us that we should expect that number to be representative of the 1970 pennies in circulation only within $\pm\sqrt{13} = \pm 3.6$.

Let us return to my original question. My sample of pennies shows over twice as many 1970 pennies as 1971 pennies. Does this mean I should reasonably conclude that there really were twice as many 1970 pennies in circulation as 1971 pennies? Or can this difference instead be accounted for simply by the statistical nature of counting a limited sample? We can make a better judgment if we put error bars—vertical bars that encompass the likely uncertainty in the number of counts—on the figure 40.3 data. Figure 40.5 shows the same data with these error bars in place, calculated simply by taking the square root of the number of counts. Careful inspection of the graph shows that the lower error bar for 1970 almost meets up with the upper error bar for 1971. This means that it is possible but not extremely likely that the differences in those counts can be accounted for by Poisson statistics alone.

Analyzing the expected statistical uncertainty in measurements is of utmost importance whenever we use measured data to form reasonable conclusions about models or theories. For example, can we form a reasonable theory to explain the ups and downs in the penny spectrum? The red line in figure 40.5 shows an attempt at explaining the data with the most obvious theory. I looked up the statistics from the US Mint, and simply plotted the number of pennies put into circulation each year—scaled to match up with the total number in my tiny sample. I scaled the US Mint data to match my penny counts at 1960—but then the red line clearly falls well below the data in more recent years. More importantly, this 'theory' lies well outside the error bars of the data.

The blue line represents a slightly more sophisticated attempt at a theory for the data. It is the same as the red line but also includes a model of how pennies might be taken out of circulation. My model assumes that pennies are always taken out of circulation at the same percentage rate. Thus much of the reason there are so few 1960 pennies compared to 1990 pennies is that 1960 pennies have been going out of circulation for many more years. This model clearly 'fits' the data very well up to about 1985, but after that, it lies at the edges or just outside the error bars of the data points.

Because the model does not completely fall within the error bars of the data, can we then conclude that the theory is incorrect? It would be nice if life were so simple. It is also possible that the measured data is the problem. In particular, I could be mistaken in my assumption that I collected the pennies randomly; the data could be biased in various ways, meaning it does not directly represent the real distribution of pennies in circulation.

The Poisson distribution is named after the French mathematician and scientist Baron Siméon Denis Poisson (1781–1840). Only one of his many accomplishments in mathematics and fields as diverse as celestial mechanics and electricity and magnetism, Poisson's analysis of the statistics of counting randomly occurring events was published in 1837. There is some irony to this, as Poisson is also well-known for his public opposition to the new advances in the wave theory of light presented by his contemporary, Augustin-Jean Fresnel. Poisson was a firm believer in the corpuscular theory of light—that it was essentially a stream of particles, and Fresnel's work did much to lay the corpuscular theory to rest. Yet, a century later, it was well established by quantum physics that in addition to its wave-like nature, light does in fact act as a stream of particles (photons). When photons arrive at a detector, they are either detected or they are not—just like counting pennies. Thus Poisson's statistics of counting plays a key role in the measurement of light.

40.3.2 Photons, signal and noise

Because a particular photon either arrives at the detector and is detected, or it is not, Poisson statistics apply to the detection of light. This is not to say there are no other factors that add to this uncertainty—but those factors can often be mitigated by altering the design of the detector. One cannot, however, eliminate the uncertainty described in section 40.3.1; it is a fundamental part of the measurement process due to the inherently statistical nature of the arrival of photons.

In practice then, there is an inherent random element to the physical detection of light. When exactly the same light falls on many nearby pixels, one might expect the detector to record exactly the same measurement for all of those pixels. But instead, they will randomly record a range of values from pixel to pixel. Only the average of these randomly-fluctuating values may correspond closely with the actual brightness of the light. We call these random fluctuations *noise*, and in a photograph, it shows up as a random speckled pattern where one would expect a uniform shade of gray (or a uniform color). The 'true' value that the noise deviates plus or minus from is called the signal.

This also means that just because one pixel records more light than another, it is not necessarily true that the light was brighter there. Consider this example: pixel A detects nine photons, and pixel B detects seven. When we combine these with their respective uncertainties ($\sqrt{9} = 3$ and $\sqrt{7} = 2.6$), it is clear it is possible that the light was actually brighter at pixel B than at pixel A.

Thus, on the one hand, we have the actual brightness of the light and, on the other, the often small and always finite sample of that light as actually measured by physical interactions with the detector. The measurement contains real information about the actual brightness of the light, but it always contains random fluctuations (noise) due to the statistical nature of the detection of light. In practice we can (and do) use our actual measurement as our best estimate of the signal, but we must always be aware that any signal comes with noise as part of the package.

40.3.3 Signal-to-noise ratio

Which has more noise, a signal of 10 000 photons detected or a signal of only 100 photons detected? Regarding the noise due to Poisson statistics, the answer is clear: the larger signal also has greater noise, for $\sqrt{10\,000} = 100$ while $\sqrt{100} = 10$. We can be reasonably confident in our smaller signal to ± 10, while we only know the larger measurement to ± 100.

But in most circumstances, we care about something very different. It is usually not the noise itself that is important but rather how that noise compares to the signal. Let us look at our example again in this new light. We have, on the one hand, a measurement of $10\,000 \pm 100$ and, on the other hand, 100 ± 10. In the first example, the noise of ± 100 is only 1% of our signal (10 000). In the second example, however, the smaller noise of ± 10 is 10% of our also smaller signal of 100. Although the smaller signal has less noise that comes along with its measurement, that noise is a larger percentage of the signal itself.

This example suggests an important but slightly-counterintuitive result: more signal means more noise, but it means less noise compared with the signal itself. A useful way to describe this numerically is with the *signal-to-noise ratio* (s/n), or simply the signal divided by the noise. Thus our larger signal would have an s/n of $10\,000/100 = 100$, while the smaller signal has an s/n of $100/10 = 10$. The s/n represents then, how many times larger the 'true measurement' is than the random statistical fluctuations in that measurement. The 'noisy' or 'grainy' look of a photograph is closely related to s/n; a higher s/n means a more uniform, less noisy look to the picture.

For Poisson statistics, since the noise (n) is simply the square root of the signal (s), we have the following simple result.

$$s/n = \frac{s}{n} = \frac{s}{\sqrt{s}} = \frac{s}{\sqrt{s}} \cdot \frac{\sqrt{s}}{\sqrt{s}} = \frac{\cancel{s} \cdot \sqrt{s}}{\cancel{s}} = \sqrt{s} \tag{40.1}$$

And so for Poisson statistics, not only is the noise the square root of the signal, but so too is the signal-to-noise ratio. Thus for a greater s/n, one simply needs more signal. This means that for a photograph, the shadow areas (where less light was detected) are the parts that will always look the most noisy or grainy. The saving grace is that if we leave those shadow areas dark on our final print or display of the photograph, then, although those areas have more noise (compared to the signal), it will be equally difficult to see that noise.

The problem arises when one alters the photograph. What if one wants to use digital photo editing software to 'brighten up' those shadows, to see detail that is otherwise hidden? A consequence is that when the signal for those shadow areas is rescaled to make it look brighter on our photograph, it is inevitable that so too is the noise that comes along with that signal. And since an underexposed shadow area is likely to have a very low s/n, that brightened region will be made to appear very noisy. See figure 40.6.

We can make this effect a little more concrete with actual numbers. Say, for example, that in my two exposures of the tree in figure 40.6, a particular dark part of the image was recorded with a signal of 1000 photons detected in the left-hand picture, but that same part detected only 10 photons for the underexposed version on the right. I then rescale the right-hand image so it looks as bright as the left one. To do this, I would have to multiply the values by 100, so the value of ten photons would also be 1000. Now let us look at what happens to the noise. The left-hand picture has an s/n of $\sqrt{1000} = 32$, while the underexposed version has an s/n of only $\sqrt{10} = 3.2$. The problem is when I rescale the low s/n image so it looks as bright as the high s/n version, the s/n remains the same. This is because I will multiply both the signal and the noise by the same factor (100), so the ratio of the two remains the same. Now my altered underexposed image seems to have a signal of 1000 just like the unaltered image on the left. But it still has an s/n of only 3.2, and so it still looks 'noisy.'

For any real detector, there are many steps between the physical interaction of the photons and the ultimate realization of an image. A CCD array detector, for example, must transfer the signal from one pixel to the next, and pass it through an amplifier. All of the these physical interactions contribute noise in some way; the detailed analysis of all of the factors that contribute to noise is complex indeed and well beyond the scope of this book. With these other sources of noise taken into account, the signal-to-noise ratio no longer has such a simple relationship to the signal as equation (40.1). But in most cases, the same lesson applies—a larger signal

Figure 40.6. The result of different s/n. The pictures on the **left** were taken with a higher s/n than the underexposed pictures on the right. The pictures on the **right** were rescaled so they appear the same overall brightness as those on the left, but doing so increased the noise proportionally, leaving the low s/n unchanged. Compare the two magnified views at the bottom; the low s/n version on the right appears grainy and with much less contrast. The contrast could be digitally enhanced, but doing so would make the picture appear even more grainy still.

comes with more noise, but there is less noise compared to the signal. In most circumstances, more signal means greater s/n.

References

Bevington P R 1969 *Data Reduction and Error Analysis for the Physical Sciences* (New York: McGraw-Hill)

Stroebel L, Compton J, Current I and Zakia R 2000 *Basic Photographic Materials and Processes* 2nd edn (Waltham, MA: Focal Press)

IOP Publishing

Photography (Second Edition)
Physics and art in focus
John Beaver

Chapter 41

Digital photography and astronomy

Digital detector arrays such as charged-coupled devices (CCDs) are especially suited for precise, quantitative measurement of light. As such, astronomers were some of the first to widely use digital imagery, and it has transformed observational astronomy utterly. The graphs and images in this section are made from raw data used for the preparation of Beaver *et al* (2013). In what follows, I describe why these digital detectors are so suited to the quantitative measurement of light.

41.1 Digital detectors are reusable

Load a roll of 35 mm film into a camera, put the camera on a tripod, focus on a still subject under controlled lighting and take 36 exposures exactly the same way. Commercial photography has striven for a particular outcome: all 36 pictures are identical. That is true, up to a point. However, every single exposure was, in fact, made with a different light detector. The pictures only seem to be identical because of the great pains film manufacturers have taken to insure that the film is uniform, from one frame to the next and from roll to roll, in its chemical and physical properties.

Of course, if one looks more closely—under a microscope, for example—then the differences become evident. At the finest level of microscopic detail, the arrangement of particular silver-halide grains on the film is random. And if one measures with enough precision, it becomes clear that the sensitivity of the film varies ever so slightly from one place to the next.

Much of this is true for a digital detector as well. Although there is nothing quite the same as film grain, the pixels on any real digital detector vary in sensitivity in a random way as compared to their neighbors. Furthermore, most digital detectors have manufacturing flaws that make certain regions of many pixels slightly more or less sensitive than other areas.

But here is the difference. Many of the flaws and imperfections of a digital detector are repeatable—the same flaws persist, and in the same way, from one

doi:10.1088/978-0-7503-3703-8ch41

© IOP Publishing Ltd 2022

picture to the next. But for film, the specific imperfections and random details in one piece of the film are unrelated to those in another. Light interacts with a new detector every time an exposure is made with film. But in a digital camera, light interacts with the same physical silicon chip for every picture. And this means that it is possible to measure that particular digital detector's imperfections and correct for them.

Some of these corrections are performed automatically by many digital cameras, and the stored data file is most often preprocessed to take into account the features of the particular detector. More sophisticated corrections can be performed after the picture has been taken by gathering additional data and using image-processing software. Specific examples are discussed in section 41.5.

41.2 Linear response

Many photoelectronic digital detectors—a CCD array is a good example—have over a broad range of exposures a true linear response. That is, if one plots the measured effect (electrons accumulated at a pixel for example) versus exposure, it is a straight line on an ordinary graph (not a log–log plot). One consequence of a linear response is that it is possible to calibrate the image—correct for systematic effects so that the measured signal for each pixel in the image is directly proportional to the intensity of light that was incident during the exposure. This allows one to combine separate exposures of the same subject in order to increase the signal-to-noise ratio. Figure 41.1 shows graphs measuring pixel values along a particular row in two CCD images. The left-hand graph is from a single CCD exposure, and the right-hand image is from many separate exposures added together, pixel by pixel. The random noise, compared to the signal, is greatly reduced.

For nearly every photochemical detector, there are many complex steps between the original formation of the image and the final print. In the traditional silver-gelatin process, for example, the latent image formed by light in the negative is first

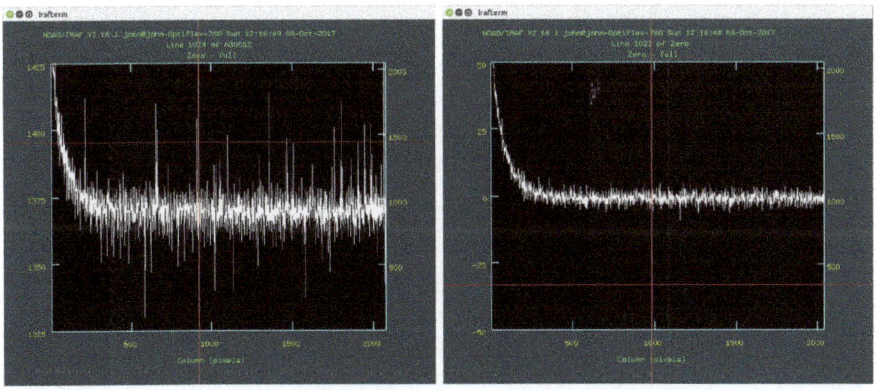

Figure 41.1. **Left:** A graph of pixel values across a single row of a low-signal CCD image. **Right:** The same, but here many low-signal images have been added together to form a new image with a much higher signal-to-noise ratio.

chemically amplified with a developer, then the remaining silver halide is removed chemically with a fixer. Then light is passed through the negative to expose the light-sensitive paper to make a photographic print, and steps of development and fixing follow once again. All of these steps add complexity to the overall relation between exposure to light and the measurable effect of density. The end result is a nonlinear response to light, as discussed in chapter 33. Recall that even over the 'straight-line part' of the log–log characteristic curve, the relation between density and exposure is better described by a power law than a straight line.

A linear response is greatly desired if one wants to use a light detector for precise, quantitative measurements. But the advantage for ordinary photography—where the goal is to instead accurately reproduce human visual perception—is less clear. The S-shaped log–log characteristic curve of traditional film works well for its ordinary purpose—reproducing images from life, to be viewed by humans. The shallow slope of the bottom and top parts of the characteristic curve compresses a lot of brightness levels in the world to relatively few in the picture. Regarding the shadows, that fact often makes little difference—the shadows of the picture are too dark to see much anyway. But regarding the highlights, this is a lucky thing—it seems to be something like how we humans see high levels of brightness (Adler *et al* 2014).

41.3 Dynamic range

The *dynamic range* of a detector is the total range of brightness levels that it can record in the same image while still retaining the ability to distinguish one brightness level from another in a useful way. There is no obvious precise definition for dynamic range, since what is a 'useful' range of brightnesses depends greatly upon the nature of the use. For astronomical applications, however, the dynamic range refers to the brightness levels for which the detector response is linear. Figure 41.2 shows two graphs taken from different rows of an image of the star cluster M 11 (shown in figure 41.4). The left-hand image is across a particular row that contained only very faint stars. Much of the low-level up and down in this graph is a uniform signal from the dim sky background and the statistical noise that comes with that

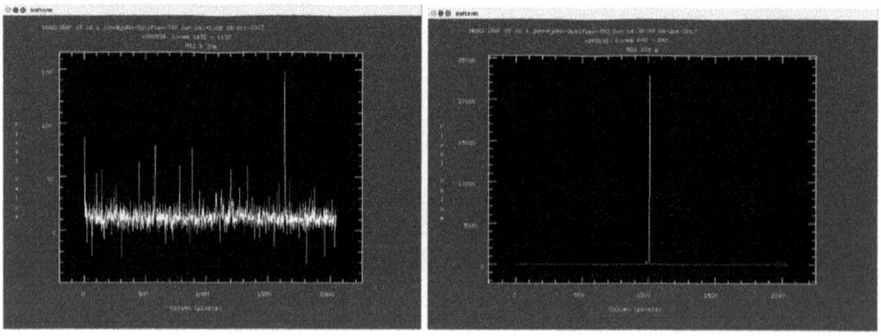

Figure 41.2. CCD detectors can have a large linear dynamic range. These two slices across different parts of the same calibrated image show usable data over a large range of values.

signal. Most of the spikes that shoot up to much higher levels on the graph are where the row cuts across individual stars. On this scale, the sky background has a value of about ten (the units are unimportant for this discussion), and that is a meaningful measurement for this fully calibrated image.

The graph on the right shows the pixel values for a row that happens to cut across a bright star near the center of the image, and it produces a huge spike in the graph. Notice that at the scale of this graph, the sky background is negligibly small in comparison. The star's brightness on the same scale is nearly 25 000. Both of these measurements in this calibrated image are meaningful, and they can be compared to each other quantitatively. On this single image, we have a sky background measurement of ten, and a bright-star measurement of 25 000 (2500 times greater), and both measurements are meaningful. Thus we can say that this image has a dynamic range of at least 2500.

Without this large dynamic range, one would need to take many exposures of different lengths in order to adequately measure the brightnesses of both the faint and the bright stars. The impressive linear dynamic range of CCD detectors is one of their principle attractions for astronomers.

A factor of 2500 corresponds to over 11 exposure steps ($2^{11} = 2048$). So, what is the dynamic range of film? What range of exposure steps can be meaningfully recorded with a single exposure of black-and-white or color film? The answer is not so straightforward for two reasons. Firstly, the characteristic curve for film is very different from the linear response of a CCD. Secondly, the intentions are different—we expose black-and-white film not for its quantitative data, but rather to make a picture that is to be viewed by humans. Since we perceive brightnesses in a way that mirrors somewhat the S-shaped characteristic curve of film, we can use a large part of that highly nonlinear response to make a good picture. And so the useful dynamic range of film is a more-or-less thing, and answers vary. For a typical example, Stroebel and Compton (2000, p 101) present a film characteristic curve with a dynamic range of about 500, or about nine exposure steps. Outdoor scenes, on the other hand, typically have a range of only about 100 in brightness, so the dynamic range of film is more than adequate.

41.3.1 Dynamic range and bit depth

For a digital photoelectric detector, the signal at a given pixel is encoded by a computer or microprocessor (built into the camera) as a number. This computer encoding of the number must be accomplished according to some prearranged scheme—and that scheme sets a limit to the numerical information that can be recorded.

Digital information is stored by a computer in binary form—a string of ones and zeros. The *bit depth* of an image can be described in terms of the number of binary ones and zeros that are used to encode the level of signal at a given pixel. A single bit has only two possibilities—zero and one. If only one bit per pixel is used to encode the image, then each part of the image has only two possibilities—black or white. But add a second bit, and the possibilities double; each bit added multiplies the possibilities by two. Thus, four different signal values could be recorded with two bits, but eight values could be recorded with three bits. The bit depth, then, is the total number of bits used to encode all of the possible signal levels that can be recorded per pixel. If N is the bit depth, then 2^N different levels can be recorded.

And so, for example, an eight bit black-and-white image can record $2^8 = 256$ different gray levels. Is that enough? What about 16 bits, $2^{16} = 65\,536$ possible shades of gray? Is that enough, or is it more than enough? The answer is that it depends on several factors. But first, we must consider another complication. The typical color image is essentially three black-and-white images that work together to synthesize color. Thus, if eight bits are enough for a black-and-white image, it follows that we would need three times this many—24 bits, to record an equivalent color image. The separate red (R), green (G) and blue (B) portions of an RGB color digital image are often called *channels*, so the terms '24 bit RGB image' and 'eight bits per channel RGB image' mean the same thing[1].

So, how many bits are enough? The answer depends on to what use the digital image is put. At one extreme, an image of simple text is often digitally encoded with only a single bit per pixel, since only black-and-white is needed. For an image that will serve no other purpose than to be directly displayed on a screen or printed, eight bits per channel is plenty. That corresponds to 256 possible shades per channel, and the human eye cannot even discern that many.

It is another matter entirely, however, if this same digital image is to be adjusted. In the course of applying an adjustment curve, some of the limited number of shades of gray are compressed, while others are stretched further apart. Shades of gray that, in the unadjusted image, are too close to each other for the eye to discern as different may be spread further apart by the adjustment. What appeared as a smooth gradation of tones in the original then shows sudden visible jumps in brightness from one area to another, leading to an unattractive contoured look. A digital image that is meant to allow for adjustments must have a greater bit depth than what is needed for the final image.

There is another important side to the entire consideration of bit depth—it also matters just what information is encoded by those bits. The left-hand image in figure 41.3 is encoded as 24 bit RGB. The right-hand image was made by transforming the 24 bit image to only eight bits. This lower-bit color image has obvious contours; clearly eight bits is not enough to encode all of the subtle levels of both hue and value present in the original image. This eight bit image is not really an RGB image like the 24 bit image on the left; it does not have separate channels for red, green and blue. Notice that eight does not even divide evenly by three. It is instead an *indexed* image; each of the 256 possible numbers in the eight bit encoding is assigned its own color out of all of the full-color possibilities. The computer software decided how best to split up the vastly more colors in the original to this much-more-limited color palette. But here is the point—one cannot then go backward. One cannot get rid of the contours in the eight bit image by simply

[1] RGB is not the only method for encoding a color image, but it is the most common.

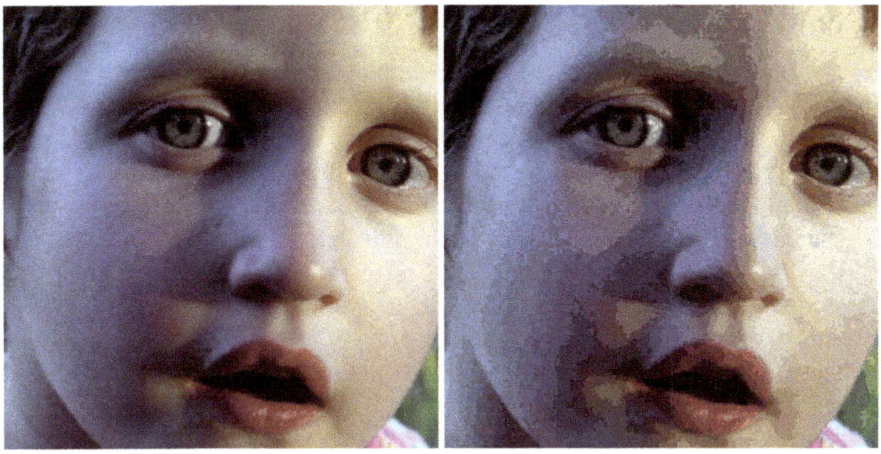

Figure 41.3. On the **left** is a digital image stored as a 24 bit RGB image file. On the **right** is the same image but after conversion to a 256 color (eight bit) indexed image. The subtle range of hues and values cannot be captured with a palette of only 256 colors, so a contoured, paint-by-number effect results.

transforming it (with computer software) to a 24 bit RGB image. The image has only 256 colors, so only eight bits are needed to encode it, and the vast majority of possibilities in the 24 bit encoding would be unused.

Thus, it also matters how much information is in the image. Choosing a greater bit depth does not, in and of itself, put more information into the image; it only allows for the possibility. The potential amount of grayscale and color information in an image ultimately depends on the physical means by which the image was originally formed. For a digital camera with a CCD detector, for example, the signal from each pixel is determined by measuring the number of electrons that accumulated at the light-sensitive site during the exposure. The bit depth that is needed to count those electrons depends on the number of electrons that can accumulate before the site is 'full.' The general rule of thumb is that for a black-and-white image on a CCD detector, eight bits (256 levels) are too few, but 16 bits (65 536 levels) are more than enough. The so-called RAW image formats on many CCD cameras are stored as 16 bits per channel—48 bits for an RGB color image.

But for most causal use, a 48 bit color image is overkill; it allows for far more adjustment than one is likely to perform. A good compromise is 24 bits—eight bits per channel, and this is the most commonly used encoding for color images not meant for scientific measurement or sophisticated image processing. The standard jpeg (.jpg or .jpeg) color image format only supports up to 24 bits, so a 48 bit image must be stored in other formats.

The CCD detectors used for astronomy usually do not have a Bayer mask to produce a color image. Instead, the astronomer takes separate images, one at a time, through different colored filters in order to record information related to color. But the astronomer needs to use all of the information that the CCD is capable of recording, so a 16 bit encoding is most often used.

41.4 Quantum efficiency

Many different physical processes can be employed to make a light detector. But whatever the method, a crucial question is what percentage of the light falling on the detector is actually detected? For it is inevitable that some of the light falling on the detector will result in no usable effect at all. And so, if 1000 photons strike the detector, how many of those photons result in a measurable effect?

The percentage of photons that strike the detector and are actually detected is called the quantum efficiency of the detector. It is not difficult to see that a detector of higher quantum efficiency has some distinct advantages. Clearly a detector of 50% quantum efficiency should be able to accomplish the same exposure with, all else being equal, half the exposure time of a 25% quantum efficiency detector. This is in fact one of the big advantages of many photoelectric detectors as compared to photochemical detectors. A CCD array detector can achieve a quantum efficiency as high as 90%. The quantum efficiency of ordinary silver-gelatin film, on the other hand, is only a few percent.

Since the universe is a big place, objects of interest to astronomers are usually very far away, and, more often than not, they are exceedingly faint. To compensate for this, astronomers must make long exposures with large telescopes (so as to gather a greater area of light). The high quantum efficiency of CCD arrays as compared to film is one of the primary reasons that amateur astronomers now take pictures with backyard telescopes that rival predigital photographs made decades ago with the largest telescopes in the world.

41.5 Image calibration

Because photoelectronic array detectors such as CCDs have the ability to produce a truly linear response over a wide dynamic range, and because—unlike photochemical detectors—it is the same detector used again and again, it is possible to calibrate them. The left-hand image in figure 41.4 is a raw, uncalibrated exposure of the star cluster M 11, in the constellation of Scutum, taken by astronomer Michael Briley with a CCD camera on a telescope at Kitt Peak National Observatory, as part

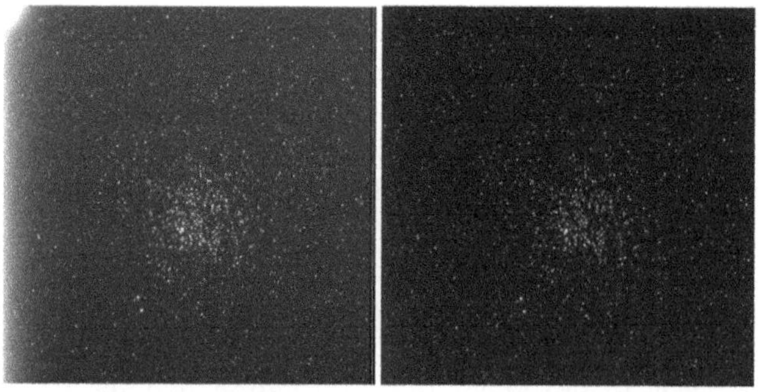

Figure 41.4. **Left**: Raw CCD image of the star cluster M 11. **Right:** The same image after calibration.

of the raw data set for Beaver *et al* (2013). The right-hand image is a calibrated version of the same. To the eye, only slight differences are apparent between the raw and calibrated images, but the raw image by itself is unusable if the goal is to make accurate quantitative measurements of the brightnesses of the stars.

The linear response means that it is practical to *calibrate* the image from a CCD detector—correct for systematic effects so that an accurate quantitative measure of the exposure to light can be determined from the measured signal. A nonlinear response makes this process far more difficult; if the deviation from linearity is too drastic, accurate calibration may be impossible in practice. A linear response between the signal, s, and the exposure, E, implies that the following is true.

$$s = aE + b \tag{41.1}$$

Here a and b are constants. Notice that the constant a is multiplied by the signal, while b is added. If this relationship holds, it means that some systematic effects are additive, while others are multiplicative.

A good example of an additive effect is the CCD *bias signal*; even with no exposure to light, a signal is produced by the very act of applying a voltage to the pixels, reading out the CCD and passing the result through amplifiers. This bias signal varies from pixel to pixel on the CCD; it must be determined separately for each pixel. The first image in figure 41.5 is an example of a *bias frame*—a zero second exposure used to directly measure the bias of each pixel. If the response is linear, this image can simply be subtracted—pixel by pixel—from the image to be calibrated, in a process called *bias correction.*

The thermal motions of the atoms in the CCD also produces a signal that is independent of the exposure to light. This *dark signal* is (hopefully) proportional to the exposure time if the CCD is kept at a constant temperature. The linear response means that the dark signal for a given exposure time is simply additive and can be measured and simply subtracted. A so-called *dark frame* is simply an exposure with the shutter closed. Typically a dark frame would be taken immediately after the image to be calibrated and would have the same exposure time. One then simply subtracts the dark frame from the 'object frame.' Some consumer digital cameras automatically do this for long-time exposures; one can tell because a 30 s exposure keeps the camera busy for a full minute. Astronomers instead cool their CCD cameras with liquid nitrogen, so as to render the dark signal negligibly small.

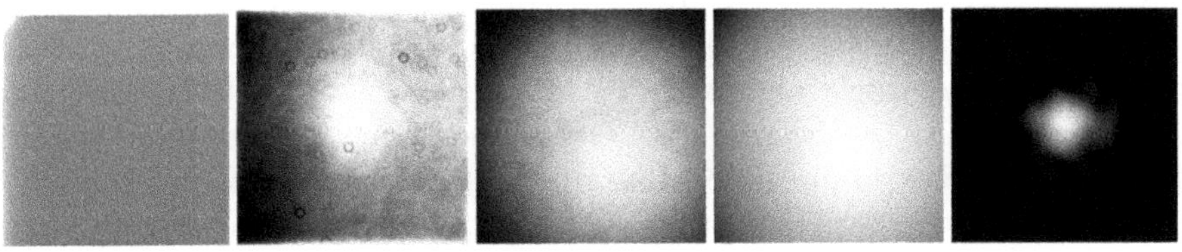

Figure 41.5. Some of the images used to calibrate the image in figure 41.4.

Once additive effects have been removed, the linear relationship becomes simply the following.

$$s = aE \tag{41.2}$$

That is, signal is simply proportional to exposure. The constant of proportionality, a, then is related to the sensitivity of the detector; if a is larger, then a greater signal results from the same exposure. But the pixels of a CCD detector are essentially separate light detectors; they only have the same sensitivity insofar as they are manufactured exactly alike. Thus the sensitivity of a real CCD varies significantly from pixel to pixel.

One can directly measure the differences in pixel sensitivity by taking a *flat field image*—an unfocused image of a uniformly illuminated, featureless white screen. If a particular pixel produces, for example, 1.37 times the signal of its neighbor, then it means that pixel is 1.37 times more sensitive. The second image in figure 41.5 is an example of a flat field image that was used for the calibration of the image in figure 41.4. Since sensitivity is a multiplicative effect, the image to be calibrated can simply be divided by the flat field image, in order to correct for these pixel-to-pixel variations in sensitivity.

It should be clear that if the response of the CCD were nonlinear, this simple flat field correction would not work. The sensitivity factor, a, in equation (41.2) is the slope of the response. But if the response is nonlinear, then there is no single slope, even for a single pixel. A nonlinear relationship between signal and exposure is a curve, not a straight line—and different parts of a curved line have different slopes. To put it differently, if a particular pixel produces 1.37 times more signal than its neighbor on a flat field image with a nonlinear response, it only means that pixel is 1.37 times more sensitive for the flat field exposure but not for the image to be calibrated, which received a different exposure for those pixels.

The third, fourth and fifth images in figure 41.5 are also multiplicative calibration images applied to the image in figure 41.4. They correct for such effects as uneven illumination of the CCD by the telescope optics and scattered light inside the CCD camera. The ability to accurately correct for these systematic effects depends crucially on the linear response of the CCD detector.

It is possible to calibrate a detector that has a significant nonlinear response, but in practice it can be too difficult to bother. Crucially, to correct for nonlinear effects requires far more experimental data. To correct for a nonlinear flat-field response, for example, one would need to take many different flat-field exposures and fit a model to them. For an astronomer, this is a serious problem since Earth rotates on its axis. Night does not last forever, so only a limited number of calibration images can be recorded between the ending and beginning of astronomical twilight. A single night's data used in Beaver *et al* (2013), for example, required a total of 16 h to record 187 CCD images. Only 28 of those images (for a total of 30 min of exposure time) were of the actual astronomical objects being studied; the rest were for calibration. And that was for a detector with a mostly linear response.

Once an image has been calibrated, it can be used for measurements. And so, for example, the brightnesses of the individual stars can now be measured from the right-hand image in figure 41.4. Scientific image calibration and measurement is usually performed with specialized software that maintains the mathematical integrity of the data and keeps a running record of what calibrations have been performed. The software must also keep track of the statistical noise in the data if measurements are to be meaningfully compared to each other.

References

Adler M, Mayo A and Alon U 2014 Logarithmic and power law input-output relations in sensory systems with fold-change detection *PLOS Comput. Biol.* **10** e1003781

Beaver J, Kaltcheva N, Briley M and Piehl D 2013 Strömgren H-β photometry of the rich open cluster NGC 6705 (M 11) *Publ. Astron. Soc. Pac.* **125** 1412–20

Stroebel Leslie, Compton John, Current I and Zakia R 2000 *Basic Photographic Materials and Processes* 2nd edn (Waltham, MA: Focal Press)

IOP Publishing

Photography (Second Edition)
Physics and art in focus
John Beaver

Chapter 42

Comparison of digital and film techniques

It is usually considered an insult to say that a photograph has been 'photoshopped,' implying that it isn't a real picture but instead one that has been artificially produced. Although I often agree in specific cases, as a general statement, it is certainly unfair. All photographs are altered in some way after they are exposed; it is not clear what a pure and unaltered photograph would even be. There are always many necessary steps between exposure and a viewable image, even if they occur (as in a digital camera) automatically and in the blink of an eye.

At a more philosophical level, even the most purely representational photograph is always a picture, not a direct capture of some element of the real world. If one takes several pictures in quick succession of any moving subject, each may be quite different from the other (see figure 42.1), to the point of making very different statements about the subject. Yet the subject did not really change in any fundamental way in the one or two seconds between exposures. The photographer chooses to show us one of these and not the others. Why is that particular choice, out of many, the one that is 'real?' At one level digital (or darkroom) editing is not so different, in principle, from the editing one always must do in deciding which picture to show and which 100 to discard.

But still, the use of digital technology often is different in many ways from predigital processes. In this chapter we try to untangle, at least a little, the old from the new to take stock of where photography is as an art in this modern digital age.

Many common digital manipulations have a direct correspondence to traditional darkroom techniques. Many tools in popular software packages for digital image processing are even named directly after darkroom processes. So here we step through some of the more common digital processing techniques found in, say, Adobe Photoshop or GIMP and compare them to traditional wet chemistry darkroom techniques. Since GIMP is free and open source, I will refer to it for my specific examples. If you know how to use either one of these software packages, it is fairly easy to learn the other.

doi:10.1088/978-0-7503-3703-8ch42

© IOP Publishing Ltd 2022

Figure 42.1. Photographs taken only seconds apart. If looked at in isolation, they may make very different statements about the subject. So which one is 'real?'

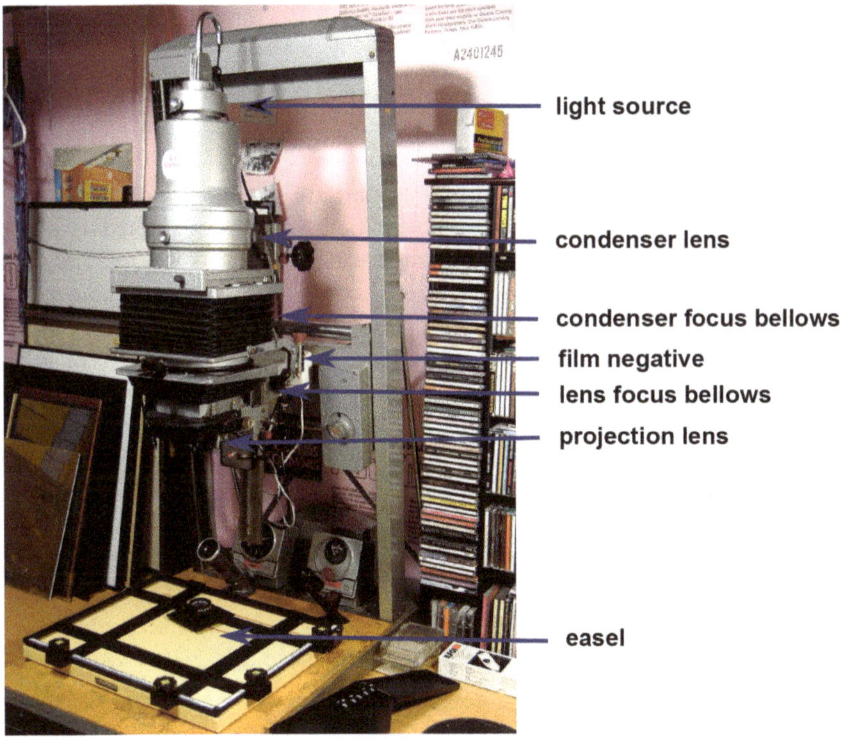

Figure 42.2. An enlarger for making black-and-white prints from negatives up to 4 × 5 in.

I will first outline the process of making a black-and-white print in the darkroom, comparing the steps at each point to how one would accomplish the same tasks in GIMP. Then I will describe how the process is different for making a color print. Detailed descriptions of these basic operations are widely available (e.g., London *et al* 2005, chapter 7, 11).

42.1 Borders and cropping

When making a print in the darkroom, the light-sensitive print paper is exposed to light from a negative, usually projected onto it by an *enlarger* (see figure 42.2). The

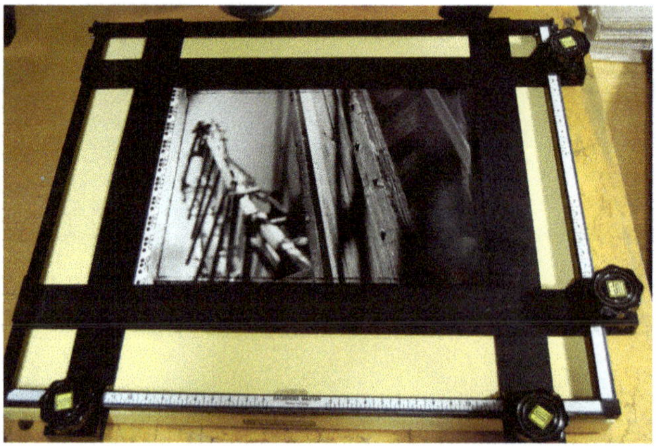

Figure 42.3. An easel is used to both hold the paper in place and to crop the borders as the paper is exposed to the image projected by the enlarger.

enlarger consists of a projection head that contains a light source, a condenser lens[1], the negative, and a projection lens. This entire assembly can be moved up and down, closer or farther from the paper, to alter the size of the projection. The lens then moves independently in order to focus the image for the particular projection distance chosen.

The paper is placed in an easel to hold it in position. The easel typically has a white or yellow background, so one can see (and thus position and focus) the projection from the enlarger head before placing the light-sensitive paper. It also has a system of adjustable metal blades (see figure 42.3) to allow one to cover parts of the print paper so they will remain unexposed by the negative. Since enlarging paper is a negative process, it gets darker wherever it has been most exposed to light. Thus areas of the paper covered by the easel blades will remain unexposed, leaving a white border around the print area. Black-and-white enlarging paper comes in standard sizes (in inches) of 5 × 7, 8 × 10, 11 × 14 and 16 × 20, and both the easel and enlarger can usually be adjusted to accommodate most of these sizes.

Sometimes an important part of composition is to crop the photograph by including only part of the original image in the final print. For example, a silhouetted tree on one side of an image can either be a vertical line cutting the image in two parts (if there is space to the right of the tree), or it can form one side of a framing element if the picture is cropped so the right edge of the picture passes through the middle of the tree.

To do this in the darkroom, one first sets the easel to accommodate the size of the print paper and to include whatever white border is desired. Then, with no enlarging paper in place, one adjusts the size of the image (by moving the enlarger head up or down) while repositioning the easel, until only the part of the picture one wants is

[1] There is a type of enlarger that does not use a condenser lens. A *diffusion enlarger* instead uses an internally reflective chamber to evenly illuminate the negative.

included in the print area. The easel blades are black, so the image projected onto them does not show up very well, so one mostly sees only the part of the image that is to be left after cropping.

One of the appeals of digital image processing is that cropping is easy. GIMP, for example, includes simple tools designed expressly for that purpose. Furthermore, digital image processing saves previous steps and allows one to instantly 'undo' an operation that did not go as planned. When cropping in the darkroom, on the other hand, there is no undo button; there is only 'start over and do it again.'

There is another fact that makes darkroom work more difficult than digital processing, even for a relatively simple task such as cropping: the image projected by the enlarger is a negative, while with GIMP one works directly with the positive image. It is difficult to compose an image—or even to know if it is worth the bother of making a print—when one sees it only as a negative. For this and other reasons, photographers often first make a set of small positive *contact prints* that are the same size as the negatives.

42.2 Brightness and contrast adjustments

When one makes a print from a negative, each shade of gray on the negative is 'mapped' to a corresponding shade of gray on the print. Of course it is a negative process, so a light shade on the negative becomes a dark shade on the print, and vice versa. But how does one tell precisely which shade from the negative ends up exactly what shade on the print?

For making a black-and-white print in the darkroom from a black-and-white negative, this question is usually answered in two parts. The overall lightness of the print is determined by controlling the exposure of the enlarging paper to the projected image. The projection lens in the darkroom enlarger has an *aperture adjustment* so that one may allow either more or less light to pass through the lens. Also, the enlarger lamp is always connected to some kind of timer (see figure 42.4) that allows one to expose the print paper for a controlled length of time. Since the enlarging paper is a negative process, a longer exposure or a larger lens aperture results in an overall *darker* print.

So how does one determine exactly what exposure to use? The short answer is that one makes an initial guess, and then adjusts that guess by trial-and-error. There are ways to make this task easier, and experience helps a lot too. One common method is to cover all but a little strip of the print while the exposure is progressing, and every few seconds, uncover a little more in steps. Thus different parts of the print are exposed for different amounts of time, and one can then pick what looks best. A related method is to make a test print exposed through a *step wedge*—a sheet of thin plastic or film printed with different gradations of transparency. It can be precalibrated such that, for example, a 60 s test exposure shows, for each step, how many seconds of exposure would be needed to make the print have that same lightness.

But both of these methods apply different exposures to different parts of the print, and the final decision must be made in the context of the whole print. And so trial

Figure 42.4. The overall lightness/darkness of a black-and-white print is controlled in the darkroom by adjusting the exposure from the image projected by the enlarger. A timer (rear left) allows one to vary the exposure time. Some enlargers also allow one to vary the brightness of the projection lamp (rear right). The two sighting devices in the foreground allow one to magnify the projected image in order to achieve a perfect focus.

and error is still an important part. This takes time[2], since each print must be exposed in the enlarger and then run through about five to ten minutes of chemical treatment in order to see the result. This is further complicated by the fact that as a wet print dries, it darkens slightly.

For a digital image on a computer, this process is, of course, much easier since one can make adjustments interactively, seeing the result for the entire image as the adjustments are made. Photo editing software packages offer a choice of methods for adjusting the lightness/darkness of an image. In GIMP, for example, there is a simple *brightness-contrast* tool. Simply slide the brightness slider to the left with the mouse, and the picture becomes darker, corresponding to a longer exposure of the print paper in the enlarger. Slide to the right and the opposite happens; see figure 42.5.

There is a sense, however, in which the process in the darkroom is more straightforward than the process of adjusting a digital image. The end result of all that work in the darkroom is a print, which is in fact the final goal of all that messing around. But when one interactively adjusts a digital image with GIMP or Photoshop, one is really adjusting how the image appears on the particular computer screen one happens to be using at that moment. It will likely appear somewhat different on another digital device. And making a digital print from that digital image is yet another step, the precise results of which depend on the make of printer and the particular inks and paper that it uses. It is a complex procedure to guarantee that the print produced will look just like the image on the photo editor's

[2] And costs money—enlarging paper costs roughly one dollar per sheet.

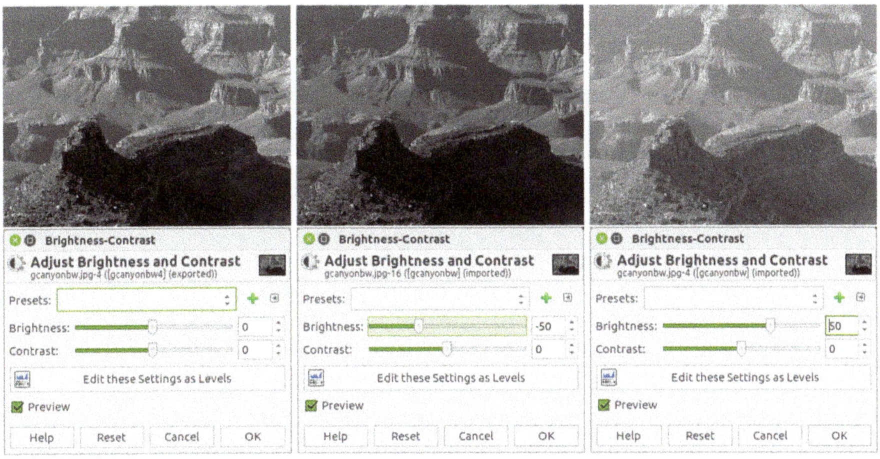

Figure 42.5. The effect of adjusting brightness in GIMP.

computer screen. But that said, modern printers and computer screens are calibrated well enough that it is easy for a not very picky person to produce a print that is at least a fair approximation of the image one adjusted with GIMP while looking at the computer screen. Technical details regarding computer monitor and printer calibration systems are available, for those who need them.

42.2.1 Digital contrast adjustments

The fact that the brightness-contrast adjustment dialog in GIMP contains an additional slider labeled contrast drives home the point that brightness, corresponding to overall exposure in the darkroom, is not the only issue. If some parts of the print are too dark, one can, if it is a digital image, slide the brightness slider to the right to make that part brighter. Or in the darkroom, one can choose a shorter exposure time, and that too will make that part of the print brighter. But doing this will also make every other part of the print brighter as well. What if some other parts of the print are now too bright?

It is at the very least a two-step process to make a connection between a particular shade of gray on the negative and a corresponding shade of gray on the print. And exactly the same is true when one wants to relate the numerical information contained in a digital image to shades of gray on the computer screen; it requires, at the very least, two separate adjustments. And so the adjustment of brightness has a complement, called contrast.

A contrast adjustment alters the relationship between the total range of brightness levels on the original and the range of shades of gray that are displayed on the adjusted image. When one increases contrast in a digital image by, for example, sliding the contrast slider to the right in GIMP, a given range of brightness levels in the unadjusted image is spread out to a larger range of brightnesses on the adjusted image. When lowering the contrast, the opposite happens; a given range of

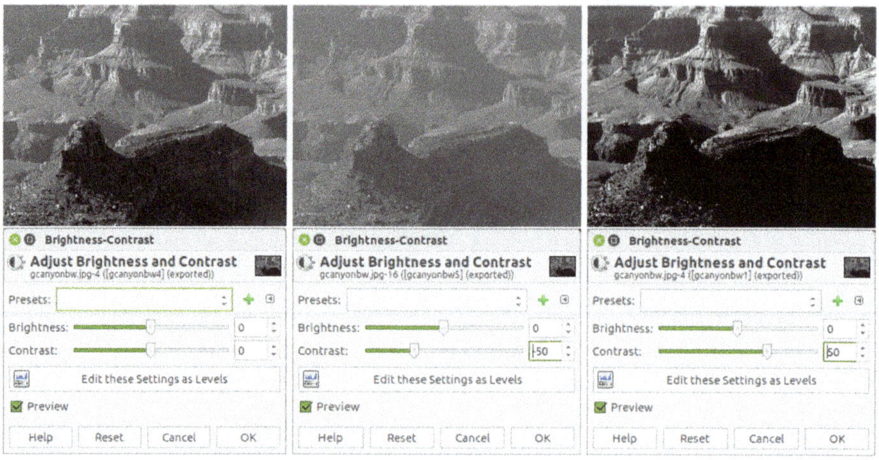

Figure 42.6. The effect of adjusting contrast in GIMP.

brightnesses on the unadjusted image is compressed to a smaller range of brightnesses on the adjusted image. See figure 42.6 for examples.

42.2.2 Contrast adjustments in the darkroom

But what about the darkroom? The print paper can be exposed either more or less. Greater exposure makes all of the printed values darker (except the ones that are already maximum black), while lesser exposure makes all of the printed values lighter (except those that are already pure white). It might seem then that it is impossible to adjust contrast at all, but in practice there are several methods for adjusting the contrast of a black-and-white print.

42.2.2.1 Preflash exposure

The contrast of a print can sometimes be lowered by flashing the print paper with diffuse light before (or after) exposing it with the projected negative. Thus one can give the paper two exposures, one with no negative in the enlarger and one with the negative in place. The first exposure is of diffuse light that darkens all of the paper evenly. The second exposure produces the image on top of this even shade of gray. This method only works for decreasing contrast, and it has the (possibly undesirable) effect that no part of the print will appear completely white.

42.2.2.2 Chemical toning

Passing the processed print through a chemical toning bath, may alter the contrast of a print (London *et al* 2005, pp 156–57). One of the most common types is selenium toning, and it slightly increases contrast while making subtle changes to the hue of the gray tones. Properly done selenium toning also increases the life of a silver-gelatin print, so it is often considered a necessary step in producing archival prints guaranteed to last at least 100 years.

42.2.2.3 Adjusting the negative

One can instead adjust the contrast of the negative, by making careful choices during exposure and development. That is to say, by carefully choosing the exposure when the picture is taken, combined with how the film is developed, one can intentionally produce a negative of either high or low contrast. Of course, one is then stuck with whatever choices were made while exposing and developing the negative.

42.2.2.4 Choosing a different negative

It may sound silly, but one can simply choose a different negative to print. To illustrate with a personal note, I have by now accumulated several times more film negatives than I will ever have time to print in three of my lifetimes—and surely I am not the only photographer my age for which this is true. Choosing which negative to print is always the first big step. Since, all else being equal, a higher-contrast negative will produce a higher-contrast print, I can simply keep that fact in mind while making the difficult choice of which negative to print and which to shun.

42.2.2.5 Graded enlarging papers

Graded photographic print papers are available to produce final prints of more or less contrast from the same negative. These are assigned numerical values of 0–5 indicating their comparative level of contrast. A grade 0 paper produces a relatively low-contrast print, while a grade 5 paper produces a print of much higher contrast when exposed with the same negative. To use graded papers effectively, one must, of course, have a selection of different grades on hand to choose from, and it can be difficult to find a source for grades 0, 1 and 5 since they are not used so much. But the very best quality print papers are usually available in graded form, and this is in fact the darkroom contrast-control method of choice for many discerning photographers.

42.2.2.6 Variable-contrast enlarging papers

Finally, one can use *variable-contrast print paper*. This is called variously by different manufacturers: *VC* (Oriental), *polycontrast* (Kodak) or *MG* (Ilford). Variable contrast black-and-white paper has two different photographic emulsions—one of high contrast and the other of low contrast—mixed together on the same paper. The trick is that although both emulsions are sensitive to blue light, the sensitivity of each emulsion to green light is different. Colored filters can be used to change how much green light, compared to the amount of blue light, is projected onto the print paper. If the color is chosen carefully, one can control how much of each emulsion is exposed relative to the other, and this has a profound effect on the overall contrast of the print. Figure 42.7 shows a box of variable contrast paper and a set of filters to go with it, along with a selection of graded papers.

With variable contrast paper, one can achieve prints of different contrast from the same paper. But to make use of it, one must have a special set of filters designed to go with the paper. These usually come in a kit, and they fit in a holder either above or below the lens of the enlarger. The filters are labeled with numbers just like those for graded paper but often with a greater range of contrast values than now available

Figure 42.7. Left: An assortment of enlarging papers for making black-and-white prints from negatives. The top box is variable-contrast paper, while the others are contrast graded papers (larger numbers are higher contrast). **Right:** A set of filters to control the contrast of prints made with the VC paper.

for graded papers. Although many photographers prefer their favorite brand of graded paper, variable-contrast paper (with the one-time investment of the proper set of filters) is so convenient and easily available that it is the most common method for controlling the contrast of a black-and-white print in the darkroom.

One complication of variable contrast paper is that increasing the contrast, by using a higher-contrast filter, also has the effect of decreasing the overall sensitivity of the paper, and so a longer exposure is needed to compensate. This means one cannot simply adjust brightness and contrast of the print separately; adjusting one affects the other. This makes the trial-and-error process of getting the right exposure all that more complicated. There are, however, many tricks one can learn to make the process more systematic and less like a game of chance (e.g., see *split-filter printing* in London *et al* 2005, p 151).

42.2.3 Levels and curves adjustments

The information contained within a digital image allows for far more sophisticated adjustments than simple brightness and contrast. A better approach is to directly control the *transformation* of the input gray levels in the unadjusted image to the output gray levels in the adjusted image. A graphical representation of this transformation is called an *adjustment curve*, and an example can be seen in the GIMP *curves* dialog shown in figure 42.8.

In the left-hand image, the different gray levels that correspond to the pixel values in the digital image—the *input*—are shown across the bottom of the graph. The ragged 'mountain range' in gray above is a histogram of those gray levels. Where the histogram is high, it means there are many pixels in the image that have that particular gray level. Where the histogram is low, there are relatively few pixels in the image that have that particular gray level.

There is another gray scale that progresses vertically on the left-hand side of the image; these are the *output* gray levels, after adjustment. The curved line cutting

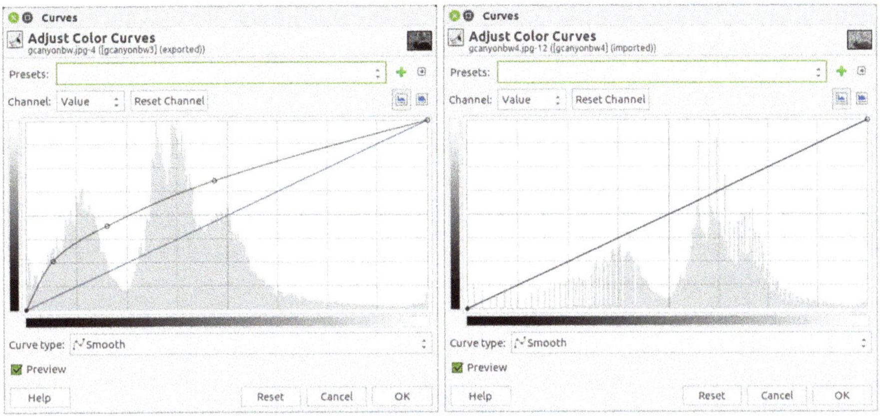

Figure 42.8. A GIMP curves dialog both before and after adjustment.

across the histogram is the *adjustment curve*. Pick a particular gray level on the input gray scale along the bottom, follow it vertically until reaching the adjustment curve and then from that point, move horizontally to the left until reaching the output gray scale. The gray level at that intersection is the result—the output—if this particular adjustment curve is chosen (by clicking 'OK').

The corner-to-corner straight line is there for comparison and represents no adjustment at all; any particular input gray level corresponds to exactly the same gray level for the output. So we can see that this particular adjustment curve makes the darkest gray levels in the image brighter on the output than on the input.

The adjustment curve has much in common with the characteristic curve of a silver-gelatin emulsion, discussed in chapter 33. It too is a graph that relates an input (photographic exposure) to an output (the resulting density of the film). But there is an important difference between the two; the density of photographic film is defined to be how dark it is, whereas the adjustment curve increases with brightness.

The right-hand image in figure 42.8 shows the result of applying the adjustment curve shown on the left. Since the adjustment curve made the darkest pixels brighter, there are now fewer dark pixels. The bulk of the pixels were darker than the middle tone in the input image, but the adjusted histogram now shows them to be more evenly distributed. Clearly the adjusted image would, overall, look brighter than the input image.

Consider another apparent feature of the output histogram shown on the right in figure 42.8. The shadow values now seem to come in discrete 'jumps.' That is, there are relatively many pixels in the image that have some of these gray values—but there are other gray values that are almost the same and have no pixels at all represented in the image. This seems to come in waves as we go along the gray scale from dark to light—the image has many pixels, no pixels, many pixels, no pixels, etc.

The reason for this is that relatively few input gray levels were transformed to a much larger range of output gray levels. A given digital image has only so many levels of gray—256 for this particular example—and there were even fewer of the dark gray levels that received the most adjustment. These few levels were 'stretched' to cover a much greater range on the output, and there were simply too few input levels to smoothly cover the much-larger range of output levels, so they appear in discrete steps instead. This may or may not be a problem. It can appear as sudden jumps in the brightness of the image, from one location to another, but if these jumps are in very dark regions of the image, they may not be noticeable. If they were, to make this adjustment successfully, one would have needed an original input image that had more than only 256 levels of gray.

Let us now look at some examples of altering the adjustment curve of an image and how this relates to the brightness-contrast adjustment in GIMP. In addition, we consider another dialog called *levels*. It represents an intermediate degree of control between brightness/contrast and curves. The levels dialog includes five adjustments and a histogram. Three of the adjustments are for the input gray levels, and two separate adjustments control the range of output gray levels. Input levels lower than the left-hand input slider are all transformed as the same (dark) gray level for the output. The right-hand input slider has a similar effect for the brightest levels. The range of output levels is set by two sliders on a separate gray scale for output. To illustrate the relation between these three ways of controlling the transformation between input and output gray levels, figure 42.9 shows both levels and curves for the adjustment of brightness portrayed in figure 42.5.

Figure 42.10 shows the examples of changing contrast from figure 42.6, in terms of both levels and curves. Notice that in all cases, if only the brightness and contrast

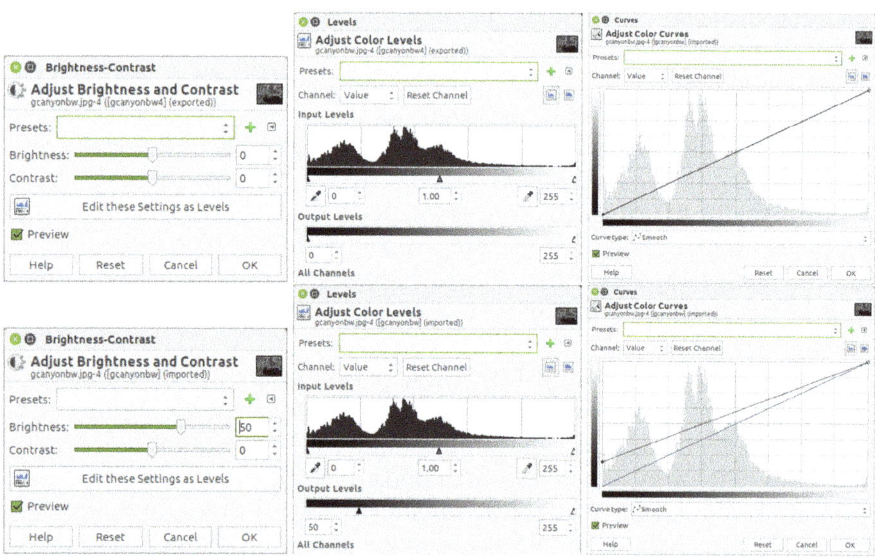

Figure 42.9. Adjusting brightness in GIMP with the brightness, levels and curves interfaces. The top dialogs are for the unadjusted image—the left-hand image in figure 42.5. The bottom dialogs show the adjustments that result in the center image.

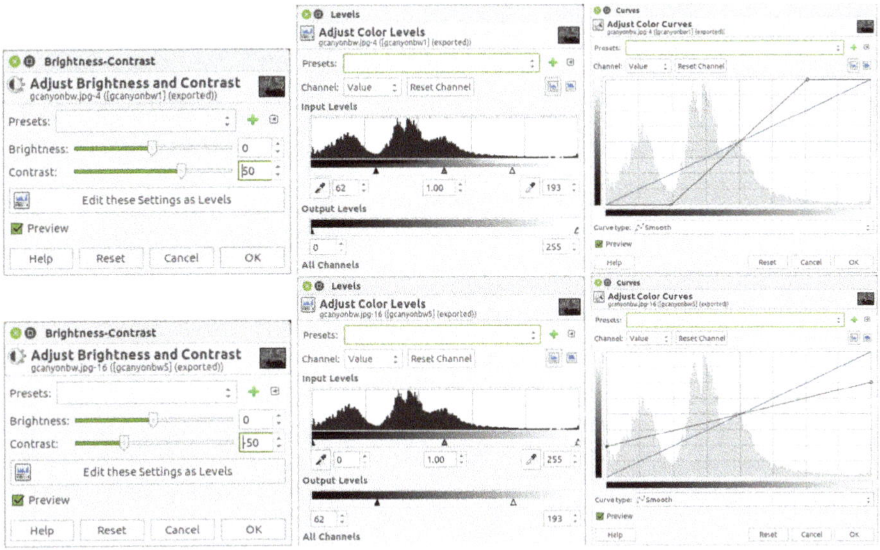

Figure 42.10. Brightness/contrast and equivalent levels and curves adjustments of the left-hand image in figure 42.6 in order to produce the center and right-hand images.

is adjusted, the adjustment curve remains a straight line. Also note that the contrast is related to the slope of the adjustment curve. A slope steeper than the no-adjustment reference line (which has a slope of exactly one) increases the contrast, while a slope that is less than unity decreases the contrast.

The center slider of the levels dialog changes the *curvature* of the adjustment curve. Moving the slider to the right results in a concave-upward adjustment curve, while sliding it to the left makes the adjustment curve concave-downward. The effect of either is to overall darken or brighten the image, but both the darkest and brightest gray levels are unaffected in either case. See figure 42.11 for examples of these adjustments and the results, both having the left-hand image in figure 42.6 as input. Such an adjustment is sometimes called a *gamma* (γ) *adjustment* because the shape of the adjustment curve on a linear scale is similar to that which alters the slope of the characteristic curve on a log–log plot. Notice the similarity between the adjustment curves in figure 42.11 and the power law curves on the right side of figure 33.5 in chapter 33.

A very useful tool is to increase contrast with an s-shaped adjustment curve. At a given point in the histogram, the *slope* of the adjustment curve indicates the change in contrast. Where the adjustment curve is sloped greater than unity (the straight-line, one-to-one preadjustment line), the contrast is increased, while the contrast is decreased wherever the adjustment curve has a slope less than one. The simple contrast adjustment tool increases or decreases the slope of the entire adjustment curve, which remains a straight line. When lowering the contrast, the result is that no part of the picture will appear either black or white. Increasing the contrast, on the

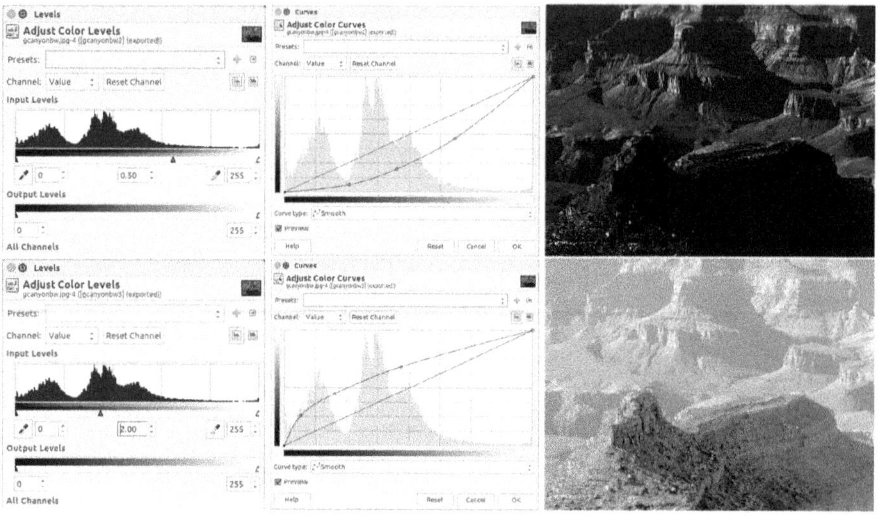

Figure 42.11. Adjusting the middle slider in the levels dialog of GIMP causes the curve to deviate from a straight line. There is no equivalent to this more-complex adjustment using only the brightness/contrast sliders.

Figure 42.12. An s-shaped adjustment curve can be used to increase the contrast of the middle tones, while decreasing the contrast (and preserving) both the highlights and shadows.

other hand, has the effect that many highlight shades of gray are compressed to pure white, while many shadow shades of gray are compressed to pure black.

An s-shaped adjustment curve, on the other hand, increases (or decreases) the slope (and thus the contrast) for the mid-tones, but preserves the tonality in the highlights and shadows. See figure 42.12 for an example of using an s-curve to increase the contrast.

For all of the adjustment curves considered so far, each output level corresponds to only one input level, but that need not be the case. Figure 42.13 shows an adjustment curve for which there are two different input levels that produce a given output level. The upward-sloping part of this adjustment curve is similar to those already considered, but for the downward-sloping part, brighter input values result

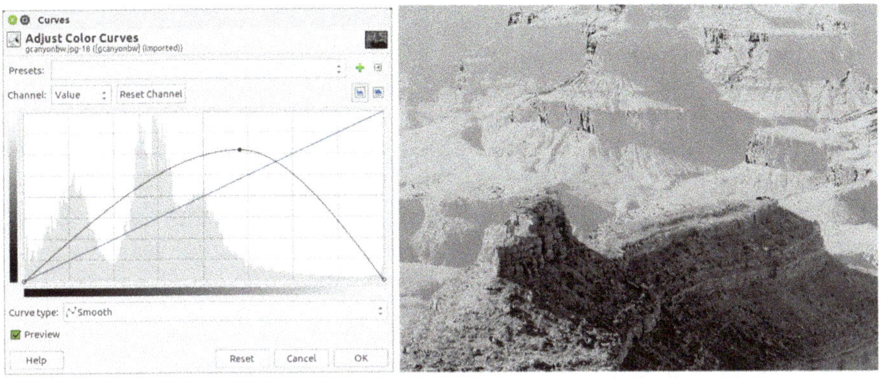

Figure 42.13. This curve adjustment has two different input values for the same output value. This causes the highlights, where the curve slopes downward, to switch to a negative, producing an image much like the Sabbatier effect (solarization).

in darker output values. Thus the brighter input values of the image have been inverted to a negative, while the darker values remained a positive. This is a digital version of the Sabbatier effect, discussed in chapter 37.

42.2.4 Levels and curves in the darkroom: the zone system

The ability to interactively change the adjustment curve between the 'RAW' digital image file and the displayed (or printed) image is one of the most important advantages of digital photography. Is there an equivalent to this with film photography? Well, of course, one can always digitally scan a negative and then use exactly the same type of level adjustments with the scanned digital image. Since digital scanners have been around for longer than digital cameras were in widespread use, photographers have been using these techniques from even before the time that digital cameras were practical.

Attempts to precisely control in a systematic way the shades of gray on a final photographic print long predate digital techniques. Probably the most important procedure is the *zone system*, formalized by Ansel Adams. The zone system relies on precise technical control at each of these steps:

1. The metering of the subject, ideally using a spot meter to measure the brightness at many different parts of the subject.
2. The exposure of the negative.
3. The development of the negative.
4. The measuring of the transparency of the negative at different locations with a *densitometer*.
5. The exposure and contrast control of the print.

Each of these steps is linked to the others in order to determine a final print that has exact reproductions of light and dark areas according to the photographer's plan. A photographer could, in theory, look at a scene and decide, before taking the picture, which areas they wanted to appear completely black, which areas completely white and which parts of the scene would appear the other shades of gray in between. See Stroebel *et al* (2000, pp 249–54) for an excellent description of precise tone reproduction with the zone system.

In practice, the zone system in its full incarnation requires an extraordinary amount of technical expertise. Apart from the mastery needed to accomplish an exact reproduction of tones according to a precise plan, one must have the artistic sense to know what is good and what is not. For both of these reasons, it is very special to see a print in the flesh by masters of printing who are also artists, such as Ansel Adams or Sally Mann.

42.3 Dodging and burning

An adjustment curve ordinarily treats all of the pixels in the image equally. That is, every pixel with a particular brightness value gets changed by the adjustment curve in exactly the same way, no matter where it appears in the picture. But what if, for example, there is a part of the image that is overall too dark, while other parts are overall too light? What if one wants a different adjustment curve for different parts of the image?

Dodging and burning are two standard techniques from the traditional black-and-white darkroom that are also used in digital image manipulation. In the darkroom, we can consider this problem in light of exposure time. What if the proper overall exposure for the print is 60 s, but that 60 s exposure makes one particular part of the print too dark? One way to correct for this is to block the light to that area only, for part of the exposure. A small circle of cardboard held on a wire can be waved around over the area to be lightened in order to reduce the exposure (remember that this is a negative process, so reducing the exposure lightens the print).

The opposite of dodging is burning. To darken one particular area of the print, a second exposure is made—but using a cardboard mask with a hole that is waved over the area that requires additional darkening. Thus the areas that project through the hole receive a greater exposure.

For both dodging and burning the mask is moved, typically in a circular motion of varying radius, in order to avoid sharp edges to the dodged or burned area of the print. The mask is held at some distance above the print to make the shadow edges somewhat blurry. Doing this well requires practice because one cannot see the result until the print is developed. Every print that has been dodged or burned is slightly different, and this adds a handmade element to darkroom prints.

42.3.1 Dodging and burning with GIMP

One can accomplish an effect very similar to dodging and burning in GIMP with digital tools of the same name, but the digital version has many advantages. In particular, one has instant feedback, so the act of dodging and burning is much like drawing with the mouse. Since prior results can be stored in computer memory, it is possible to undo what one has done, and even to 'back up' part-way and proceed from that point. This makes the digital versions of dodging and burning far easier than the corresponding darkroom methods, but there is a cost—the unique handmade feature of a darkroom silver-gelatin print is lost.

When dodging or burning in the darkroom, one achieves different effects by choosing a mask of different size or shape and by determining how close to the print one holds it. These choices are mimicked in GIMP by choosing a different brush and altering its size.

42.3.2 Spot healing and retouching

It is almost inevitable that a picture will have bad spots that need to be repaired. The most common example in the black-and-white darkroom printing is the white spots on a print produced by dust on the negative while the print was exposed. The process of removing mistakes such as this is called *retouching*, and it is a difficult process for a black-and-white print. Traditionally, one applies special dyes to the spot with a tiny paint brush, hoping to match the surrounding print area. To lighten an area is even more difficult; the spot on the print must first be bleached and then redyed to proper tone.

This can be a frustratingly difficult process, and it must be done by hand to each print. A mistake in the retouching process can easily ruin a print altogether. Apart from fixing mistakes such as dust specks, one might also want to alter the actual photographic subject, such as when a portrait photographer uses retouching to hide an unwanted skin blemish or to remove a feature that distracts from the composition. Highly skilled techniques, using both water- and oil-based paints, are needed to successfully retouch a traditional silver-gelatin or color print in this way.

42.3.3 Digital retouching

Digital image editors come equipped with many tools that make retouching far easier than traditional brush work. For one thing, the ability to work on a copy of the image file, to save work completed at any stage and to 'undo' previous steps means that one is much less likely to ruin many hours of work with a slip of the mouse. Below I list just a few of the more popular digital retouching tools (the examples are from GIMP).

- *Spot Heal*—This tool is designed to easily cover over a small imperfection. It is especially useful for a tiny spot that is nearly round, such as a dust speck. When the mouse is clicked (while holding it steady), the region is replaced by sampling a ring around the mouse position. In some versions, a different, featureless region can be used as the model for the underlying texture, and the hue, saturation and value are modeled from the region surrounding the mouse. It also can be used with a sweeping motion of the mouse, and in this case, a region on either side of the mouse sweep is sampled.
- *Cloning Tool*—Two regions are selected with the mouse. When the mouse is clicked and moved, the sample region is copied to the mouse location. This

tool is more difficult to use, but when used with care, it can accomplish complex retouching tasks not possible with the spot heal tool.

- *Drawing Tools*—There are many tools for selecting colors and values and then 'drawing' directly with the mouse, filling in selected regions with a uniform color or deleting details directly with the mouse (as if using an eraser).
- *Masks*—A region of complex shape can be selected with several different tools, and adjustments applied to that selection alone or to its inverse—everything but that selection. The selection can have fuzzy edges so that there is a gradation between where the changes are applied and where they are not.

When editing a digital image, it is easy to magnify its every detail to a level that is much greater than what is seen by someone viewing a print made from it. And so if care is taken and time is spent, the photographer can retouch the image down to the finest detail.

42.4 Color darkroom versus digital

I have so far compared digital and darkroom techniques only for black-and-white photography; color adds many complications. This chapter is not meant to be a how-to guide for digital image processing, but I will here add just a few items specific to color photography.

To make a *chromogenic* color print in the darkroom, light is projected through the negative with an enlarger, onto light-sensitive color print paper. In these ways, the color darkroom is just like the black-and-white version, but there are many important differences:

1. Since the color print paper must be sensitive to all different colors of light, one cannot use a special safe-light; all handling of the paper must be carried out in total darkness. But there are special processing tanks that allow chemicals to be poured in and out while keeping the light-sensitive paper in darkness.
2. A special light source, called a *color head*, is used to expose the paper. The color head combines three different light sources, one each for cyan, magenta and yellow. The brightness of these can be varied separately so the color of the combined light can be precisely controlled.
3. The chemical development of the light-sensitive paper requires additional steps.
4. For black-and-white printing, the primary considerations are exposure and contrast control. For color printing, it is exposure and color balance that are most important.

42.4.1 Contrast control

We saw how the contrast of a black-and-white print can be controlled in the darkroom by using graded papers or variable-contrast (VC) paper. A moment's thought should make it clear that one could not use something like VC paper for color printing. The contrast of VC paper is altered by changing (with special filters) the color of the light falling on the print paper. Obviously, for color photography, such filters would also change the *color* of the print.

One could still, in principle, design color print papers that are analogous to graded black-and-white papers in order to produce prints of higher or lower contrast. This is, however, not all that practical since the steps in the color process are so much more complex. For these reasons, it is difficult to exert as much control over the contrast of a color print in the darkroom as one can for black-and-white printing.

In fact, by now most film processing centers have abandoned altogether the traditional darkroom method of projecting light through the film negative onto the light-sensitive print paper. Instead, the processing machine scans the film negative, and the resulting digital image is then scanned onto the light-sensitive print paper with computer-controlled colored lasers. The print paper is then chemically developed in the usual way to produce what is, physically, a traditional chromogenic color print. Thus contrast (and color balance) is controlled digitally before the print paper is even exposed by the lasers, and this allows for far greater control over contrast and color balance than is possible with traditional wet-chemistry darkroom techniques. This has the added advantage that the same processing machine can produce chromogenic prints from both color film and digital image files.

Changing the contrast of a color digital image is, in principle, the same as for a black-and-white image. The only difference is that the color image contains, in effect, three separate digital images, one each for red, blue and green. Each of these three parts of the image is called a *channel*. We can control the levels and curves on each channel separately, or we can vary them all together.

42.4.2 Color balance

The default in GIMP, whenever one selects *levels* or *curves* is that all three red, green and blue channels are varied together. But one can instead select only one channel to vary at a time. This will change not only the contrast of the picture overall but also the color balance. Employing adjustment curves directly on individual color channels allows for sophisticated control of the overall color of the image. But there are simpler alternatives as well. Editing software such as GIMP also feature simple tools that allow one to interactively shift the overall color balance, with sliders that adjust the mix between the additive and subtractive primaries. And so one can separately adjust the balance between red-cyan, green-magenta and blue-yellow. As an alternative, one can adjust the hue, saturation and lightness separately with a different dialog, but this only scratches the surface of the many powerful digital tools available for adjusting color. More information on color adjustment of digital images can be found in many standard references both online and in print (e.g., London *et al* 2005, chapter 11).

The overall color balance of a chromogenic print in the darkroom is adjusted at the stage of exposing the light-sensitive paper to light. The photographic enlarger must have a special *color head* that allows for easy adjustment of the color of the projected light. This is usually accomplished with separate adjustments for the subtractive primaries—cyan, magenta and yellow. Often, a numbering system is used for these adjustments so the photographer can easily reproduce the proper color balance, once they have found it. The cyan, magenta and yellow values are simply 'dialed in' by the three knobs on the enlarger head. To make this mostly trial-and-error process easier, many color printing systems employ a set of correction filters. One looks at a finished print through these filters, selecting the one that makes the color balance look the best. The particular filter selected tells the photographer how many steps they need to add or subtract from the cyan, magenta and yellow adjustments on the enlarger head, such that the next print will come out with the correct color balance.

References

London B, Upton J, Stone J, Kobré K and Brill B 2005 *Photography* 8th edn (Englewood Cliffs, NJ: Prentice-Hall)

Stroebel L, Compton J, Current I and Zakia Z 2000 *Basic Photographic Materials and Processes* 2nd edn (Waltham, MA: Focal Press)

IOP Publishing

Photography (Second Edition)
Physics and art in focus
John Beaver

Chapter 43

Image manipulation

By image *manipulation* I refer to alterations that significantly change the content of the image, often adding elements that were not present in the original exposure to light or removing elements that were. The image *processing* outlined in chapter 42, on the other hand, mostly treats the entire image at once, adjusting overall properties such as brightness, contrast and the balance of color, usually maintaining the same recognizable content. In particular, image-processing techniques most often do not invent new elements. I call it image processing, for example, when the color balance of what was originally a color image is altered to make a final print. I call it image manipulation when different colored elements are added after the fact to an image that was originally monochrome.

In the sections that follow, I describe many techniques by which images may be manipulated. I make a distinction between techniques that directly and intentionally manipulate specific content and those that manipulate the entire image as a whole. Finally, I consider a contemporary (and perhaps curious) connection between digital image manipulation and the antiquarian, and I make some brief comments about the art and ethics of manipulating photographic images.

43.1 Direct content manipulation

In this section I describe some techniques that intentionally alter selected representational elements in a picture.

43.1.1 Cropping and retouching

The simplest and most common way to alter the content of a photograph is simply to refrain from showing all of what was recorded by the camera. *Cropping* a photograph can often greatly improve the formal properties of an image, from a design perspective, or one may remove elements that are a distraction and do not serve the larger goals of the photograph. At a more extreme level, cropping can capture a tiny

 © IOP Publishing Ltd 2022

detail of a photograph to make an entirely new image. Figure 43.1 is an example; the image on the left was cropped from only a tiny detail of the full in-camera image on the right.

Retouching may be as innocuous as the after-the-fact dyeing of white spots on a darkroom print that were caused by the magnified images of unwanted specks of dust on the negative. But it also may refer to the intentional deletion of a detail that was faithfully captured by the camera—a small feature on the face of a portrait subject, for example, that is perceived by the photographer to be a distracting blemish. Figure 43.2 shows an example; the digitally retouched version is on the right, and the original on the left.

Apart from removing the white shadows (this is a negative process) of many tiny dust specks on the scanned paper negative, I have removed the much-larger white

Figure 43.1. The image on the **left** was cropped from a tiny detail of the image on the **right**.

Figure 43.2. *Plant sphere*, John Beaver 2021. The image on the **right** was retouched from the original image on the **left**.

spot near the bottom-center. The feature—perhaps the out-of-focus glint of sunlight off a metal bolt—is a confusing distraction and adds nothing to understanding the image.

The retouching of even a small detail on a finished silver-gelatin print can be a painstaking and potentially destructive process. The digital equivalents, on the other hand, are comparatively simple, and more importantly, they are reversible. Sophisticated digital techniques even sometimes allow for direct *content-sensitive selection*. Complex algorithms attempt to identify, often with some accuracy, the actual content of the picture. One can then, for example, click on a chair in the picture, and it is recognized by the software as the image of a chair, separate from the background of floor, wall and a more-distant sofa. If the algorithm is successful, the chair can then be removed from the image with a mouse click. A reasonable background is constructed by the software to replace the parts that were covered by the chair in the original image.

43.1.2 Manipulation during exposure

From the earliest days of photography, artists have sought out tricks to manipulate images during the very taking of the picture. The appeal is obvious—to create one's own fantasy world while armed with only a camera. One simple technique is the *multiple exposure*; record two or more separate exposures, one after the other, onto the same piece of light-sensitive material. Most early cameras, in fact, had no mechanism to *prevent* this from happening, even by accident. The left side of figure 43.3 shows an example. The greatly over-exposed second exposure creates an odd, irregular bottom border to the line of trees, as if they were reflected in water—but unfaithfully so.

Figure 43.3. Left: An accidental multiple exposure produced an image very different from either of the two scenes photographed. **Right:** *Leaving my head behind*, John Beaver and Teresa Patrick 2003. Different parts of the moving subject were 'painted' with light from a flashlight during a long exposure, to give an odd multiple-exposure effect.

The right side of figure 43.3 shows a different variation on the concept of multiple exposure, using the technique of *light painting*. Instead of distinct separate exposures taken one after the other, the figure moved during a time exposure lasting many seconds. The subject was illuminated, in an otherwise darkened room, by a flashlight. Choreographed with the movement, the light was used to successively 'paint' different parts of the moving subject.

For figure 43.4, I erected a tall mirror as part of the foreground. The focus is set on the more distant background and the reflected image. The mirror, much closer than the image reflected in it, is out of focus, and thus one sees only the reflection, which appears oddly superimposed onto a background in the opposite direction. This particular image is manipulated further by use of the Sabattier effect, described in chapter 37.

Images may also be intentionally (or accidentally) manipulated in-camera by the use of internal camera and lens reflections, called *flares*. Most typically, a small, bright source of light just out of the image frame can add odd features, especially if the rest of the image is rather dark.

Finally, the scene itself may be manipulated and then photographed in such a way as to disguise the manipulations. Objects may be, for example, suspended by fine wires that are rendered invisible by clever focus control or a confusing background.

All of the in-camera techniques described in this section were used, sometimes in combination, in the nineteenth century tradition of *spirit photography*. The idea was that spirits and ghosts may be invisible to the human eye, but the new and seemingly

Figure 43.4. A tall mirror placed in the foreground gives an effect as if two views are pasted on top of each other. This image also features the Sabattier effect.

Figure 43.5. Two 'spirit photographs' by contemporary photographer Gregg McNeill. The image on the **left** was made by stacking two negatives, from two different in-camera exposures, when exposing the print. The image on the **right** was made with a simple in-camera double exposure. Gregg McNeill, 2022, used by permission.

magic technology of photography could capture them. Spirit photography has seen something of a revival by photographers who specialize in nineteenth century processes, although without the pretense of the ghosts being real. See figure 43.5 for two examples by contemporary photographer Gregg McNeill.

43.1.3 Photomontage

A *photomontage* is an image constructed from two or more separate images that occupy different parts of the picture plane, working together to seem as one single photograph. This is not quite the same as a *photocollage*, which usually refers to clearly separate photographic prints pasted together to make a larger piece of art.

David Hockney famously blurred the boundary between montage and collage with his *My Mother, Bolton Abbey, Yorkshire* (Cooke and Kinneberg 2014, p 223). It is technically a photomontage, but he has intentionally left apparent the rectangular edges of the individual prints he pieced together, giving the overall piece a collage-like look. The left side of figure 43.6 shows an attempt in that spirit, made from image transfers from instant film, the different exposures made from different portions of the same original 35 mm transparency.

The image on the right side of figure 43.6, on the other hand, is a little more in the spirit of a photocollage. I made three separate exposures onto the same piece of light-sensitive paper, but I used a hand-brushed accelerator to selectively expose only parts of the paper with each exposure. It is clearly a composite, but one could

Figure 43.6. Left: *Along the line*, John Beaver 2008. **Right:** *Worm*, John Beaver 2016. A montage made by three successive exposures onto the same piece of photographic paper. The accelerated lumen technique allowed for only selected parts of the paper to be rendered light sensitive for each exposure.

imagine the grass seed heads and caterpillar as being part of the same scene. The border was added to the exposure by hand with a laser pen.

A photomontage may be constructed in the darkroom onto a single piece of light-sensitive print paper by using masks to allow only certain parts of a given negative to expose the paper. The unexposed parts can then be exposed with a different negative and a mask that blocks the parts of the print that were previously exposed. This can be repeated many times, but it is a technical challenge to do it well. To make an image that is not obviously pieced together requires careful control of cross-fading and blurring. The 'Surrational Images' (Mutter 1992) of the US American photographer Scott Mutter (1944–2008) are good examples. He seamlessly combined everyday images to make often-whimsical, impossible scenes—a misty forest with a wood parquet floor (p 23, 24) or a skyscraper in a cityscape that is also the bottom of a Corinthian column inside a building (p 17).

Digital image processing has made photomontage far more accessible. It is easy to combine separate images in virtual 'layers,' with digital masks and automatic blurring of edges. Each layer can be processed separately, with the masks controlling which parts of an image layer receive processing and which parts do not. An image layer can then be combined with other layers, and the transparency controlled interactively. One must still carefully create the image adjustment masks, selecting only the image elements one wants to use in each case. In some cases, even this usually tedious task is made simple by context-sensitive selection and other digital processing tools.

In the darkroom, the masks and layers are physical objects, and they must be constructed either by hand or photographically. The process is done without visual input, rather than interactively; the complex set of multiple exposures produces only an invisible latent image on the light-sensitive material. To see the result of even a single, minor change, one must go through the time-consuming cycle of chemical processing.

43.1.4 Physical and chemical manipulation of image content

Many historical and alternative-process photographic techniques naturally create prints that have a lot of nonphotographic content—the brush strokes of a hand-brushed emulsion on watercolor paper, for example. But there are also some clever ways to intentionally manipulate separate elements of certain types of prints even if exposed to the negative in an otherwise-ordinary way.

The left image in figure 43.7 shows an instant-film *emulsion lift*. I took a picture in the ordinary way using integral film in a Polaroid camera. The small, rather unremarkable print was then dissected with a razor knife, separating out the different layers in the complex integral film package. The top-most layer is a sheet of transparent plastic with the developed silver-gelatin emulsion stuck to its back face.

I soaked the plastic and adhered emulsion in a shallow tray of warm water for several minutes until the gelatin layer separated from the plastic. Once lifted, it is something like a jellyfish, distorting and moving freely with any swirl of the water. But it is a jellyfish with a picture on it. Using a soft brush, I then teased it onto a sheet of watercolor paper immersed in the same tray of water. Carefully removing the water from the tray with a syringe, I manipulated the rather gooey gelatin with its silver image onto the watercolor paper, causing it to tear and fold in particular places.

Of course none of this ever happens exactly as planned, and that is much of the point. The emulsion lift technique appeals to many artists for the very reason that one has only partial control of the results. Much is still left to accident.

The Sabattier effect can also be used as a form of image manipulation. A silver-gelatin emulsion is exposed to form a latent image. But while it is undergoing chemical development, it is flashed with diffuse light, causing a partial reversal of the some of the tones, and outlines (so-called Mackey lines) around parts of the image.

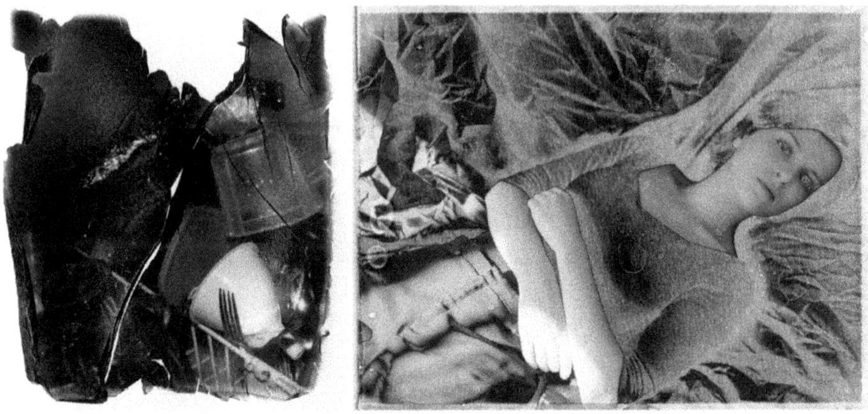

Figure 43.7. Left: An example of a transfer of an integral film image onto watercolor paper. **Right:** By flashing an instant-film negative with light before development was complete, the Sabattier effect reversed the tones on some of the elements in the image.

Unless a mask is used during the flash of diffuse light, one cannot pick and choose which image elements will reverse. But with experience, one can anticipate the results and plan the composition accordingly.

For the right-hand image in figure 43.7 I used a black cloth as the background, and the model wore dark clothes. I anticipated that the dark tones would reverse, the lighter skin tones would not and the Mackie lines at their boundary would seem to cut the face and arms out and lay them on top of the picture. One can, with experience, guess approximately what will happen, but much is still left to chance and happy accident.

The *mordençage process* allows for a complex physical and chemical manipulation of image elements in an already-finished silver-gelatin print. The technique seems to have been first described by L P Clerc in 1930, although it was mostly forgotten until recently (James 2016, pp 816–18). Challenging to master, mordençage can result in a spectacular transformation of an image. Figure 43.8 shows an example by the Edinburgh-based photographer Brittonie Fletcher.

Figure 43.8. *Nest* ©Brittonie Fletcher, Mordençage 2013. Used by permission.

43.2 Image-wide manipulations

43.2.1 Toning and split toning

The process of *toning* is considered a standard, even necessary step when making a silver-gelatin print from a negative in the darkroom. With *selenium toning*, for example, some of the developed-out silver is converted to silver selenide; it has a subtle intensifying effect on the print and makes it more stable against oxidation, but it also alters the overall color of the printed image (Stroebel *et al* 2000, p 232). This is usually a good thing, making the blacks slightly off-neutral, either warmer or cooler. One can, in fact, choose emulsions that are already off-neutral, using either warm-tone or cool-tone emulsions. There are other types of chemical toning that are far from subtle. Sepia toning, for example, turns the image a strong reddish-brown, while blue toning makes it look like a cyanotype (it uses some of the same chemistry).

Yet even a blue-toned silver-gelatin print is still *monochrome*; it is just now different values of blue instead of gray. The process of *split toning*, on the other hand, is a more significant intervention. A split-tone print has two different, often contrasting, tones—one for the highlights and one for the shadows. Chemical split toning is tricky and complicated, and the results are usually subtle. It is almost trivial, on the other hand, to split-tone a digital monochrome image. Figure 43.9 is an example, processed in GIMP from a scan of a black-and-white 35 mm negative. The important question is, as always, does this intervention improve the artistry of the photograph—or is it instead merely a distraction?

Figure 43.9. Split toning was easily applied with GIMP to a digital scan of a black-and-white negative.

43.2.2 Physical and photochemical manipulations

Many years ago I received a call from the local processing lab, regarding some color film I had dropped off earlier. 'We are sorry, but something terrible has happened to our processing chemistry today, and your film is ruined.' I picked up the negatives anyway, and they refunded my money. But I considered myself doubly lucky; not only did I get the processing for free but I also liked the pictures much better that way. See the left side of figure 43.10; odd swirls of color and light made something new and mysterious out of what would have been just another landscape photograph, so similar to hundreds of others I had taken.

The right side of figure 43.10 shows another example. This time the fault was my own, an attempt at reversal processing gone bad. Again, an otherwise unremarkable photograph was, arguably, improved by the accident.

I am far from alone in having experienced these sorts of happy accidents (many are also unhappy) that arise from errors in processing chemistry. It should be unsurprising then that some photographers arrange for this to occur 'accidentally on purpose.' One popular technique is known as *film soup*. A roll of color 35 mm film is soaked—cannister and all—for a day or two in some concoction of choice. The cannister is then rinsed under the tap and left to dry inside and out for many days or even weeks. It is then exposed in a camera, and the film processed in the usual way. Armed with both knowledge of photochemistry and wild guesses, film soup artists experiment with a wide range of often everyday kitchen chemistry for pretreating the film. Some of the recipes sound good enough to eat.

The high contrast and odd color palette of the picture in figure 43.11 arises from *cross-processing*. Film designed to be reversal processed as a positive transparency (for direct projection onto a screen) is instead processed as color negative film. It is the processing, not the emulsion itself, that decides whether the image comes out as a negative or positive. Even so, films meant to be reversal processed are designed differently than those meant to be processed as negatives. When one is used for the other, odd and possibly interesting results can be expected.

Figure 43.10. Left: An accident in the chemistry of the color processing resulted in swirling colors and odd patches of light. **Right:** In a failed attempt at reversal processing, improper bleaching and uneven, rapid development put swirling 'clouds' into what was really a blank, overexposed sky.

Figure 43.11. Film meant for 35 mm positive transparencies was instead cross-processed to make a negative, resulting in an odd shift of color and contrast.

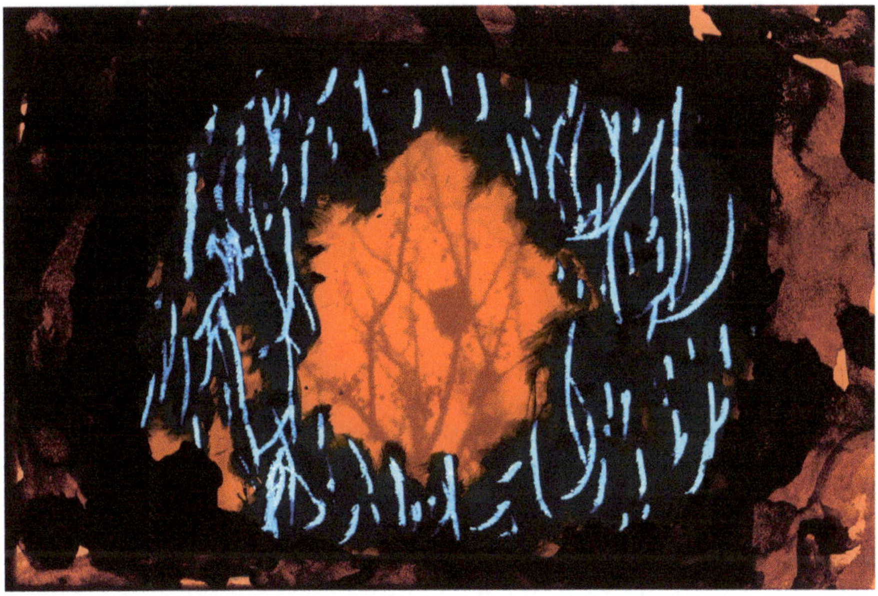

Figure 43.12. From *Rivers/bookends*, Judith Waller and John Beaver 2018. The original photographic transparency by John Beaver was over-painted and scratched by Judith Waller. Used by permission.

Finally, there is a long tradition of printing from a negative or positive transparency that has been physically altered in some way. Figure 43.12 shows a rather extreme example. The original 35 mm positive transparency was over-painted and then scratched with a needle while viewed under a low-power stereo microscope.

Figure 43.13. A mixture of several photochemical techniques was used to make this ultimately digital image.

All of the techniques described in this section, and many others as well, may be mixed. Think of them as artistic tools, either useful or not in a given circumstance. The image in figure 43.13 began as a 35 mm black-and-white negative. I then made a positive contact print onto lithography film, using an ephemeral-process accelerator. After fixing the resulting positive transparency, I toned it with blue toner. Finally, I rephotographed the transparency with a digital camera and stretched the contrast of the resulting image with GIMP.

43.2.3 Digital filters and effects

A *digital filter* acts on all of an image according to the same set of rules, or *algorithm*, but such an algorithm can have an added level of complexity that is difficult to achieve with photochemical processes; a given part of the image can be modified according to the parts around it. This allows for wide-ranging transformations of an image.

Some filters are *convolutions*—a sort of two-dimensional running average that replaces each pixel with a mathematical combination of the pixels around it according to a preset formula. In the simplest case, this blurs detail, perhaps preferentially in a particular direction.

But a convolution may instead replace pixels with a more complex aspect of its surroundings—a *gradient* for example. A gradient is a measure of how rapidly the pixel values change in a particular direction. This is sometimes called *edge detection* since only those parts of the image with sharp edges show detail.

The inverse of a convolution is called a *deconvolution*, and such a filter may act so as to sharpen the image or at least give an illusion of increased sharpness. Increased

Figure 43.14. The image on the **left** was first blurred (detail **upper right**), and then the blurred image sharpened (detail **lower left**). The sharpening process inevitably introduces random noise into the image. For the image at **lower right**, an edge-detection convolution was applied.

random image noise is the inevitable consequence. The process of blurring, however, does destroy image information that cannot be recovered by deconvolution. See figure 43.14 for examples of blurring, sharpening and edge detection.

A digital filter may also bring in its own image detail from the outside, for example, superimposing preprogrammed textures onto image details, or it may incorporate masks in its algorithm, applying the filter differently to different x-y positions on the image.

A script of successive filters can be applied to an object to produce a complex *digital effect.* An effect might be designed, for example, to emulate a particular artistic technique, and may involve many individual filters. See figure 43.15 for examples of different digital effects applied to the same original image. The intent may be subtle, or it may be the opposite of subtle.

43.3 The art and ethics of image manipulation

We have seen in chapter 40 that digital image processing techniques are in many ways not so new as they are sometimes made out to be. Much of it is simply a different technical means to alter photographs in pretty much the same way photographers have been doing for a century.

Figure 43.15. The image on the upper left has been transformed by five different digital effects from the popular G'MIC package of GIMP.

Yet, there is still something that rankles many. When certain pictures are disclosed to have been digitally manipulated, one sometimes feels cheated or let down, so here I try to lay out a few of the issues as best I can, to better identify where, when and how that reaction arises, to better judge under what circumstances it is or is not justified.

43.3.1 Analog antecedents

One could argue that the only 'pure' image is a direct-positive photograph that uses the full image circle of the lens, with the same physical object hanging on the wall as was in the camera for the exposure. One might further claim that it is only truly 'pure' if the artist hangs on the wall every photograph they have ever taken; why should the artist be allowed to decide what is and what is not a good picture?

But of course this deciding—being an intermediary—is very much a part of the essence of the artist. A photograph has some extra baggage that a painting does not. We have developed something of an expectation that it is a literal 'capture' of a moment of reality, so we sometimes feel deceived when we find that it is not.

Since the earliest days of photography, photos have been manipulated in the simplest of ways—the subject itself is altered, arranged or even staged before the photograph is taken. Hippolyte Bayard's direct-positive photograph on paper, *The Drowned Man*, is thought to be the first photographic image to portray a fictional event; the photograph is a self-portrait (Cooke and Kinneberg 2014, p 48). The year was 1840, only three years after the first daguerreotype.

Some of Sally Mann's most 'candid' family portraits were unapologetically staged; *Gorjus* is a good example. The young girl playing dress-up and the titular license plate on the abandoned automobile did not really come together by chance. Shana and Robert ParkeHarrison built elaborate sets and props for their photographs, assembled in a studio even though the visual pretense is decidedly outdoors in nature.

Even in that most naturalistic branch of photography—photojournalism—scenes are sometimes staged while still maintaining a commitment to both art and truth. The composition of the elements in Don McCullin's famous Vietnam War photograph of a dead North Vietnamese soldier, *Hue, Vietnam 1968* (Cooke and Kinneberg 2014, p 315) is too good to be true. As McCullin explains, he did not simply stumble across the scene to photograph it just as it was (McCullin 2012).

> I never use flash, I don't believe in cropping, and there's no way I'm going to manipulate a street photograph other than to get the composition right. I've only ever staged one picture in my life, in 1968, when I found a dead North Vietnamese soldier whose possessions had been rifled through by some American marines during the Tet offensive. They walked away making derogatory remarks about this man, calling him a 'gook' and so on.
>
> I decided this wasn't right, so I shovelled his possessions together—his pictures of his family, his pathetic little medical kit and his bag of bullets—and photographed them at the foot of his dead body (pictured left). I know it was staged, and I did it as a statement. I have to own up to that. I just thought it was important that this man's voice could somehow be heard.

In the end, the staging of this photograph is essential to both its artistry and the truth of its message. It is not just the taking of a picture; it is a memorial, with the elements carefully placed, as one arranges flowers at the grave site of a loved one.

Even after the photograph is taken, it can happen that a tiny detail ruins the overall composition of a picture—'It would be perfect if not for that edge of a sign in the upper left corner…' It can be tempting to fix a picture by removing or altering minor details. This has always been a part of photography, and individuals differ as to whether (or to what degree) *retouching* destroys some important notion of purity for a photograph. But even some very famous photographers retouched shamelessly, not only removing detail they didn't like, but sometimes adding in new detail as well.

As an example, I have seen an original print of one of Bill Brandt's photographs, which features a gull caught in mid-flight. The gull is slightly blurred by motion and just barely out of perfect focus, but Brandt used retouching to enhance the sharpness of the leading edge of its wing, so it forms a high-contrast sharp line. In my opinion, it was necessary for that particular detail to be sharp, so Brandt's aggressive retouching improved the photograph greatly. Brandt in fact goes far beyond this retouching with paint; the photograph itself is a montage. The seagull was photographed separately, and added later to a landscape photograph that was taken under lighting conditions that would have rendered impossible the sharp and distinct capture of the bird.

Was it cheating? I think not. For one thing, he was not trying to trick us into believing he achieved this with his camera alone. Anyone curious as to how he managed to get that particular detail to be so sharp and high-contrast would do just what I did—move up close to the print and look at it carefully. From a close inspection, it is obvious that it was retouched with paint. Furthermore, he did not hide the fact that the photograph was a montage, for he unabashedly superimposed exactly the same gull photograph onto other landscapes as well.

But his techniques were considered controversial by some at the time, as they did not square with the purist conceptions of photography advanced by other influential photographers. Brandt's response to these criticisms was that, 'I am not interested in rules and conventions… photography is not a sport' (Brandt 1948). It should be clear from my own photography that I essentially agree with his response, but it should also be clear that I would not take it to be *literally* true that 'anything goes,' and I doubt that Bill Brandt meant it that way either. Rather, he meant (I suppose) that whatever rules there may be, they are always open for debate, and it is the art that matters in the end.

Most digital retouching techniques have their antecedents in the traditional black-and-white darkroom, but many find digital retouching to be even more controversial, mostly for the reason I already discussed: it is too easy. Retouching a silver-gelatin print with paint brushes and dyes requires a lot of skill and patience, and there is the very real risk that one will instead destroy the print. Digital editing is, in a sense, cost-free; if you don't like the result, simply click the 'undo' button. One must always keep in mind that just because you can do something, it doesn't mean you should.

43.3.2 Digital mimicry of physical processes

The easy accessibility of digital image filters over the past two decades has led us to an ironic place—a nostalgia for strictly physical and photochemical processes that make their distinctive mark on the image. Even more strangely, these same digital processes have made it easy to alter any digital image so as to mimic the 'look' of these same historical processes. Figure 43.16 shows an example. I have used a filter in Gimp to emulate the look of a cross-processed film photograph. It is just one of many such options in the popular FxFoundry set of digital tools.

Figure 43.16. Left: Original image. **Right:** A digital filter was used to instantly give an image (left) a look as if it had been recorded with cross-processed color film (compare to figure 43.11).

We have already seen how many alternative and historical photographic processes accentuate the physical interaction between light and the physical surface that responds to it during the exposure. When this interaction is uneven in complex ways, a kind of picture plane is put into the image itself, even when it is capture digitally from, for example, a scanned paper negative.

It has now become common to emulate this look, creating a mask overlay to superimpose upon an otherwise-ordinary image the uneven edges and other non-photographic content of, for example, a wet collodion tintype.

43.3.3 The question of cliché

Digital filters have the ability to greatly alter the look of a photograph, sometimes in ways that are quite interesting, but since a digital filter is a mathematical algorithm that always does the same thing when applied in the same way, too-heavy use of a digital filter can give a look of sameness to one's photographs. This is not to say that it is invalid for a photographer to specialize in a particular kind of look. Far from it! But there must be room for originality, and the particular look must have depth to it; ideally it should give the appearance that there is more there than what one literally sees in the picture.

For example, when one makes an instant film transfer, it usually looks like an instant-film transfer. But every transfer looks different in unpredictable ways. It is possible to design a digital filter to make any digital image look like an instant film transfer, but when that same filter is applied to different images, they will likely have a sameness to them that will not stand up well when compared with a set of real instant-film transfers. Furthermore, the instant film transfer is not only an image; it is an object, made of dyes and silver on good watercolor paper.

When someone looks at a photograph and instantly recognizes the technique that was used to make it look as it does, different things can happen. If the technique is difficult and is the result of a not completely controllable physical process, recognition of that fact by the viewer can sometimes add to the power of the photograph. If, on the other hand, the viewer suddenly realizes that much of the photograph's appeal was achieved by simply clicking a button in GIMP, then the opposite effect is likely. It goes from 'That's so cool!' to 'Oh, that's the Van Gogh filter.' See figure 43.17 for an example. *Cool* can easily turn into cliché if there is too much of it and it always looks the same.

43.3.4 Paying one's dues

As we have seen in chapter 40, many of the techniques for digital image processing are far easier to accomplish than their traditional film counterparts. When a lot of effort is required to accomplish a task, that fact provides something of a brake. One thinks long and hard before beginning a task that is sure to be difficult. When performing a task is too easy, one is not forced to think so much about whether it should be done, even if done well. Thus a common criticism of some examples of digital manipulation is a lack of good taste. The digital adjustment may have been

Figure 43.17. An unremarkable image from a digital camera, run through the GIMP digital filter called 'Van Gogh.' Digital filters can alter the look of a photograph in interesting ways, but sometimes the result is more cliché than captivating.

done with expertise, but it makes the photograph worse and should not have been done at all.

This basic point can hold in a larger sense, however, regarding a photographer's body of work rather than just a specific picture. A lot of time is required to master anything beyond basic film and darkroom technique. The digital analogs of many of those same techniques, on the other hand, can be mastered quite easily. With much less knowledge, one can produce photographs that are acceptable from a technical standpoint. This may mean that one has spent less time contemplating one's own work before showing it to others. One might say that digital photography in general is too easy.

But of course it need not turn out that way, and for many photographers it does not. Anything done well, including digital image processing, requires care and hard-earned skill. One can argue that the ease of many digital processes frees one to spend time on what really matters for a photograph. Many photographers who use digital exclusively have paid their dues, just in a different way than the traditional photographer.

43.3.5 Honesty

When one sees a picture and knows that it is a photograph, many expectations come with this. We often expect a photograph to be a literal record of an event—the taking of the photograph itself. We want to believe that a photographer was holding a camera and clicked a shutter while stuff was happening. So how is one supposed to

Figure 43.18. *Iceberg*, Ralph A Clevenger 1999. A composite of different photographs. As Clevenger explained upon the publication of the photograph, it would be impossible to view this scene in real life. Ralph A. Clevenger, 1999, used by permission of Ralph A. Clevenger.

look at a photograph that could in theory have been taken just as shown but was not?

As an example, there is a much-reproduced photograph from 1999, by the nature photographer Ralph A. Clevenger, of an iceberg showing both the parts above and below the waterline (see figure 43.18). A few years after the original image was released, in a classic example of an urban legend in the form of an internet hoax, a version of the image appeared that was accompanied by a completely fanciful story of how the photograph had come to be. This hoax version of the picture was widely distributed, without the photographer's permission, on the internet along with the false description that the photograph had been snapped as-is by a rig manager from an oil drilling company. Every few years the picture reappears on social media with a similar false description (Snopes 2018).

In fact, Clevenger made a composite of multiple photographs to portray a dramatic scene that could not have occurred in real life. The digital image processing used to make the composite photograph was an impressive technical feat at the time. Although the final image is digital, it was all done with scans from film photography. Clevenger was always completely honest about this; he described repeatedly and in detail that the photograph had been extensively 'photoshopped' and just how it had been done.

And so in this particular case, we really have two photographs—an honest one distributed by the photographer and a dishonest one. The pixels are identical but we see them very differently.

To my eye, the iceberg picture, as presented honestly by the photographer, is an excellent photograph. I am not at all put off by the fact that the photographer 'faked' real life. If this is somehow illegitimate in principle, then much of painting is also illegitimate. It is important to keep in mind that whatever else a photograph might be, it is still a bunch of marks on a flat surface that makes stuff happen in our brains when we look at it. No matter how representative the subject, we always 'make stuff up,' at least in part, when we make any picture. And although Clevenger's picture shows something that could not have been seen from any vantage point, it does nonetheless depict something real. It is an excellent picture partly because it manages to capture the essence of the subject (an iceberg), even as it shows a particular view that is not possible in a literal sense.

In a way, what Clevenger did with the iceberg has much in common with the long history (predating photography) of artistic illustration of scientific works. Look at the paintings of birds in the field guides of Roger Torey Peterson or David Sibley (e.g., see Sibley 2000, Peterson 2002). They look unlike any particular individual view of a bird; there is no sense of a light source, for example. Yet any one of those paintings captures the essence of what that particular species of bird looks like in a way that no single straight photograph could, which is why many birdwatchers prefer these illustrated guides to ones that use actual photographs from life.

For me, not only is the internet-hoax version of the picture dishonest about its origin, the all-too-perfect story encourages the viewer to *stop* thinking. And it does worse than tell a lie about nature; it implicitly states that neither nature nor the truth matter a whit. The natural scientist in me finds this not only dishonest but also unspeakably arrogant. Clevenger's original version, on the other hand, is *evocative.* Even with no accompanying description of how the picture was taken, one cannot see it without immediately asking questions. And it is this asking of questions that leads us to see the world in a deeper and richer way.

References

Brandt B 1948 A statement on photography *Camera in London* (Waltham, MA: Focal Press)

Cooke T and Kinneberg C (ed) 2014 *The Photography Book* 2nd edn (London: Phaidon Press Limited)

James C 2016 *The Book of Alternative Photographic Processes* 3rd edn (Boston, MA: Cengage Learning)

McCullin D 2012 The art of seeing https://www.theguardian.com/artanddesign/2012/nov/15/don-mccullin

Mutter S 1992 *Surrational Images: Photomontages* (Champaign, IL: University of Illinois Press)

Peterson R T 2002 *Birds of Eastern and Central North America* 5th edn (Boston, MA: Houghton Mifflin Company)

Sibley D A 2000 *The Sibley Guide to Birds* (New York: Alfred A. Knopf)

Snopes 2018 Giant Newfoundland iceberg https://www.snopes.com/fact-check/giant-newfoundland-iceberg/

Stroebel L, Compton J, Current I and Zakia R 2000 *Basic Photographic Materials and Processes* 2nd edn (Waltham, MA: Focal Press)

Chapter 44

The image, the object and the process

44.1 Some preliminary ideas

Part of the goal of *Photography: Physics and Art in Focus* is to explore the possible connections between physics and the physicality of a piece of visual art. Artists use the term *physicality* in a variety of ways, but it often relates to evidence of the physical in the medium itself. Throughout, I have more often than not depended on examples of photographic processes that result in a rather blatant physicality.

But the *image* is important too, especially in photography, in a sense that is wholly separate from the physical embodiment of the medium. If the physical object *Mona Lisa* were to be tragically destroyed in a fire and cease to exist, would the Mona Lisa cease to exist? Only a small percentage of the human population has ever seen the Mona Lisa, but it is also true that a rather large percentage of the human population has seen the Mona Lisa. The image is real too, apart from the physical object.

Finally, the *process* is intimately connected to both the image and the object, especially whenever either of the two has an obvious physicality. This is nowhere more evident than in the community of photographers who identify with so-called *alternative process* photography. And it is part of the motivation for my own term 'ephemeral process' (EP) as a (perhaps pretentious) name for the techniques described in chapter 25. The process is the means, and both the image and the object are the ends. An explicit focus on process recognizes the inevitable connection between means and ends—what a physicist might call cause and effect.

The following sections in this chapter can be seen as a set of *Gendankenexperiments* on the connections between process, the image, and the object in art photography. The German term translates in English to 'thought experiment,' and it is often associated with the birth of modern physics in the late nineteenth and early twentieth centuries. Sometimes one can learn much from carefully posing a question as a detailed description for doing something—even if that thing is impossible to do in practice—

doi:10.1088/978-0-7503-3703-8ch44

© IOP Publishing Ltd 2022

because the question itself is thus clarified. This technique was used famously in the pedagogy of Einstein, Schrödinger and others in explaining the development of relativity and quantum physics.

In what follows, I provide no answers, but I hope the examples call attention to, and better clarify, the questions.

44.1.1 Photographic and representational content

Some of the elements of a photograph relate directly to the straight line rays of light passing from things in the world, redirected in an organized way by the lens and arriving on the surface of the film or other light detector. I call this *photographic content.*

But photographs can and often do contain elements that are not directly related to the photographic image. There may be *flares* of light that come from reflections and scatterings within the lens itself. The detector may not respond to light in a uniform way. Other processes may have added elements after the detector was exposed, as when we solarize a negative by flashing it with light.

- **Photographic content:** elements of the photograph that correspond to rays of light from the world, redirected in an organized way to an image on the detector.
- **Nonphotographic content:** elements of the photograph that do not correspond directly to rays of light from the world, redirected in an organized way to an image on the detector.

This is not the same thing as 'abstract' versus 'representational.' I prefer the terms *representational* and *nonrepresentational,* because the term *abstract* has a more general meaning. For example, one can identify abstract elements in any two-dimensional image, as discussed in chapter 13. By representational content and nonrepresentational content I mean the following.

- **Representational content:** elements of the photograph that correspond in a recognizable way to forms in the real world.
- **Nonrepresentational content:** elements of the photograph that do not correspond in a recognizable way to forms in the real world.

Photographic content may or may not be representational. For example, an out-of-focus background is photographic, but it may represent only abstract shapes that are unrecognizable as forms in the world. Nonphotographic content, on the other hand, is usually nonrepresentational, since it does not arise from rays of light connected to objects in the world. But accidents happen, and it is possible even for nonphotographic elements to make, or at least alter, representational elements. See figure 44.1 for an example. The 'sun' in the picture is actually from a flaw in the hand-brushed emulsion.

Nonphotographic content can be introduced at the level of the detector, at the level of the printmaking or at both. Part of the appeal of alternative-process

Figure 44.1. The 'sun' in this photograph is representational, since it seems to represent a form in the world. But it is nonphotographic; it did not arise from rays of light redirected by a lens, but rather from an imperfection in the hand-coated emulsion.

techniques is that they naturally introduce nonphotographic elements in ways that are only partially under the photographer's control, and this introduces an element of chance and the possibility of happy accidents. The photographer must still, of course, have a good enough eye to recognize the many bad accidents when they see them and throw them away.

44.1.2 The picture plane

The picture plane is the flat, two-dimensional surface of the image or print itself and the marks thereon, independent of what forms in the world those marks might represent. It has long been recognized that there is a power in calling attention to the picture plane in a photograph, as a painter does with obvious brush strokes. Inexpensive plastic cameras have been popular among art photographers in part for this reason. They often leak light, for example, and this imparts a look that is as if some source of light is shining from a low angle onto a flat surface—which is exactly what happened to the negative inside the leaky camera. Another trick is to file the machined edges of the film frame in a 35 mm camera, so the image has an irregular edge and maybe extends onto the sprocket holes of the film. An even simpler trick is

Figure 44.2. *Prairie path*, John Beaver 2011.

to simply include the sprocket holes and edge markings of the film in the print itself, instead of cropping them off. See figure 44.2 for a rather cliché example.

Some photographers would intentionally damage the negative after the fact, scratching it with sandpaper or placing pieces of cellophane tape onto it. All of these techniques put into the image itself a reminder that a flat surface was used to record the image. Not only is there the image, but there is also a superimposed image of the flat negative itself.

An emphasis of the picture plane calls attention to the fact that there was a physical interaction between light and matter. This may or may not always be a good thing. But for the times when it is a wise choice, it is one of the goals of this book to offer some new tools for doing so, even if these new tools sometimes introduce the virtual into our aesthetic of the physical.

44.1.3 Control and happy accidents

It is possible to seek out processes that take away some of the photographer's control, that lead to 'happy accidents.' The idea of an accident implies that something random has been introduced (usually as nonphotographic detail) to the picture. Sometimes this is a good thing, sometimes not, but there can be value in allowing for the possibility of the unplanned. Figure 44.3 shows two examples. The image on the left was made from a transfer of Polaroid instant film onto watercolor

Figure 44.3. Left: *Pig and goat*, John Beaver 2006. **Right:** *The doorway*, John Beaver 2018.

paper; the bands at the top and bottom of the picture come from the fact that the Polaroid emulsion is larger than the print area. In a regular Polaroid print, the edges of the emulsion are masked off to form an even edge; for this picture I removed that mask before I made the transfer. It wasn't until later that I realized how well that feature works for this particular image, as it is echoed in the pattern of the coats of the two animals.

The EP photograph on the right was a long time exposure. I was unlucky enough to start the exposure right before the church service ended, so people began to stream out the door and down the sidewalk, right past the camera. Their motion made them invisible, but one figure apparently stood almost perfectly still, right in the doorway for nearly the entire ten minute duration of the exposure.

Not all accidents are happy; most are, at best, uninteresting. But with an emulsion or digital detector that produces a 'perfect' image every time, there are no accidents at all.

44.1.4 Negative versus positive

It is often said that the invention of the first photographic negative process was important because it allowed one to make many identical prints from the same in-camera original. It is true that a negative of a negative is a positive—so light can be projected through a transparent negative, in a second step, to make a positive print. But the key here is that the in-camera original is *transparent*, not that it is a negative. For it is also true that a positive of a positive is a positive, so a positive transparency can also be used in a second step to make a positive print. It turns out, however, that the early direct-positive processes—daguerreotype, ambrotype and tintype—were all opaque, so two-step photographic processes have mostly been negative-on-negative.

Apart from the ability to make multiple prints from the same original, there are two additional important features of two-step negative processes. First, the step of transforming from in-camera original to the final print allows one to make adjustments; the print can be better than the original. Secondly, and less well-understood,

Figure 44.4. Left: *The red table*, John Beaver 2013. Instant film transfer on watercolor paper. **Right:** *Valeria*, John Beaver 2007. Digital image made by scanning an instant film transfer on watercolor paper of a *negative*. The digital image was then inverted to a positive.

the process of transforming from a negative to a positive adds some psychological distance between the in-camera original and the print.

It is this last feature that I am most concerned with for this particular discussion. Consider figure 44.4. Both images use the technique of transferring peel-apart instant film directly to watercolor paper. The transfer process introduces much nonphotographic content, with all of the possibilities for happy accidents, and an emphasis of the picture plane.

The image on the left is from a digital scan of an instant-film transfer, directly from the camera to watercolor paper. Seen in the flesh, it is an intriguing (but small) image/object. The example on the right, however, was made by exposing instant film to a projection of a color film negative. The exposure was then transferred to watercolor paper in the same manner as the image on the left. But this image transfer was of a *negative*—so I scanned the negative image transfer and used GIMP to invert the resulting digital image to a positive.

One could make a large digital print from either of these images. But the result for the left-hand image would likely look just like what it is—a photographic reproduction of a small image transfer on nice paper. But for the right-hand image, the act of moving from a negative to a positive establishes psychological distance from the original (negative) image transfer on paper. It has many of the features of the left-hand image—nonphotographic content, accidental features, emphasis of the picture plane. But it is now its own thing; it no longer looks like a magnified reproduction of a piece of paper with an image on it.

The image on the right is just an image, and the original instant film transfer negative used to make it is not particularly interesting in and of itself. But on the other hand, there exists an object in a frame, on the wall of a friend's house, that looks much like the image on the left. And that object is a nice piece of Arches Aquarelle, underneath an image made of the very molecules of silver and dyes that were in the piece of Fuji FP100 film that was in my Polaroid 195 camera, when I pointed that camera at a red table and tripped the shutter. Which is best?

44.1.5 Order, complexity and randomness

It would seem that there is a range of possibilities between the completely uniform and predictable on the one hand and the completely random and unpredictable on the other. Consider the silver-gelatin emulsion that is used for traditional film photography. It is engineered and manufactured to be completely uniform in sensitivity and thickness at the scale of the photographic image. But when seen under a microscope, it is comprised of individual grains of silver halide that are arranged randomly.

But it is not so simple as that. Figure 44.5 shows four patterns. The first example is simple in the sense that it would require few instructions to describe how to reproduce it. The second example was generated by a particular mathematical algorithm—a set of step-by-step instructions—called a Julia-set fractal. It has the appearance of complexity, but the algorithm used to generate it is surprisingly simple, when expressed in the appropriate mathematical language. A kind of complexity arises from a set of simple rules.

The third example in figure 44.5 is fundamentally different from the first two; it has an element of randomness. A simple algorithm takes as an input a list of random numbers in order to generate the specific details. If one were to use the same algorithm but with a different input list, a different pattern would result. There is a sense in which this pattern is more complex than the others; to reproduce it, one would need not only the algorithm but also that particular long list of random numbers.

The fourth example was generated with the same algorithm as the third; only the list of random numbers was different. The third and fourth patterns are very different from each other; to reproduce each in detail would require a long set of instructions, and most of those instructions—the long list of random numbers used—would

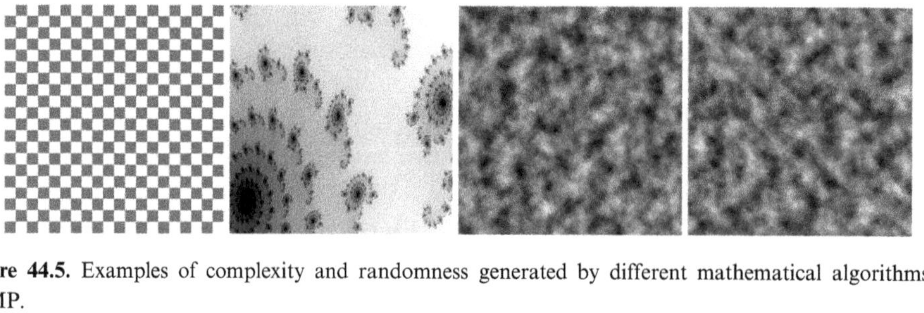

Figure 44.5. Examples of complexity and randomness generated by different mathematical algorithms in GIMP.

be different for the two patterns. But we see those differences as unimportant, so the patterns appear to us as essentially the same, even though the overwhelming majority of the details are different.

Randomness is associated with unpredictability—but the relation between the two ideas is subtle. In the last two examples in figure 44.5, the lists of random numbers were generated by a computer. Given one of the lists, one cannot predict the other, and in that sense, they are random. But given exactly the same instructions, the computer will generate exactly the same list.

Nature, however, can produce true randomness via the rules of quantum physics, as described in chapter 4. It is possible (but quite difficult) to make a device that takes advantage of the laws of quantum mechanics to generate a string of truly random numbers (e.g., see Chen 2018). But it is easy to produce a near-randomness —simply find a physical system for which the end result depends on many overwhelmingly complex critical details along the way. This is called *sensitive dependence on initial conditions*, and flipping a coin is a good example. The result of a coin flip is not truly random. If all of the forces acting on the coin were reproduced in precise detail every time, the same side of the coin would come up with every flip. But those details are so complex and, in practice, uncontrollable that the result is, at casual inspection, as good as random. Thus a complex physical system can be used to produce a seeming randomness even when the strange, mostly subatomic, world of quantum mechanics is not a significant factor.

The left-hand image in figure 44.6 shows a tiny portion of the image on the right-hand side of the same figure (changed to black and white for comparison with figure 44.5). There is certainly a kind of randomness to the complex pattern of light and dark, but that randomness comes not from computer-generated random numbers but rather from the almost unimaginable—and thus unpredictable—complexity of the arrangement of the fibers in this piece of high-quality watercolor paper.

The center image in figure 44.6 is also cropped from a tiny portion of the right-hand image. It has added elements that arose from the hand-brushing of the cyanotype sensitizer used to make the picture. Although these elements of the brushwork are unpredictable in detail, they clearly depart significantly from 'random.' In their complex connection to a physical act (human-controlled in this

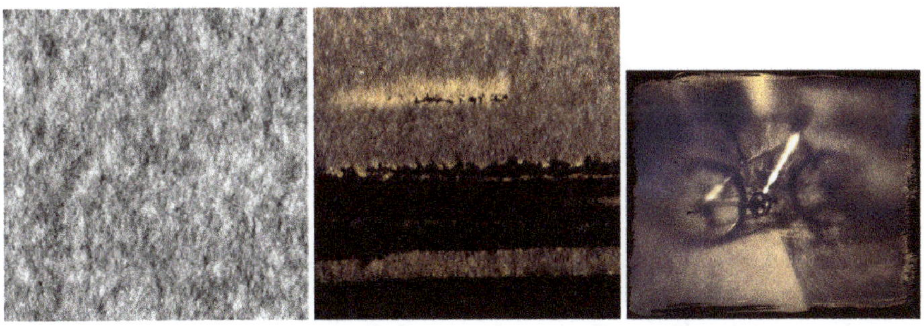

Figure 44.6. Magnified details from the cyanonegative image at far right, showing nonphotographic detail that is unpredictable but not quite random.

case), they could not have been predicted exactly as shown. But they can be understood after the fact.

44.1.6 The new antiquarian movement

Much of the history of photography has consisted of attempts to exert greater and greater conscious control and to eliminate randomness and nonphotographic content as much as possible. Digital photography has taken this process to even higher levels, using the same detector over and over to give precisely repeatable results. But for photography as an art, this is a mixed blessing.

Lyle Rexer in his important book, *Photography's Antiquarian Avant-Garde: The New Wave in Old Processes* (Rexer 2002), describes how the standard gelatin silver negative photographic process became, over the decades, more and more flawless and exactly reproducible. His book showcases contemporary art photographers who consciously seek out techniques that subvert that perfection and exact reproducibility—a modern renaissance in pre–gelatin-silver techniques from the earliest days of photography. Thus we have a *new antiquarian* movement, seeking out quirky and partially uncontrollable techniques that can provide not only a new look but also the happy accidents that subvert the inherent sameness of the photographic process.

Compared to digital photography, even the traditional silver-gelatin process has a physical elegance to it; there is a continuous flow of interactions between light and matter. The paths of light rays from the world are altered by geometry and the glass of the camera lens, to fall in an ordered manner onto a flat surface. And that flat surface undergoes an invisible physical change as a result in the formation of a latent image. The direct action of chemistry brings these hidden changes forth as a visible image, and a transparent negative is made. In the darkroom, this negative alters light passing through it from the photographic enlarger—a sort of camera in reverse. Modulated by the negative, this light is focused by a lens to form an image onto another piece of light-sensitive material. And it too is physically changed by that process, and a print—marks of dark silver on white paper—is made. Part of the appeal of the new antiquarian is that this continuous interplay between artist, medium and process is even more direct still.

44.1.7 The archival ethos

Any particular photographic print or direct positive is an *object*. But what kind of object will it be in 200 years? The idea that a photographic print at its best should be archival—relatively permanent—has long been a value among photographers. There is no official definition of archival; there is no standard answer to the question how permanent is permanent enough. But great pains are often taken to ensure that a photographic print can last as long as possible. The silver-gelatin printing process, for example, requires special materials and processing techniques to produce a print that is considered archival (e.g., see London *et al*, 2005, p 155).

Some of the appeal of alternative processes is undoubtedly the relatively permanent physical objects they make. I was fortunate enough to see firsthand (thanks to the wonderful staff of the Royal Botanic Garden in Edinburgh, Scotland) one of the 17 known copies of *Photographs of British Algae: Cyanotype Impressions*, published between 1843 and 1853 by Anna Atkins. Widely held to be the first photographically-illustrated book, these direct cyanotype photograms from the mid nineteenth century are still achingly beautiful, comparable in tone and contrast to many cyanotypes made today.

Some alternative processes do not follow this ethos of the archival. Or perhaps instead, they subvert that ethos, and therein lies the appeal. The anthotype, for example, is faded by the very light one must use to see it, and this gives it an endearing, ephemeral quality.

Archival considerations are also considered to be important for digital printing. After all, no one wants to buy an expensive art print only to have it fade or fall apart in a few years. But one can also consider the archival nature of the digital image itself. The central idea of digital storage is that the physical storage mechanism is only used to save a code—for a computer it is usually a string of ones and zeros. This means that the *physical* storage medium can be allowed to degrade up to a point—so long as the ones are still recognizable as ones and the zeros as zeros. Since the image is in the code, not in the physical medium itself, a degradation of the medium may produce no degradation at all in the image.

Of course if the medium degrades *too* much, then it may be impossible to tell ones from zeros, and then the image is lost. But unlike an image that is also its own physical medium—a cyanotype, for example—it is possible for a digital image to be immortal. Simply copy the digital code to a fresh medium every now and then, before the medium is allowed to degrade too much. If one were to do this often enough, the image could be preserved forever.

This is nothing new. Text that is hand-printed on paper, for example, is essentially digital storage; the content depends not on the precise shape of any given character, but rather on which letter of the alphabet it represents. Paper and ink degrade, so any individual piece of written text will not last forever, but one can simply recopy the text by hand onto fresh paper before the old piece becomes unreadable. This process of regular hand-copying is in fact why many ancient texts still exist today.

44.2 Examples from EP and lumen photography

The EP technique described in chapters 25 and 31 allows for many ways to investigate the interaction between the image, the object and the process. It also has a deep connection to lumen prints, and the similarities and differences are worth exploring, in light of issues raised in section 44.1.

44.2.1 EP pictures from pictures

Like many photographers my age, I have boxes and boxes of small chromogenic prints, printed commercially from the hundreds of rolls of color film I have taken over the years. EP techniques allow me to mine these prints as source material for new images by making EP contact negatives from the positive prints. The result is almost certain to be different from the original. The print is placed face-down, in contact with the accelerated enlarging paper, and the light source passes through the back of the color print for the exposure. A brief illustrated guide to making EP images in this way can be found in chapter 48.

As described in more detail in chapter 31, printed-out enlarging paper can result in a duotone color image, even though black-and-white paper is used. With the right source of light and choice of paper and accelerator, there can be an odd synergy between the printed-out scattering colors that may appear in the enlarging paper and the colored dyes used in the chromogenic printing process. The resulting color palette, although clearly related to the colors of the source print, has quite a different feel to it.

Since the exposure is only intense wherever the accelerator is applied, the original picture can be recomposed to make something new. Accidents are easy to come by, and some of them are happy (many are not). The accelerator must be hand applied to the paper in dim working light, working quickly, so it is easy to end up with something that is not quite as planned. Figure 44.7 shows two examples. For the image on the left, I intended the accelerated area to be entirely underneath the color print during the exposure, but some strayed beyond that boundary, making a stark white mark at the bottom of the picture from the directly exposed accelerated region (remember, this is a negative process). That feature may or may not be a good thing; if I really wanted to find out, I could do it again many times, and each time would be different. But its unexpected addition to the image forced me to see the picture in a completely different way than I otherwise would have.

The original print for the image on the right was very different. It had a portrait orientation, not landscape, and it was of a preserved frog upside down in the bottom of a jar. By selective placement of the accelerator, I used only the bottom portion of the original picture for this image and then turned it upside down.

The photogram (shadow print) technique can be combined with the exposure of the print. Before beginning the exposure for the left-hand image in figure 44.8, I sandwiched blades of grass between the color print and the light-sensitive paper. The colors come out as a duotone with the hues correct on the negative, which makes them backward upon inversion to a positive. They can be switched back, as in both images in figure 44.7, or they can be left reversed as in those of figure 44.8.

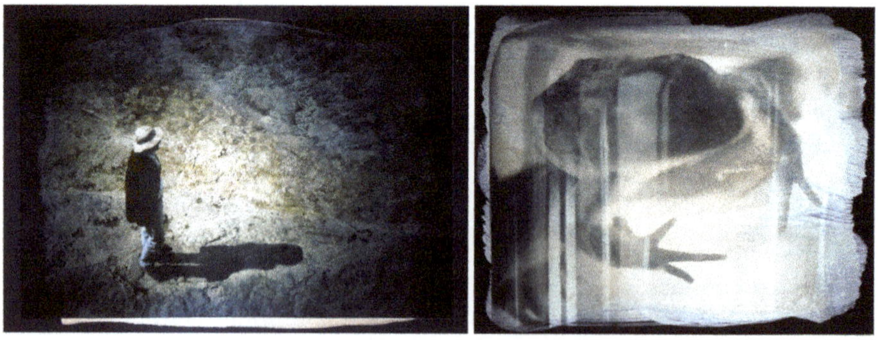

Figure 44.7. **Left:** *Doug on Mars*, John Beaver 2018. **Right:** *Preserved frog*, John Beaver 2017.

Figure 44.8. Left: *St. Mary's Church #2*, John Beaver 2018. **Right:** *Blue dandelion*, John Beaver 2016.

Figure 44.9. Left: *The wildcat*, John Beaver 2016. **Right:** *Vali and the leaves*, John Beaver 2015.

A good picture can sometimes be made from a bad print. The left-hand example in figure 44.9 was made from a disappointing picture that I had stashed away and nearly forgotten. But none of its flaws mattered for the result shown here; in particular, nonphotographic content is added to what was a featureless, overexposed sky in the original. On the other hand, a color print that is excellent on its own terms may not produce anything interesting when used as a source material for an EP image.

The original print used as source for the right-hand image in figure 44.9 also had its problems. There was much I liked about it, but the composition did not quite work. This EP image derived from it has more unity, with the simplified color palette better drawing the connection between the leaves and the clothes of the figure, and the overall shape of the illuminated region (from the hand-brushing of the accelerator) echoing the pose of the subject. Neither of those features were part of my conscious intent when I made the picture.

The color prints from commercial photo finishers typically have back printing—the date and watermark of the brand of paper used, perhaps, or maybe a code that makes it easier for them to find a negative if asked for additional prints. Since the exposure is made by shining light through the print (from the back), these features will end up in the EP image; close inspection reveals a faint example in the right-hand image of figure 44.9. The result may simply be distracting, but in certain cases, it may add to the image, grounding it by calling attention to the picture plane and the world of thoughts and words that exists outside the image.

Figure 44.10 was made from a contact print with a drawing, rather than a print. A photographic contact print made with a transparent (or translucent) handmade drawing is often called a *cliché verre* (James 2016, p 506), and the technique can be used with any contact-printing process. But there is a twist in this case; the drawing was made directly from a black-and-white photographic film negative, using the drawing-from-negatives technique described in section 44.3. So it is a print in a book (or an image on a screen if on the e-version) of a digital positive made from a scanned paper negative made from a drawing made from a film negative exposed in a camera. So, is it a photograph?

Figure 44.10. An EP image made in the same manner as those of figures 44.7–44.9, but with a pencil drawing on paper, instead of a color print, used as the source material.

44.2.2 Limited edition prints from EP negatives

The right side of figure 44.11 shows an EP photograph made from a scan of an in-camera medium format negative, exposed in a homemade camera, in the manner described in chapter 25. Or rather, it is a reproduction of that digital image on either a printed page or a lighted screen, depending on whether you are reading a printed copy or an ebook. Digital image capture is the only way I have found to effectively reproduce the subtle range of values and hues to make an image from these unfixed paper negatives. Because the unexposed silver-halide crystals have not been removed from the emulsion by fixing, the light areas of the negative reflect a lot of light, but they do not transmit very much.

Thus one cannot simply place the EP negative in an ordinary photographic enlarger and project it onto light sensitive paper to make a traditional black-and-white silver-gelatin print. But there is a way, and I have tried it; one of my most successful results can be seen on the left in figure 44.11, made directly in the darkroom from the same EP negative after it was scanned to produce the image on the right.

I was able to make a silver-gelatin print directly from the EP negative by retrofitting my enlarger for opaque projection. I removed the light source and condenser from the enlarger and placed the paper negative on top of the condenser focus bellows, instead of beneath it as ordinary film is intended to be placed. I then mounted small light-emitting diode (LED) lights inside the condenser bellows,

Figure 44.11. The image on the **left** is a reproduction in this book made from a scan of a silver-gelatin print made in the darkroom from an EP negative. The image on the **right** is a digital image from a direct scan of that same EP negative.

pointing up toward the negative and away from the lens. The light source reflected off the EP negative, instead of passing through it.

Once the exposed EP negative is washed and dried, it is far less sensitive than the developed-out enlarging paper used to make the print. Although the negative is exposed to much brighter light than is the paper for the print, it is still possible to make a print without damaging the still-light-sensitive negative too much. This is true, at least, so long as one makes only a few prints total, including the test exposures that are usually necessary to determine the proper exposure and contrast. This is made far more difficult by the fact that it is a moving target. The second print is made with a negative that has been altered by the light used to make the first print. Even the first print is made from a negative that has been altered by the light from the test exposures.

My initial experiments proved that it is possible to make at least a few silver-gelatin prints from an EP negative. For the example in figure 44.11, I succeeded in making two prints, which are at least not awful, before the negative became too fogged by the light of the exposures. I know that with more work, I could do better. But to make a series of very good prints in this way would require much planning and experimentation; it is surely beyond my modest skills as a darkroom technician. One intriguing feature, however, is that it would be—by necessity—a limited edition; the act of making the edition of prints destroys the negative used to make them. Even with the best technique, each print would almost certainly be different.

The version on the right side of figure 44.11 is only an image, while that on the left is of a physical object, made of silver and gelatin on pressed wood pulp (paper). Which is better, and why? Can one confidently answer that question without both holding the physical version in hand and seeing the digital version rendered in various contexts other than this book? Does it matter that only two physical prints exist and that I have no way to make more (because the act of making them destroyed the negative)? Does the very existence of the two (and only two) physical silver-gelatin prints render a print made from the digital version to be less valuable because it would be only a reproduction of the two 'originals?'

44.2.3 Ephemeral prints

EP techniques can be used to make positive prints, by exposing them in contact with negatives. These EP prints could be scanned to produce a positive digital image that could be printed much larger than the original negative. One can, however, instead think of the resulting EP positive print as the final image—an image that is also an ephemeral object. See figure 44.12 for an example.

I sometimes imagine, half tongue-in-cheek, an exhibition of these ephemeral image/objects of various sizes and subjects. The visitor to the gallery is greeted by nothing but blank walls and a scattering of simple boxes on floor pedestals. Next to each *sealed* box is a label—with the usual listing of title, medium, artist and date—and a framed critique of what is inside. But to see the picture, one must buy it.

It is possible to fix an EP print to make it permanent, and an example can be seen on the right side of figure 44.13, with an unfixed example from the same negative on

Figure 44.12. An EP print made from a digital negative made from a scanned 35 mm film negative.

Figure 44.13. Left: Unfixed EP print made from a color digital negative. **Right:** A similar print, but after fixing. The exposure was greater than for the left-hand print because the fixing process reduces the density considerably.

the left. We could call this an 'accelerated, unfixed lumen,' instead of a 'fixed EP.' In all of my tests, the image hues disappear as a result of the fixing process, although some papers retain colors that are unrelated to the colors in the negative. Fixed EP prints are better suited to black-and-white negatives. These have their own appeal, and it is an easy technique if one wants to produce a positive print with the appearance of a hand-brushed emulsion.

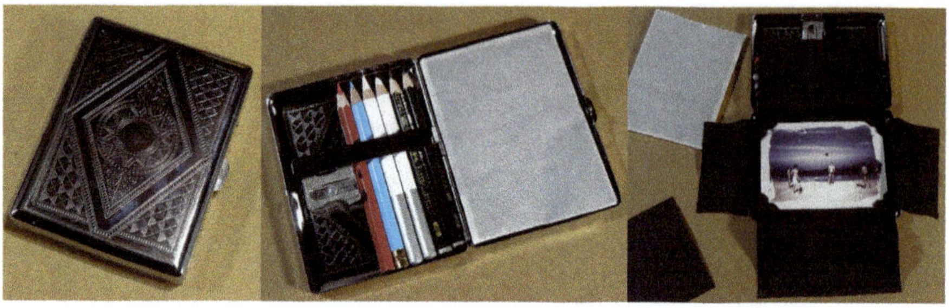

Figure 44.14. A protective case for a still-sensitive-to-light EP print.

Figure 44.14 shows one solution to displaying a still-light-sensitive EP print. It is small, framed inside a closed case like an eighteenth century daguerreotype or tintype. Meant as a gift, the extra space in the case includes drawing materials, in the event the recipient dislikes the photograph.

44.2.4 To …, or not to …

There are many possible variations on the theme of EP, lumen process and combinations thereof. We can choose to

- Use an accelerator or not,
- Expose with a positive to produce a negative or expose with a negative to produce a positive,
- Produce normal color hues or inverted color hues,
- Scan the result to make a digital image or let the exposed enlarging paper be its own image/object or
- Fix the printed-out enlarging paper to make it permanent, or leave it unfixed.

Which combination is best? It should be clear that I believe there is no correct answer, in general. Rather, we should look at these possibilities as artistic tools—some more appropriate, some less so—for whatever task may be at hand.

Consider figure 44.15. The same EP photogram has been simply scanned (left), inverted to a positive with the contrast stretched (center) and fixed (right). The image on the left is of a physical object that looked like this. But the original object would need to be kept in darkness most of the time. It would be changed by the act of viewing it, because it is still sensitive to light. For the center image, subtle variations in hue and value appear that were not obvious in the original photogram. But this is only a digital image. The right-hand image is now a permanent image/object, but the act of fixing it reduced the subtle range of hues and tones to a much flatter, near monochrome. And because of the fixing process, the image/object reproduced at the far left no longer exists; there is only the digital image of it in this book. Which of the three versions is best, and why?

In figure 44.16, I show several iterations of the same photogram concept, both with and without a hand-brushed accelerator. For each I compare the result of

Figure 44.15. Three versions of the same original EP photogram exposure. **Left:** A faithful scan of the original, unfixed photogram. **Center:** The previous digital scan inverted to a positive and with the contrast appropriately stretched. **Right:** A faithful scan of the original photogram, but now it has been chemically fixed so it is no longer sensitive to light.

Figure 44.16. The same photogram concept executed eight different ways, all using printed-out enlarging paper. For the **top** row, I used no accelerator, but for the **bottom** row, I hand brushed vitamin C accelerator onto the paper before exposing. From **left** to **right:** (1) after exposure, (2) after fixing, (3) enhancement of prefixing scan and (4) enhanced negative version of scan.

fixing, not fixing and making contrast-enhanced digital images that are either inverted to a positive or left as a negative. Which is best, and why?

44.2.5 EP accelerator transfers

Another EP approach is to use objects in the world to transfer accelerator to only selective parts of the enlarging paper, which can then be simply exposed to light with no object in place. Wherever the accelerator was applied, it turns dark. And so, for example, the accelerator can be brushed onto a leaf, and the leaf pressed onto the paper. If done carefully, the accelerator can be transferred to the paper according to the shape and texture of the leaf. When the paper is held up to the light, the pattern of transfer appears.

An intriguing variation is to mist an abandoned spider web with accelerator, and then let the web stick to the enlarging paper. Since the web is sticky, it efficiently transfers the accelerator to make a direct and unique print of the web (which is destroyed in the process). Since I usually do this to the webs of orb weaver spiders (which typically build new webs each day), I call them *araneitypes,* after the Latin name for this family of spiders (araneidae). Three examples can be seen in figure 44.17. The araneitypes can be made permanent with fixer; even though there is an inevitable loss in density, the high contrast of the technique yields satisfactory results. Depending on the paper used, the colors tend to shift toward the brown, but that too can be appealing. But I usually leave mine unfixed to better draw a connection to the ephemeral nature of the spider web.

44.3 Drawing from negatives

Figure 44.18 demonstrates a technique I call *drawing from negatives.* A single frame of a 35 mm black-and-white *negative* is mounted in a glass slide mount. In a darkened room it is projected with a slide projector directly onto a piece of drawing paper. Graphite or charcoal is then used to make the light areas of the projected negative dark, until all of the picture appears the same dark value. Thus one draws in until the picture disappears. When the slide projector is turned off and the room lights turned on, a positive appears.

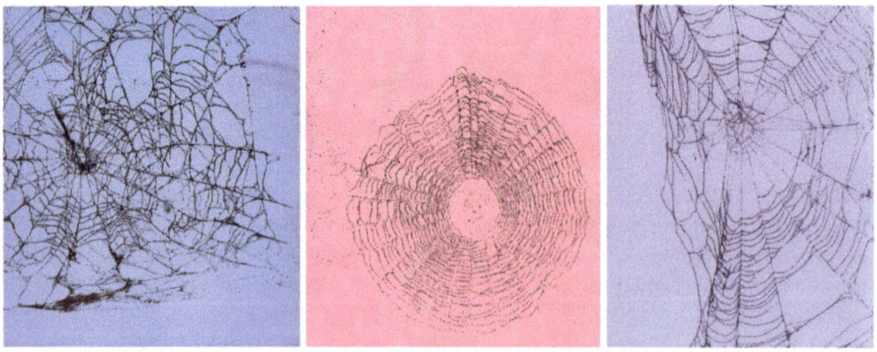

Figure 44.17. Araneitypes: accelerator was sprayed onto spider webs, and the wet webs stuck to the paper immediately before exposure to sunlight. The background colors are a result of the different varieties of enlarging paper used.

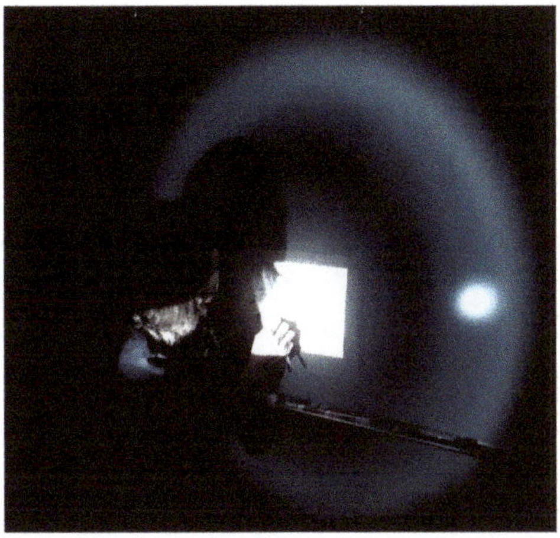

Figure 44.18. Artist Judith Baker Waller drawing from a negative. The projected black-and-white film negative is darkened in by hand, wherever it is light, resulting in a positive image.

I have done many of these (and held workshops to teach others to do it), and the experience is always odd. At one level it is very tedious. While the projector is turned on, it is almost impossible to tell how the picture is progressing, since the areas already darkened with the pencil simply merge together with neighboring areas that are dark because the negative is dark there. One starts with an image of a negative, and one tediously, bit-by-bit, obliterates it with a pencil. It is always tempting to turn off the projector to check one's progress, for only then can the actual drawing be seen.

Yet, one can make choices. For the example in figure 44.19, I spent a lot of time on the human figure, carefully coloring in the tiniest details in the light parts of the negative. For the trees, on the other hand, I made the marks very quickly and with a lot of gesture, trying to darken the light areas as best I could with fast movements of the pencil, but I didn't go back and touch up every little detail. For the field in the foreground, I made marks that were only in the same spirit as the original negative (which consisted of a sharply focused field of dandelions). I made quick gestures with the pencil in an attempt to mimic the ball-shaped flowers and vertical stems and to agree with the broader areas of light and dark, but I did not try at all to line up my dark marks with the specific light areas of the negative. If I had tried to do that, I would probably still be there to this day, tediously drawing.

Clearly, I could not make a second identical 'print' from this same negative, even if I tried. In making a second print, I could make different choices regarding the making of marks, with the intention of producing very different results. Figure 44.20 shows two drawings, both made by the artist Judith Waller from one of my negatives. This particular negative is such that it is very difficult to tell, simply by looking at it, what the picture even represents. Waller made the second drawing without having seen the result of her first drawing. That is, I made her turn away while I turned off the projector, removed the drawing and then attached a second piece of paper.

Figure 44.19. A negative was projected onto a piece of drawing paper, and the light areas darkened in by hand. Is it a drawing or a photographic print? Does it matter?

Figure 44.20. Two pictures drawn from one of my negatives by the artist Judith Waller. (Judith Waller, 2019, use by permission.)

Waller made two drawings, each with different choices regarding how to make her 'mark' on the paper. But she did not know what the drawings even represented as she made them. So are these drawings or prints? Was she simply an underpaid and overworked, sometimes complaining but often cheerful, self-propelled and expensive-to-feed inkjet printer? Or was there more to it than that?

Part of the central idea of 'the print' is that once the negative (or image file or printer's block) is chosen, one can no longer decide which parts of the paper will be light and which parts will be dark. The process of making the print is mechanical, so it can be done again and again, producing multiple versions of the same picture. A painter, on the other hand, gets to look at the whole picture at once and alter their decisions based on the work taken as a whole.

But these thought experiments show, I think, that it is not quite so simple. The matter of human agency is important, even if it is at the level only of the mark and not regarding the entire picture. It matters that a human mind is controlling a hand that is making marks and that that mind is responding to what it has done, thus altering future marks. This happens only a little when one makes a straightforward silver-gelatin print in the darkroom, and it happens not at all when one clicks 'print.'

44.4 The *camera stupida*

The *camera lucida*, as originally conceived by William Hyde Wollaston (see section 32.5), is in many ways misnamed. The name means, literally, 'light box.' Clearly, the name was a play on *camera obscura*, the 'dark box' to form images widely known since the 16th century that evolved into the modern camera. But in Wollaston's design, there is no camera, no box.

To the contemporary ear, the name would seem to imply that when one uses a camera lucida, one is basically 'taking a picture'—only the image must be drawn to be captured, since there is no light-sensitive material to record it directly. This is how I imagined the device, when I first heard it described in *Herschel at the Cape*—the annotated letters and diaries of photography pioneer John Herschel, describing his intellectual exploits between 1834 and 1838 in South Africa (Evans *et al* 1969). I pictured it as a box with some arrangement of lenses and beam splitters, which one could carry around and point like a camera. When looking into the box one would see, simultaneously, the paper and the image, and thus one could simply mechanically trace over the image.

For anyone who has used (or tried to use) a real camera lucida[1], however, it is clear that this is wishful thinking. The traditional camera lucida is a drawing aid more than it is a device for 'capturing an image' by drawing. For one thing, it must be set up and clamped to a drawing table, and the paper and device must be carefully positioned before one can even begin to draw. Secondly, it is difficult to see both the image and the drawing paper simultaneously because in most circumstances, the lighting is vastly different for both. One usually compensates for this by holding some kind of screen in the non-drawing hand, to cast just the right shadow onto the drawing paper so it does not drown out the virtual image superimposed upon it, or by flipping a filter in or out of the light path when needed.

Another difficulty is that the drawing paper is only a foot or so away, while the subject may be at effective infinity. Thus one cannot necessarily focus the eye on both simultaneously. This is particularly troublesome if one is, like me, not young and needs reading glasses to focus on a nearby object.

The camera lucida is not a camera-by-drawing. It is really a device that helps an artist, someone who already knows how to draw, get the proportions, locations and angles exactly correct in their drawing. The typical use is to employ the camera lucida to mark the locations of key points in the drawing—the ends of lines, points to mark the rough borders of objects. Then the drawing is drafted in the conventional

[1] There are at least a couple of companies that manufacture and sell modern versions.

way, using these markings as guides. Two of photography's pioneers—John Herschel and William Henry Fox Talbot—both experimented with Wollaston's camera lucida. Herschel made many excellent drawings with the device, but Fox Talbot's frustrating attempts were instrumental in his quest for what became some of the earliest photographic processes (James 2016, p 118).

I was disappointed when I first sought out the details of the design, as there was a practical reason for my wishful thinking. I have had no training in drawing, and neither have most of my photography students. But I wanted a device that would give them (and me) a visceral connection between an image and its permanent rendering. Toward this end, I developed a variation on the camera lucida that really is somewhat like what I ignorantly first imagined. The result borrows from both the camera lucida and the camera obscura and adds a few new ideas of my own. It is simple and inexpensive to build, and because the end product is in some ways a bit silly, I dub it the *camera stupida.*

1. I have put the camera back in the camera stupida. The device is built around a box that one can put on a tripod or rest on a tabletop and point at the subject like an ordinary camera.
2. It is designed for one size of drawing paper, and other design features necessitate that it is a small piece of drawing paper. The drawing paper is analogous to the film format in any particular camera. My prototype model uses sheets of paper that are 4 × 4 inch square.
3. The small format means that one can quickly sketch the drawing, because there is not that much to draw, so the intent is completely different. One is not making an art drawing from life. One is using the device and a pencil to quickly capture an image in a format not unlike the picture spit out by a Polaroid SX-70 of decades ago. The result is more likely to appear pinned under a magnet on the refrigerator than framed, wired and hung on the wall of an art gallery.
4. Establishing a fixed format and putting it all in a box allows other technical problems to be solved. In particular, the relatively dimly lit box (open only on the side facing the drawing hand) means that a simple piece of thin glass or clear plastic can be used in place of an expensive, partially silvered beam splitter, and *polarization* can be used to make the brightness of the subject image adjustable by simply rotating a circular filter. The fixed format also allows one to use inexpensive lenses to ensure that the images of both the nearby drawing paper and distant subject are in focus simultaneously. With this arrangement of lenses, one sees a relatively wide-angle view of the subject superimposed upon a magnified view of the drawing paper.
5. The entire device is easily portable. Ideally, it would close up into its own box with all of the optical parts stowed safely inside. There would be room left for spare precut sheets of drawing paper, as well as pencils, erasers, etc. The box would close with bi-parting doors, which provide adjustable sun-shade 'wings' when the device is in use.

Figure 44.21. The working prototype for my 'camera stupida' design. One looks through the lens at the top and sees a virtual image from the forward direction superimposed upon a direct (but magnified) view of the drawing paper inside. Both images are simultaneously in focus, and one can adjust their relative brightnesses by rotating the polarizer over the viewing lens.

One simply puts my camera stupida onto a tripod and points it at the subject. Peering through a small, simple converging lens at the top of the box, one sees both the drawing paper and the subject, and both are in focus[2]. As one traces over the image, a polarizing filter over the lens can be rotated with the other hand to brighten or dim the projected image. The image can be dimmed for relatively bright parts of the subject, and then brightened for relatively dim parts of the subject, with a simple twist of a dial. If the light is too dim inside the box to see the paper well, the built-in LED lamp can be turned on. See figure 44.21 for images of my working prototype, constructed easily with simple and inexpensive materials. More details on the optical design can be found in appendix F.

I find the camera stupida very satisfying to use if my goal is to quickly capture the essence of an image. I usually get the best results when I try to make a particular drawing as fast as I can; usually, only about five minutes is required to make the 4×4 inch sketch. I find this exercise enjoyable, but also it forces me to think hard about just what it is in a given picture that makes it a picture. Anyone who ever makes any drawing from life must do this too. But it is my contention that, for the photographer, it is very instructive to draw while seeing like a camera sees. See figure 44.22 for two examples of drawings I made with the prototype device.

The camera stupida brings up the issue again regarding what is the distinction between a drawing (or painting) and a photograph? What if your best friend were to lead you blindfolded to a camera stupida they had already set up and pointed at the subject. You then look only through the device and not at the world, and you carefully and exactly trace what you see. Did you make a drawing? Or did your best friend take a photograph—using you as the light detector?

[2] This is true, but only within limits in my design, which is optimized for relatively distant objects. An acceptable focus is achieved for objects more distant than about 1 meter.

Figure 44.22. Two drawings made with the camera stupida. Each 4 × 4 inch sketch was made in about five minutes and completed entirely while I was looking through the camera stupida. The traditional camera lucida was primarily used only to mark key points in the drawing in order to establish scale, proportions and angles; the drawing was then drafted by eye in the usual way, using these marks as guides.

References

Chen S 2018 Quantum mechanics could solve cryptography's random number problem https://www.wired.com/story/quantum-mechanics-could-solve-cryptographys-random-number-problem/

Evans D S, Deeming T J, Evans B H and Goldfarb S (ed) 1969 *Herschel at the Cape: Diaries and Correspondence of Sir John Herschel* (Austin, TX: University of Texas Press)

James C 2016 *The Book of Alternative Photographic Processes* 3rd edn (Boston, MA: Cengage Learning)

London B, Upton J, Stone J, Kobré K and Brill B 2005 *Photography* 8th edn (Englewood Cliffs, NJ: Prentice-Hall)

Rexer L 2002 *Photography's Antiquarian Avant-Garde: The New Wave in Old Processes* (New York: Harry N. Abrams)

IOP Publishing

Photography (Second Edition)
Physics and art in focus
John Beaver

Chapter 45

Four photographers and a musician

I here consider five artists whose work I admire, in light of issues raised in section 44.1. All of these artists do work that is outside the commercial mainstream of photography, and they employ techniques that highlight the interplay between the image, the object and the process.

Throughout *Photography: Physics and Art in Focus*, I have mostly illustrated the concepts with my own photography. The artists in this chapter take it to another level. The technical expertise, attention to detail, and innovative use of unusual and experimental processes is not, for these five artists, a gimmick—it is essential to their art.

45.1 Ky Lewis

Ky Lewis (http://www.kylewis.co.uk, @kylewis1) is an artist and photographer who lives in London. With prior experience in design and illustration, she now uses analog photographic processes. Often working with landscapes, her work focuses on the environment and cycles of life, death and decay. Lewis uses a variety of self-made and modified devices to capture her images, including pinhole and large format cameras, as well as camera-less techniques (all quotes are from http://www.kylewis.co.ukhttp://www.kylewis.co.uk, April 2022):

> …many are made of simple waste items such as old tins, film pots, boxes and the like, from such disposable items come dreamy images, resonant of past memories.

She is a prolific presenter of workshops, sharing her experience and knowledge of traditional and alternative photographic processes. Lewis's multi-award-winning work has been published and shown widely in solo and group exhibitions in the United Kingdom, United States, Poland, Spain, Germany, India, Russia, Austria and Canada. Lewis's process comes in at both the time of recording an image with a

doi:10.1088/978-0-7503-3703-8ch45

© IOP Publishing Ltd 2022

camera and later in making unique hand-manipulated prints from the negative:The

> emotional response to a specific site is what is important to me, feelings based on light, fleeting moments barely noticed, the essence of a place held at the edges of awareness. I have a strong interest in *wabi-sabi*, and the ethos of imperfection, history, growth, decay, even smells influence my work. Using a variety of processes to achieve my work, I manipulate light and chemicals, working with my subject matter both on site and in my darkroom.

Figure 45.1 shows images from *The Blackberry Path*, part of a much-larger series called *Ness*. Here Lewis has used the chromoskedasic sabattier process (section 35.7) to make multiple unique prints, all from the same large-format negative. Each handmade print evokes its own universe from this same path she has walked many

Figure 45.1. *The Blackberry Path*, from the series *Ness*, Ky Lewis 2021 (used by permission). Unique chromoskedasic sabattier prints from the same large format negative, made with a pinhole camera.

times. A photograph from her *Ness* series was recognized in 2021 for *Outstanding Work* from the prestigious Denis Roussel Award.

> …'Chance–Mistake–Repetition' …is definitely the core to my practice, especially with the way I use my process. The emotive response to each print is different despite it being fundamentally the same image, the same as the landscape captured the change from moment to moment, it is the same in the darkroom. Muscle memory, emotion reflection, it can all change the print. Each print is a unique iteration, a discourse with that particular landscape, an understanding of its form and what it means to me. The decisions made regarding the choice of the final chosen image are primarily my gut response and an aesthetic balance.

> *The Blackberry Path* was really one of the key images for the 'Ness' series that brought about its reason for being. The Blackberry Path references the walks I had with my late father. We would go on hunts for blackberries, up on what used to be the cliffs, now softened with time and erosion. Most of these places have fallen to the sea, but this reminded me so much of the adventures we had all those years ago.

45.2 Brittonie Fletcher

Brittonie Fletcher (MFA, brittoniefletcher.com, @etherwaves) is an artist, educator and curator from Boston, Massachusetts (USA), based at the time of this writing in Edinburgh, Scotland. Her work is political—mixed with personal ideas of the environment, resources and community. Fletcher works in many forms of visual art but is best known for her alternative photography. She is an enthusiastic and passionate educator working primarily in North America and the United Kingdom, as well as online.

Brittonie's work has been exhibited internationally and published with Focal Press/Routledge and Random House as well as other publications online and in print. Her work can be found in public collections such as the Preus Museum: Norwegian National Museum of Photography and the National Library of Scotland. She has received awards, nominations and fellowships for her work, including the seventh Julia Margaret Cameron Award for Women Photographers, a nomination for the Royal Photographic Society's '100 Heroines' and an Arctic Circle Fellowship. Fletcher has been a repeat curator alongside members of the Wellcome Trust for Imperial Colleges Art of Research and founder and director of the ACTINIC Festival.

Known for her experimentation and mastering of difficult, nearly lost historical techniques, it is still the art that matters for her in the end. Fletcher uses these experiments as ways for the process to be part of the art and appropriate to it. She is

Figure 45.2. *Folly* from Material Threat ©Brittonie Fletcher, Toned Cyanotype 2020. Used by permission.

deeply interested in material culture. One can see this in her description of her process in making the images in figures 45.2 and 45.3, from her work-in-progress series, *Material Threat* (quotes from https://www.brittoniefletcher.com/material-threat, June 2022).

> Purple is an ancient and globally recognized colour of power (it is also a colour of passion, independence and mourning) from Asia to South America. It is the most rare colour to occur in nature and the most powerful of the visible light spectrum. Purple served to preserve the 'natural order' and political power of the ancient Elite. In the Late Bronze Age or early Classical Antiquity 10 000 crushed predatory sea snails were needed to dye just one cloak. Phoenicia translates as 'land of the purple' and lends its name to the word 'phoenix' the mythological bird that rises from the ashes. The word 'tyrant' comes from the ancient greek word 'tyrannos' for one who rules with absolute power unrestrained by law... I wonder what they were wearing.

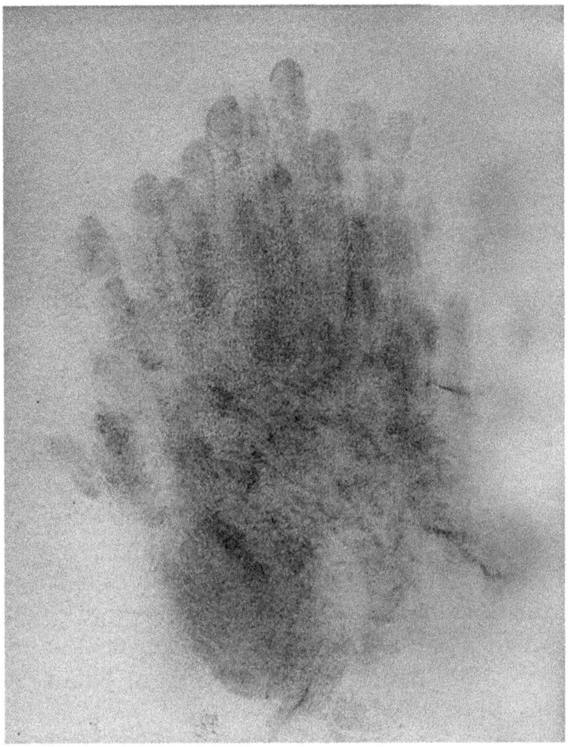

Figure 45.3. *Bruise No. 2 (Lapse)* from Material Threat ©Brittonie Fletcher, Toned Cyanotype 2020. Used by permission.

In ancient Celtic areas archaeologists have found middens purely composed of the crushed shells. The dyes were supplemented with that of Orchil made from lichen—an alleged symbiotic partnership between fungi and cyanobacteria which typically is a good indicator of air lacking nitrogen pollutants. Tea fields are often fertilized with ammonium—a nitrogen rich resource.

I built a fire with a friend to watch it burn. Afterwards I dragged the charred remains to the beach and built a henge which I photographed with my large format camera. I also used a piece of burnt wood to draw, my hands covered in soot—using them to grind and rub pieces of charcoal into the paper. Later, slapping and hitting the paper as hard as I could past the point of hurting until my hands were clean(er), creating artifacts of process. Energy transference.

A bruise is a visual sign of impact, some sort of damage, internal bleeding. The colours are blue, purple, brown, yellow and sometimes black. My prints are toned in the colour-scheme of bruising using store-

> bought and hand-made teas, wind-blown found lichens, naturally occurring and sustainably sourced ammonia and urea, calcium carbonate, red wine turned to vinegar and the tiniest amount of high carbon impact muriex.
>
> Power dynamics, and consent, consequence management and sustainability. Relationships. Resources. Abuse. Waste. Industrial production/personal consumption. Destroyed environments/displaced life. Caught up in a moment without thinking about the longer term effects. We don't need to burn everything to the ground.
>
> Material Threat is an active and unresolved WIP [work in progress].

45.3 Chrystal Lea Nause

Chrystal Lea Nause (linktr.ee/chrystal.lea.nause, @chrystal_nause_photography) is a New Orleans, Louisiana–based award-winning visual artist and anthropologist who does both commercial and art photography. Her photography has been juried into many exhibitions, both online and in galleries in Illinois, Vermont, Oregon, Kentucky, Tennessee, Missouri and China.

Her fine art photography usually starts with a traditional 4 × 5 large-format view camera. She uses ordinary large-format film and (no longer made) Type 55 Polaroid instant film. She includes the edges of the instant film negative in her prints, and the image detail there is overlaid with many imperfections. This not only provides an interesting frame but also draws attention to the picture plane, a fact noticed by many other photographers who used (and now very much miss) this film (see O'Brien and Waits 2011, for another good example).

But Nause sometimes makes atypical use of this film. For the image in figure 45.4, the slight tilting of the photographic image and the irregular corner resulted from a flaw in the instant film processing as it was pulled through the metal rollers of the special film holder—a not-uncommon occurrence with this type of film. But also, the film partially and unintentionally solarized. In this case, both of these accidents are happy ones, as she explains (Nause 2016),

> I think that the image of McAndrew Stadium is a great study of transitory places, since it was demolished less than a week after the image was created. The negative solarized spontaneously, so I was never able to do a direct print of it. I'm guessing that the solarization was a by-product of the film's age.

Nause uses traditional wet-chemistry darkroom techniques to make, directly from her negatives, archival, toned silver-gelatin prints of the finest quality. But this particular negative was essentially unprintable in this way, so a print of this negative

Figure 45.4. Chrystal Lea Nause, *McAndrew Stadium, Southern Illinois University. Carbondale, Illinois*, 2011, used by permission of the artist.

is available only by way of the intermediary of digital scanning and printing. She has this to say about the image in figure 45.5 (Nause 2016):

> I made the photo about 10 minutes before my train left the station. From there, I was rushing back to my apartment to gather all of my possessions and begin the less-than-glamorous journey back to Rome for my departing flight. The negative sloshed in the Sodium Sulfite fix and unfiltered water for about three days. All of the chemical deterioration and scratches are a result of these 'rough' travels. The perfectionist in me cringed at my mistakes—but I've really come to love that this image alone can represent my exploration of southern Italy: traveling in rough conditions, missing connections, sleeping on my backpack in empty stations, exploring the rural mundane of the south and really getting to know myself as a photographer.

45.4 Erin Woodbrey

Erin Woodbrey's visual art (http://www.erinwoodbrey.com, @erinwoodbrey) includes a range of media—sculpture, printmaking, arrangements of found objects, photography and time-based media. She often works in the installation format,

Figure 45.5. Chrystal Lea Nause, *Vecchia Stazione Ferroviaria. Manfredonia, Italia Old Train Station. Manfredonia, Italy*, 2007, used by permission of the artist.

with many such pieces working together in a single room, often with both two-dimensional (2D) and three-dimensional elements. Based in New England, she uses her art in part to explore the interrelated aspects of the current global environmental crisis.

Woodbrey's art has been shown widely in the United States and many other countries, including the United Kingdom, Germany, Iceland, Denmark, the Netherlands and Estonia. Erin Woodbrey studied at the School of the Museum of Fine Arts at Tufts University (BFA, 2007) and the School of the Art Institute of Chicago (MFA, 2014).

Woodbrey's work is wide-ranging and evocative, and she dives deeply into her physical materials to give them an intimate connection to such intertwined themes as time, material, nature, the body, ecology, architecture and the tension between permanence and transience. She often draws ideas from literature; The Carrier Bag Series, for example, draws its name from a 1989 essay, *The Carrier Bag Theory of Fiction*, by Ursula K LeGuin (Le Guin 1989, pp 165–170).

When taken as a whole, her work is like a universe. Each piece provides hints to all of the others, and this gives a refreshing solidity and depth to her 2D work, even for that flattest of photographic techniques, the photogram. Figure 45.6 shows (at left), one of her anthotype photograms, made by allowing a spinach-based dye to fade in sunlight. At right is a view of one of her installations of found objects: *The Continuing and Spreading Results of an Event or Action (Carrizozo, New Mexico)*.

She has interesting things to say about her process (Woodbrey 2022).

> I often work on different things simultaneously and it can take years to finish a body of work. (I also have a lot of material experiments that just take time to develop.) Sometimes I'll pick something up and work intensely on it, move on to something else, and then pick up the other piece again. Kind of like accumulating different kinds of time at the same time. I somehow think that creates a kind of continuity between different pieces/bodies that may seem quite different from each other but are held together by a similar idea/thought/experience.

Like her installation work, Erin Woodbrey's photographic objects seem almost as if they are artifacts. Look at the photographic prints in her series Time Beings; figure 45.7 shows one example. The hands in *Time Beings (Thank You)* are literally dirty, but so is the gum transfer process she used to make the picture.

Figure 45.6. Left: Erin Woodbrey, *Fragments of Clear and Green Glass Organized by Color and Opacity from the Center to the Edge*, 2021, Shadowgram Anthotype, spinach on paper, 40.6 × 35.6 cm (used by permission of the artist). **Right:** Erin Woodbrey, *The Continuing and Spreading Results of an Event or Action (Carrizozo, New Mexico)*, 2021. Found objects and fragments of plastic, paper, glass, foam, aluminum, tin and organic matter, variable dimensions (used by permission of the artist).

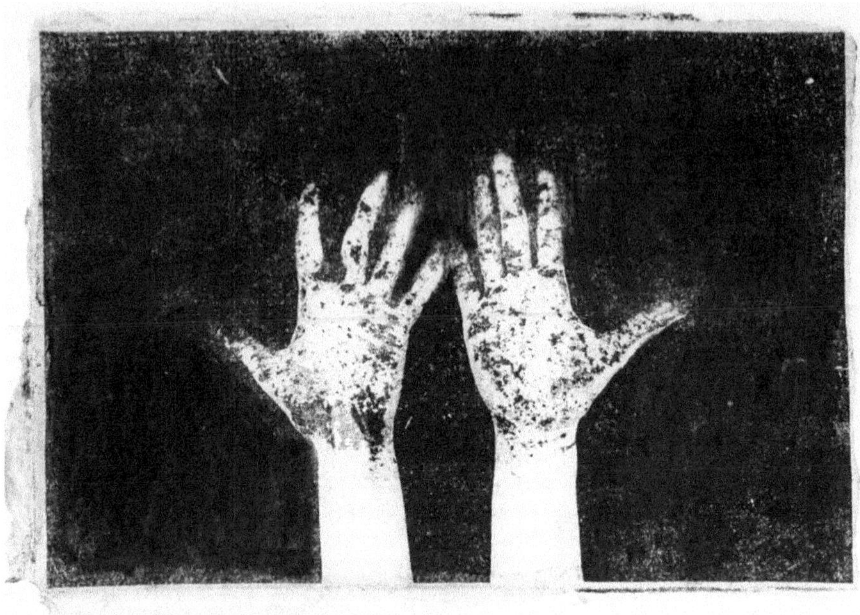

Figure 45.7. Erin Woodbrey, *Time Being (Thank You)*, 2020, Gum transfer of images derived from video stills on handmade paper. Unframed: 30.5 × 43.2 cm (used by permission of the artist).

45.5 Matt Turner

Matt Turner, cellist and pianist, teaches in the Jazz & Improvisational Music Department at Lawrence University, in Appleton, Wisconsin (USA). Improvising in a myriad of styles, Turner appears on over 100 recordings and has performed in Canada, Europe and Asia. Turner is a Yamaha and D'Addario Performing Artist and has dozens of pieces published with Carl Fischer, Alfred, Latham and Alliance. He currently performs with Bill Carrothers, Randy Sabien and Hal Rammel.

Widely regarded as one of the world's leading improvising cellists, Turner completed his undergraduate studies at Lawrence University and his Master of Music degree in Third Stream Studies (now the Contemporary Improvisation program) at the New England Conservatory of Music, where he studied with Dave Holland, Geri Allen and Joe Maneri and where he was the recipient of a Distinction in Performance Award.

I first met Matt in 1998, while working on the first of many collaborative ventures with artist Judith (Baker) Waller. *Keeping Time: A Sight and Sound Poem* was a live planetarium multimedia performance piece; it was Judith's idea, and her original visual art. Matt improvised cello, performing live and in the dark, with the projected visuals surrounding him and the audience. I was the technical director, having previously worked in a planetarium.

I profile Matt Turner in this chapter because of his collaborative (see figure 45.8) and improvisational spirit and because my own development as an artist parallels my exposure to Matt's approach to his own art. Many of my ideas about art that

Figure 45.8. *Karmadog*, Matt Turner (cello, left) and Greg Pagel (keyboard, right) in a live improvisational performance (photograph by John Beaver).

form the subtext of this book are greatly influenced by my long-time collaborations with Waller and Turner. But it is doubly appropriate because Matt has often welcomed collaborations that integrate his music with the visual arts. I interviewed Matt on 6 May 2022, and I here quote some of his insights, edited from that interview.

Turner is also a teacher. I am fortunate to have been present for several performances by students in his improvisation course, who make up the Improv Group at Lawrence University, which he directs. They improvise live as an ensemble, usually with some outside nonmusical impetus that provides a framework—a simultaneous screening of a silent film, for example, or an improvised performance by their colleagues in the dance department. Many of his remarks in our conversation reflect his role as a teacher.

Regarding improvisation, a lot of people think it's a free for all. And it can be, but it usually doesn't sound good that way. It's like throwing things at the wall, hoping something sticks. But even then, it could be good—and that's the dilemma.

With my students, I often try to set some kind of structure. It's like an invisible framework, but you're allowing yourself to be flexible within the framework. Maybe you're also avoiding something comfortable—tonality, major triads, whatever. We are always trying to modulate, going to an uncomfortable place and embracing the discomfort. Experienced improvisers are going for that.

Solo live improvisation is the most difficult thing to do because you are so vulnerable. The temptation is that you fall into what is comfortable and that you know people will like. But instead, maybe I find a particular spot on the cello or piano, I just don't know what's going to happen, and I just throw myself into it. Sometimes it works really well, and sometimes it's a disaster, and you just get out as best you can.

The best moments are when you discover something that is completely new, and you work on it to create something beyond what you discovered. Performing live, you are taking those risks. It's like an onion. You know this particular layer is working, but can you then peel it and find something deeper.

It is really about being able to listen to yourself. When you are improvising as part of an ensemble, you have to also listen to the others, and it takes training. I often stop my students when they are performing and ask, 'So what did you do one minute ago, and why did you do that?' or 'What was John playing two minutes ago?' For the beginning students (the more experienced students know this is coming), the answer to both questions is often 'I have no idea.'

When improvising to visual art, you can look at it and use it as a score, in a literal sense. For me, I try to avoid that; instead I try to see what space it puts me in—it's hard to describe. I see things when I improvise, it's almost like a film … colors.

I like to use the word 'avoid.' I often avoid what people think I would play. I do this with my students too. Say we are improvising to a silent film, and Buster Keaton is running around crazy, and you expect the music to do [imitates sound], and that's probably what the pianists or organists did at the time. But I tell my students, 'Now I want you to avoid playing what you would expect to hear; instead play something sustained, and slow and morose.' My students say 'Ugh,' but then when they do it, they realize 'oh my gosh, this changes *everything*.' It changes what you see.

You have all those years of being taught this is how a cello sounds. Sometimes I bow on the tailpiece. It's a cool sound because it has this low hiss, but I can't control everything. Sometimes when I'm bowing between the bridge and tailpiece, there's not a lot of control there. If I play really close to the bridge where all the harmonics come together, if you experiment with that, moving between a heavy and light bow, all these harmonics start popping out, and all of a sudden there is something I've never heard before. It just *explodes*. I've used a hairless bow [bowing with just the wood part] on a junk cello. When I play the recording for people, they don't know its a cello, and they think I had two or three tracks going on.

What does it mean? It can be evocative; you don't have to bang the audience over the head with, 'This is what it means.' Let them work at it. It is okay for audiences to work at things.

References

Le Guin U K 1989 *Dancing at the Edge of the World: Thoughts on Words, Women, Places* (New York: Grove Press)

Nause C L 2016 private communication

O'Brien M and Waits T 2011 *Hard Ground* (Austin, TX: University of Texas Press)

Woodbrey E 2022 private communication

IOP Publishing

Photography (Second Edition)
Physics and art in focus
John Beaver

Chapter 46

Toward an art and science of nature: a personal note

Rays of light from the world are altered—blocked or redirected—by matter and then allowed to fall onto some flat surface that is *light-sensitive*. That is, the light striking that surface disappears, and the surface is forever changed as a result. These two processes taken together—the altering of the paths of light rays and the interaction of light with matter—might reasonably be taken to be the defining principals of *photography*.

But photography is also a form of two-dimensional art, and as such, a photograph is a set of marks on a flat surface—and stuff happens in our minds when we look at those marks on that surface. Thus it is of crucial importance to recognize that there is also an *image*—something that exists apart from its physical representation, presumably identifiable only because of the common nature of our human minds. If Martians, with their obviously green blood and differently structured minds, were brought into the mix, who knows what the 'image' would mean to them?

Throughout *Photography: Physics and Art in Focus*, I have tried to find ways to make explicit the connection between the physical part of photography—the part that is separate from the image—and the image itself. Part of the trick has been to look for processes and techniques that place evidence of this physicality directly into the photographic medium or into the image.

Much of the reason for this emphasis is personal. I do not want perfect control over my art. I want instead to make use of the tension that naturally arises when the artist's conscious control is tempered by technique that is at least in part uncontrollable. Or rather, I want my art to be partly controlled not consciously by myself, but by the world itself. Perhaps part of this comes from the natural scientist in me. The natural sciences are human endeavors, and there is often a particular 'style' that

 © IOP Publishing Ltd 2022

comes with that. But on the other hand, the scientist doesn't get to just make stuff up; there is *The World*, and it would still be there if we were not.

Perhaps this is why I am drawn to an art that allows me to make my mark in my own way but only within limits. I can arrange things as in a still life, but I am still photographing something that exists in the world, not only in my mind. And the nonphotographic content I seek out is a way to let nature (through complex physical processes only partly under my control) touch the surfaces of my pictures directly; they are not made solely by my own hand. It comes down to two beliefs.

1. My mind is much like the minds of those who look at my photographs.
2. Nature is much bigger than any of our minds.

The first point brings with it the possibility of art, for it is the commonality of our minds that allows us to see something of the same thing when two of us look at the same picture. Thus we can use our art to communicate with each other.

But it also brings a dilemma. For if our minds are alike, then when I use my art to say to you only what is in my mind—then I am, in a sense, telling you only what you already know. So I find myself bored with a photography that allows me to portray whatever is in my mind with perfect precision. Why would anyone want to see something so mundane as that?

I believe the second point is the way out of this dilemma, for it means I have nature—infinitely bigger and older than all of us—to both talk about and listen to. I can thus let go for a bit and leave some of the details of a photograph to the world. Seen in this light, alternative and historical photographic processes, with their rather blatant physicality, are ways to capture, at least a little, some of the grand processes of nature. This unimaginable complexity adds chance but without the banality of pure randomness. It allows for the possibility of accidental discovery.

Perhaps as photographers, we are simply stuck here, trying to arrange lenses and boxes and light-sensitive materials such that nature makes interesting marks on flat surfaces, in the hope that someone will see those marks and, in so doing, new thoughts and feelings—new things—are created. So we change a tiny part of our only world.

IOP Publishing

Photography (Second Edition)
Physics and art in focus
John Beaver

Chapter 47

Make your own pinhole fisheye camera

47.1 Basic principles

Figure 47.1 lays out the dimensions of the principle parts of a Wood fisheye pinhole camera, as described in chapter 10. I have oriented the camera such that it is pointing upward, as in R W Wood's original experiments. Light from the subject enters through a transparent window at the top into the upper fluid-filled chamber, through a pinhole at the center and into the lower chamber (also filled with fluid), arriving at a detector located at the bottom. Thus in my terminology, I refer to the fluid-filled chamber facing the subject as the *upper chamber* and that holding the light detector as the *lower chamber.*

The most important constraint is that both fluid-filled chambers must allow for the critical angle, θ_C, between the fluid and air. Note that it is the critical angle for the fluid inside the camera that is important, even if there is a layer of some other transparent material (glass, for example) on top of the upper chamber (see section 10.4.3).

The two chambers need not be the same size, but their relative proportions must hold to the following constraint regarding the diameter (D) of the chamber compared to its thickness (T) for a given critical angle (θ_C) .

$$\theta_C = \arcsin(1/n) \tag{47.1}$$

$$D > 2\tan\theta_C T \tag{47.2}$$

$$D > 2\tan[\arcsin(1/n)]T \tag{47.3}$$

For a water-filled camera, with $n = 1.33$, we must have $D > 2.28T$. If this condition is unmet for either chamber, the camera will image less than a full 180° view.

doi:10.1088/978-0-7503-3703-8ch47

© IOP Publishing Ltd 2022

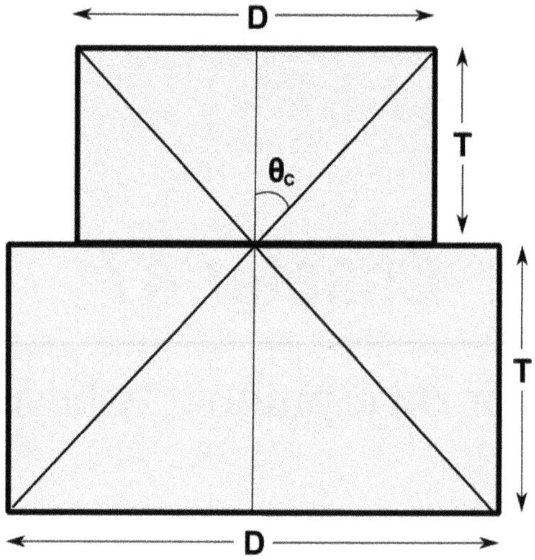

Figure 47.1. Basic constraints on the dimensions of a fluid-filled pinhole fisheye camera. Light from the subject above enters through a transparent window at the top of the upper chamber, through a pinhole at the center and into the lower chamber, which contains the light detector at its bottom. θ_C is the critical angle between air and the fluid inside the camera.

The size of the lower chamber decides our image scale, but the upper chamber could in principle be as small as we like. We should, however, stick to the following for the thickness of the upper chamber.

$$T_{\text{upper chamber}} \gg D_{\text{pinhole}} \tag{47.4}$$

That is, we should be sure the size of the upper chamber is at least many times greater than the size of the pinhole itself.

We also have the basic constraints for any pinhole camera, in that there is an optimal pinhole diameter given the image distance (Appendix C).

$$D_{\text{pinhole}} = \frac{\sqrt{T_{\text{lower chamber}}}}{27} \tag{47.5}$$

Here both $T_{\text{lower chamber}}$ and D_{pinhole} are expressed in millimeters.

In section 47.3, we dispense with the fluid in the lower chamber by placing the pinhole at the center of a transparent hemisphere that extends into the upper chamber, as outlined in section 10.4.4, but the constraints on the chamber dimensions remain the same, as the light rays emerge from the pinhole at the same angle on their way to the light detector.

In section 47.4, we consider the use of acrylic casting resin as the refracting medium, instead of water. The index of refraction of acrylic is higher than that of water, typically about $n = 1.49$. This lessens somewhat the constraints on the dimensions: $D > 1.81$ T.

47.2 A design for a wet camera

A medium-format R W Wood camera can be made from inexpensive and readily-available PVC plastic plumbing components, which are of course already meant to be waterproof. I built my camera around three parts purchased at the local hardware store—a *reducing bushing*, a *female adapter* and a *cleanout plug*. See figure 47.2 for top and bottom views of each part.

The reducing bushing fits into the unthreaded end of the female adapter, and the cleanout plug threads into the other end. The left-hand view of the reducing bushing shows the camera's 1.5-inch diameter upper chamber. The small lip at the bottom of the small chamber provides a shelf upon which to rest the pinhole (located at the center of a circle of thin brass sheet). A circular platform is attached, via spacers, to the inside of the cleanout plug. The detector attaches to this platform, so the cleanout-plug side of the female adapter forms the lower chamber of the camera. Figure 47.3 shows top and bottom views of the assembled camera. The left-hand image shows the upper chamber, with the pinhole at its bottom. A clear plastic window has been glued to the top of the reducing bushing, and the area around the upper chamber has been blackened. A wooden block with a tripod-mount threaded hole is glued to the outside of the camera body with hand-moldable epoxy.

The cleanout plug has a circular platform (cut from sheet plastic and painted black) that sits on spacers; magnets are glued to its surface with epoxy. Under safe-light conditions, a piece of light-sensitive paper is cut to shape and attached (with double-sided adhesive tape) to a round sheet-metal plate. Several of these metal plates can be preloaded with light-sensitive paper and placed in a dark bag. To load the camera, one of these metal plates is affixed to the magnets, and the cleanout plug is then threaded into the back of the camera using Teflon plumber's tape on the plastic threads to ensure a water-tight seal. This can be accomplished by feel in a changing bag, so the camera can be loaded (and unloaded) in the field.

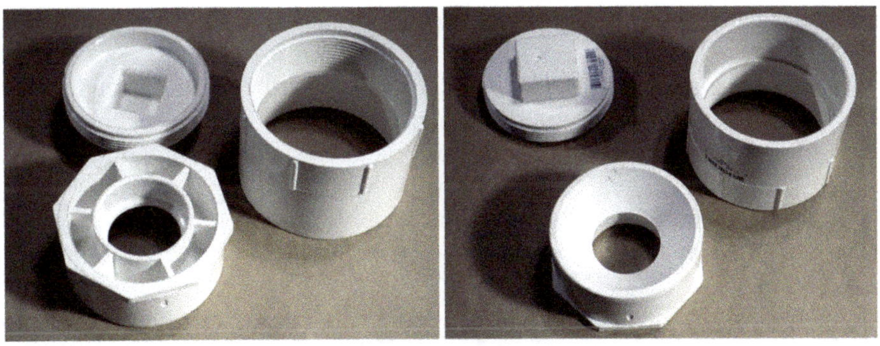

Figure 47.2. Top and bottom views of three plumbing parts used for the construction of a water-filled pinhole fisheye camera. **Top left:** Four-inch cleanout plug; **top right:** Four-inch female adapter; **bottom:** Four-inch to 1.5-inch reducing bushing.

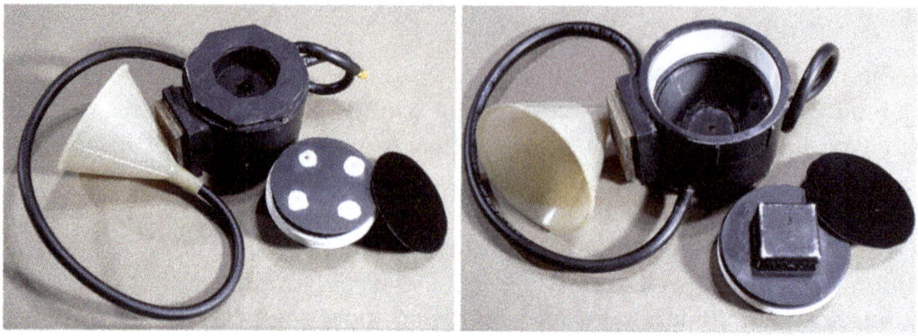

Figure 47.3. Top (**left**) and bottom (**right**) views of the assembled pinhole fisheye camera. The light-sensitive paper is taped to the metal plate at right, which is then held to the cleanout plug with magnets. The assembly is then threaded into the back of the camera, using Teflon plumbing tape on the plastic threads to ensure a waterproof seal.

The camera is loaded while empty and then filled with water via the hose and funnel that attaches at the bottom; a second short hose at the top acts as an air outlet. The small openings surrounding the upper chamber in the reducing bushing (figure 47.2, left-hand view) serve a convenient purpose. A series of holes drilled between those openings and into both the upper and lower chambers allow for water to enter (and air to exit) from the hoses into both chambers. All of the openings are painted flat black, so they act as light traps. Instead of drilling holes for the hose attachments directly into the camera chambers, the hoses empty first into those surrounding spaces, passing from section to section and finally into the two camera chambers.

To load the camera, one first wraps Teflon plumber's tape around the threads of the cleanout plug. The camera parts and the light-tight bag of preloaded sheet-metal detector holders are then placed inside a changing bag. While inside the changing bag, a metal plate holder is removed from its light-tight envelope and affixed to the cleanout plug, which is then threaded tightly into the back of the camera. A light-tight cover is placed over the upper chamber, and the camera is then removed from the dark bag.

Once out of the dark bag, the loaded camera is positioned on a tripod so the air outlet hose is at the very top of the camera; the water enters the bottom of the camera. Plugs are removed from both hoses, and the water hose with its funnel is held higher than the camera while water is poured into it. The camera is full when water begins to emerge from the air outlet hose. Both hoses are then plugged, the camera is pointed at the subject, and the cover removed for the duration of the exposure.

After the cover is put back in place, ending the exposure, the camera can be drained simply by removing the plugs from both hoses, allowing water to flow out of the bottom hose (the end of which is now kept lower than the camera) while the top hose allows air to flow in. Once the camera is completely drained of water, it may be placed back in the changing bag, the cleanout plug removed and the exposed (and wet!) photo paper with its metal plate placed in a second light-tight bag, for later processing in the darkroom.

The camera can then be removed from the dark bag and dried off with a towel. After applying new Teflon tape to the threads of the cleanout plug, the camera is ready to be loaded for a second exposure.

47.3 A self-contained fisheye water 'lens'

It is only the upper chamber, where the incoming rays refract from air to fluid, that is essential to produce a fisheye effect. The reason Wood's camera has a lower chamber filled with water rather than air is so the rays do not revert back to their original angles when they encounter the water-air interface. But we can avoid this with a trick; we place the pinhole at the center of a small, air-filled hemisphere, as described in section 10.4.4. The rays on their way to the pinhole thus exit from water to air through an interface surface that is normal to their propagation—so no refraction occurs. Thus we place a small, clear-plastic hemisphere over the pinhole, protruding into the upper chamber.

With this arrangement, the lower chamber with its light detector can now be dry, rather than fluid filled, greatly simplifying the construction of the camera. We can, in fact, allow the fluid-filled upper chamber with its pinhole to be a self-contained unit —a separate 'lens' if you will—that can be attached to any ordinary box camera.

It is easy to make such a fisheye pinhole lens attachment out of inexpensive and easy-to-find materials. Figure 47.4 shows top and bottom views of an example, mounted on a 4 × 4 ft wooden lens board.

Figure 47.4. Top (**left**) and bottom (**right**) views of a self-contained, water-filled fisheye pinhole lens, mounted on a 4 × 4 inch board for attachment to a camera. The pinhole is removable, so the optimal pinhole size can be used for any choice of camera image distance.

Figure 47.5. **Top:** The parts for making an inexpensive, detachable, water-filled pinhole fisheye 'lens.' **Bottom:** The finished assembly, ready to be filled with water.

Since this unit can be attached to any box camera, the pinhole is on a separate small sheet of thin brass, which is affixed to the back of the assembly with adhesive tape. This allows for the pinhole to be changed for one with a diameter appropriate to the image distance.

The lens attachment is constructed around a small, black polyethylene terephthalate (PET) plastic jar with a threaded watertight lid. Similar jars can be easily purchased online or at some craft supply stores; they are marketed to, for example, at-home makers of scented oils and creams. High-quality, small, transparent hemispheres are sold as weatherproof housings for outdoor surveillance cameras. But a lower-quality example will also work well; distortions are minimized by the tiny pinhole aperture. I purchased inexpensive, small, transparent spheres that separate in half, so they can be filled with tiny objects to make hanging decorations. Figure 47.5 shows the basic parts. A hole is cut into the lid, and the hemisphere sealed over the hole with waterproof epoxy. The bottom of the plastic jar is cut off so it has the appropriate thickness, in accordance with equation (47.3) (but still, of course, thick enough to fully contain the transparent hemisphere). A circle is cut from thin sheet plastic and epoxied to the cut-off jar. The pinhole is temporarily covered with waterproof tape, and the two parts are threaded together underwater.

If the lens is filled with ordinary water directly from the tap, it will likely fill with bubbles some hours afterward because of the air dissolved in tap water. Thus, one should first boil the water for several minutes and allow it to cool before filling the lens assembly.

In principle, the assembly can be mounted on any large-format or box camera. But in order to allow for a full 180° fisheye image, the camera portion must still

Figure 47.6. Top (**left**) and bottom (**right**) views of a self-contained pinhole fisheye lens using acrylic casting resin, rather than water, for the refracting medium.

conform to the constraint of equation (47.3), with (for a water-filled lens) the light detector at least 2.28 times wider than the pinhole-detector distance.

47.4 Using casting resin instead of water

An interesting variation is to follow the instructions of section 47.3, but fill the jar with acrylic casting resin instead of water. One then has a permanent, solid fisheye 'lens' that will not freeze during long exposures in subzero weather.

Obviously one cannot fill the container with resin in the same way as with water. Instead we dispense with the sheet plastic cover, and simply fill the assembled container with resin, pinhole-side-down.

It is tempting to place sheet plastic on top of the poured resin, allowing it to cure while flattened by the sheet of clear plastic. But this requires caution. First, the curing process can produce a myriad of tiny bubbles, and they must be allowed to escape. One technique is to gently heat the surface with a small flame torch. But even so, one must take care in placing the sheet plastic on the sticky resin so as not to trap an air bubble in the process. Finally, even if one avoids bubbles, the resin typically shrinks slightly as it cures, thus producing an unwanted concave surface that will restrict the angle of view of the image.

A better alternative, in my view, is to simply allow the resin to cure uncovered. The surface can then be sanded smooth with sandpaper, in a series of increasingly fine grits, and then polished with polishing compound meant for plastic. A small power sander or polisher makes this task fairly easy. Figure 47.6 shows front and rear views of my example.

Figure 47.7. An image made with a casting resin fisheye lens. The surface of the resin was allowed to be uneven. Tiny bubbles trapped in the resin produced the dark spots. The dark form at the very bottom is the end of one the tripod legs used to support the camera.

The resin surface can also be left as is; the inevitable uneven surface will likely not produce significant distractions in the already-distorted image. More problematic (if one wants a pristine image) are tiny bubbles trapped within the resin. Figure 47.7 shows an image made with the same lens as in figure 47.6, but with an uneven surface.

IOP Publishing

Photography (Second Edition)
Physics and art in focus
John Beaver

Chapter 48

Some technical notes on 'ephemeral process' photography

48.1 Introduction

What I call *ephemeral-process* (EP) photography uses printed-out enlarging paper, and in that sense, it is similar to the *lumen process* described in detail in chapter 35. But for EP photography, the paper is left unfixed, so it is still sensitive to light; this puts the 'ephemeral' in EP. Furthermore, in most circumstances I apply an *accelerator* immediately before exposure in order to greatly increase the speed at which the paper prints out. Details about the sensitivity and photochemical mechanisms involved with this accelerated printing out can be found in chapters 25 and 35.

48.2 EP photography: general considerations

We will look at several ways to use the basic concept of EP photography. In this section I step through the parts of the process that the different techniques have in common.

48.2.1 EP accelerator formula

The key ingredient of a printing-out accelerator is plain water, which has a significant accelerating effect on most enlarging papers. This is, for most papers, greatly enhanced over plain water by adding sodium sulfite or ascorbic acid (vitamin C).

I recommend ascorbic acid, as it is safe and can be easily purchased in powdered crystals, sometimes even at the grocery store. Sodium sulfite is also an excellent, inexpensive and easily-available oxygen scavenger, and it is considered to be mostly nonhazardous. It has some advantages over ascorbic acid, but it does cause an allergic response in some people, and direct contact to the skin (or inhalation of the dry powder) should be generally avoided.

 © IOP Publishing Ltd 2022

In order to make the accelerator brush more easily onto the paper, I use xanthan gum as a binder. It works better (for this purpose) and is far less expensive than a traditional art-medium binder such as gum arabic. It can be easily found online or in the gluten-free baking section of many grocery stores; the smallest package is far more than you will need.

The mixed accelerator should be kept in an air-tight container; only a small amount should be mixed at a time. My strategy is to premix the dry ingredients to the correct ratio. I then use a small water-tight container to mix a small amount of accelerator with water. The relative amounts of premixed dry ingredient and water can be predetermined for the particular container.

I keep the mixed solution for only about a week; it may lose its oxygen-scavenging potency if kept much longer. Immediately before application, I pour a small amount of the accelerator solution into a separate plastic container that is wide and shallow enough for a brush. After applying accelerator to a piece of paper, I then simply pour the unused portion back into the water-tight bottle. This keeps the accelerator out of contact with air as much as possible and allows for tidy use in the field, even while under a dark cloth.

The powdered xanthan gum binder does not easily dissolve in water, as clumps quickly seal themselves off. This is greatly mitigated if it is first mixed dry with the powdered sodium sulfite or ascorbic acid. When the dry combination is mixed with water, the powdered accelerator helps keep the tiny grains of xanthan gum separate from each other, and there are fewer clumps. This strategy is very effective for the sodium sulfite formula, since the xanthan gum is only a small portion of the total. It is less successful for the ascorbic acid formula, but even so, most of the clumps dissolve after a few minutes of vigorous shaking. If left to rest for 24 hours, the remaining clumps will disappear on their own.

Ascorbic acid will stain some papers brown, especially if either the exposure time or concentration is too high. The ascorbic acid solution also dries fast, so it can be problematic for very long exposures. And it can turn into something like a glue as it dries, and so the acetate (see below) must be carefully peeled off under running water.

Sodium sulfite has none of these problems, but it has the disadvantage that its use is much less benign. Also, sodium sulfite will fog some types of papers; prolonged contact darkens the paper even with no exposure to light. This tendency can be counteracted by adding a little citric acid to the accelerator. But like ascorbic acid, citric acid may stain some papers brown. So far, I have not found a paper that both needs citric acid to prevent fogging but also is stained by it.

Below I outline my own recipes for both sodium sulfite and ascorbic acid accelerators. One should consider them as starting points for experimentation. Some papers require more (or less) of the ascorbic acid or sodium sulfite, and the amount of xanthan gum can be adjusted to make the solution either thicker or more watery. In my experience, the ascorbic acid recipe brushes onto the paper more smoothly than does the sodium sulfite version.

48.2.1.1 Ascorbic acid formula

1. Dry ingredients:
   (a) Measure xanthan gum powder and dry crystals of ascorbic acid (vitamin C) in a ratio of 1:1 by volume, enough to make roughly ten grams (about one tablespoon or 15 ml).
   (b) Put the dry ingredients in a small container with a tight-fitting lid and shake vigorously.
2. Mix 360 mg (0.65 ml or 1/8 teaspoon) of dry ingredients for every 60 ml of water in a small container with a water-tight lid and shake vigorously. The clumping can be mitigated somewhat by carefully sprinkling tiny bits onto the surface of the water, shaking, and then repeating the process. Mix only enough that it will likely be used in one week.

48.2.1.2 Sodium sulfite formula

1. Dry ingredients:
   (a) Measure xanthan gum powder and dry crystals of sodium sulfite (Na_2SO_3, not Na_2SO_4) in a ratio of 1:8 by volume, enough to make roughly 60 g (about four tablespoons).
   (b) Put the dry ingredients in a small container with a tight-fitting lid and shake vigorously.
2. Mix 1.2 ml (1/4 teaspoon) of dry ingredients for every 30 ml of water in a small container with a water-tight lid and shake vigorously until the clumps disappear. Mix only enough that it will likely be used in one week.

   If the accelerator fogs that particular enlarging paper, try adding citric acid to the dry accelerator mixture immediately before mixing with water, in the ratio one part dry citric acid crystals to five parts of the dry mixture of sodium sulfite and xanthan gum.

48.2.2 Choosing the paper

There are many varieties and sizes of black-and-white enlarging paper that can be purchased online or possibly at a nearby camera store. The cost is usually about one dollar per 8 × 10 sheet—less if bought in larger quantities and more if bought in larger sizes. There are two basic categories:

1. **Resin-coated papers (RC):** The light-sensitive silver-gelatin emulsion is coated onto paper that is waterproof, as it is sealed with a plastic resin. This is usually the least expensive type of paper, and it is the easiest to use. But for this purpose it may be less satisfying, as the paper surface has a plastic-like perfection. Furthermore, of the RC papers I have tested, they all are more sensitive while dry than the fiber based papers I describe below. But they are, in general, no more sensitive when the accelerator is applied. And so they tend to show less contrast between the unaccelerated and accelerated parts of the paper. RC papers also tend to have a very high unaccelerated sensitivity

when the humidity is high, and this means one must then take extra care when scanning the exposed but still light-sensitive paper.

2. **Fiber-based papers (FB):** FB papers are the go-to choice for the art photographer. The silver-gelatin emulsion is applied directly to good quality paper, with all of its subtle microtexture. It is usually more expensive and difficult to handle than RC paper. For EP photography, I have found FB papers, in general, to be more interesting and useful than RC papers. But the results vary widely from one type of paper to another.

For EP photography, it is not necessary to use newly-purchased enlarging paper. My favorite papers, in fact, have been unavailable for decades outside of the used market. Papers that are long expired and nearly useless for their original purpose may give outstanding results for EP photography. Almost any black-and-white enlarging paper will produce results that are at least interesting in some way. Experiment!

Two different enlarging papers that are nearly identical when developed out may be strikingly different when printed out. Below I lay out some important considerations and define symbols to make it easier to compare different papers to each other.

1. Paper type, T (RC, FB): Whether the paper is FB or RC.
2. Unaccelerated dry sensitivity, S_{UD} (L, M, H): The sensitivity of the paper to printing out when dry, in low humidity (L, M or H for low, medium or high).
3. Unaccelerated wet sensitivity, S_{UW} (L, M, H): The sensitivity of the paper to printing out when wet with plain water. This also gives some indication as to how sensitive the paper is when dry, but under conditions of high humidity.
4. Accelerated sensitivity, S_A (L, M, H): The sensitivity of the paper when a sodium sulfite or ascorbic acid accelerator is applied.
5. Accelerated color without fixing, C_{AU} (Pi, Br, Bl, Pu, Gr): The color tone of the paper when printed out with an accelerator but left unfixed. Pi, Br, Bl, Pu and Gr mean, respectively, pink, brown, blue, purple and gray.
6. Accelerated color after fixing, C_{AF} (Pi, Br, Bl, Pu, Gr): The color tone of the paper when printed out with an accelerator, after it has been fixed.
7. Dry color without fixing, C_{DU} (Pi, Br, Bl, Pu, Gr): The color tone of the paper when printed out dry, with no accelerator, and left unfixed.
8. Dry color after fixing, C_{DF} (Pi, Br, Bl, Pu, Gr): The color tone of the paper when printed out dry, with no accelerator, after it has been fixed.
9. Does the paper stain with ascorbic or citric acid, *Stn* (Y, N)?
10. Does the paper fog with sodium sulfite, *Fog* (Y, N)?
11. Wavelength discrimination, W (L, M, H): Does the color tone at which the accelerated paper prints out bear a relation to the wavelength of the light used to expose it? Ratings of L, M and H represent low, medium and high degree of separation of printed out color tone by wavelength of incident light.

Table 48.1. Printing out characteristics of several types of enlarging papers. The symbols are defined in the text.

Paper	T	S_{UD}	S_{UW}	S_A	C_{AU}	C_{AF}	C_{DU}	C_{DF}	Stn	Fog	W
Kodak polyfiber	FB	L	L	H	Pu	Br	Gr	Br	Y	N	H
Kodak polyprint	RC	M	H	H	Pu	Br	Gr	Br	N	N	H
Kodak ektamatic SC	FB	M	M	H	Br	Br	Br	Br	N	N	M
Kodak polycontrast F	FB	L	L	M	Pi	Br	Bl	Br	N	N	L
Ilford multigrade IV	FB	L	L	H	Pu	Br	Gr	Br	N	Y	L
Ilford multigrade IV	RC	M	H	H	Gr	Br	Gr	Br	N	Y	L

Table 48.1 shows my data for a few papers I have tested. Most of these are vintage papers, but it is not difficult to find them in online auctions. A few of the papers I list are still produced and available as of this time of writing. There are, of course, many other subtleties and complications that are not captured by my categorizations. You will discover many interesting variations if you choose to experiment with different papers under a wide variety of circumstances.

The 1980s-era Kodak polyfiber is my all-around favorite. One can see from table 48.1 that it has high sensitivity when accelerated, but low sensitivities when unaccelerated, or even when wet with only water (or dry but in high humidity). This means it has a high contrast between the accelerated and unaccelerated parts, and it is easy to handle when dry, even in moderate lighting. Yet it is very sensitive when accelerated. It also responds differently to different wavelengths of light, so it can be used for the color EP processes described in chapter 31.

The color response brings up another qualification. Since papers like Kodak polyfiber print out different colors depending on the wavelength of the light that fall upon them, then obviously the colors described in the table for many of the entries also depend somewhat on the color of the light used for the exposure. I do not try to untangle those complications here; you will have to experiment.

There are other important features of some papers not captured by table 48.1. As an example, it would appear that Kodak ektamatic SC is a poor choice, in particular because it is still moderately sensitive even when dry, leaving much less contrast between accelerated and unaccelerated parts. And it appears to mostly print out with only brown tones. But in practice this paper, usually made with a low-quality paper stock, exhibits interesting nonphotographic content that can sometimes make the picture[1]. And it does in fact show slightly different tonalities under various circumstances. When used as a negative, with that extra step of translating the scanned negative into a printable digital positive, all of this can sometimes lead to an

[1] Ektamatic SC was intended for use by photojournalists. It was designed primarily for quick, nonarchival processing, with only partial fixing, to make a print that was only stable enough to be used as an image source for a newspaper edition.

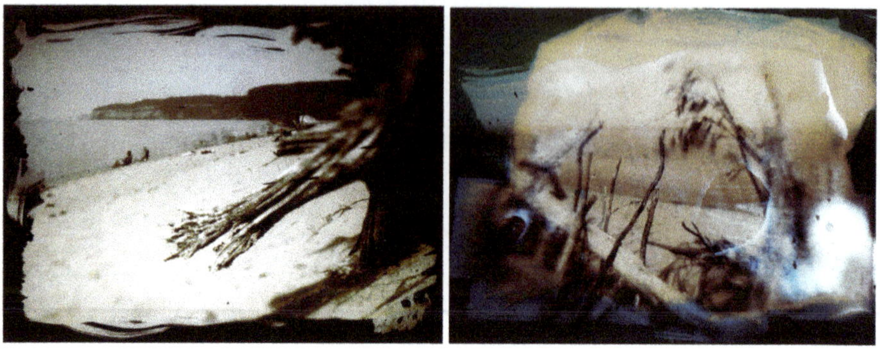

Figure 48.1. Both of these images were made with EP negatives exposed in a camera, but different kinds of paper were used, leading to very different effects. **Left:** *Pictured rocks #5* (John Beaver 2017) exposed with Kodak polyfiber. **Right:** *Kohler-Andrea State Park #1* (John Beaver 2020), exposed with Kodak ektamatic SC.

extraordinary image. I often seek out my oldest box of this paper; the low-quality, thin paper stock has turned brown. This makes for poor rendition of photographic content, but the interesting texture of the crude paper is exaggerated in the image. Figure 48.1 shows examples of EP photographs made with polyfiber and ektamatic SC in-camera negatives.

As a final point, which features are most important depends upon what one is doing. Are you using the paper as a negative in a camera, or for a photogram exposed with sunlight, for example?

48.2.3 Preparing the paper for exposure

Once accelerator is applied, the paper is far more sensitive to light, and one must be very careful about unwanted exposure. But before applying the accelerator, one can be far more casual about this. How much so depends on the paper and the humidity; most papers print out faster in higher humidity.

Once the accelerator is applied, the enlarging paper should be handled for as little time as possible and in light that is just barely bright enough to work under. Incandescent lighting will produce less exposure than daylight, fluorescent or white light-emitting diode (LED) lighting of the same brightness. A red LED lamp can be used with little fear of exposure at all.

Brush the sensitizer onto the emulsion side of the paper. Any type of paintbrush can be used, but I prefer either inexpensive foam brushes or flat Japanese hake brushes. Gently lay a thin sheet of acetate on it, and smooth out the bubbles. In order to determine how the final image will be affected by air bubbles and unevenness in the application of the sensitizer, you will have to experiment. The answer depends on too many details to describe here, but that is much of the fun.

48.2.4 Washing and drying the paper

Once the exposure is complete, washing should be carried out in dim, indirect light, and the paper should be left to dry in total darkness. Before washing, separate the acetate from the paper. Do this carefully under running water so as not to damage the emulsion, especially with the combination of FB paper and ascorbic acid accelerator. Wash the paper under gently running water until it no longer feels slippery from the accelerator solution.

A smooth, clean surface and a shower-stall squeegee can be very helpful for removing most of the water from the paper. RC paper can be simply hung to dry from a corner with a spring clothespin. FB paper should be left to dry upside down on a clean porous surface (a plastic window screen works well).

FB papers will wrinkle somewhat as they dry. To flatten, place the dry paper upside down on a sheet of kraft paper, and iron the back with a medium-low temperature setting. Place a flat weight on top while the print is still warm and allow it to cool.

The emulsion layer on some papers is sticky until dry, so a small hair dryer can be handy to dry the surface until it is smooth to the touch, before placing the emulsion in contact with, for example, a drying rack[2].

48.2.5 Scanning

Once your EP negative is dry, it will be much less sensitive to light than when the accelerator was applied. But it will still gradually turn dark if exposed to ordinary room light. At what rate this happens depends mostly on the particular type of paper, the brightness and spectrum of the light it is exposed to and the humidity of the air. But even under the worst of circumstances, a high-quality digital version of the image can be captured with any good-quality scanner.

Scan the paper only after it is dry. The original EP image will almost certainly be darkened—damaged if you will—by the light of the scanner. A higher-resolution scan usually means a greater exposure by the scanning light, as does repeated scanning. And so it is prudent to practice first before scanning that one-of-a-kind perfect exposure you just made. With my Epson V700 scanner, I usually scan EP negatives at 1200 dpi, which is enough resolution to easily see the grain of FB papers. Most papers (especially RC papers) will suffer more damage from scanning if the humidity is high.

Instead of scanning, one can simply rephotograph the EP negative with a digital camera. I say 'simply' with some irony; unless you are already set up with the proper camera, lens, copy stand and lighting, you will likely find it easier to make a high-quality digital capture with a scanner. But with the right equipment it can have an advantage—if one uses strobes to illuminate the EP negative for the exposure, then it can be rephotographed with significantly less damage to the unfixed negative. But in the end, I have not found this to be a significant problem.

48.2.6 The option of fixing

It is possible to make your original EP negative permanent by applying the same chemical fixing process that is used when making silver-gelatin prints in the

[2] This is particularly true for ektamatic SC.

darkroom. For a black-and-white print in the darkroom, exposed and chemically developed in the way intended by the manufacturer, the fixing process barely alters the image. Not so for EP photography. Results vary widely, but in general, the act of fixing your EP negative is likely to greatly lighten the image, and the overall color will likely shift to brown. Subtle differences in hue and value that are easily visible in the original will likely be lost when it is fixed.

If the goal is to produce a negative to be scanned in order to make a digital image, there is nothing to be gained by fixing the EP paper negative. Shadow tones will be greatly reduced and the digital image is easily captured from the unfixed negative. But I describe below some variations on EP photography where one might consider fixing the printed out negative as the final piece.

While in the fixer, the print will fade considerably. But it will darken upon drying —likely to within roughly two steps of its prefixing density. One can partially compensate for this overall loss in density by using a longer initial exposure. Expose the print for as long as is needed to bring it to the desired prefixing density and then continue to expose for two to three times longer.

48.3 EP photograms

Because accelerated silver-gelatin paper prints out with much greater speed than cyanotype, let alone anthotype, and it is sensitive to visible light as well as ultraviolet, nearly any light source can be used. In particular, an ordinary flood light or even a household lightbulb can be used to expose an EP photogram. Direct, full sunlight may require only a few seconds of exposure.

An ordinary old-style incandescent lightbulb is probably the worst choice. Bright photo flood lights work well, both the hot quartz type and the newer and much-easier-to-use cool LED versions. An inexpensive LED 'blacklight' bulb may also work well, even if it does not look very bright to the eye. They emit light mostly at approximately 400 nm, the very wavelength to which the enlarging paper is most sensitive. The shortest exposures will almost certainly be with direct sunlight, whenever it is available. Depending on the choice of paper and accelerator, one may find that different colors of photogram objects, and different sources of light, produce somewhat different colors in the printed-out image.

If you are to expose your photogram in sunlight, then cover everything with something opaque and carry it outside for the exposure. If you are using artificial lighting, then it is convenient to have some way to position the light source directly over the photogram in the same dim room you use to prepare it before exposure.

With a bright source of artificial light, you will likely find that an exposure time of only a minute or two (or even less than a minute) is required. For some combinations of paper and accelerator, only a few seconds may be required when exposed to the direct light of the sun when it is high overhead on a clear day. Since, like cyanotype, this is a printing out process, you can monitor the exposure as it happens. See figure 48.2 for a step-by-step illustration of the process and figure 48.3 for the scanned result.

Figure 48.2. The EP photogram process. Except for the actual exposure, the entire process was carried out in a room with dim, indirect lighting from a single lightbulb. **Upper left:** The accelerator is brushed onto the paper, and it is covered with a thin sheet of acetate. **Upper right:** Objects are placed on the paper and held down by glass (some objects were placed on top of the glass in this example). **Lower left:** The photogram was exposed to an LED photo-floodlight for 90 s. **Lower right:** The objects are removed.

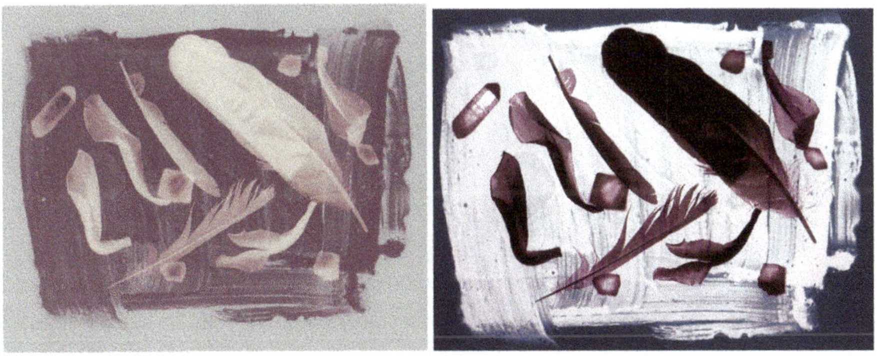

Figure 48.3. Left: A digital scan of the finished photogram from figure 48.2. **Right:** The same scan with the values digitally inverted from a negative to a positive and the levels and white balance adjusted.

48.4 A camera for wet paper negatives

The basic design of the EP/Cyano camera I present here is that of a 7.6 × 10.2 cm format press camera, like a small Crown Graphic camera from the 1940s. Constructed out of hardwood and hobby plywood, the camera with film back measures 9 × 13 × 18 cm closed and weighs only about 0.5 kg. The bellows is folded by hand from opaque paper, and the camera has a focus range of approximately 100–200 mm. The front standard is held in place by a thumb screw inserted through the bottom of a single slot running the length of the bed, and into the bottom-center of the front standard. Since a single screw holds the front standard in place, the front standard has swing; in fact care must be taken if one does not want swing. Focus and swing adjustment is achieved simply by loosening this screw, moving the front standard, and then tightening the screw. This same thumb screw is used to hold the hinged bed in place when the camera is closed. A second thumb screw is inserted from the back to lock the bed in place when open. The camera bed opens in the portrait-mode direction. A second tripod mount on one side is used to orient the camera in landscape mode (at which point, front swing becomes front tilt).

These features so far described differ from those found in an ordinary press camera only in their relative simplicity, mechanical crudeness, light weight and low cost. One can get away with this here because the goals of the photographer are different and precise mechanical alignment and stability are often neither necessary nor a virtue for the EP or cyanonegative photographer. Furthermore, the camera need only be merely light tight; it need not be really most sincerely light tight.

This particular design is a convenient size; it is small and lightweight but still large enough to capture a significant amount of detail. Most of the EP pictures in this book were taken with this very camera. The small format means that one standard sheet of enlarging paper is enough for six exposures, and this translates to about only pennies per exposure (the cost of the accelerator solution is trivial).

The Graflex Optar 135 mm $f/4.7$ lens shown in figure 48.4 is a good choice for this camera. These relatively low-cost but excellent lenses were used for many years on the popular Speed and Crown Graphic press cameras, and there are many

Figure 48.4. Cameras and film holders for cyanonegative and EP photography. The cameras are of the simple design described in this article, with paper bellows and crude mechanics. The film backs are modified from $3\frac{1}{4} \times 4\frac{1}{4}$ film holders. The sensitized paper is sandwiched between the focus screen (mounted on a hinged door) and a sheet of glass. Also pictured (**right**) is a 4 × 5 film holder modified in the same way.

available on the used market, at a typical cost (2022) of less than \$100—much less if the shutter is broken (not a concern for EP photography!). But even a simple magnifying glass can produce interesting results. Aperture stops can be made out of black paper; if the magnifying glass is stopped down to $f/5.6$ or $f/8$, daylight exposures are still easily within reach, but the image shows far fewer lens aberrations than if used with its full aperture (typically about $f/3$).

48.4.1 The film back

The heart of the EP/cyano camera is a special combination film back and focus screen that is designed like a frame for contact printing. The example described here is modified from a standard, double-sided $3\frac{1}{4} \times 4\frac{1}{4}$ inch film holder. But this same kind of back can be made from any 4×5 inch or 8×10 inch film holder, thus converting any view camera into an EP camera. A good choice is the oldest version of the Speed Graphic (sometimes called a 'pre-anniversary' Speed Graphic). It is lighter—and cruder—than later versions and is usually much less expensive on the used market. The extra features on the later models are not needed for EP photography. If the bellows are damaged (thus lowering the price), a little bit of tape is all that is needed since they do not need to be perfectly light tight for EP photography.

As for the case of making photograms, the sensitizer is brushed onto the paper, and a precut sheet of thin acetate is laid on top before placing it in the film back. This must, of course, be carried out as quickly as possible and in subdued lighting. It can be accomplished in the field underneath a large dark sheet, preferably while in the shade.

For my modified EP film back, the accelerated paper with its acetate cover is sandwiched between two pieces of glass, thus ensuring the wet paper stays flat during the exposure. The rearward piece of glass is on a hinged door, and it is frosted (it is covered with strips of matte-finish cellophane tape). Thus it becomes the focus screen when no paper is present. A hinged cover is attached to the door and swings 90° to it. During an exposure this prevents light from getting through the focus screen to the back of the light-sensitive paper. During focusing and composition, it props up to form a sun shade.

All of the printing frame mechanism is housed on one side of the double-sided film holder. The other side still retains its dark slide, which can be used as a shutter. The film can be loaded on site, directly into the film holder while it is attached to the camera (under a dark cloth, for example). Alternatively, the film holder can be removed from the camera after focusing and composing, loaded separately and then reattached for the exposure.

Unlike FB paper, RC paper does not curl when wet, so it is possible to cut such paper to fit in a standard unmodified sheet film holder. The accelerator can then be applied once the paper is in the holder, a protective sheet of acetate placed on top and the dark slide put in place. Any large format camera can be used unmodified for EP photography. The paper may still curl a little (it should be cut slightly undersized to give it room to expand) and possibly affect the focus.

48.5 EP negatives from color prints

Figure 48.5 illustrates the steps in making an EP negative from a chromogenic print. Brush the sensitizer onto the emulsion side of the paper; any type of paintbrush can be used. Gently lay a thin sheet of acetate on it, and smooth out the bubbles. Lay the print image-side down on top of the accelerated paper and acetate, and place a sheet of glass on top to hold it in place. Alternatively, one can dispense with the piece of acetate and simply stick the print directly to the accelerated paper. Depending on the combination of enlarging paper, accelerator formula and whether the print has a glossy or matte finish, they may be difficult to separate without soaking in warm water. But a typical chromogenic print is not permanently harmed by the process.

Expose through the back of the chromogenic print. Sunlight works well, and an exposure time of roughly one minute should be expected. But indoor flood lights can also be used, although the exposure time is likely to be longer. The colors produced depend critically on the choice of enlarging paper, print subject, accelerator and light source. Experiment!

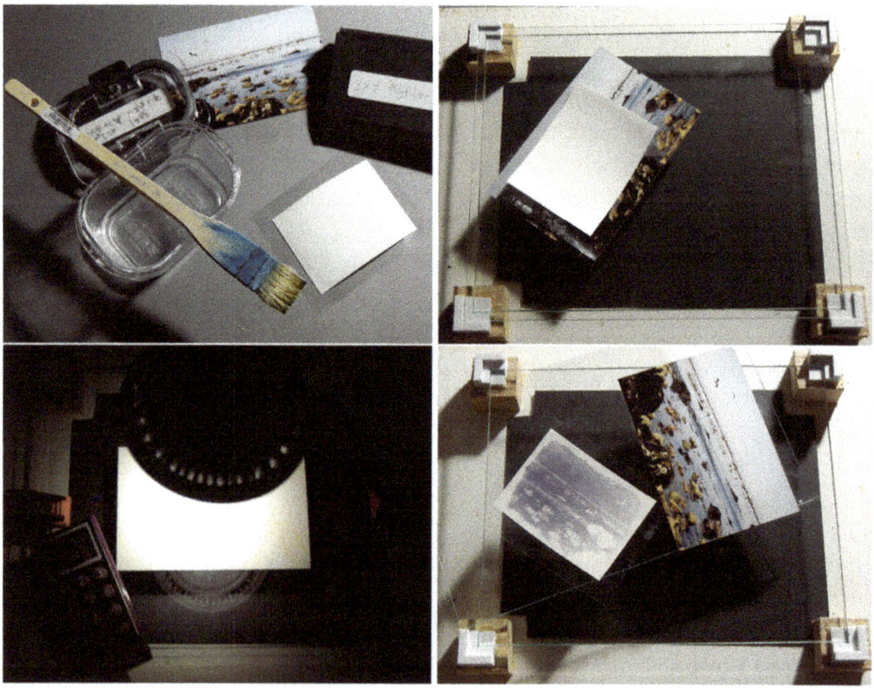

Figure 48.5. Upper left: The accelerator is brushed onto the enlarging paper and overlaid with acetate. **Upper right:** The enlarging paper is positioned face down on the print. **Lower left:** The print with enlarging paper is turned upside down and held flat under glass, and the enlarging paper is exposed through the back of the print to light from an LED flood. **Lower right:** The resulting negative. The entire procedure (except for the exposure) was carried out under indirect light from a single 60 W lightbulb.

Figure 48.6. The image resulting from the negative shown in figure 48.5, after washing, drying, scanning and simple digital processing with GIMP.

See figure 48.5 for an outline of the procedure. Figure 48.6 shows the positive image that resulted, once it was scanned and inverted digitally from a negative to positive.

IOP Publishing

Photography (Second Edition)
Physics and art in focus
John Beaver

Appendix A

Units, dimensions, and scientific notation

A.1 Units and dimensions

When we refer to a physical quantity, it must always have associated with it a set of *dimensions* and also, in many circumstances, a set of *units*.

In this context the word 'dimension' refers not to spatial dimensions but rather to the *type* of physical quantity. For example, length is a fundamentally different type of quantity than time. One cannot add a length to a time, nor can one subtract one from the other, because that would equal nonsense. Note that this is not the same thing as apples and oranges. Unlike length and time, one *can* add apples and oranges (it equals fruit salad).

But on the other hand, it is just fine to multiply or divide a length by a time. This produces something with different dimensions that are a combination of the two. For example, if one divides a length by a time, the result is something that has dimensions of length/time ('length per time'). Often these combined dimensions have special names. This example of length/time has the special name of velocity or speed. And so whenever one divides a length by a time, something with dimensions of length/time results.

But what about the actual numbers one plugs into the calculator in a specific case? What if one has a specific length, and a specific time, and wants to calculate a specific speed? Whenever actual numbers are involved, there must also be *units*.

A length of 12.0345 is ambiguous. Is it 12.0345 meters or 12.0345 furlongs? The meter and the furlong are examples of *units*, which are agreed-upon standards for attaching a numerical value to a particular physical quantity. And so the meter is a unit of the dimension of length, and so is a furlong. One can convert between units of the same dimension by establishing an equivalence between them. And so 1 meter = 3.280 feet = 39.37 inches = 0.00497 furlongs, etc.

In the physical sciences we mostly use a particular international system of units, called *SI*, which stands for 'International System' (in French). The SI unit of length is the meter, while the SI unit of time is the second. Every SI unit has an official abbreviation. The abbreviation for the meter is *m*, and for the second it is *s*

doi:10.1088/978-0-7503-3703-8ch49

© IOP Publishing Ltd 2022

(it matters that they are lowercase). Table A.1 lists some common SI units with their dimensions and official abbreviations.

Just as we can derive new dimensions by multiplying or dividing dimensions by each other (length/time, for example), we can do the same for units. And so we can divide meters by seconds to get a new derived unit, which we write m/s (called 'meters per second'). What if we want to divide m/s by seconds? We can do that just fine, and we get $m/s/s = m/s^2$ (called 'meters per second squared'). Many of the units in table A.1 are actually derived combinations of other units. For example, the newton is actually a combination of kilograms, meters and seconds:

$$1\ \mathrm{N} = 1\ \mathrm{kg}\ \frac{\mathrm{m}}{\mathrm{s}^2} \tag{A.1}$$

These base units can be modified by any one of a number of official prefixes, which then multiplies the unit by some power of 10. These prefixes and their abbreviations are listed in table A.2, although some are more commonly used than

Table A.1. Common SI units.

Dimension	Unit	Abbreviation
Length	meter	m
Time	second	s
Mass	kilogram	kg
Temperature	kelvin	K
Force	newton	N
Energy	joule	J
Power	watt	W

Table A.2. Prefixes for SI units.

Prefix	Abbreviation	Meaning
Femto	f	$\times 10^{-15}$
Pico	p	$\times 10^{-12}$
Nano	n	$\times 10^{-9}$
Micro	μ	$\times 10^{-6}$
Milli	m	$\times 10^{-3}$
Centi	c	$\times 10^{-2}$
Deci	d	$\times 10^{-1}$
Hecto	h	$\times 10^{2}$
Kilo	k	$\times 10^{3}$
Mega	M	$\times 10^{6}$
Giga	G	$\times 10^{9}$
Tera	T	$\times 10^{12}$

others. For example, 'milli' means '×1/1000'. And so a millimeter (abbreviated mm) is one thousandth of a meter.

A.2 Scientific notation

We have used scientific notation for the values in table A.2. Physical quantities in nature can vary by many powers of 10. And so for example the light given off by the Sun, its power, P, is many times greater than the light given off by a 60 W lightbulb:

$$P_{\text{Sun}} = 667\,000\,000\,000\,000\,000\,000\,000\,000 P_{\text{lightbulb}} \tag{A.2}$$

After the 667, there are 24 zeros there. What if I had mistyped (or you miscounted) and you found 23 zeros instead? Well, that number would be *ten times* too small. And so clearly, when dealing with numbers like this, we need a better way. And so we use what is called scientific notation. Written this way, the above equation becomes

$$P_{\text{Sun}} = 6.67 \times 10^{26} P_{\text{lightbulb}} \tag{A.3}$$

The $\times 10^{26}$ part means ×100 000 000 000 000 000 000 000 000. But in practical terms this also means 'take the decimal point in 6.67 and move it 26 places to the right, filling in with zeros as needed'.

Raising something to a negative power means the same thing as dividing 1 by that same thing, but raised to the same *positive* power. For example,

$$27^{-3} = \frac{1}{27^3} \tag{A.4}$$

And so we can also use negative numbers in scientific notation; it means simply *divide* by the power of 10 instead of multiplying by it. And as with positive powers, we can also express this as a decimal equivalent:

$$3.27 \times 10^{-5} = 3.27 \times \frac{1}{10^5} = \frac{3.27}{10^5} = 0.000\,0327 \tag{A.5}$$

Here we can see that 3.27×10^{-5} means 'take the decimal place in 3.27 and move it five places to the *left*, filling in with zeros as needed'.

This has a couple of advantages. For one thing, we can see at a glance the most important part numerically: how many powers of ten. Secondly, when we write it this way, we do not need the zeros for placeholders. And so if I put them there, it means I believe that they are significant.

And so 6.67×10^{26} and 6.670×10^{26} are not really the same number, although they will both appear the same on a calculator. 6.67×10^{26} could possibly be 6.673×10^{26} or even 6.668×10^{26}. If I do not include any more decimal places, then I am making a statement that, based on my uncertainty in the measurement of that quantity, I have no idea what the value of the next decimal place would be. If on the other hand I write 6.670×10^{26}, then I am saying that I believe (even if with some uncertainty) that it really is 6.670×10^{26} and not, say, 6.673×10^{26}.

Note that one *could* use scientific notation to write the same number in several different ways. You should verify for yourself that the following is true:

$$9.75 \times 10^7 = 975 \times 10^5 = 0.009\,75 \times 10^{10} = 97\,500\,000\,000 \times 10^{-3} \qquad \text{(A.6)}$$

Clearly, the last two possibilities look a bit silly, but we try to avoid even the second version. When using scientific notation, it is customary to pick whatever power of 10 is needed in order to have one and only one digit to the left of the decimal place.

IOP Publishing

Photography (Second Edition)
Physics and art in focus
John Beaver

Appendix B

Notes on the golden rectangle

The lengths of the sides of the golden rectangle are in the proportion of approximately 1:1.618 034, a proportion known as the *golden ratio*. It has been claimed that this ratio of lengths is inherently pleasing to the eye and that a rectangle thus proportioned is the most aesthetically pleasing. Part of the fascination with the golden ratio and the golden rectangle arises from its fascinating mathematical properties.

We can easily work out, with simple algebra, what must be the proportions of such a rectangle. Let us label the rectangle as in figure 13.8, with the short side of the original rectangle a and the long side $a + b$. The ratio of the long to the short side of the original rectangle is $\frac{a+b}{a}$, while the same ratio for the sectioned-off rectangle is $\frac{a}{b}$. Thus, for the two rectangles to have the same proportions, we must have

$$\frac{a}{b} = \frac{a+b}{a} \tag{B.1}$$

Doing a bit of algebra on this, we have

$$a^2 = b(a + b) \tag{B.2}$$

$$a^2 - ab - b^2 = 0 \tag{B.3}$$

$$\frac{a^2}{b^2} - \frac{ab}{b^2} - \frac{b^2}{b^2} = 0 \tag{B.4}$$

$$\left(\frac{a}{b}\right)^2 - \left(\frac{a}{b}\right) - 1 = 0 \tag{B.5}$$

If we define a new variable, ϕ (the Greek letter phi), to be $\frac{a}{b}$, the ratio of the sides of the golden rectangle, then the last equation becomes

$$\phi^2 - \phi - 1 = 0 \tag{B.6}$$

doi:10.1088/978-0-7503-3703-8ch50

© IOP Publishing Ltd 2022

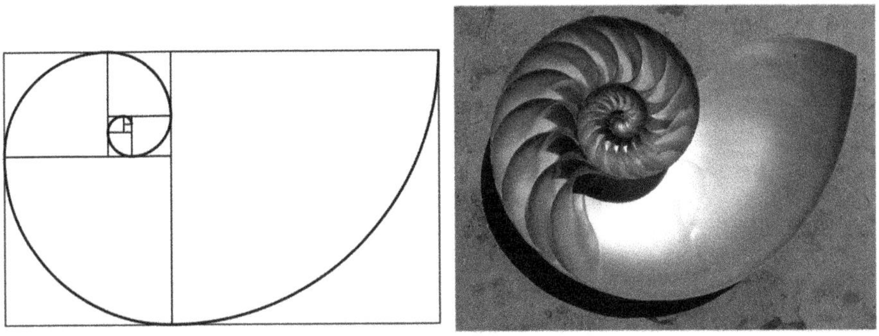

Figure B.1. Left: The golden spiral can be constructed from a series of golden rectangles subdivided into squares and golden rectangles. (Copyright © Dicklyon – Own work, Public Domain, https://commons.wikimedia.org/wiki/File:Fibonacci_spiral_34.svg, Fibonacci spiral 34.svg.) **Right:** The shell of the chambered nautilus shows an example of such a form in nature. (Copyright © Chris 73, Attribution-ShareAlike 3.0 Unported (CC BY-SA 3.0), https://commons.wikimedia.org/wiki/File:NautilusCutawayLogarithmicSpiral.jpg, NautilusCutawayLogarithmicSpiral.jpg.)

This is called a *quadratic equation*, and as such, it has two solutions:

$$\phi = \frac{1 \pm \sqrt{1 + 4}}{2} \tag{B.7}$$

This gives, to five decimal places, $\phi = 1.618\,034$ or $\phi = -0.618\,034$. The positive root says that the length of the long side of the golden rectangle is 1.618 034 times the length of the short side. If you want to look at this the other way, the length of the short side is $\frac{1}{1.618\,034} = 0.618\,034$ times the length of the long side (this is the meaning of the negative root).

It is an odd thing that ϕ is a number for which its inverse is equal to itself minus one. That is, $\frac{1}{\phi} = \phi - 1$; it (and its negative sibling) is the only number that has that peculiar property. This number, 1.618 034 (or its inverse, 0.618 034), is known as the *golden ratio* or the *golden section*.

There are many other peculiar and fascinating mathematical properties of the golden rectangle, but perhaps even more interesting for two-dimensional art are its geometrical properties. If one starts with a golden rectangle, it can be subdivided with a square, and this leaves another golden rectangle. This smaller rectangle can also be subdivided with a square, and it too will leave a golden rectangle, etc.

If one connects the corners of these squares, it makes a special type of spiral, known as the golden spiral. This spiral sometimes appears in nature, most directly and famously in the shell of the chambered nautilus; the nautilus and the golden spiral can be seen together in figure B.1.

Appendix C

Optimal pinhole size for a pinhole camera

A pinhole camera uses simple geometry to restrict rays of light coming from the object to their proper position at the image. This is accomplished simply with a tiny hole on one side of a dark box. The idea is that since it is only a tiny hole, only one ray from each part of the object can get through the hole to the image. There are two problems with this:

1. This could work perfectly only if the hole were infinitesimally small, in which case, only an infinitesimally small amount of light would enter the camera.
2. If the hole is small enough that its size is comparable to the wavelength of the light, then the laws of diffraction dictate that the light will spread out as it goes through the hole. And furthermore, the laws of diffraction say that a *smaller hole* will make the light spread out *more*.

And so we are faced with the fact that the laws of geometry alone indicate that a smaller hole would give a sharper image, but the laws of diffraction say the opposite. This means that there is a trade-off between these two factors. And so there is a particular size hole—neither too large nor to small—that yields the sharpest image.

Let us see what this best pinhole size would be. Imagine that a point source of light, located very far away, is imaged in our pinhole camera. The perfect image would be a point, since the source of light is a point. But how large would the image be in practice?

A full analysis is beyond the scope of this book, but the simpler analysis I present here gives nearly identical results. It turns out that there is more than one way to define the problem in the first place; when one looks in fine detail, it is not obvious what one means by 'the best'. It is not such a bad thing to have a slightly vague answer when the question itself is imprecisely defined.

One simple way to approach the problem is to add the effects of both geometry and diffraction. For a point source of light very far away, geometry would indicate that the size of the image would be the same size and shape as the pinhole. And so, if our pinhole has a diameter, D, then the image in our pinhole camera would also be a circular spot of that same diameter.

doi:10.1088/978-0-7503-3703-8ch51

© IOP Publishing Ltd 2022

But diffraction plays the opposite role. From section 3.7 we have seen that a smaller hole produces a larger diffracted image. The exact relation, for the *angular* diameter, θ, of the brightest part of the diffracted image, is given by equation (C.1):

$$\theta = 2.44\frac{\lambda}{D} \tag{C.1}$$

where λ is the wavelength of the light. But this is the angular diameter, not the physical diameter, x, (in units of length) on the detector of our pinhole camera. To calculate that, we must use the small-angle formula:

$$x = \theta F = 2.44\frac{\lambda F}{D} \tag{C.2}$$

where F is the focal length, the distance between the pinhole and the light detector. And so, combining all of this and adding the effects of both geometry and diffraction, the size, x, of the image on the light detector would be

$$x = D + 2.44\frac{F\lambda}{D} \tag{C.3}$$

The question here is, for a given value of D, how big is x? And what is the particular value of D that gives the *smallest* value of x? We can approach this in a couple of ways. The most direct and accurate way is to use calculus: finding the value of a variable that gives the minimum (or maximum) value of some quantity is one of the fundamental uses of calculus.

But we can also make a graph of x for different values of D and look for the minimum value. This approach is especially illustrative if we also graph the two parts—geometry and diffraction—separately to see how they individually affect the sum. One complication is the focal length, F, of our pinhole camera. How might that affect the answer? We will need calculus to answer that question, but we can still use a simple graph to see why there is a best value of D for a particular choice of F. Figure C.1 shows the size, x, of our pinhole-camera image, for different values of pinhole diameter, D. The three curves show, respectively, the result for geometry alone, diffraction alone and the sum of the two. It is clear from the graph that both effects combined result in a particular value of D that results in the smallest value of x—and thus the sharpest image. For figure C.1, I have chosen a pinhole-camera focal length of 1 m, and for this example, it appears that the best-possible pinhole size (the lowest point in the graph) is a diameter of about 1 mm.

It would be much more useful, however, if we had a formula that would allow us to calculate the best pinhole diameter for *any* given focal length. To find the value of D that gives the minimum value of x, we take the derivative of D with respect to x, dD/dx. We then determine for what value of D is this derivative equal to zero. Taking the derivative of equation (C.3), we find

$$\frac{dD}{dx} = 1 - \frac{2.44F\lambda}{D^2} \tag{C.4}$$

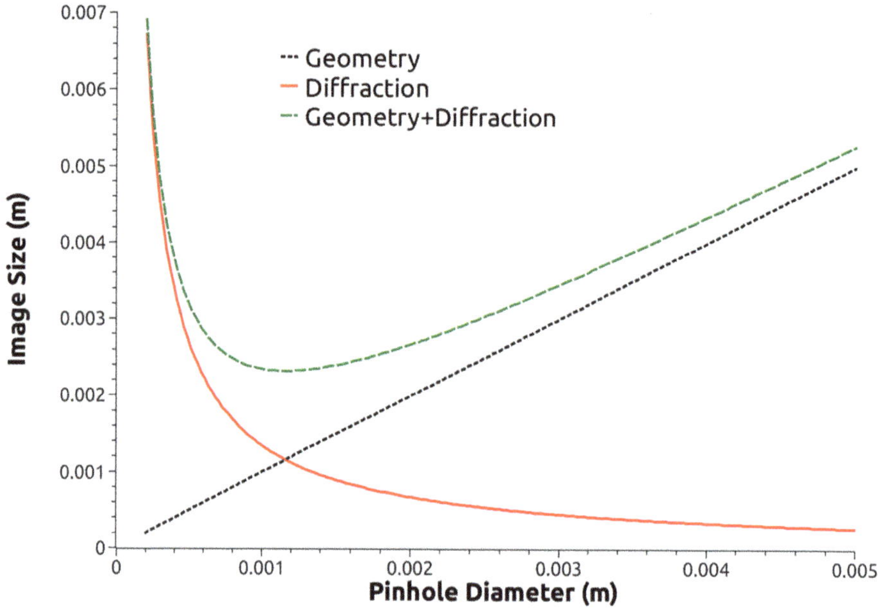

Figure C.1. The two factors, geometry and diffraction, that contribute to the size of the image of a point in a pinhole camera. Geometry alone indicates that a larger pinhole results in a larger (and thus blurrier) image of a point source. Diffraction indicates the opposite; a smaller hole yields a larger and blurrier point image. The two effects combined yield a particular pinhole size that gives the smallest (and thus sharpest) image. For a camera focal length of 1 m (assumed for this graph), it is clear from the minimum in the graph that the best pinhole diameter is about 0.001 m or 1 mm. To determine a formula for the best pinhole diameter for other focal lengths, one must analyze the problem with calculus.

If we set this derivative equal to zero, and solve for D, we get

$$1 - \frac{2.44F\lambda}{D^2} = 0 \tag{C.5}$$

$$D = \sqrt{2.44F\lambda} \tag{C.6}$$

If we pick a wavelength of 550 nm, in the middle of the visible part of the spectrum, and put in a conversion factor to convert from meters to millimeters, then we get, for the best value of D,

$$D(\text{mm}) \approx \frac{\sqrt{F(\text{mm})}}{27} \tag{C.7}$$

We can compare the result of this formula with our graph. Putting in $F = 1000$ mm gives $D = 1.16$ mm.

In practice, when making a pinhole it is prudent to err on the side of a larger-than-best-sized hole, rather than a smaller one. A hole that is slightly too large at least has the advantage that it admits more light into the camera, and so the exposure time is correspondingly shorter. This is not the only possible way to analyze this problem, and other approaches yield slightly different formulae. The most sophisticated approaches use a more accurate model of diffraction, and a slightly larger best size results. In practice, however, the process of making a pinhole is trial and error. And so my advice is to make a pinhole that has a size that is *no smaller than* that given by equation (C.7); your best attempt will likely be slightly larger than, and close to, the best value.

Appendix D

Lambertian reflectors and emitters

D.1 Lambertian versus isotropic reflectors

Here we derive results used in chapters 19–21. We consider a small circular spot that reflects light diffusely to a point directly in front and a distance, d, away that is much greater than the dimensions of the reflecting spot. We show that if the spot reflects as a Lambertian reflector, the intensity directly in front is twice that of an isotropic reflector.

Imagine that a certain power, P, is reflected from our circular spot. Then, the intensity, at a given point distant from the spot, is that power per unit area crosswise to the ray. So let us imagine we are at a distance d from that spot, off in some direction making an angle of θ to the normal.

For the simple but unrealistic case of an isotropic reflector, the intensity would be the same for any θ, so long as we maintain the same distance, d. For a Lambertian reflector, on the other hand, *the intensity is proportional to the projected area* of our circular spot. If θ is large, then from that perspective one sees a spot of small area, and so the intensity of the light is proportionally less. Since the intensity as seen from a particular distance is proportional to the projected surface area, *the Lambertian reflector maintains the same apparent surface brightness, while the isotropic reflector maintains the same total brightness.* We can summarize these results as follows:

$$B_{\rm I}(\theta) = B_{\rm I0} = \text{constant} \tag{D.1}$$

$$B_{\rm L}(\theta) = B_{\rm L0}\cos(\theta) \tag{D.2}$$

where the subscripts refer to isotropic and Lambertian reflectors and the zero subscript refers to the intensity for $\theta = 0°$.

Since power equals intensity times area, the integral of the intensity over the area of the forward hemisphere of radius d must equal the total power reflected, $P_{\rm R}$, for both the isotropic and Lambertian cases:

doi:10.1088/978-0-7503-3703-8ch52

© IOP Publishing Ltd 2022

$$P_R = \int B_I(\theta)dA \tag{D.3}$$

$$P_R = \int B_L(\theta)dA \tag{D.4}$$

where dA is an infinitesimal element of surface area at distance d, and the integral is carried out over the forward-facing hemisphere. We can think of the area element, dA, as a thin ring on that hemisphere, defined by the space between θ and $\theta + d\theta$. This ring would have an area given by the product of its thickness ($d \times d\theta$) and its circumference ($2\pi d \sin\theta$). And so we have

$$dA = 2\pi d \sin\theta \times d \times d\theta \tag{D.5}$$

$$dA = 2\pi d^2 \sin\theta d\theta \tag{D.6}$$

To integrate over the forward hemisphere, the integral over θ would be carried out from $0 \longrightarrow \pi/2$, and thus equations (D.3) and (D.4) become

$$P_R = 2\pi d^2 B_{I0} \int_0^{\pi/2} \sin(\theta)d\theta = 2\pi d^2 B_{I0} \tag{D.7}$$

$$P_R = 2\pi d^2 B_{L0} \int_0^{\pi/2} \cos(\theta)\sin(\theta)d\theta = \pi d^2 B_{L0} \tag{D.8}$$

Note that the two definite integrals evaluate to 1 and $\frac{1}{2}$, respectively. Thus we have

$$\boxed{B_{L0} = 2B_{I0}} \tag{D.9}$$

D.2 A flat, circular Lambertian emitter

Here we consider the variation of intensity versus distance directly in front of a circular, Lambertian *emitter* with radius, R, and total emitted power, P_0. We calculate the flux, B, passing through point p directly in front, at a distance d from the center of the circle.

We can use our results in section D.1 for a Lambertian reflector to describe light emitted by a flat *source* of light and arriving some distance, x, away, at an angle θ with the normal. We simply equate our emitted power, P, to the reflected power, P_R, of equation (D.8). But equation (D.8) in this context would apply not to the entire circular emitter but rather to only an infinitesimally small part of it. Doing so, and combining equations (D.2) and (D.8), we have, for the case of a small piece of emitting flat surface at a distance, x, and seen from an angle θ,

$$B = \frac{P}{\pi x^2}\cos\theta \tag{D.10}$$

And so a bit of area dA from some portion of our nearby circular disk emits a differential bit of power, dP. We can see from equation (D.10) that dA must contribute an amount of flux, dB, given by

$$dB = \frac{dP \cos\theta}{\pi x^2} \tag{D.11}$$

Given our assumption that the circle emits with uniform intensity over its surface, the power emitted by a given area is simply the product of the total power, P_0, and the *fraction* of the total area ($A = \pi R^2$) represented by dA. And so we have

$$dB = \frac{P_0 \cos\theta}{\pi R^2 \pi x^2} dA \tag{D.12}$$

The area element dA is a ring of circumference $2\pi r$ and width dr:

$$dA = 2\pi r dr \tag{D.13}$$

$$dr = x d\theta \tag{D.14}$$

$$r = x \sin\theta \tag{D.15}$$

$$dA = 2\pi x^2 \sin\theta d\theta \tag{D.16}$$

Thus we can combine equations (D.12) and (D.16) to get an expression for dB in terms of angle alone:

$$dB = \frac{P_0 2\pi x^2 \sin\theta d\theta \cos\theta}{\pi^2 R^2 x^2} \tag{D.17}$$

$$dB = \left(\frac{2P_0}{\pi R^2}\right) \sin\theta \cos\theta d\theta \tag{D.18}$$

We can now integrate equation (D.18) over the circular emitter, from $\theta = 0$ to $\theta = \theta_{\max}$, noting that $\theta_{\max} = \arctan(R/d)$:

$$B = \left(\frac{2P_0}{\pi R^2}\right) \int_0^{\theta_{\max}} \sin\theta \cos\theta d\theta \tag{D.19}$$

$$B = \left(\frac{2P_0}{\pi R^2}\right) \frac{1}{2} [-\cos^2\theta_{\max} - (-\cos^2 0)] \tag{D.20}$$

$$B = \left(\frac{P_0}{\pi R^2}\right) (1 - \cos^2\theta_{\max}) \tag{D.21}$$

$$B = \left(\frac{P_0}{\pi R^2}\right) \sin^2\theta_{\max} \tag{D.22}$$

where we have used the trigonometric identity $1 - \cos^2\theta = \sin^2\theta$. And thus we have

$$\boxed{B = \left(\frac{P_0}{\pi R^2}\right)\sin^2[\arctan(R/d)]} \tag{D.23}$$

We can explore the cases of $d \gg R$ and $d \ll R$ to find, respectively, B_{far} and B_{near}. For very small d, R/d approaches infinity and the arctangent approaches $\pi/2$. Thus we have

$$\boxed{B_{\text{near}} = \frac{P_0}{\pi R^2}} \tag{D.24}$$

This is simply the surface brightness of our uniformly emitting circle—the total power emitted over its total surface area.

To find an approximation for B_{far} we note that the following is true when $\theta \ll 1$:

$$\tan\theta \approx \sin\theta \approx \theta \tag{D.25}$$

But if $\tan\theta \approx \theta$, then it must also be true that $\arctan\theta \approx \theta$. And so we have, for $d \gg R$,

$$\arctan(R/d) \approx R/d \tag{D.26}$$

$$B_{\text{far}} = \frac{P_0}{\pi R^2}\sin(R/d)^2 \tag{D.27}$$

$$B_{\text{far}} = \frac{P_0}{\pi R^2}\left(\frac{R}{d}\right)^2 \tag{D.28}$$

which gives

$$\boxed{B_{\text{far}} = \frac{P_0}{\pi d^2}} \tag{D.29}$$

And so we have that for distances much larger than R, the intensity is the result we derived in section D.1 for the case of a Lambertian reflector. The Lambertian intensity is an inverse-square law with twice that of a small disk emitting isotropically in the forward direction only, with P_0 spreading out uniformly over the forward hemisphere of area $2\pi R^2$.

Appendix E

A practical way to denote very-low sensitivity detector 'speed'

Since exposure times for very-low sensitivity (VLS) photography are measured in minutes rather than fractions of a second, it makes sense to abandon the idea of detector 'speed' and instead define a detector *slowness*, which I will denote by *S*. There is no reason for this to have any connection at all to ISO speed, and so we can define it in whatever manner is convenient.

The ordinary ISO speed uses numbers that make sense when applied to ordinary film or digital detectors, for which the typical exposure time is a small *fraction* of a second. And so it is natural for the ISO speed to be linked to typical *reciprocals* of exposure times. This is the foundation of the sunny 16 rule: the ISO speed is roughly equal to the reciprocal of the exposure time on a sunny day at $f/16$ (a typical aperture setting for sunny-day film photography). And so, for example, with ISO 200 speed film a proper exposure would be 1/200 second at $f/16$. And so the definition of ISO speed for film is linked in an intuitive way to its typical use. And furthermore, ISO speed, when applied to film, gives us easy-to-use numbers such as 100, 200, 400, etc.

But for VLS photography, typical sunny-day exposure times are measured in minutes, not *fractions* of a second. Corresponding ISO speeds are tiny fractions requiring scientific notation (or careful counting of zeros) to express. Furthermore, one is more likely than not to use small focal ratios in VLS photography, even on a sunny day. And so instead of defining a *speed* based on the reciprocal of the sunny-day exposure time in seconds at a large focal ratio, I choose to do the opposite. I define instead a *slowness*:

> The slowness, *S*, of a VLS detector is defined according to a 'sunny 5.6' rule. Under sunny-bright, $EV_{100} = 15$ conditions, the slowness equals the proper exposure time *in minutes* at $f/5.6$. And so, under sunny-bright lighting, a detector with a slowness of 15 will require a 15 minute exposure at $f/5.6$.

doi:10.1088/978-0-7503-3703-8ch53

© IOP Publishing Ltd 2022

I choose $f/5.6$ as the standard focal ratio because it is a typical largest aperture of a medium-fast large-format lens.

We can relate this definition of slowness to incident metering, via equation (20.1) in chapter 20, simply by defining more clearly what we mean by a 'proper exposure.' In ISO-standard photography, a reflectivity of $R = 0.18$ is taken to be the standard. But for VLS photography that may not be the best choice.

Part of the point of VLS photography is that non-photographic content, related to the paper detector surface itself, can be intentionally incorporated into the image. This happens most with what would ordinarily be considered an *under*-exposure, which leads to shadow detail dominated by non-photographic content. Thus I argue that it is more often the case that we want a mid-range of detector density to correspond to a subject reflectivity significantly *greater* than 18%.

But to stay somewhat in line with traditional metering, I propose to use instead a standard of 20.9% reflectivity, or $R = 0.209$. I make this odd choice, instead of an even 20%, because it leads to a simple numerical connection between S and incident-light metering time, t_{SUB}. If we choose $R = 0.209$ and also incorporate our $f/5.6$ standard, while applying a conversion from seconds to minutes, equation (20.1) then becomes

$$t_{\mathrm{D}} \text{ (minutes)} = 10.00\ t_{\mathrm{SUB}} \text{ (seconds)} \qquad \text{(E.1)}$$

This leads to the following connection between incident light metering and my definition of slowness:

> A detector that reaches a mid-range density in t_{SUB} seconds, when illuminated normally by sunny-bright, $\mathrm{EV}_{100} = 15$ light, has a slowness given by the following:

$$S = 10\ t_{\mathrm{SUB}} \text{ (seconds)} \qquad \text{(E.2)}$$

We can extend this definition to measurements at other levels of incident light with the following equation:

$$S = \frac{10\ t_{\mathrm{SUB}} \text{ (seconds)}}{2^{(15-\mathrm{EV}_{100})}} \qquad \text{(E.3)}$$

And so, for example, if one finds that under lighting conditions of $\mathrm{EV}_{100} = 13$ (cloudy-bright) the VLS detector reaches a mid-range density in 8 seconds, then we have

$$S = \frac{10 \times 8}{2^{(15-13)}} = \frac{80}{2^2} = 20 \qquad \text{(E.4)}$$

Note that this same detector would presumably take only two seconds to reach that same density under sunny-bright lighting conditions of $\mathrm{EV}_{100} = 15$, and equation (E.3) would still give the same answer of $S = 20$. And so for this particular detector, we should expect a proper exposure to take 20 minutes at $f/5.6$ under sunny-bright lighting.

And so this simple test of our VLS detector to measure S then informs us what exposure time, t_D, we should expect, given the focal ratio and EV_{100} lighting conditions, if we simply count steps in the proper direction from both $f/5.6$ and $EV_{100} = 15$. Gathering all of this into one equation, we have

$$t_D \text{ (minutes)} = S \times (f/5.6)^2 \times 2^{(15-EV_{100})} \tag{E.5}$$

Notice that equation (E.5) means that, at $f/5.6$ and sunny-bright lighting, the exposure time in minutes is simply equal to the detector slowness.

If you disagree with my choice of $R = 0.209$ as a standard, the procedure and equations described here can still be used. Simply make your own choice of what you mean by 'mid-range density' when you take your incident light reading. If your incident-light test density is *darker* than mine, then your application of these equations will result in exposures that are *longer* than mine. Associating a more-than-mid-range density to $R = 0.209$ is equivalent to associating a mid-range density to a reflectivity that is less than $R = 0.209$.

This is not a laboratory test standard, such as ISO. It is a rough standard that would in practice be interpreted by each photographer in their own subjective way. It gets us all in the same book, and in the same chapter—but probably not exactly on the same page.

IOP Publishing

Photography (Second Edition)
Physics and art in focus
John Beaver

Appendix F

The optics of the camera stupida

What I call the *camera stupida*, as described in section 44.4, uses a mirror–beamsplitter arrangement similar to that used in some versions of the *camera lucida*. One looks directly at a sheet of drawing paper, but with a superimposed view of the subject to be drawn; the view of the subject undergoes two reflections, so the image is not reversed. One of these reflections is off a normal mirror, while the other is off a beamsplitter, which allows some light to pass through and some to reflect. A beam splitter that divides this light into equal parts uses a semi-transparent layer of vacuum-deposited metal. In the camera stupida, I use a simple piece of thin glass, taking advantage of the partial reflection off the surface of an air-glass interface.

If the mirror and glass beamsplitter are angled at 45° to each other, then the straight-through and reflected images are 90° apart. Thus, one can look straight down at the drawing paper and, simultaneously, straight forward at the subject. The difference is that I angle the mirror–beamsplitter combination at an angle of 11° to the vertical. This allows the final reflection off the beamsplitter to be at an angle of 34° to the surface of the glass, or 56° to its normal.

When light reflects off the interface of a sudden change in index of refraction between two transparent materials, it is strongly *polarized* if the reflection is at an angle to the normal called the *Brewster angle*, given by

$$\tan\theta = \frac{n_2}{n_1} \tag{F.1}$$

where n_1 is the index of refraction of the material for the incident ray and n_2 is the index of refraction of the material the ray reflects off. For light in air ($n_1 = 1$) reflecting off a glass surface with a typical index of refraction of about 1.5, this corresponds to an angle of about 56° with the normal, or 34° with the surface of the glass. See figure F.1 for the geometry of the camera stupida, incorporating this angle into the design.

And so one looks from above, directly through the glass beamsplitter at the drawing paper, but one also sees through two reflections to the subject. The reflection off the glass is at the Brewster angle, and so the light from the subject is strongly polarized while that from the drawing paper is not.

doi:10.1088/978-0-7503-3703-8ch54

© IOP Publishing Ltd 2022

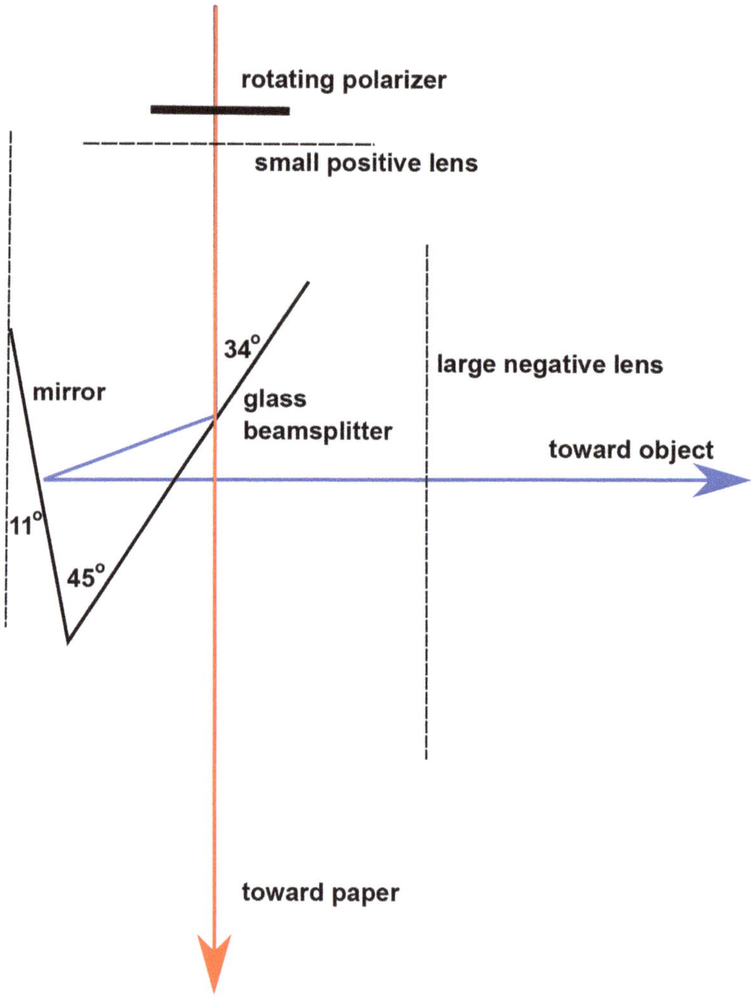

Figure F.1. The arrangement of the optics for the camera stupida. One looks downward through the polarizer and small lens at the top to see a simultaneous view of the drawing paper directly below and the subject in the distance horizontally.

All that is needed, then, is a polarizing filter at the top viewing lens to adjust the brightness of the subject image, while leaving the view of the drawing paper unaffected. This works because a polarizing filter passes only light of a particular polarization. If the filter is oriented so that it passes only light polarized perpendicularly to that of the polarized light of the subject, then no light is passed.

The fixed format allows one to view the paper through a small positive focal length (convex) magnifier lens. For the view of the subject, a large negative focal length (concave) lens is placed in front of the beamsplitter–mirror assembly. When looking at the subject, one sees through *both* lenses, while the view of the drawing paper is only through the convex lens. The particular arrangement of lenses for the subject view is a reverse-Galilean viewfinder (section 11.9.1), and it shows a somewhat wide-angle view of the subject. The focal lengths of the two lenses, combined with the fixed distance to the drawing paper, can be chosen such that both images are in focus.

www.ingramcontent.com/pod-product-compliance
Ingram Content Group UK Ltd.
Pitfield, Milton Keynes, MK11 3LW, UK
UKHW051911100926
8556IPUK00005B/847

9 780750 337014